# NUBA Conference Series-1: Nuclear Physics and Astrophysics 2014



Journal of Physics: Conference Series Volume 590

## Antalya, Turkey
## 15 – 21 September 2014

ISBN: 978-1-5108-0260-5
ISSN: 1742-6588

**Printed from e-media with permission by:**

Curran Associates, Inc.
57 Morehouse Lane
Red Hook, NY 12571

**Some format issues inherent in the e-media version may also appear in this print version.**



Printed by Curran Associates, Inc. (2015)

For permission requests, please contact the Institute of Physics at the address below.

Institute of Physics
Dirac House, Temple Back
Bristol BS1 6BE UK

Phone: 44 1 17 929 7481
Fax: 44 1 17 920 0979

techtracking@iop.org

**Additional copies of this publication are available from:**

Curran Associates, Inc.
57 Morehouse Lane
Red Hook, NY 12571 USA
Phone: 845-758-0400
Fax: 845-758-2634
Email: curran@proceedings.com
Web: www.proceedings.com

# TABLE OF CONTENTS


**NEUTRINOS, RARE ISOTOPES OF EXOTIC NUCLEI AND NUCLEAR ASTROPHYSICS** ..... 1
*A B Balantekin*

**CLUSTER STATES AND CONTAINER PICTURE IN LIGHT NUCLEI, AND TRIPLE-ALPHA REACTION RATE** ..... 7
*Yasuro Funaki*

**NEW PROBE TO ML STRENGTHS IN GDR REGION FOR SUPERNOVA NEUTRINO INTERACTIONS** ..... 13
*T Hayakawa, S Miyamoto, T Mochizuki, K Horikawa, S Amano, K Imazaki, D Li, Y Izawa, S Chiba, K Ogata*

**BOHR HAMILTONIAN WITH DEFORMATION-DEPENDENT MASS** ..... 19
*Dennis Bonatsos, N Minkov, D Petrellis, C Quesne*

**FUTURE EXPERIMENTS WITH INTENSE LASER BEAMS AND BRILLIANT GAMMA BEAMS AT THE ELI-NP FACILITY** ..... 25
*Dimiter L Balabanski*

**THE PYGMY DIPOLE RESONANCE – EXPERIMENTAL STUDIES OF ITS STRUCTURE AND NEW DEVELOPMENTS** ..... 33
*A Zilges*

**STRUCTURE AND DYNAMICS OF WEAKLY-BOUND SYSTEMS: A ONE-DIMENSIONAL MODEL** ..... 39
*Andrea Vitturi, Laura Moschini*

**FOUR-BODY EFFECTS ON $^{9}$BE+$^{208}$PB SCATTERING AND FUSION AROUND THE COULOMB BARRIER** ..... 45
*P Descouvemont, T Druet, L F Canto, M S Hussein*

**FULLY DAMPED MOTT OSCILLATIONS IN SUB-BARRIER ELASTIC SCATTERING OF IDENTICAL HEAVY IONS AND THE NUCLEAR INTERACTION** ..... 51
*M S Hussein, L F Canto, W Mittig*

**NUCLEAR STRUCTURE STUDIES WITH STABLE AND RADIOACTIVE BEAMS: THE SPES RADIOACTIVE ION BEAM PROJECT** ..... 57
*G de Angelis , G Prete, A Andrighetto, M Manzolaro, S Corradetti, D Scarpa, M Rossignoli, A Monetti, M Lollo, M Calderolla, J Vasquez, D Zafiropoulos, L Sarchiapone, D Benini, P Favaron, M Rigato, R Pegoraro, D Maniero, L Calabretta, M Comunian, M Maggiore, A Lombardi, L Piazza, A M Porcellato, C Roncolato, G Bisoffi, A Pisent, A Galatà, M Giacchini, G Bassato, S Canella, F Gramegna, J Valiente, J Bermudez, P F Mastinu, J Esposito, J Wyss, A Russo, S Zanella*

**RECENT RESULTS ON THE PYGMY DIPOLE RESONANCE** ..... 63
*Deniz Savran*

**RECENT RESULTS ON REACTIONS WITH RADIOACTIVE BEAMS AT RIBRAS (RADIOACTIVE ION BEAMS IN BRAZIL)** ..... 69
*A Lépine-Szily, R Lichtenthäler, V Guimarães, A Arazi, A Barioni, E A Benjamim, P N de Faria, P Descouvemont, L R Gasques, E Leistenschneider, D R Mendes Jr, M C Morais, V Morcelle, A M Moro, R Pampa Condori, K C C Pires, M Rodriguez-Gallardo, V Scarduelli, J M B Shorto, J C Zamora*

**NUCLEAR ASTROPHYSICS WITH THE TROJAN HORSE METHOD** ..... 75
*Claudio Spitaleri*

**THE STRUCTURE OF $^{12}$C AND $^{16}$O** ..... 81
*Martin Freer*

**NUCLEAR STRUCTURE AND SETTING CONSTRAINTS ON THE R-PROCESS?** ..... 87
*Ani Aprahamian*

**STUDIES OF THE SHAPES OF HEAVY NUCLEI AT ISOLDE** ..... 93
*Peter A Butler*

**COMPLETE ELECTRIC DIPOLE RESPONSE IN $^{120}$SN AND $^{208}$PB AND IMPLICATIONS FOR NEUTRON SKIN AND SYMMETRY ENERGY** ..... 99
*Peter von Neumann-Cosel*

**KNOWN AND EXOTIC LIGHT HYPERNUCLEI STUDIED WITH THE $^{6}$LI+$^{12}$C REACTION AT 2 A GEV** ..... 105
*Take R Saito, Vakkas Bozkurt, Christophe Rappold*

**SPIN RESPONSES IN NUCLEI AND NUCLEAR WEAK PROCESSES IN STARS** ..... 111
*Toshio Suzuki*

**SUBBARRIER FUSION OF CARBON ISOTOPES: FROM RESONANCE STRUCTURE TO FUSION OSCILLATIONS** ........ 117
*K Hagino, N Rowley*

**SINGLE-SHELL STATES FOR MULTI-SHELL DESCRIPTION OF NUCLEI** ........ 123
*Erdal Dikmen*

**COULOMB AND NUCLEAR COMPONENTS OF THE BREAKUP, THEIR INTERFERENCE AND EFFECT ON THE FUSION PROCESS** ........ 129
*P R S Gomes, J Lubian, L F Canto, D R Otomar, M S Hussein*

**PHOTONEUTRON REACTIONS IN NUCLEAR ASTROPHYSICS** ........ 135
*H Utsunomiya, S Goriely, D M Filipescu, O Tesileanu, I Gheorghe, T Glodariu, H-T Nyhus, T Renstrøm, Y-W Lui, T Shima, K Takahisa, S Miyamoto*

**PHOTONUCLEAR REACTIONS INDUCED BY A CLINICAL LINAC** ........ 141
*I Boztosun, H Dapo, M Karakoc, S F Ozmen, Y Cecen, A Coban, A Cesur, T Caner, E Bayram, T R Saito, T Akdogan, V Bozkurt, D Kaya, Y Kucuk*

**COLLECTIVE ENHANCEMENT OF NUCLEAR STATE DENSITIES BY THE SHELL MODEL MONTE CARLO APPROACH** ........ 147
*C Özen, Y Alhassid, H Nakada*

**THE EFFECT OF PROTON HALO ON FUSION REACTIONS** ........ 151
*J A Lay, Raj Kumar, A Vitturi*

**MONTE-CARLO STUDY OF THE INFLUENCE OF BACKSCATTERED ELECTRONS ON THE TRANSMISSION OF A MINI-ORANGE ß SPECTROMETER** ........ 155
*Pavel Detistov, Dimiter L Balabanski*

**NUCLEAR LEVEL DENSITY AS A TOOL FOR PROBING THE INELASTIC SCATTERING OF 6HE** ........ 159
*Bora Canbula, Halil Babacan*

**STUDY OF INTERSTRIP GAP EFFECTS AND EFFICIENCY FOR FULL ENERGY DETECTION OF DOUBLE SIDED SILICON STRIP DETECTORS** ........ 163
*D Torresi, J Forneris, L Grassi, L Acosta, A Di Pietro, P Figuera, L Grilj, M Jaksic, M Lattuada, T Mijatovic, M Milin, L Prepolec, N Skukan, N Soic, D Stanko, V Tokic, M Uroic, M Zadro*

**EXPLORING THE $^{12}$C($^{18}$O,$^{16}$O)$^{14}$C TWO-NEUTRON TRANSFER REACTION AT ENERGIES FAR ABOVE THE COULOMB BARRIER** ........ 167
*D Carbone, F Cappuzzello, C Agodi, A Bonaccorso, M Bondì, M Cavallaro, A Cunsolo, M De Napoli, A Foti, R Linares, J Lubian*

**RECONCILIATION OF MASS-ASYMMETRY SYSTEMATICS FOR INCOMPLETE FUSION** ........ 171
*Pushpendra P Singh, Abhishek Yadav, Vijay R Sharma, R Kumar, Manoj K Sharma, B P Singh, R P Singh, S Muralithar, R K Bhowmik, R Prasad*

**INTERPLAY BETWEEN THE PAIRING AND QUADRUPOLE INTERACTIONS IN THE ALGEBRAIC REALIZATION OF THE MICROSCOPIC SHELL MODEL** ........ 175
*Kalin Drumev, Ana Georgieva*

**ANALYSIS OF LOW LYING COLLECTIVE STATES IN THE SYMPLECTIC INTERACTING VECTOR BOSONS MODEL** ........ 179
*A I Georgieva, V P Garistov*

**FUSION HINDRANCE AND EXTRA PUSH IN FUSION REACTIONS WITH HEAVY NUCLEI** ........ 183
*Kouhei Washiyama*

**MEASUREMENTS OF $^{12}$ C ION FRAGMENTATION ON THIN CARBON TARGET FROM THE FIRST COLLABORATION AT GSI** ........ 187
*M Toppi*

**LIFETIME MEASUREMENTS AND THE HIGH-SPIN STRUCTURE OF $^{36}$CL** ........ 191
*S Aydin, M Ionescu-Bujor, F Recchia, A Gadea, S M Lenzi, S Lunardi, C A Ur, D Bazzacco, P G Bizzeti, A M Bizzeti-Sona, M Bouhelal, G de Angelis, I Deloncle, E Farnea, A Gottardo, F Haas, T Huyuk, H Laftchiev, D Mengoni, R Menegazzo, C Michelagnoli, D R Napoli, D Sahin, P P Singh, D Tonev, J J Valiente-Dobón*

**EXPLORING THE $^{10}$LI STRUCTURE BY THE D($^{9}$LI,P)$^{10}$LI TRANSFER REACTION** ........ 195
*M Cavallaro, M De Napoli, F Cappuzzello, C Agodi, M Bondì, D Carbone, A Cunsolo, B Davids, T Davinson, A Foti, N Galinski, R Kanungo, H Lenske, S E A Orrigo, C Ruiz, A Sanetullaev*

**$^{12}$C+$^{12}$C AND $^{16}$O+$^{16}$O FUSION REACTIONS AT LOW ENERGIES** ........ 199
*M Assunção, P Descouvemont*

**HALF-LIFE MEASUREMENTS OF THE $^{144}$PM ISOTOPE WITH $\gamma$-SPECTROSCOPY** ........ 203
*C Yalçin, Gy Gyürky, Z Halász, T Szücs, R T Güray, G G Kiss, Zs Fülöp, N Özkan, E Somorjai*

**COUPLING TO CONTINUUM EFFECTS IN THE $^{6,7}$LI +$^{64}$ZN REACTIONS AT ENERGIES AROUND THE COULOMB BARRIER.** ........ 207
*J P Fernández-García, M Zadro, A Di Pietro, P Figuera, M Fisichella, M Lattuada, A M Moro, D Torresi*

**NUCLEAR G-FACTOR MEASUREMENT WITH TIME-DEPENDENT RECOIL IN VACUUM IN RADIOACTIVE-BEAM GEOMETRY** ..... 211
*A Kusoglu, A E Stuchbery, G Georgiev, A Goasduff, L Atanasova, D L Balabanski, M Bostan, M Danchev, P Detistov, K Gladnishki, J Ljungvall, I Matea, D Radeck, C Sotty, I Stefan, D Verney, D T Yordanov*

**DETERMINATION OF NATURALLY OCCURRING RADIONUCLIDES IN SOIL SAMPLES OF AYRANCI, TURKEY** ..... 215
*Osman Agar, Canel Eke, Ismail Boztosun, M. Emin Korkmaz*

**COMPARISON OF DIFFERENT INC PHYSICAL MODELS OF MCNPX TO COMPUTE SPALLATION NEUTRONICS OF LBE TARGET** ..... 219
*Seyed Amir Hossein Feghhi, Zohreh Gholamzadeh, Claudio Tenreiro, Zahra Alipoor*

**DEPENDENCE OF ACTIVITY CONCENTRATION OF NATURAL AND ARTIFICIAL RADIONUCLIDES ON DEPTH IN SOIL SAMPLES FROM ANTALYA IN TURKEY** ..... 225
*Canel Eke, Osman Agar, Ismail Boztosun*

**THE INFLUENCE OF NUCLEAR DEFORMATIONS ON THE EXOTIC CLUSTER DECAY HALF-LIVES** ..... 229
*A Soylu, O Bayrak, S Evlice*

**THE EFFECTS OF DOUBLE FOLDING CLUSTER MODEL POTENTIAL ON SOME ASTROPHYSICAL REACTIONS** ..... 233
*Azni Abdul Aziz, Norhasliza Yusof, Muhammad Zamrun Firihu, Hasan Abu Kassim*

**PHOTONUCLEAR REACTION CROSS SECTIONS FOR GALLIUM ISOTOPES** ..... 237
*S Akkoyun, T Bayram, S O Kara*

**THE LONG RANGE POTENTIAL EFFECT ON THE $^{20}$O+$^{12}$C SYSTEM FOR NEAR AND SUB-BARRIER FUSION CROSS SECTION** ..... 241
*G Kocak, R Kabak*

**ISOMERIC AND GROUND STATE ENERGY LEVEL MEASUREMENTS OF NATURAL TELLURIUM ISOTOPES VIA (γ,N) REACTION** ..... 245
*M Tamkas, O Akcali, A Durusoy*

**THICKNESS MEASUREMENT USING ALPHA SPECTROSCOPY AND SRIM** ..... 249
*C Yalçin*

**ENERGY LEVELS AND HALF-LIVES OF GALLIUM ISOTOPES OBTAINED BY PHOTO-NUCLEAR REACTION** ..... 253
*F Dulger, S Akkoyun, T Bayram, H Dapo, I Boztosun*

**DYNAMICAL DIPOLE MODE IN HEAVY-ION FUSION REACTIONS IN THE $^{192}$PB MASS REGION** ..... 257
*C Parascandolo, D Pierroutsakou, R Alba, A Del Zoppo, C Maiolino, D Santonocito, C Agodi, V Baran, A Boiano, M Colonna, R Coniglione, E De Filippo, M Di Toro, U Emanuele, F Farinon, A Guglielmetti, M La Commara, B Martin, C Mazzocchi, M Mazzocco, C Rizzo, M Romoli, C Signorini, R Silvestri, F Soramel, E Strano, D Torresi, A Trifiró, M Trimarchi*

**INVESTIGATION OF THE PHOTO-ACTIVATION ANALYSIS ON CONCENTRATIONS OF VARIOUS MATERIALS** ..... 261
*A Cesur, M Karakoc, I Boztosun, H Dapo*

**QUALITATIVE ELEMENTAL ANALYSES OF A METEORITE SAMPLE FOUND IN TURKEY BY PHOTO-ACTIVATION ANALYSIS METHOD** ..... 265
*C Ertugay, I Boztosun, S F Ozmen, H Dapo*

**INTERACTION OF 1P NUCLEI: CASE OF $^{14}$N+$^{12}$C ELASTIC SCATTERING AT 21.0 MEV** ..... 269
*N Burtebayev, D Alimov, I Boztosun, J Burtebayeva, Zh K Kerimkulov, M Nassurlla, N Amangeldy, A K Morzabayev, S K Sakhiev, Sh Hamada*

**OPTICAL AND COUPLED-CHANNELS DESCRIPTION OF $^{20}$NE+$^{16}$O ELASTIC SCATTERING** ..... 274
*N Burtebayev, M Nassurlla, D Alimov, I Boztosun, J Burtebayeva, A K Morzabayev, S K Sakhiev, S V Artemov, J M Mussaev*

**REACTION DYNAMICS STUDIES FOR THE SYSTEM $^{7}$BE+$^{58}$NI** ..... 279
*D Torresi, M. Mazzocco, L Acosta, A Boiano, C Boiano, A Diaz-Torres, N Fierro, T Glodariu, L Grilj, A Guglielmetti, N Keeley, M La Commara, I Martel, C Mazzocchi, P Molini, A Pakou, C Parascandolo, V V Parkar, N Patronis, D Pierroutsakou, M Romoli, K Rusek, A M Sanchez-Benitez, M Sandoli, C Signorini, R Silvestri, F Soramel, E Stiliaris, E Strano, L Stroe, K Zerva*

**ANALYSIS OF ELASTIC SCATTERING OF $^{8}$HE+$^{208}$PB SYSTEM AT AROUND THE COULOMB BARRIER ENERGIES** ..... 283
*M Direkci, Y Kucuk, I Boztosun*

**PHOTOFISSION OF $^{238}$U NUCLEI** ..... 289
*S Selek, A Çoban, I Boztosun, H Dapo, M Karakoç*

**Author Index**

# PREFACE

The international conference series "NUBA Conference Series 1: Nuclear Physics and Astrophysics" was held on September 14-21 2014 in Antalya-Turkey. Akdeniz University hosted the conference and the Adrasan Training and Application Centre was chosen as a suitable venue to bring together scientists from all over the world as well as from different parts of Turkey. The conference was supported by the Scientific and Technological Research Council of Turkey (TÜBİTAK) and Akdeniz University Nuclear Sciences Application and Research Center (NUBA).

Based on the highly positive remarks received from the participants both during and after the conference, we believe that the event has proven to be a fulfilling experience for all those who took part. The conference provided an opportunity for the participants to share their ideas and experiences in addition to exploring possibilities for future collaborations. Participants of the conference focused on:

- Nuclear Structure and Interactions
- Nuclear Reactions,
- Photonuclear Reactions and Spectroscopy
- Nuclear and Particle Astrophysics
- Nuclear Processes in Early Universe
- Nuclear Applications
- New Facilities and Instrumentation

Participants included a number of distinguished invited speakers. There was significant interest from the international nuclear physics community and numerous abstracts and papers were submitted. The scientific committee conducted a careful and rigorous selection process, as a result of which 75 contributions were accepted. Of those, 65 of them were given as oral and 10 as poster presentations. The superb quality of the papers ensured fruitful discussion sessions. We thank all the participants for their efforts and also for promptly sending in their papers for publication. This issue of the Journal of Physics: Conference Series was peer-reviewed by expert referees and we also thank them for peer-reviewing the papers.

Journal of Physics: Conference Series **590** (2015) 011001 doi:10.1088/1742-6596/590/1/011001

The national and international advisory committee also deserve appreciation for their involvement in the shaping of the conference programme. The local organizing committee, Mesut Karakoç, Haris Djapo, Fatih Ozmen and Deniz Kaya worked diligently and ensured that the programme ran smoothly. We sincerely thank them all.

Our final thanks go to IOP for publishing the proceedings in a most timely and meticulous manner.

We hope to see the participants again in Turkey, in the second conference of this series.

Ismail Boztosun

Akif Baha Balantekin

Yasemin Kucuk

Journal of Physics: Conference Series **590** (2015) 011001 doi:10.1088/1742-6596/590/1/011001

Conference Photographs taken at Akdeniz University Adrasan Training and Application Centre.

Journal of Physics: Conference Series **590** (2015) 011002 doi:10.1088/1742-6596/590/1/011002

NUBA Conference Series 1: Nuclear Physics and Astrophysics

Scientific Programme

**MONDAY (September 15)**

| | |
|---|---|
| 08:50 I. Boztosun | Opening Ceremony |
| **Chairman: A. Lepine-Szily** | |
| 09:00 B. Balantekin | Neutrinos, Rare Isotopes of Exotic Nuclei and Nuclear Astrophysics |
| 09:40 T. Kajino | Big-Bang and Supernova Neutrinos and Nucleosynthesis |
| 10:20 Break | |
| 10:40 Y. Funaki | Cluster states and container picture in light nuclei and triple-alpha reaction rate |
| 11:20 T.Hayakawa | New probe to M1 strengths in GDR region for supernova neutrino interactions |
| 12:00 Lunch | |
| **Chairman: T.Hayakawa** | |
| 13:30 D. Bonatsos | Bohr Hamiltonian with deformation-dependent mass |
| 14:10 D. Balabanski | ELI-NP: Status and Perspectives for Nuclear Physics Studies with Intense Laser Beams and Brilliant Gamma Beams |
| 14:50 A.Zilges | The Pygmy Dipole Resonance-status and new experimental developments |
| 15:30 Break | |

**PARALLEL SESSIONS**

**Parallel A, Chairman: P. Von Neumann-Cosel**

| | |
|---|---|
| 16:00 C. Özen | Statistical Properties of Heavy Deformed Nuclei |
| 16:20 P. Detistov | Investigation of the backscattered electron events influence on the transmission of the Mini-orange type beta Spectrometr Performance by Monte-Carlo simulations |
| 16:40 E. Yüksel | Tensor and temperature effects on the nuclear properties |
| 17:00 İ. Albayrak | Study of nucleon structure through ınclusive and exclusive reactions: measuring structure functions |
| 17:20 N. Benouaret | Pygmy mode evidence in the stable odd-proton nucleus $^{205}$Tl |

**Parallel B, Chairman: T. Saito**

| | |
|---|---|
| 16:00 J. A. Lay Valera | The effect of proton halo on fusion reactions |
| 16:20 B. Canbula | Nuclear level density as a tool for probing the inelastic scattering of $^{6}$He |
| 16:40 D. Torresi | Study of interstrip gap effects and efficiency for full energy detection of Double Sided Silicon Strip Detectors |
| 17:00 D. Carbone | Exploring the ($^{18}$O,$^{16}$O) two neutron transfer reaction at energies far above the Coulomb barrier |
| 17:20 A. Dehghanniri | The effect of nucleon-nucleon correlations on neutrino emissivity |

Journal of Physics: Conference Series **590** (2015) 011002 doi:10.1088/1742-6596/590/1/011002

**TUESDAY (September 16)**

| | |
|---|---|
| **Chairman: B. Balantekin** | |
| 09:00 A.Vitturi | Pairing interaction and reaction mechanism for two-particle transfer reaction: a one-dimensional model |
| 09:40 P. Descouvemont | Four-body effects on $^{9}Be$+$^{208}Pb$ scattering around the Coulomb barrier |
| 10:20 Break | |
| 10:30 M. S.Hussein | Disappearance of Mott oscillations in sub-barrier elastic scattering of identical heavy ions and the nuclear interaction |
| 11:10 L. C. Chamon | Realistic analysis of the alpha-nucleus scattering at low energies |
| 12:00 Lunch | |
| **Chairman: D. Bonatsos** | |
| 13:30 G. de Angelis | Exploring nuclear structure by binary reactions with stable and radioactive nuclear beams |
| 14:10 D. Bucurescu | Alpha decay as a nuclear spectroscopy tool for heavy nuclei |
| 14:50 D. Savran | Recent results on the Pygmy Dipole Resonance |
| 15:30 Break | |

**PARALLEL SESSIONS**

**Parallel A, Chairman: S. Shimoura**

| | |
|---|---|
| 16:00 P. Singh | Reconciling the famous mass asymmetry systematics for incomplete fusion |
| 16:20 G. Zimba | Octupole correlations and Collective Couplings in the rare earth nucleus $^{154}Dy$ |
| 16:40 K. Washiyama | Fusion hindrance and extra push in fusion reactions with heavy nuclei |
| 17:00 S. Aydın | Lifetime measurements and high spin structure of $^{36}Cl$ nucleus |
| 17:20 M. Assunção | The $^{12}C$+$^{12}C$ and $^{16}O$+$^{16}O$ fusion reactions at low energies |
| 17:40 J. P. Fernández-García | Coupling to continuum effects in the $^{6,7}Li$ +$^{64}Zn$ reactions at energies around the Coulomb barrier. |

**Parallel B, Chairman: M. Freer**

| | |
|---|---|
| 16:00 K. P. Drumev | Interplay between the Pairing and Quadrupole Interactions in the Algebraic realization of the Microscopic Shell Model |
| 16:20 A. Georgieva | Analysis of Low Lying Collective States in the Symplectic Interacting Vector Bosons Model |
| 16:40 M. Toppi | Measurements of Carbon ion fragmentation on thin C target from the FIRST collaboration at GSI. |
| 17:00 M. Cavallaro | Probing the $^{10}Li$ structure by the $^{9}Li(d,p)^{10}Li$ transfer reactionabove the Coulomb barrier |
| 17:20 C. Yalçın | Half-life measurements of Pm-144 isotope with gamma-spectroscopy |
| 17:40 N. Benouaret | Isospin mixing extraction from beta delayed proton decay strengths |

Journal of Physics: Conference Series **590** (2015) 011002 doi:10.1088/1742-6596/590/1/011002

**WEDNESDAY (September 17)**

| **Chairman: T. Kajino** | |
|---|---|
| 09:00 A. Lepine-Szily | Recent results on reactions with radioactive beams of RIBRAS |
| 09:40 C. Spitaleri | Nuclear Astrophysics with the Trojan Horse Method |
| 10:20 Break | |
| 10:30 S. Nishimura | Study of R-process Nucleosythesis via Decay Spectroscopy at RIBF |
| 11:10 E. Khan | Crystal, cluster and quantum liquid states |
| 12:00 Lunch | |
| | |
| 16:30 POSTER SESSİON | |

**THURSDAY (September 18)**

| **Chairman: T. Suzuki** | |
|---|---|
| 09:00 M. Freer | Filling in the gaps in the structure of light nuclei |
| 09:40 A. Aprahamian | Nuclear Structure and constraints on the r-process |
| 10:20 Break | |
| 10:30 S. Shimoura | Charge exchange reactions of RI-beams using SHARAQ spectrometer |
| 11:10 P. Butler | Studies of the Shapes of Heavy Nuclei at ISOLDE |
| 12:00 Lunch | |
| **Chairman: A. Vitturi** | |
| 13:30 P. Von Neuman-Cosel | Complete electric dipole response in $^{120}$Sn and $^{208}$Pb and implications for neutron skin and symmetry energy |
| 14:10 T. Saito | Mysteries in few-body nuclei with strangeness |
| 14:50 C. Segebade | Photon Activation Analysis – A Suitable Method For Studies of Lunar Material |
| 15:30 Break | |

**PARALLEL SESSIONS**

**Parallel A, Chairman: P. Descouvemont**

| | |
|---|---|
| 16:00 V. Bozkurt | Hypernuclear Spectroscopy with Heavy Ion Collision The HypHI Phase 0.5 Experiment |
| 16:20 A.Kuşoğlu | Nuclear g-factor measurement with Time-Dependent Recoil In Vacuum in radiactive-beam geometry |
| 16:40 M. Firoozabadi | Simulating produced Cherenkov emission from reactor with eliminating cosmic Gama by geant4 software |
| 17:00 Z.Gholamzadeh | Comparison of Different INC Physical Models of MCNPX to Compute Spallation Neutronics of LBE Target |

Journal of Physics: Conference Series **590** (2015) 011002 doi:10.1088/1742-6596/590/1/011002

**Parallel B, Chairman: C. Segebade**

| | |
|---|---|
| 16:00 E. Şahiner | An alternative approach for the experimental total-to-peak (TTP) efficiency curve used in the calculation of true coincidence correction factors in a HPGe detector |
| 16:20 G.S. Polymeris | Luminescence detectors for setting bounds on dark matter particle characteristics |
| 16:40 O. Agar | Determination of naturally occurring radionuclides in soil samples of Ayrancı, Turkey |
| 17:00 C. Eke | Dependence of activity concentration of natural and artificial radionuclides on depth in soil samples from Antalya in Turkey |

**FRIDAY (September 19)**

| | |
|---|---|
| **Chairman: M. Hussein** | |
| 09:00 T. Suzuki | Spin responses in nuclei and nuclear weak processes in stars |
| 09:40 K. Hagino | Subbarrier fusion of carbon isotopes: From resonance structure to fusion oscillations |
| 10:20 Break | |
| 10:30 E. Dikmen | Full Shell Model Calculations of $^{113,115}$Sn |
| 11:10 P. Gomes | Nuclear and Coulomb breakup of $^{6}$Li at near barrier energies, their interferences and their effect on fusion |
| 12:00 Lunch | |
| **Chairman: A. Aprahamian** | |
| 13:30 H. Utsunomiya | Photoneutron Reactions in Nuclear Physics and Astrophysics |
| 14:10 S. Noji | Stellar Electron-Capture Rates to $^{46}$Sc via the $^{46}$Ti(t,$^{3}$He+γ) Reaction |
| 14:50 H. Đapo | Photonuclear reactions induced by a clinical linac |
| 15:30 Break | |
| **Chairman: G. De Angelis** | |
| 16:00 A.Soylu | The influence of nuclear deformations on the exotic cluster decay half-lives |
| 16:20 M. Koçak | Parametrization in Description of Prompt Fission Neutron Spectrum in Spontaneous Fisson of $^{252}$Cf |
| 16:40 G. Koçak | The long range potential effect on the$^{20}$O+$^{12}$C system for near and sub-barrier fusion cross section. |
| 17:00 A. A. Aziz | The effects of Double Folding Cluster Model Potential on some astrophysical reactions |
| 17:00 S. Akkoyun | Photonuclear Reaction Cross Sections for Gallium Isotopes |

NUBA Conference Series - 1: Nuclear Physics and Astrophysics IOP Publishing
Journal of Physics: Conference Series **590** (2015) 011002 doi:10.1088/1742-6596/590/1/011002

# LIST OF POSTER CONTRIBUTIONS

**Chairman: I. C. Chamon**

| | |
|---|---|
| M Tamkas, O Akcalı and A Durusoy | Isomeric and ground state energy level measurements of natural tellurium isotopes via (γ,n) reaction |
| C Yalçın | Thickness measurement using alpha spectroscopy and SRIM |
| F Dulger, H Đapo, S Akkoyun, T Bayram and I Boztosun | Energy Levels and Half-Lives of Gallium Isotopes Obtained by Photo-Nuclear Reaction |
| C Parascandolo, D Pierroutsakou, R Alba, A Del Zoppo, C Maiolino, D Santonocito, C Agodi, V Baran, A Boiano, M Colonna, R Coniglione, E De Filippo, M Di Toro, U Emanuele, F Farinon, A Guglielmetti, M La Commara, B Martin, C Mazzocchi, M Mazzocco, C Rizzo, M Romoli, C Signorini, R Silvestri, F Soramel, E Strano, D Torresi, A Trifir and M Trimarchi | Dynamical Dipole mode in heavy-ion fusion reactions in the $^{192}$Pb mass region |
| A Cesur, M Karakoc, I Boztosun and H Đapo | Investigation of the Photo-activation Analysis on Concentrations of Various Materials |
| C Ertugay, I Boztosun, S F Ozmen and H Đapo | Qualitative Elemental Analyses of a Meteorite Sample Found in Turkey by Photo-activation Analysis Method |
| N Burtebayev, D Alimov, I Boztosun, J Burtebayeva, Zh K Kerimkulov, M Nassurlla, N Amangeldy, A K Morzabayev, S K Sakhiev and Sh Hamada | Interaction of 1*p* nuclei: Case of $^{14}$N+$^{12}$C Elastic Scattering at 21.0 MeV |
| N Burtebayev, M Nassurlla, D Alimov, I Boztosun, J Burtebayeva, A K Morzabayev, S K Sakhiev, S V Artemov and J M Mussaev | Optical and Coupled-Channels Description of $^{20}$Ne+$^{16}$O Elastic Scattering |
| D Torresi, M. Mazzocco, L Acosta, A Boiano, C Boiano, A Diaz-Torres, N Fierro, T Glodariu, L Grilj, A Guglielmetti, N Keeley, M La Commara, I Martel, C Mazzocchi, P Molini, A Pakou, C Parascandolo, V V Parkar, N Patronis, D Pierroutsakou, M Romoli, K Rusek, A M Sanchez-Benitez, M Sandoli, C Signorini, R Silvestri, F Soramel, E Stiliaris, E Strano, L Stroe, K Zerva | Reaction dynamics studies for the system $^{7}$Be+$^{58}$Ni |
| M Direkci, Y Kucuk and I Boztosun | Analysis of Elastic Scattering of $^{8}$He+$^{208}$Pb System at around the Coulomb Barrier Energies |
| S Selek, A Çoban, I Boztosun, H Đapo and M Karakoç | Photofission of $^{238}$U Nuclei |

Journal of Physics: Conference Series **590** (2015) 011003 doi:10.1088/1742-6596/590/1/011003

# INTERNATIONAL ADVISORY COMMITTEE

for the NUBA Conference Series 1: Nuclear Physics and Astrophysics

| | |
|---|---|
| T. Akdoğan | Boğaziçi University, Turkey |
| D. Bonatsos | National Center for Scientific Research Demokritos, Greece |
| M. Harakeh | Kernfysisch Versneller Instituut, Netherlands |
| T. Kajino | Tokyo University, Japan |
| N. Meriç | Ankara University, Turkey |
| T. Motobayashi | RIKEN, Japan |
| T. Otsuka | Tokyo University, Japan |
| S. Reddy | Intstitue for Nuclear Theory, USA |
| K. Rusek | Warsaw University, Poland |
| B. Sherrill | Michigan State University, USA |
| C. Spitaleri | Catania University, Italy |
| A. L. Szily | Sao Paulo University, Brasil |
| J. A. Tostevin | Surrey University, UK |
| Ş. Türköz | Ankara University, Turkey |
| A. Vitturi | Padova University, Italy |
| O. Yılmaz | Middle East Technical University, Turkey |

# ORGANIZING COMMITTEE

for the NUBA Conference Series 1: Nuclear Physics and Astrophysics

| | |
|---|---|
| M. Karakoç | Akdeniz University, Turkey |
| H. Djapo | Akdeniz University, Turkey |
| D. Kaya | Akdeniz University, Turkey |
| S. F. Özmen | Akdeniz University, Turkey |
| E. Açıksöz | Akdeniz University, Turkey |
| A. Cesur | Akdeniz University, Turkey |
| A. Çoban | Akdeniz University, Turkey |
| S. Selek | Akdeniz University, Turkey |
| F. Dülger | Akdeniz University, Turkey |

# EDITORIAL BOARD

for the Proceedings of NUBA Conference

| | |
|---|---|
| I. Boztosun (Chairman) | Akdeniz University, Turkey |
| B. Balantekin | University of Wisconsin-Madison, USA |
| Y. Kucuk | Akdeniz University, Turkey |

# Peer review statement

All papers published in this volume of *Journal of Physics: Conference Series* have been peer reviewed through processes administered by the proceedings Editors. Reviews were conducted by expert referees to the professional and scientific standards expected of a proceedings journal published by IOP Publishing.

Journal of Physics: Conference Series **590** (2015) 012001 doi:10.1088/1742-6596/590/1/012001

# Neutrinos, Rare Isotopes of Exotic Nuclei and Nuclear Astrophysics

**A.B. Balantekin**

Physics Department, University of Wisconsin, Madison, WI 53706 USA

E-mail: baha@physics.wisc.edu

**Abstract.** The connection between neutrino physics, nucleosynthesis of elements in astrophysical sites, laboratory measurements with rare exotic nuclei and astronomical observations is discussed. The key role played by neutrinos is emphasized and the close connection between neutrino physics and nucleosynthesis is highlighted.

Understanding where and how various nuclei are synthesized during the evolution of the Universe is one of the key questions of modern science. Element synthesis is thought to be a multi-site and multi-epoch process as depicted in Figure 1. Tackling the question of the origin of elements requires a multitude of tools: High-quality observations of stellar spectra, laboratory atomic physics data, modeling stellar photospheres as well as theoretical and experimental investigations of the relevant nuclear processes. In addition, typically copious amounts of neutrinos are present in most nucleosynthesis sites. This feature makes neutrino physics and neutrino-nucleus interactions salient components of many nucleosynthesis scenarios. This very close link between astronomical observations, neutrino physics, and nucleosynthesis is depicted in Figure 2 and was described in detail in a recent special issue of Journal of Physics G where the reader is referred to for a more complete explanation of many details [1].

In nuclear physics on the theoretical side there have been significant accomplishments during the last decade in calculations with microscopic ab initio methods, driven in part by advances in computational capabilities. The most salient of these results is the realization of the necessity of incorporating three-body forces, in addition to the conventional two-body forces, to describe a wide range of nuclear phenomena. On the experimental side several facilities, already operating or under construction, with intense beams of rare isotopes now have the ability to explore nuclear physics far from stability, at the r-process nucleosynthesis path [2].

There is considerable debate about the astrophysical site of the r-process nucleosynthesis, the two most favorite sites being core-collapse supernovae and neutron-star mergers. Neutrinos not only play a crucial role in the dynamics of these sites, but they also control the value of the electron fraction, the parameter determining the yields in the r-process. (For a recent brief summary see Ref. [3]). Understanding a core-collapse supernova requires answers to a variety of questions some of which need to be answered by nuclear physics, both theoretically and experimentally: Analysis of the formation and subsequent cooling of the neutron star involves high-density equation of state for nuclear matter. Studies of the oscillations of neutrinos as they travel from the neutrinosphere to where r-process or neutrino heating of the shock may take place need knowledge of neutrino properties. One should keep in mind that there are still many puzzles in neutrino physics. For example the near symmetry between second and third generations

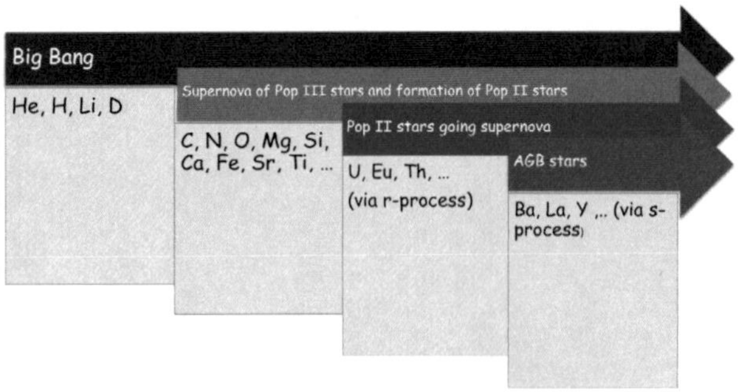


**Figure 1.** The sites of nucleosynthesis for various nuclei.

of neutrinos [4], resulting from an almost maximal mixing angle $\theta_{23}$ and comparatively small mixing angle $\theta_{13}$ [5], seems to mimic astrophysical sites where muons and tau leptons do not exist in bulk and cannot be produced by neutrinos because of the energy considerations. We demonstrate this in Figure 3 where a mixing-matrix triangle ($|U_{\beta 2}|^2$ vs $|U_{\beta 3}|^2$) is plotted. (Note that since $|U_{\beta 1}|^2 + |U_{\beta 2}|^2 + |U_{\beta 3}|^2 = 1$ for all $\beta = e, \mu, \tau$ only two of the mixing matrix entries are free parameters if one ignores possible sterile neutrinos that may mix with the active ones). Finally neutrino-nucleon and neutrino-nucleus interactions determine the electron fraction at the r-process site, element formation rate at the $\nu$-process site, and eventually detection rates at the terrestrial detectors.

New and novel effects come into play when one considers neutrinos in a supernova. Energy released in a core-collapse supernova is about $10^{53}$ ergs which is $10^{59}$ MeV. 99 % of this energy is carried away by approximately $10^{58}$ neutrinos and antineutrinos. This necessitates including the effects of neutrino-neutrino interactions in the Hamiltonian describing neutrino propagation throughout the supernova:

$$H = \sum a^\dagger a + \sum (1 - \cos\theta) a^\dagger a^\dagger a a, \tag{1}$$

where the first term describes neutrino oscillations as well as interaction of neutrinos with matter (including the MSW effect), and the second term describes neutrino-neutrino interactions. The second term makes the physics of a neutrino gas in a core-collapse supernova a very interesting and non-trivial many-body problem, driven by weak interactions. (The angle in the second term is the angle between momenta of the two interacting neutrinos), The resulting so-called collective neutrino oscillations gave rise to an active research area (for recent reviews see Refs. [6] and [7]). Neutrino-neutrino interactions lead to interesting collective and emergent effects, such as conserved quantities and striking features in the neutrino energy spectra (spectral swaps or splits). Symmetries also seem to play an important role in delineating those phenomena [8, 9, 10]. Of particular importance to the element production is the impact of collective neutrino oscillations on the electron fraction. In many, although not all, calculations of the collective neutrino oscillations the angle in Eq. (1) is averaged over (called "single-angle" approximation). Indeed in the first calculation of the electron fraction resulting from the collective neutrino oscillations such an approximation was employed [11]. Subsequent work, however, highlighted the importance of "multi-angle" calculations [12].

Journal of Physics: Conference Series **590** (2015) 012001 doi:10.1088/1742-6596/590/1/012001

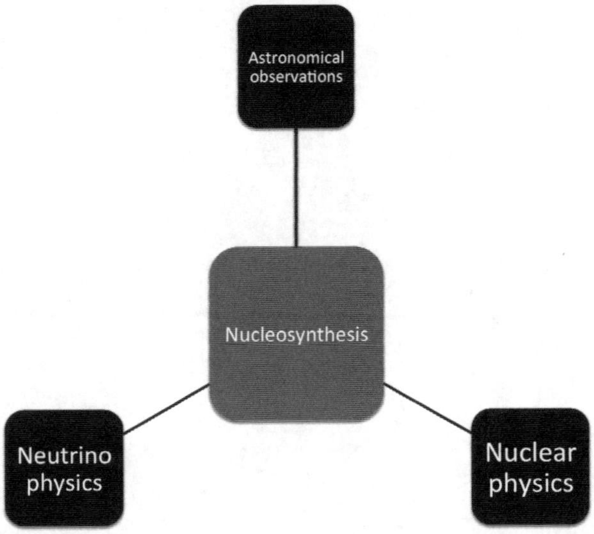


**Figure 2.** Three major physics input areas to describe nucleosynthesis.

It turns out that, because of practical limitations, calculations of the collective effects are carried out in an approximation where the neutrino in question travels in a mean field generated by the other neutrinos. The adequacy of this approximation is an open question, explored in Refs. [13] and [14]. However, at least in the single-angle case, the recognition that the invariants of the exact many-body Hamiltonian reduce to the appropriate constants of motion in the mean field limit [9, 10] suggests that the mean-field approximation appears to capture most of the relevant physics.

It should be remarked that the astronomically observed fossil record of the r-process element abundances resulting from supernova explosions may also give us clues about the dynamics of the collapse event. For example if the progenitor star is sufficiently heavy, following the collapse a black hole may be formed at the center instead of a proto-neutron star. This would truncate emission of the neutrinos that cool the proto-neutron star, possibly altering the abundances of the elements formed during the r-process nucleosynthesis [15]. Current data, however, do not seem to support the possibility of the r-process truncation [16].

Part of the research program with exotic beams is to better understand the r-process nucleosynthesis. This requires understanding beta-decays of nuclei both at and far-from stability. One needs half-lifes at the r-process ladders (N = 50, 82, 126) where abundances peak, accurate values of initial and final state energies as well as the spin-isospin responses of the nuclei. Clearly it will not be possible for the exotic beam facilities to measure those quantities for all the nuclei in the r-process path; the task at hand is to identify particular measurements which will contribute the most to our comprehension of the r-process.

Understanding the spin-isospin response of a broad range of nuclei to a variety of probes is crucial for astrophysics applications [2, 3, 17, 18]. For this purpose one needs to know not only the matrix elements of the Gamow-Teller operator $\sigma\tau$ between the initial and final states, but, in a good number of cases, contributions from the forbidden transitions. Charge-exchange reaction experiments both with direct and inverse kinematics will help. Recently there have been significant developments in this area, particularly in the high-resolution study of the Gamow-Teller transitions with the ($^3$He, t) reactions [19, 20]. On the theoretical side a persistent problem is the need to quench the axial-vector coupling strength $g_A$ in nuclei as compared to the free neutrons. Part of this quenching comes from the limited size of the model space

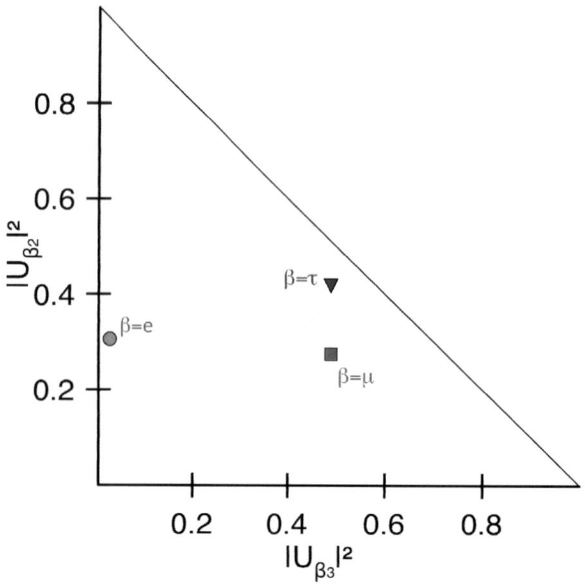


**Figure 3.** The mixing matrix triangle for the three flavors of neutrinos where $\beta$ is any of the In this plot, ignoring any possible mixing between active and sterile neutrinos, the identity $|U_{\beta 1}|^2 + |U_{\beta 2}|^2 + |U_{\beta 3}|^2 = 1$ is used.

and the effective interactions used. A recent development towards relaxing these limitations is a shell-model Hamiltonian, referred to as SFO. SFO Hamiltonian, enhancing monopole terms of the matrix elements in the $p_{1/2}$ and $p_{3/2}$ orbitals, includes tensor components consistent with the general sign rule for the tensor-monopole terms [21, 22, 23]. Calculations with this Hamiltonian reproduces the measured neutrino-$^{12}$C cross sections with a reduced quenching of $g_A$, as compared to the previous calculations [24]. Motivated by this success, SFO Hamiltonian is used to calculate neutrino-$^{13}$C cross sections [25, 26], which are helpful to know in analyzing data from carbon-based scintillators.

Core-collapse supernovae are not the only astrophysical phenomena to explore which information about radioactive, unstable isotopes of nuclei is needed. A different kind of supernovae, called type IA, are set in motion by mass transfer onto the surface of a white dwarf from a giant companion star. This mass transfer is very rapid making white dwarf to approach Chandrasekhar limit and initiating a thermonuclear explosion. Sometimes the mass transfer is slow enough to cause explosive novae on the surface of a white dwarf star. Accreting neutron stars from a binary companion can result in X-ray bursts similarly powered by nuclear reactions. To understand all those phenomena information to be gained from nuclear radioactive beam facilities is essential.

Although data radioactive-beam facilities are crucial to obtain a complete picture of element synthesis, one should not ignore the need for more precise data from stable-beam experiments. Arguably three most critical such reactions for nuclear astrophysics are $^3$He$(\alpha,\gamma)^7$Be, $^{12}$C$(\alpha,\gamma)^{16}$O, and $^{14}$N(p,$\gamma$)$^{15}$O [27]. The first reaction is the source of largest uncertainty in solar modeling and also Big Bang nucleosynthesis (albeit at different energy scales). The second reaction determines the ratio of $^{12}$C to $^{16}$O in the Universe, the determining parameter which makes the existence of living organisms for example on Earth possible. The third reaction is the key component of the CNO cycle. Although the Sun shines primarily with pp-cycle, CNO-cycle is the dominant energy generating mechanism for heavier stars, which comprises most of the past and present stars.

Clearly many aspects of the close connection between neutrino physics and nucleosynthesis still need to be explored using observational, theoretical, computational, and experimental tools available for a better understanding of the origin of elements.

## Acknowledgments

I would like to thank Akdeniz University nuclear physics group for its hospitality and my collaborators G. Fuller, W. Haxton, T. Kajino, Y. Pehlivan, and T. Suzuki for many enlightening discussions. This work was supported in part by the U.S. National Science Foundation Grant PHY-1205024 and in part by the University of Wisconsin Research Committee with funds granted by the Wisconsin Alumni Research Foundation.

## References

[1] Volpe C and Balantekin A B 2014 *J. Phys. G* **41** 040301; Jacobson H R and Frebel A 2014 *J. Phys. G* **41** 044001 [arXiv:1309.0037 [astro-ph.GA]]; Qian Y Z 2014 *J. Phys. G* **41** 044002 [arXiv:1310.4462 [astro-ph.SR]]; Frohlich C 2014 *J. Phys. G* **41** 044003; Caballero O L, Malkus A C, McLaughlin G C and Surman R A 2014 *J. Phys. G* **41** 044004; Arcones A and Bliss J 2014 *J. Phys. G* **41** 044005 [arXiv:1312.0434 [astro-ph.SR]]; Surman R, Caballero O L, McLaughlin G C, Just O and Janka H T 2014 *J. Phys. G* **41** 044006 [arXiv:1312.1199 [astro-ph.SR]]; Kajino T, Mathews G J and Hayakawa T 2014 *J. Phys. G* **41** 044007; Martnez-Pinedo G, Fischer T and Huther L 2014 *J. Phys. G* **41** 044008 [arXiv:1309.5477 [astro-ph.HE]].

[2] Balantekin A B, Carlson J, Dean D J, Fuller G M, Furnstahl R J, Hjorth-Jensen M, Janssens R V F and Li B A *et al.* 2014 *Mod. Phys. Lett.* A **29** 1430010 [arXiv:1401.6435 [nucl-th]].

[3] Balantekin A B and Fuller, G M 2013 *Prog. Part. Nucl. Phys.* **71** 162 [arXiv:1303.3874 [nucl-th]].

[4] Balantekin A B and Fuller G M 1999 *Phys. Lett. B* **471** 195 [hep-ph/9908465].

[5] For a recent brief review of the neutrino properties see e.g., Balantekin A B and Haxton W C 2013 *Prog. Part. Nucl. Phys.* **71** 150 [arXiv:1303.2272 [nucl-th]].

[6] Duan H and Kneller J P 2009 *J. Phys. G* **36** 113201 [arXiv:0904.0974 [astro-ph.HE]].

[7] Duan H, Fuller G M and Qian Y Z 2010 *Ann. Rev. Nucl. Part. Sci.* **60** 569 [arXiv:1001.2799 [hep-ph]].

[8] Duan H, Fuller G M and Qian Y Z 2009 *J. Phys. G* **36** 105003 [arXiv:0808.2046 [astro-ph]].

[9] Pehlivan Y, Balantekin A B, Kajino T and Yoshida T 2011 *Phys. Rev. D* **84** 065008 [arXiv:1105.1182 [astro-ph.CO]].

[10] Pehlivan Y, Balantekin A B and Kajino T 2014 *Phys. Rev. D* **90** 065011 arXiv:1406.5489 [hep-ph].

[11] Balantekin A B and Yuksel H 2005 *New J. Phys.* **7** 51 [astro-ph/0411159].

[12] Duan H, Friedland A, McLaughlin G C and Surman R 2011 *J. Phys. G* **38** 035201 [arXiv:1012.0532 [astro-ph.SR]].

[13] Balantekin A B and Pehlivan Y 2007 *J. Phys. G* **34** 47 [astro-ph/0607527].

[14] Volpe C, Väänänen D and Espinoza C 2013 *Phys. Rev. D* **87** 113010 [arXiv:1302.2374 [hep-ph]]; Väänänen D and Volpe C 2013 *Phys. Rev. D* **88** 065003 [arXiv:1306.6372 [hep-ph]].

[15] Sasaqui T, Kajino T and Balantekin A B 2005 *Astrophys. J.* **634** 534 [astro-ph/0506100].

[16] Aoki W, Suda T, Boyd R, Kajino T and Famiano M 2013 *Astrophys. J.* **766** L13 [arXiv:1302.6573 [astro-ph.SR]].

[17] Litvinova E, Brown B A, Fang D L, Marketin T and Zegers R G T 2014 *Phys. Lett. B* **730** 307 [arXiv:1402.0877 [nucl-th]].

[18] Cole A L, Anderson T S, Zegers R G T, Austin S M, Brown B A, Valdez L, Gupta S and Hitt G W *et al.* 2012 *Phys. Rev. C* **86** 015809 [arXiv:1204.1994 [nucl-ex]].

[19] Fujita Y 2008 *Nucl. Phys. A* **805** 408.

[20] Fujita Y, Adachi T, Fujita H, Algora A, Blank B, Csatlos M, Deaven J M and Estevez-Aguado E *et al.* 2013 *Phys. Rev. C* **88** 014308.

[21] Suzuki T, Fujimoto R and Otsuka T 2003 *Phys. Rev. C* **67** 044302.
[22] Otsuka T, Suzuki T, Fujimoto R, Grawe H and Akaishi Y 2005 *Phys. Rev. Lett.* **95** 232502.
[23] Suzuki T, Chiba S, Yoshida T, Higashiyama K, Honma M, Kajino T and Otsuka T 2007 *Prog. Part. Nucl. Phys.* **59** 486.
[24] Suzuki T, Chiba S, Yoshida T, Kajino T and Otsuka T 2006 *Phys. Rev. C* **74** 034307 [nucl-th/0608056].
[25] Suzuki T, Honma M, Balantekin A B, Kajino T and Chiba S 2014 *EPJ Web Conf.* **66** 07025.
[26] Suzuki T, Balantekin A B and Kajino T 2012 *Phys. Rev. C* **86** 015502 [arXiv:1204.4231 [nucl-th]].
[27] Adelberger E G *et al.* 2011 *Rev. Mod. Phys.* **83** 195 [arXiv:1004.2318 [nucl-ex]].

Journal of Physics: Conference Series **590** (2015) 012002 doi:10.1088/1742-6596/590/1/012002

# Cluster states and container picture in light nuclei, and triple-alpha reaction rate

**Yasuro Funaki**[1]

[1] RIKEN, Nishina Center, Wako 351-0198, Japan

E-mail: funaki@riken.jp

**Abstract.** The excited states in $^{12}$C are investigated by using an extended version of the so-called Tohsaki-Horiuchi-Schuck-Röpke (THSR) wave function, where both the $3\alpha$ condensate and $^8$Be $+ \alpha$ cluster asymptotic configurations are included. We focus on the structures of the "Hoyle band" states, $2_2^+$, and $4_2^+$ states, which are recently observed above the Hoyle state, and of the $0_3^+$ and $0_4^+$ states, which are also quite recently identified in experiment. We show that the Hoyle band is not simply considered to be the $^8$Be$(0^+) + \alpha$ rotation as suggested by previous cluster model calculations, nor to be a rotation of a rigid-body triangle-shaped object composed of the $3\alpha$ particles. We also discuss the rate of the triple-alpha radiative capture reaction, applyng the imaginary-time method. Results of the triple-alpha reaction rate are consistent with NACRE rate for both high ($\approx 10^9 K$) and low ($\approx 10^7 K$) temperatures. We show that the rate of the imaginary-time calculation in coupled-channels approach has a large enhancement for low temperatures if we truncate the number of channels.

## 1. Introduction

The Hoyle state, the second $J^\pi = 0^+$ state at 7.654 MeV in $^{12}$C, is a typical example of cluster states. For this state, the aspect of the $\alpha$ condensate, where the $3\alpha$ clusters occupy an identical $S$-orbit, has triggered a great interest for recent years, since the so-called Tohsaki-Horiuchi-Schuck-Röpke (THSR) wave function [1], which has the $\alpha$ condensate character, was shown to be equivalent to the Hoyle state wave function obtained by solving the equations of the full $3\alpha$ resonating group method (RGM) or generator coordinate method (GCM) [2, 3].

The excited states above the Hoyle state have also recently been highlighted by experiments. The $2_2^+$ state, which theoreticaly had been predicted at a few MeV above the Hoyle state as a member of the Hoyle band, was recently confirmed by many experiments [4, 5, 6, 7, 8]. A strong candidate of the member of the Hoyle band with $J^\pi = 4^+$ was also reported by Freer *et al.* [9]. Besides the $2^+$ and $4^+$ states, quite recently Itoh *et al.* insisted that a broad width $0^+$ state at 10.3 MeV should be decomposed into the $0_3^+$ and $0_4^+$ states [4, 10]. This observation of the two $0^+$ states is consistent with theoretical predictions done by using the OCM combined with the CSM [11, 12].

On the other hand, the Hoyle state is known to play a key role in a synthesis of $^{12}$C in stellar evolution. At a temperature above 0.1 GK, the Hoyle resonance contributes dominantly [13]. Below 0.1 GK, however, non-resonant contribution is considered to be significant [14, 15, 16]. In a standard compilation NACRE [17], which is used to describe various astronomical phenomena, the non-resonant triple-alpha rate is evaluated by extending the sequential model at high temperatures to low temperatures. Recently, Ogata *et al.* [18] reported that the radiative capture rate was surprisingly large below 0.1 GK in comparison with NACRE rate ($\approx 10^{26}$ at 0.01 GK) by using the continuum-discretized coupled channels (CDCC) method [19]. We should also mention that some other different techniques are also applied to the calculation of the reaction rate and give large increases of the magnitude at low temperature regions compared to NACRE, for example, $\approx 10^{18}$ in Ref. [21] and $\approx 10^{6}$ in Ref. [20] at 0.01 GK, whereas the Faddeev calculation gives very similar result to NACRE [22].

The first half of this contribution is devoted to the structure study of the positive parity excited states above the $3\alpha$ threshold by using an extended version of the THSR wave function [23] so as to include $^8\mathrm{Be}+\alpha$ asymptotic configurations with a treatment of resonances. In particular, we focus on the "Hoyle band" (the $0_2^+$, $2_2^+$, and $4_2^+$ states, and the $0_3^+$ and $0_4^+$ states, together with the corresponding experimental data. In the latter half, we discuss the triple-alpha reaction rate by using the imaginary-time method where the radiative capture rate as a function of temperature can be calculated directly by solving an equation that looks like a time-dependent Schrödinger equation along the imaginary-time axis [24, 25].

## 2. Extended THSR wave function and Hoyle-family states in $^{12}$C

The extended version of the THSR wave function is written as follows:

$$\Phi_{JM}^{\mathrm{THSR}}(\boldsymbol{\beta}_1,\boldsymbol{\beta}_2)=\widehat{P}_{MK}^{J}\mathcal{A}\Big[\prod_{i=1}^{2}\exp\Big\{-\mu_i\sum_{k=x,y,z}\frac{2(\xi_{ix})^2}{(B_{ik})^2}\Big\}\prod_{i=1}^{3}\phi(\alpha_i)\Big], \tag{1}$$

where $B_{ik}^2=b^2+2\beta_{ik}^2$ with $i=1,2$ and $k=x,y,z$. This wave function is characterized by the parameters $\boldsymbol{\beta}_1$ and $\boldsymbol{\beta}_2$, which correspond to the sizes of the $^8$Be core and the remaining $\alpha$ particle c.o.m. motion, respectively. The case of $\boldsymbol{\beta}_1=\boldsymbol{\beta}_2$ results in the original THSR wave function. Thus this new THSR wave function is a natural extension of the original version, so that taking $|\boldsymbol{\beta}_1|\le|\boldsymbol{\beta}_2|$ allows for the $^8\mathrm{Be}+\alpha$ cluster structure, deviating from the identical $3\alpha$-cluster motion for $\boldsymbol{\beta}_1=\boldsymbol{\beta}_2$. We should note that this THSR wave function includes the $^8\mathrm{Be}+\alpha$ asymptotic feature for $|\boldsymbol{\beta}_1|\ll|\boldsymbol{\beta}_2|$, and is also feasible for approaches to resonances like the complex scaling method (CSM) and the analytic continuation of the coupling constant (ACCC) method. In order to obtain the excited states, we solve the Hill-Wheeler equation by this new THSR wave function taking both $\boldsymbol{\beta}_1$ and $\boldsymbol{\beta}_2$ as the generator coordinates, where axial symmetric deformation is assumed, i.e. $\beta_{i\perp}\equiv\beta_{ix}=\beta_{iy}$. In this calculations we effectively remove spurious continuum components by using a method of the so-called radius constraint method [26, 23]. In Fig. 1(left), the calculated energy spectra are shown. One can see that besides the ground state band, there are many $J^\pi$ states obtained above the Hoyle state. All these states have large r.m.s. radii (3.7 fm $\sim$ 4.7 fm), and therefore may have gas-like cluster structures. These states seem to appear as the excited states of the Hoyle state that looks like the "ground state" of the gas-like cluster states.

In Fig. 1(right), the $E2$ transition strengths between $J$ and $J\pm 2$ states and monopole transition strengths between $J^\pi=0^+$ states are also shown with the corresponding arrows. We can see the very strong $E2$ transitions inside the Hoyle band, $B(E2;4_2^+\to 2_2^+)=591\ e^2\mathrm{fm}^4$ and $B(E2;2_2^+\to 0_2^+)=295\ e^2\mathrm{fm}^4$. The transition between the $2_2^+$ and $0_3^+$ state is also very large, $B(E2;2_2^+\to 0_3^+)=104\ e^2\mathrm{fm}^4$. We should note that the $0_3^+$ state is strongly excited from the Hoyle state by monopole transition, whose strength is calculated to be $M(E0;0_3^+\to 0_2^+)=35$ fm$^2$. In Fig. 1(right), the calculated energy levels are plotted as a function of $J(J+1)$, together with the experimental data. We can say that roughly the $0_2^+$, $2_2^+$, and $4_2^+$ states both in theory and experiment follow a $J(J+1)$ trajectory, and therefore, on the one hand, the rotational picture is very good. On the other hand, the $J^\pi=0^+$ bandhead in experiment seems to be fragmented into the Hoyle state and the $0_3^+$ state, and the calculated levels also have similar tendency, where the Hoyle state is located slightly below the $J(J+1)$ line.

Journal of Physics: Conference Series **590** (2015) 012002 doi:10.1088/1742-6596/590/1/012002

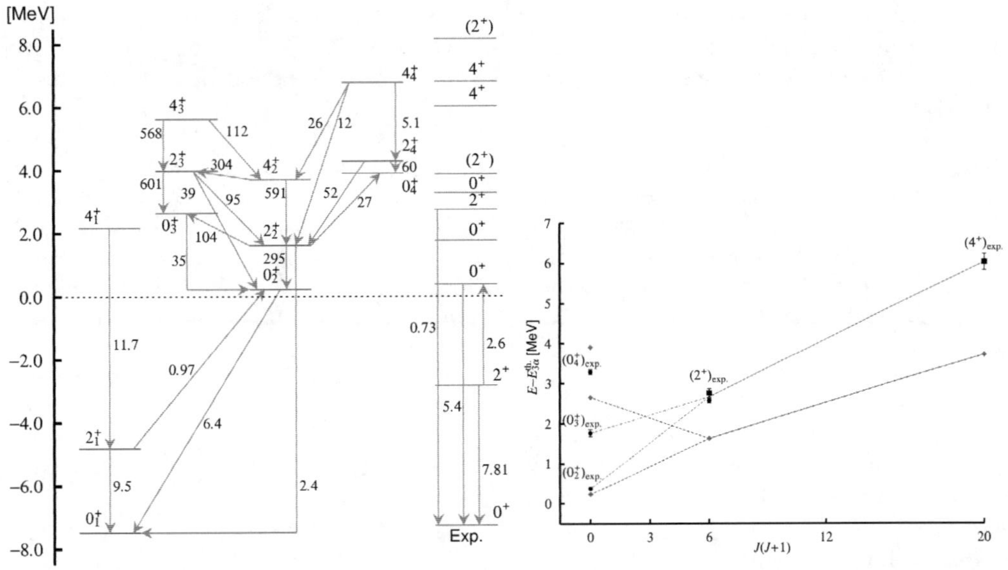


**Figure 1.** (color online). (Left): Calculated energy levels and electric transition strengths are shown and compared with experiments. (Right): The observed energy levels for the $0_3^+$, $0_4^+$, and $2_2^+$ states in Ref. [4], and the $2_2^+$ [5] and $4_2^+$ [9] states are denoted by black circles and black squares, respectively. The calculated energy levels for the five states are denoted by red diamonds. Figure (right) taken from [23].

**Table 1.** The $S$-factors $S^2_{[I,l]}(J^\pi)$ for the five $J^\pi$ states.

| $J^\pi$ | | $S^2_{[I,l]}(J^\pi)$ | | | | | |
|---|---|---|---|---|---|---|---|
| | $[I,l]$ | $[0,0]$ | | $[2,2]$ | | $[4,4]$ | |
| $0_2^+$ | | 1.85 | | 0.13 | | 0.04 | |
| $0_3^+$ | | 1.09 | | 0.06 | | 0.03 | |
| $0_4^+$ | | 0.25 | | 1.47 | | 0.05 | |
| | $[I,l]$ | $[0,2]$ | $[2,0]$ | $[2,2]$ | $[2,4]$ | $[4,2]$ | $[4,4]$ |
| $2_2^+$ | | 1.24 | 0.53 | 0.07 | 0.03 | 0.08 | 0.02 |
| | $[I,l]$ | $[0.4]$ | $[4,0]$ | $[2,2]$ | $[2,4]$ | $[4,2]$ | $[4,4]$ |
| $4_2^+$ | | 0.88 | 0.24 | 0.65 | 0.06 | 0.02 | 0.0 |

We then discuss the $S$-factor of the $^8$Be + $\alpha$ channel, which is defined as $S_{[I,l]}(J) = \int dr \left(r\mathcal{Y}_{[I,l]_J}(r)\right)^2$. Here $\mathcal{Y}_{[I,l]_J}(r)$ is the reduced width amplitude (RWA) of the $^8$Be$+\alpha$ channel. We show in Table 1 the calculated $S$-factors for these excited states. We can see that except for the $0_4^+$ state, all the states have the largest contribution from the $[0,J]_J$ channel. This supports the idea of $^8$Be + $\alpha$ rotation for the Hoyle band, where the $^8$Be core is in the $0^+$ ground state. On the other hand, the Hoyle state is considered to be the $3\alpha$ condensate state, where the $3\alpha$ clusters mutually move in an identical $S$-wave. Since the ground state of $^8$Be is composed of weakly interacting $2\alpha$ clusters coupled loosely in a relative $S$-wave, it is natural that the Hoyle state, with the $\alpha$ condensate structure, also has a large overlap with the $^8$Be$(0^+) + \alpha$ structure.

However, the $3\alpha$ condensate structure in the Hoyle state is not the same as the usual $^8$Be$(0^+)+\alpha$ rotation, in which the remaining $\alpha$ cluster orbits outside the $^8$Be core. Namely in the Hoyle state, the remaining $\alpha$ cluster also orbits inside the $^8$Be core and the independent $3\alpha$-cluster motion in an identical $0S$-orbit is realized. As a result, the Hoyle state gains an extra binding, and hence its energy position is considered to be made lower than the $J(J+1)$ line, as shown in Fig. 1(right). The same effect is also argued in the study of the $4\alpha$ condensate and $^{12}$C$(0_2^+)+\alpha$ rotational band [27, 28], where the $4\alpha$ condensate is mentioned as "complete condensate" and the $^{12}$C$(0_2^+)+\alpha$ state as "local condensate". Due to the existence of the "complete condensate", a higher $0^+$ excited state, which has the prominent $^8$Be$(0^+)+\alpha$ structure, with the remaining $\alpha$ cluster orbiting outside the $^8$Be core, appears as a higher nodal state, that is the $0_3^+$ state, excited by strong monopole transition from the Hoyle state. In Table 1, the $0_4^+$ state is shown to have the component of $[2,2]_0$ channel dominantly. This is consistent with the $0_3^+$ state calculated by GCM [2], AMD [29], and FMD [30]. In particular, in AMD and FMD calculations this state is shown to have linear-chain-like intrinsic density. We also mention that the $2_2^+$ state also includes non-negligible mixture from the $[2,0]_2$ channel and $4_2^+$ states from the $[2,2]_4$ channel and smaller amount from the $[4,0]_4$ channel. These mixtures also deviate the $2_2^+$ and $4_2^+$ states from a pure $^8$Be $+\ \alpha$ rotational structure.

### 3. Triple-alpha reaction rate

We start with the following expression for the triple-alpha reaction rate [24, 25],

$$N_A^2\langle\alpha\alpha\alpha\rangle = 6\cdot 3^{3/2}N_A^2\left(\frac{2\pi\beta\hbar^2}{M_\alpha}\right)^3\frac{8\pi(\lambda+1)}{\hbar\lambda((2\lambda+1)!!)^2}\sum_{M_f\mu}\langle\Phi_f|M_{\lambda\mu}e^{-\beta H}\left(\frac{H-E_f}{\hbar c}\right)^{2\lambda+1}PM_{\lambda\mu}^\dagger|\Phi_f\rangle, \tag{2}$$

where $\beta$ is the inverse temperature $\beta = 1/k_BT$, $M_\alpha$ is the mass of the $\alpha$-particle, $H$ is the Hamiltonian of three $\alpha$ particles, $\Phi_f$ is the wave function of $^{12}$C in the final state after the $\gamma$-ray emission, $E_f$ is the energy of the final state measured from the three $\alpha$ threshold, and $M_f$ is the magnetic quantum number of the final state. $M_{\lambda\mu}$ is the multipole transition operator of $\gamma$-ray emission with multipolarity $\lambda$. We consider reactions of three $\alpha$-particles in the total angular momentum $J=0$, emission of $\lambda=2$ $\gamma$-ray, and $2^+$ state of $^{12}$C at 4.44 MeV for the final wave function $\Phi_f$. $P$ is the projection operator which eliminates any bound eigenstates of the three-body Hamiltonian. We may easily show that Eq. (2) exactly coincides with the expression for the triple-alpha reaction rate, Eqs. (2) $\sim$ (5) of [31], if we insert a completeness relation of a functional space of three $\alpha$-particles in Eq. (2).

To evaluate the rate in practice according to Eq. (2), we need to calculate $\Psi(\beta) = e^{-\beta H}M_{\lambda\mu}^\dagger\Phi_f$. We treat it by evolving the wave function along an imaginary-time axis,

$$-\frac{\partial}{\partial\beta}\Psi(\beta) = H\Psi(\beta), \tag{3}$$

starting with the initial wave function, $\Psi(\beta=0) = M_{\lambda\mu}^\dagger\Phi_f$. The rate at the temperature specified by $\beta$ is then evaluated using the wave function at $\beta/2$,

$$\langle\alpha\alpha\alpha\rangle \propto \sum_{M_f\mu}\left\langle\Psi\left(\frac{\beta}{2}\right)\middle|\left(\frac{H-E_f}{\hbar c}\right)^5\middle|\Psi\left(\frac{\beta}{2}\right)\right\rangle. \tag{4}$$

We have achieved numerical calculations using the model pace and the three-body Hamiltonian described below. We treat $\alpha$-particles as point particles, and consider reactions of three $\alpha$ particles in the space of total angular momentum $J=0$ as mentioned above,

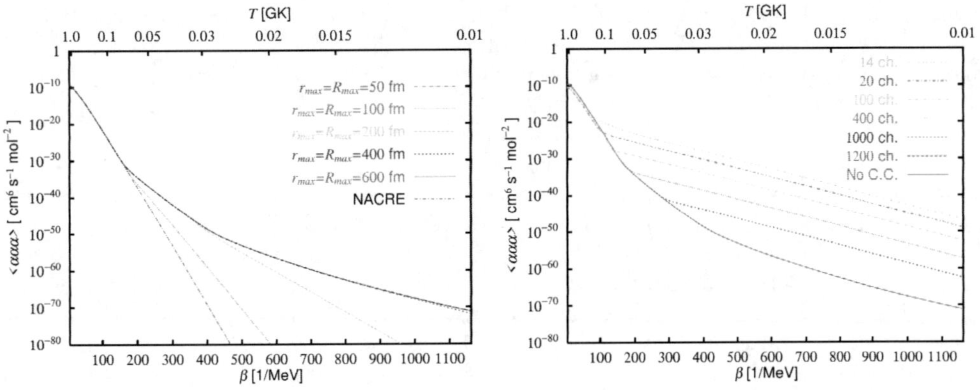


**Figure 2.** (Color online) (Left): Calculated triple-alpha reaction rates for several choices of radial cutoff distances, $r_{\max}$ and $R_{\max}$, in comparison with the NACRE rate. (Right): Triple-alpha reaction rates employing the coupled-channels expansion. Results of different number of employed channels specified by $n_{\max}$ are compared. Figures taken from Ref. [25].

which should be sufficient below $T < 1.0$ GK [31]. We use a Jacobi coordinate system of two relative coordinates, $\boldsymbol{r} = \boldsymbol{r}_1 - \boldsymbol{r}_2$ and $\boldsymbol{R} = (\boldsymbol{r}_1 + \boldsymbol{r}_2)/2 - \boldsymbol{r}_3$, with $\boldsymbol{r}_i$ $(i = 1-3)$ being coordinates of three $\alpha$-particles. The three-body wave function is expanded in partial waves, $\Psi(\beta) = \sum_L (u_L(R,r,\beta)/Rr)[Y_L(\hat{R})Y_L(\hat{r})]_{J=0}$. In the present work, we only treat $L=0$ component which should be the most important at low temperature. We note that the same model space is adopted in the CDCC calculation [18].

The Hamiltonian $H$ of three $\alpha$-particles is constructed as follows. For the potential between two $\alpha$-particles, we use the Ali-Bodmer potential of angular momentum $l=0$ channel. The potential parameter is slightly modified so that it reproduces the resonance corresponding to the ground state of $^8$Be at 92.08 keV accurately. We add a three-body potential among three $\alpha$ particles, $V_{3\alpha}(\boldsymbol{r}_1, \boldsymbol{r}_2, \boldsymbol{r}_3) = V_3 e^{-\mu(r_{12}^2 + r_{23}^2 + r_{31}^2)}$ with $\mu = 0.15$ fm$^{-2}$. The value of $V_3$ is chosen so that the resonance energy of the Hoyle state, $0_2^+$ state of $^{12}$C, is reproduced at 379.8 keV above the three $\alpha$ threshold. As for the final wave function, $\Phi_f$, of $^{12}$C $J^\pi = 2^+$ at 4.44 MeV, we use the one constructed by the orthogonality condition model [32].

In Fig. 2(left), calculated triple-alpha reaction rates adopting different spatial areas specified by $R_{\max}$ and $r_{\max}$ are compared. We also show the NACRE rate [17] for comparison. The figure shows that a convergent reaction rate in the entire temperature region is obtained if we adopt $R_{\max}$ and $r_{\max}$ larger than 400 fm. Calculations in smaller spatial areas provide the rate which is correct in the limited higher temperatures.

We next examine the origin of the discrepancy between the theories: why different theoretical approaches result in so different rates at low temperatures. In particular, the CDCC approach [18] which employs almost the same model space as adopted here, differs by $10^{26}$ order of magnitude at low temperature from the NACRE rate and ours. To clarify the origin of this discrepancy, we have achieved the imaginary-time evolution of Eq. (3) using the coupled-channels method. In the coupled-channels approach, we first solve the eigenvalue problem of $\alpha$-$\alpha$ relative motion described by the Hamiltonian $h_{\alpha\alpha}(r)$. Discretizing the radial variable $r$ in 0.5 fm step up to 600 fm, we have 1200 grid points for this coordinate. Diagonalizing the radial Hamiltonian, we obtain 1200 eigenfunctions, $w_n(r)$, $(n = 1-1200)$. Eigenstates belonging to low eigenvalues are characterized by a large $\alpha$-$\alpha$ separation outside the Coulomb barrier, except for the resonant state corresponding to the $^8$Be ground state which appears as 14th eigenstate. In the coupled-channels approach, we expand the wave function as $u(R,r,\beta) = \sum_n v_n(R,\beta)w_n(r)$ and calculate the imaginary-time evolution of $v_n(R,\beta)$ in the form of matrix-differential equation. We numerically confirmed that, employing all 1200 eigenstates in the expansion, calculated rate exactly coincides with the result shown in Fig. 2(left), as it should be. However, if we truncate the number of basis in the expansion, results depend on the truncation. Convergence with respect to the number of employed basis functions is shown in Fig. 2(right).

Taking only the lowest state ($n_{\rm max} = 1$), the reaction rate is much higher than the fully convergent calculation by $10^{16}$ order of magnitude at $T = 0.01$ GK. This enhancement may be easily understood. In the lowest eigenstate of $h_{\alpha\alpha}(r)$, two $\alpha$-particles are much apart to each other and stay outside the Coulomb barrier. Then the Coulomb barrier for the third $\alpha$-particle is very small in this channel, giving large reaction rate. The rate first increases as the number of channels increases. In Fig. 2(right), the rate shows maximum when we include the channel of $\alpha$-$\alpha$ resonance, the 14th channel, $10^{24}$ times larger than the convergent value at $T = 0.01$ GK. The artificial enhancement in small-channel calculations should be compensated by off-diagonal coupling terms as we increase the number of channels. However, as is seen from the figure, the convergence is extremely slow. Even with 400 channels, the reaction rate is still $10^{13}$ times larger than the fully convergent result. We consider it manifests a numerical difficulty of expressing exponentially small function in the tunneling process using the basis expansion method.

## References

[1] Tohsaki A, Horiuchi H, Schuck P, and Röpke G 2001 *Phys. Rev. Lett.* **87** 192501
[2] Fujiwara Y, Horiuchi H, Ikeda K, Kamimura M, Katō K, Suzuki Y, and Uegaki E 1980 *Prog. Theor. Phys. Supple.* **68** 29
[3] Funaki Y, Tohsaki A, Horiuchi H, Schuck P, and Röpke G 2003 *Phys. Rev.* C **67** 051306(R)
[4] Itoh M *et al.* 2004 *Nucl Phys. A* **738** 268; Itoh M *et al.*, 2011 *Phys. Rev. C* **84** 054308
[5] Zimmerman W R *et al.* 2013 *Phys. Rev. Lett.* **110** 152502
[6] Freer M *et al.* 2009 *Phys. Rev. C* **80** 041303(R)
[7] Zimmerman W R *et al.* 2011 *Phys. Rev. C* **84** 027304
[8] Fynbo H O U and Freer M 2011 *Physics* **4** 94
[9] Freer M *et al.* 2011 *Phys. Rev. C* **83** 034314
[10] Itoh M *et al.*, 2013 *J. Phys. Conf. Ser.* **436** 012006
[11] Kurokawa C and Katō K 2005 *Phys. Rev. C* **71** 021301; 2007 *Nucl. Phys. A* **792** 87
[12] Ohtsubo S, Fukushima Y, Kamimura M, and Hiyama E 2013 *Prog. Theor. Exp. Phys.* 073D02
[13] Hoyle F, Astrophys. 1954 *J. Suppl.* **1** 121
[14] Nomoto K, Thielemann F-K, Miyaji S 1985 *Astron. Astrophys.* **149** 239
[15] Langanke K, Wiescher M, and Thielemann F-K 1986 *Z. Phys. A* **324** 147
[16] Descouvemont P and Baye D 1987 *Phys. Rev. C* **36** 54
[17] Angulo C *et.al.* 1999 *Nucl. Phys.* **A656** 3
[18] Ogata K, Kan M, Kamimura M 2009 *Prog. Theor. Phys.* **122** 1055
[19] Kamimura M, Yahiro M, Iseri Y, Sakuragi Y, Kameyama H, Kawai M 1986 *Prog. Theor. Phys. Suppl.* **89** 1
[20] Garrido E, de Diego R, Fedorov D V, and Jensen A S 2011 *Eur. Phys. J. A* **47**, 102
[21] Nguyen N B, Nunes F M, Thompson I J, Brown E F 2012 *Phys. Rev. Lett.* **109**, 141101
[22] Ishikawa S 2013 *Phys. Rev. C* **87**, 055804
[23] Funaki Y, arXiv: 1408.5855.
[24] Yabana K, Funaki Y 2012 *Phys. Rev. C* **85**, 055803
[25] Akahori T, Funaki Y, and Yabana K, arXiv: 1401.4390
[26] Funaki Y, Horiuchi H, and Tohsaki A 2006 *Prog. Theor. Phys.* **115** 115
[27] Ohkubo S, Hirabayashi Y 2010 *Phys. Lett. B* **684** 127
[28] Funaki Y, Yamada T, Horiuchi H, Röpke G, Schuck P, and Tohsaki A 2012 *Prog. Theor. Phys. Supple.* **196** 439
[29] Kanada-En'yo Y 2007 *Phys. Rev. C* **75** 024302
[30] Chernykh M, Feldmeier H, Neff T, von Neumann-Cosel P, and Richter A 2007 *Phys. Rev. Lett.* **98** 032501
[31] de Diego R, Garrido E, Fedorov D V, Jensen A S 2011 *Phys. Lett. B* **695** 324
[32] Yamada and T, Schuck P 2005 *Eur. Phys. J. A* **26** 185

Journal of Physics: Conference Series **590** (2015) 012003 doi:10.1088/1742-6596/590/1/012003

# New probe to M1 strengths in GDR region for supernova neutrino interactions

**T. Hayakawa[1], S. Miyamoto[2], T. Mochizuki[2], K. Horikawa[2], S. Amano[2], K. Imazaki[3], D. Li[3], Y. Izawa[3], S. Chiba[4], K. Ogata[5]**

[1]Japan Atomic Energy Agency
[2]University of Hyogo
[3]Institute for Laser Technology
[4]Tokyo Institute of Technology
[5]RCNP, Osaka University

E-mail: hayakawa.takehito@jaea.go.jp

**Abstract.** The M1 strength (or level density of $1^+$ states) is of importance for estimation of interaction strengths between neutrinos and nuclei for the study of the supernova neutrino-process. We have proposed a method using (γ, n) reactions with linear polarized laser Compton scattering γ-rays to measure M1 strength. In 1957, Agodi predicted theoretically angular distribution of neutrons emitted from states excited via dipole transitions with linearly polarized γ-ray beam at the polar angle of $\theta$=90° can be described by a simple function, a + b cos($\phi$2), where $\phi$ is azimuthal angel. However, this theoretical prediction has not been verified over the wide mass region. We have measured neutron angular distributions with (polarized gamma, n) reactions on Au, NaI, and Cu. We have verified the Agodi's prediction for the first time over the wide mass region. This suggests that (polarized gamma, n) reactions may be useful tools to study M1 strength in giant resonance regions.

## 1. Introduction

During core-collapse supernovae copious amounts of energetic neutrinos are emitted from the proto-neutron star. As these neutrinos pass through the outer layers of the star they can induce nuclear reactions on ambient nuclei. This process is referred to as the neutrino process. This has been proposed as the mechanism responsible for the origin of some rare isotopes of light and heavy elements [1]. The neutrino process is also of importance for constraining the neutrino spectra from SNe [2, 3] and for studying neutrino oscillations [2, 4]. Although many nuclides can be synthesized by the neutrino process, the produced abundance is usually negligibly small compared to production by other major nucleosynthesis processes such as s-process or r-process. Thus, the neutrino process can only play a significant role in the synthesis of isotopes which are bypassed by other major processes. The astrophysical origin of $^{180}$Ta has been unresolved problem for long time. Hayakawa et al. [5] presented that the neutrino process reproduces systematically the solar abundances of $^{138}$La and $^{180}$Ta with a new isomer model, whereas other scenarios can explain partly their abundances. $^{92}$Nb is a beta unstable isotope with a half-life of about $3.5\times10^7$ y, which does not exist at the present solar system. The abundance of $^{92}$Nb at the solar system formation has been measured by primitive meteorite analysis. However, its origin has been also unresolved problem. Hayakawa et al. [6] has proposed the neutrino process origin for $^{92}$Nb and calculated a theoretical abundance at the solar system formation using a supernova model with neutrino-nucleus interactions calculated by a QRPA model. The origin of $^{92}$Nb can be explained by the supernova neutrino process and the duration time between the time when the supernova explosion occurred and the time of the solar system formation is $10^6$–$3\times10^7$ y. In this way, the supernova neutrino process has important role for understanding of origin of rare isotopes, the cosmological history before the solar system formation, and neutrino physics.

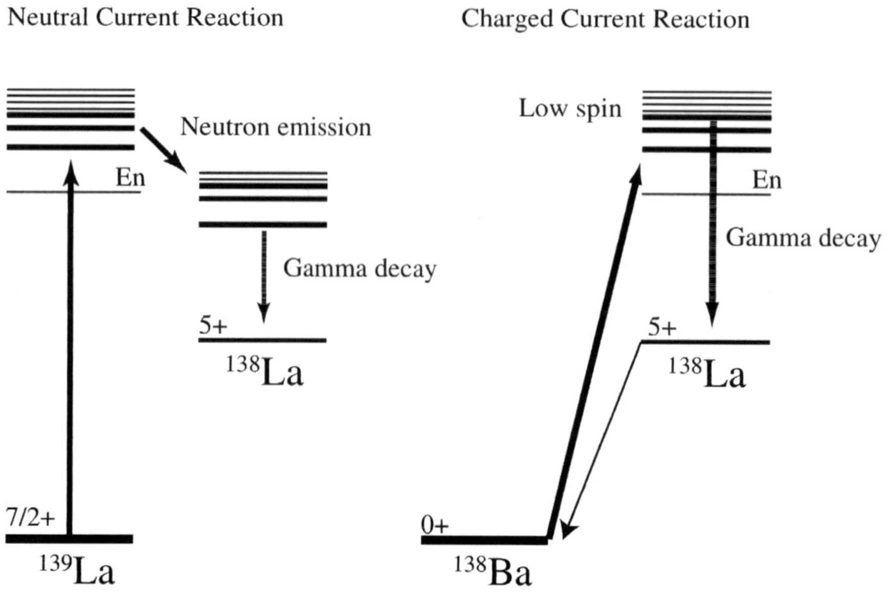


FIGURE 1. Schematic view of neutrino-induced reactions. Neutral current reactions (left) and Charged current reaction (right). In the neutral current reactions, nuclei are firstly excited by neutrino-induced reactions and subsequently decay by a nucleon such as a neutron, whereas, in the charged current reactions, a change exchange occurs and decays though emission of a gamma-ray or a nucleon.

Neutrino-nucleus interactions are key physics for understanding the neutrino process. However, individual neutrino reaction cross sections depend on the detailed nuclear structures of the nuclei involved [7]. Experimental measurements of neutrino–nucleus interactions for heavy nuclei are almost impossible because the associated weak reaction cross-sections are extremely small. Thus, one should calculate cross sections using nuclear structure models based on experimental data for detailed nuclear structure. The neutrino-process isotopes are predominantly synthesized by neutral current reactions and charged current reactions (see Fig. 1). In the neutral current reactions, the pre-existing seed nuclei are firstly excited by neutrino-induced reactions and subsequently de-excited by emission of nucleons such as neutron or $\gamma$-ray. On the other hand, in the charged current reactions, excited states on neutrino-isotopes are populated by electron neutrinos and subsequently decay to the ground state. The M1 strengths (or level density of 1+ states) are of importance for the estimation of the interaction strengths between neutrinos and atomic nuclei for the study of the supernova neutrino-process [7, 8]. However, there is no effective method to measure the M1 strength in the GDR region because of its large E1 strength.

The progress of relativistic engineering provides a new generation of photon beam, laser Compton scattering (LCS) gamma-rays. The energy tunable quasi-monochromatic LCS gamma-ray beam with MeV energy has been used for studying fundamental science and various applications at Duke University [9] and NewSUBARU [10] at present. The LCS gamma-rays have advantages that the maximum energy is sharply determined by the basic QED process and that the brightness near the maximum energy is relatively high. In addition, an advantage of the LCS gamma-ray is that it is almost 100% linearly (circularly) polarized, since the polarization of the laser is directly transported to the scattering photons.

The polarized gamma-ray beams are powerful tools to study nuclear physics. For example, electric-dipole (E1) and magnetic-dipole (M1) transition strengths from the ground state are measured directly with parity assignments using nuclear resonance fluorescence. On the other hand, (γ, n) reactions with linear polarized γ-rays have not been studied well. In 1957, Agodi [11] predicted theoretically angular distribution of neutrons emitted from states excited via dipole transitions with linearly polarized gamma-ray beam at the polar angle of $\theta=90°$ should be followed by a simple function, $a + b\cos(2\phi)$, where $\phi$ is azimuthal angel. However, this theoretical prediction has not been verified over the wide mass region except for light nuclei as deuteron for long time. The Agodi's paper had been cited by previous papers in 50~60' but it has been almost forgotten at present. We have proposed a method using (γ, n) reactions with linear polarized laser Compton scattering (LCS) γ-rays to measure M1 strength [10]. We have measured neutron angular distributions with (polarized gamma, n) reactions on Au, NaI, Cu targets to examine the Agodi's prediction [12].

## 2. Method

We have performed a nuclear experiment using laser Compton scattering γ-rays with liner polarization at NewSUBARU in SPring-8. The γ-ray beam was generated by Compton scattering laser photons with electrons stored in the electron storage ring NewSUBARU. The details of the LCS γ-ray source and typical experimental procedures are described in previous papers [12-14]. The storage ring NewSUBARU was operated with a mode of "Top-up", in which electron bunches were supplied continuously from an electron linac without extra acceleration into NewSUBARU. We irradiated three targets of Au, NaI, and natural Cu. Their sizes were 10 $mm^{\phi} \times 40$ mm, 10 $mm^{\phi} \times 50$ mm, and 5 $mm^{\phi} \times 50$ mm for Au, NaI, and Cu, respectively. Figure 2 shows a schematic view of photonuclear reactions on the nuclide $^{197}Au$ as an example. Excited states on $^{197}Au$ are populated by induced γ-rays. Individual populated states decay predominantly to the ground state or an excited state in the neutron deficient isotope $^{196}Au$ by emitting a neutron.

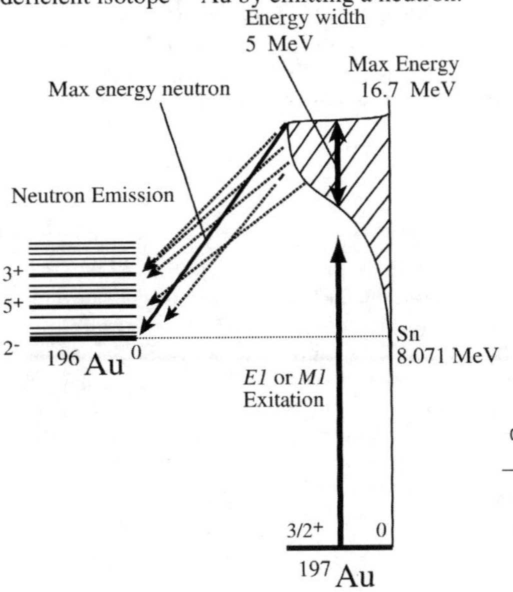


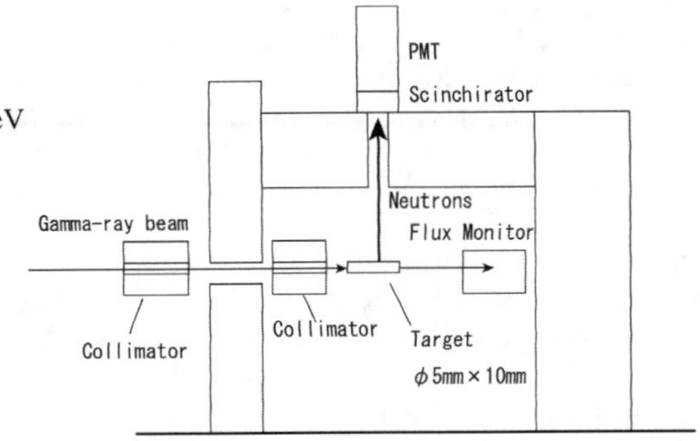


FIGURE 2. Schematic view of (γ, n) reactions. FIGURE 3. Schematic view of experimental setup.

The energy of generated γ-rays is determined by the wave length of laser and the energy of electron beam. The energy of the electron beam was 974 MeV. A Q-switch Nd:$YVO_4$ laser system provided laser with a wave length of 1064 nm with a power of 4 W. The maximum energy of the generated LCS γ-ray beam was 16.7 MeV. We used pulsed electron beams at a repetition rate of 2.5 MHz with a pulse width of 60 ps. The pulse width of the laser was 8 ns. The energy spread of the γ-ray beam depends on the collimator size and the emittance of the electron beam. We used two collimators as shown in Fig. 3. The first collimator was located inside of the storage ring, whereas the second collimator was located in front of the target. The γ-ray energy width was about 5 MeV for Au and 3 MeV for NaI and Cu. The estimated γ-ray flux was $(1\text{-}2)\times10^6$ photons/s in an energy range from 12 MeV to 16.7 MeV. The second collimator with a diameter of 6 mm was located before the target position. The diameter of the incident beam was about 6 mm on the target NaI or Cu. This collimator was not used for the Au measurement and the beam diameter was larger than the Au target diameter.

The neutrons were measured using a time-of-flight (TOF) method. The detector was set at a polar angle of $\Phi = 90°$ outside the experimental room (see Fig. 3). Each target was located inside the irradiation room with a concrete shield with a thickness of 540 mm. Neutrons were guided to the detector through a hole with a diameter of 80 mm and length of 970 mm. A lead shield with a thickness of 2 mm was located in front of the detector. A time-to-amplitude converter (TAC) was used to measure neutron energies. A start signal is generated from an output of the neutron detector. The synthesizer controlled the frequency of electron bunch with 2.5 MHz. A stop signal was generated from this synthesizer signal. Since this signal was also used as the external trigger signal to generate the laser pulse with 25 kHz by the divider, the trigger signal can be used as the stop signal. The generation of the LCS γ-ray was synchronized with the electron bunch. The pulse height of TAC signals were recorded using a multi-channel analyzer.

To measure the angular distribution of neutrons, we changed the linear polarization plane angle of the LCS γ-rays, whereas the detector system was fixed. The advantage of this method is that it deduces the systematic error. The angle of the linear polarization plane of the γ-ray beam was tuned by changing the linear polarization plane of the incident laser. We measured the neutron energy spectra as a function of the laser polarization angle in a range from $\phi = 0°$ to 360° in 30° steps for NaI and Cu, where the $\phi = 0$ was defined as the electric polarization vector being in the plane of the detector. A key point of this experiment is how to verify the angle of linear polarization of the incident LCS γ-ray beam. The angle of the linear polarization plane was determined by the linear polarization plane of the incident laser. However, the linear polarization plane of the incident laser may be changed by mirror. Thus, the laser beam was extracted to an area outside of the electron storage ring after LCS without additional mirror and the laser polarization angle was measured.

## 3. Result

Prompt γ-rays and neutrons can be separated by the TOF method. As shown in Fig. 4, the neutrons and prompt γ-rays are clearly separated. The measured time width of the prompt γ-ray is about 3 ns (FWHM) which is longer than the expected duration of 60 ps from the electron pulse width because of time fluctuation of the slow rising-time photomultiplier and the time jitter between the laser generation time and the external trigger signal from the electron storage ring. The energy spectra of the neutrons were derived from the TOF signals. The neutrons with energies lower than 2 MeV were not measured because of the detection efficiency. The measured maximum energy of the neutrons is almost 8 MeV for Au. This energy is almost same as the energy of 8.6 MeV, which is the difference between the maximum photon energy of 16.7 MeV and the neutron separation energy of 8.1 MeV. The time resolution, originated from the ratio of the detector depth to the flight pass length, is about 5%. However, the energy width of the incident LCS γ-ray beam is about 3~5 MeV, which is much higher than typical excitation energy differences near the ground state. Thus, the level structure of the residual nucleus cannot be observed in the present TOF spectra.

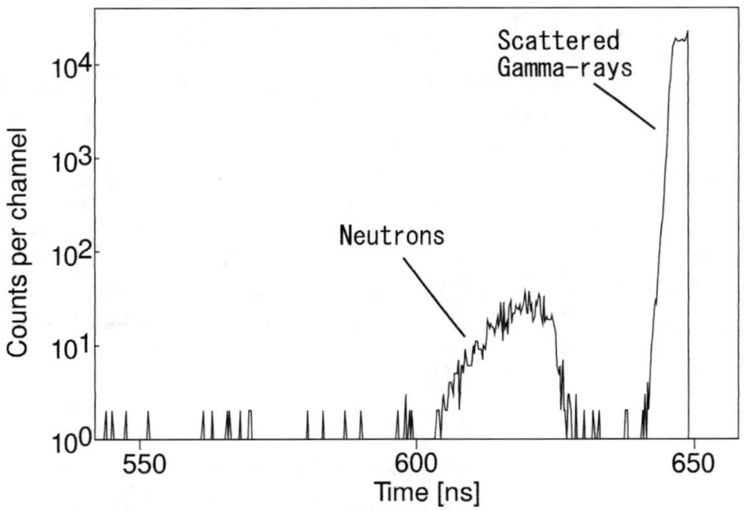


FIGURE 4. Time-Of-Flight spectrum for the Au target. The peaks corresponding to γ-rays and neutrons are clearly separately.

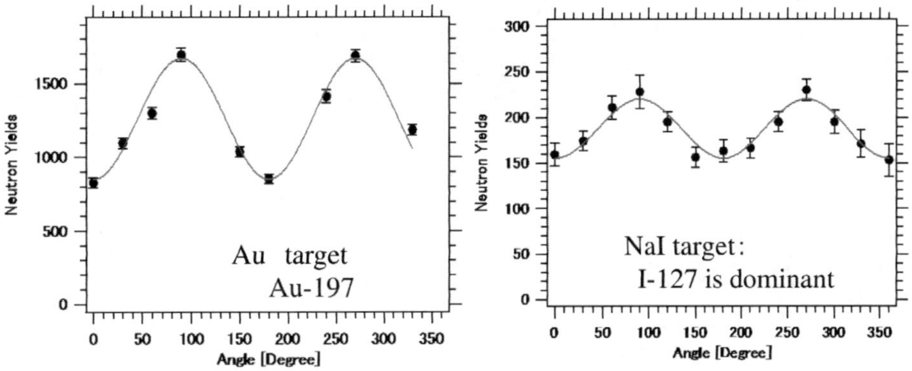


FIGURE 5. Angular distributions

## 4. Discussion

The neutron yields are presented as a function of azimuthal angle $\phi$ for An and NaI (see Fig. 5). The solid lines in Fig. 5 show functions in form of a + b cos($2\phi$) obtained by $\chi^2$-fitting. The NaI target includes $^{23}$Na and $^{127}$I. The neutron separation energies of these nuclides are 12.4 MeV ($^{23}$Na) and 9.1 MeV ($^{127}$I) and the contribution of $^{127}$I is dominant for the NaI target. The experimental result for Cu is also reproduced by the function a + b cos($2\phi$) although the anisotropy, b/a, is very low. The three neutron angular distributions are described as the function of a + b cos($2\phi$) independent of nuclides. Therefore, the theoretical prediction by Agodi is for the first time verified over the wide mass region.

The targets used in the present experiments are odd-A isotopes. The residual isotope is an odd-odd nucleus. The level density of the odd-odd nuclei is higher than even-even nuclei and odd-A nuclei. There are many combinations of the transitions as shown in Fig. 2 and it is difficult to calculate realistic values for neutron angular distribution for the odd-A isotopes. In contrast, in the case of an even-even nucleus target, the spin and parity of states populated by the incident photon is in general limited only to $J^{\pi} = 1^{\pm}$ and $2^{+}$ since E1, M1, and E2 transitions are dominant. Thus the theoretical calculation is easy comparing with those of odd-A targets.

Agodi [11] indicated that the sign of the parameter b for the M1 transitions is different from that of E1 when the initial and final states are same. This suggests that if we select a transition of which initial and final states are known we can measure, in principle, M1 strength in this transition. For example, the M1/E1 mixing can be evaluated from a polarization asymmetry if by gating with the highest neutron energy to select the transition to the ground state of the residual nucleus (see Fig. 2). For such a purpose, the key point is the energy resolution of neutron measurements. In the present experiment, the energy resolution is finally determined by the energy width of the incident gamma-ray beam, which is either 3 or 5 MeV. In the present available LCS γ-ray facilities, the typical energy spread of the photon beam is 3-10%. In this case, with 17-MeV γ-rays, the neutron energy resolution is about 0.5-1.7 MeV. The progress in laser and accelerator physics enables us to realize the next generation of the LCS γ-ray sources including ELI-NP [15], MEGA-ray [16], and the ERL-LCS [17].

The energy spread of these γ-ray beams is expected to be lower than $dE/E = 10^{-3}$. If we couple the present experimental method with such high energy resolution γ-beam, it is possible to study the detailed nuclear structure of the GDR with an excellent resolving power of the order of keV.

## 5. Conclusion

Agodi predicted theoretically the anisotropic angular distribution of neutrons emitted from states excited via dipole transitions with linearly polarized γ-ray beam at the polar angle of $\theta=90°$ should be followed by a simple function, $a + b\cos(2\phi)$, but it has not been verified experimentally over half a century. We have measured neutron angular distribution using linear polarized γ-ray beam generated by Compton scattering at NewSUBARU. We have verified this Agodi's prediction for the first time in the wide mass region. In the near future, the next generation of the LCS γ-rays will be available. The lost Agodi's prediction will have a more precious role to study nuclear physics in the GDR region and applications using photonuclear reactions.

## References

[1] Woosley S E *et al* 1990 *Astrophys. J.* **356** 272
[2] Heger A *et al* 2005 *Phys. Lett. B* **606** 258
[3] Yoshida T *et al* 2005 *Phys. Rev. Lett.* **94** 231101
[4] Yoshida T *et al* 2006 *Phys. Rev. Lett.* **96** 091101
[5] Hayakawa T *et al* 2010 *Phys. Rev. C* **81** 052801(R)
[6] Hayakawa T *et al* 2013 *Astrophys. J. Lett.* **779** L9
[7] Langanke K *et al* 2004 *Phys. Rev. Lett.* **93** 202501
[8] Cheoun M *et al* 2010 *Phys. Rev. C* **82** 035504
[9] Litvinenko V N *et al* 1997 *Phys. Rev. Lett.* **78** 4569
[10] Miyamoto S *et al* 2007 *Radiation Measurements* **41** S179
[11] Agodi A 1957 *Il Nuovo Cimento* **5** 21
[12] Horikawa K *et al* 2014 *Phys. Lett. B* **737** 109
[13] Hayakawa T *et al* 2006 *Phys. Rev. C* **74** 065802
[14] Hayakawa T *et al* 2008 *Phys. Rev. C* **77** 068801
[15] Habs D Tajima T and Zamfir V 2011 *Nuclear Physics News* **21** 23
[16] Albert F *et al* 2011 *Phys. Rev. ST-AB* **14** 050703
[17] Hajima R *et al* 2008 *J. Nucl. Sci. Technol.* **45** 441

Journal of Physics: Conference Series **590** (2015) 012004 doi:10.1088/1742-6596/590/1/012004

# Bohr Hamiltonian with deformation-dependent mass

Dennis Bonatsos[1], N Minkov[2], D Petrellis[1], and C Quesne[3]

[1]Institute of Nuclear and Particle Physics, National Centre for Scientific Research "Demokritos", GR-15310 Aghia Paraskevi, Attiki, Greece

[2]Institute of Nuclear Research and Nuclear Energy, Bulgarian Academy of Sciences, 72 Tzarigrad Road, 1784 Sofia, Bulgaria

[3] Physique Nucléaire Théorique et Physique Mathématique, Université Libre de Bruxelles, Campus de la Plaine CP229, Boulevard du Triomphe, B-1050 Brussels, Belgium

E-mail: bonat@inp.demokritos.gr, nminkov@inrne.bas.bg, petrellis@inp.demokritos.gr, cquesne@ulb.ac.be

**Abstract.**
The Bohr Hamiltonian with a mass depending on the nuclear deformation is solved using the techniques of supersymmetric quantum mechanics (SUSYQM). Analytical expressions for spectra and wave functions are obtained. Spectra and B(E2) transition rates are calculated for more than 50 $\gamma$-unstable nuclei and more than 60 prolate deformed nuclei using the Davidson potential and the Kratzer potential. In addition to solving the long standing problem of the too rapid increase of the moment of inertia with deformation, the method reveals a conformal factor in the Bohr Hamiltonian, embedding the Bohr space in six dimensions.

## 1. Introduction

Bohr Hamiltonians with mass dependent on the nuclear deformation have been introduced in [1, 2, 3] and further considered in [4]. Here we are going to briefly review why these are needed (Section 2), how they can be solved (Section 3), why they are useful (Section 4), and how they are related to other approaches to nuclear collectivity (Section 5).

## 2. Why is deformation dependent mass (DDM) needed

In the framework of the Bohr Hamiltonian [5], nuclear collective motion is described in terms of the collective coordinates $\beta$ (expressing the size of the deformation) and $\gamma$ (related to the shape of the deformed nucleus), plus the three Euler angles (showing the orientation of the nucleus in space). The mass, $m$, is usually considered to be a constant.

Within this framework, the moments of inertia of atomic nuclei are found to increase as $\beta^2$, i.e. too fast [6] in comparison to the experimental data. We are going to see that within the deformation dependent mass (DDM) framework, this increase is slowed down (see Fig. 1), while the one parameter entering in the dependence of the mass on the deformation can be related to the curvature of the 5D space, as well as to parameters of the Interacting Boson Model (IBM) [7], the complementary model describing nuclear collective motion in algebraic language.

Effective masses depending on the coordinates are not unheard of. They have been used for a long time in semiconductor theory [8, 9, 10, 11, 12, 13], in compositionally graded crystals [14], quantum dots [15], and liquid crystals [16], as well as in energy-density functional theory applied to nuclei [17], quantum liquids [18] and metal clusters [19].

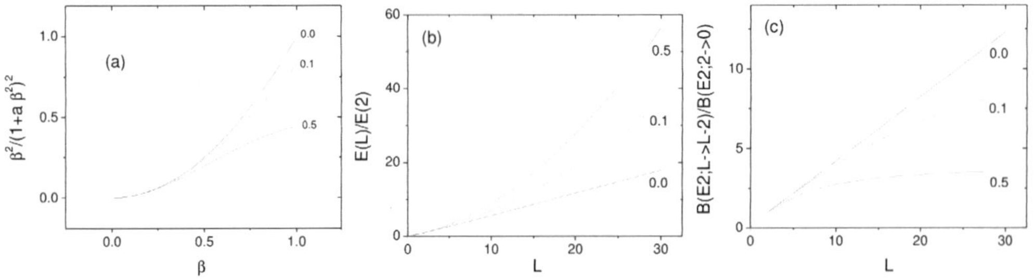


**Figure 1.** The function $\beta^2/f^2(\beta) = \beta^2/(1+a\beta^2)^2$, to which moments of inertia are proportional in the DDM Davidson model, plotted (a) as a function of the nuclear deformation $\beta$ for different values of the parameter $a$ [1]. Energy levels $E(L)$ [normalized to the energy of the first excited state, E(2)] (b) and $B(E2; L \to L-2)$ transition rates [normalized to the transition from the first excited state to the ground state, $B(E2; 2 \to 0)$] (c) are shown for the ground state band as functions of the angular momentum $L$ in the DDM Davidson model for $\beta_0 = 1$ and varying values of $a$. Figures taken from Ref. [1].

Additional motivation for introducing DDM in the Bohr Hamiltonian comes from the following points:

1) In the framework of the Interacting Boson Model (IBM) [7], it is known that in its geometrical limit [20, 21, 22], obtained through the use of coherent states [7], terms of the form $\beta^2\pi^2$ and/or more complicated terms appear [23], in addition to the usual term of the kinetic energy, $\pi^2$. Thus it might be appropriate to search for a modified form of the Bohr Hamiltonian, in which the kinetic energy term will be modified by terms containing $\beta^2$ and/or more complicated terms.

2) Detailed comparisons to experimental data have recently pointed out [24, 25, 26] that the mass tensor of the collective Hamiltonian cannot be considered as a constant and should be taken as a function of the collective coordinates, with quadrupole and hexadecapole terms present in addition to the monopole one. This approach has been recently extended [27, 28] to odd nuclei as well.

## 3. How can the DDM Bohr Hamiltonian be solved

If the mass depends on the coordinates, then mass and momentum do not commute. One then has to modify the kinetic energy operator, however keeping it Hermitian. The most general solution of this problem, involving two free parameters, has been given by von Roos [12].

In order to derive a DDM Bohr Hamiltonian, one has to work in the 5-dimensional (5D) space of the Bohr Hamiltonian, involving the Bohr–Wheeler coordinates $\beta$, $\gamma$ and the three Euler angles, and allowing the mass to depend on the deformation $\beta$. Using the usual Pauli–Podolsky prescription [29], one arrives at the DDM Bohr Hamiltonian given explicitly in Refs. [1, 2, 3].

This Hamiltonian can be solved using Supersymmetric Quantum Mechanics (SUSYQM) [30, 31], as pointed out in Ref. [32]. The basic steps of the solution, described for simplicity in 1D, are:

1) Factorize [33] the Hamiltonian $H_1$ in the form $A^\dagger A$. Its supersymmetric partner, $H_2$, is $AA^\dagger$. The two Hamiltonians share the same kinetic energy, but their potentials are $V_1 = W^2 - W'$ and $V_2 = W^2 + W'$, where $W$ is called the superpotential and $W'$ is its first derivative with respect to the relevant variable ($\beta$ in the present case). The basic feature of these two Hamiltonians is that they share the same spectrum, the only difference being that the ground state of $H_1$ has

Journal of Physics: Conference Series **590** (2015) 012004 doi:10.1088/1742-6596/590/1/012004

no partner in the spectrum of $H_2$. Using the operator $A$ one can jump from the eigenfunctions of $H_1$ to the eigenfunctions of $H_2$ with one node less, while using the operator $A^\dagger$ one can jump from the eigenfunctions of $H_2$ to the eigenfunctions of $H_1$ with one node more [30, 31].

2) Write a hierarchy of Hamiltonians [30, 31]. Each Hamiltonian $H_{i+1}$ is defined as $A_i A_i^\dagger$, in terms of the operators $A_i$, $A_i^\dagger$ of the previous Hamiltonian $H_i = A_i^\dagger A_i$, and is subsequently factorized into $H_{i+1} = A_{i+1}^\dagger A_{i+1}$, where $A_{i+1}$, $A_{i+1}^\dagger$ are new operators.

3) Impose shape invariance conditions [34] (integrability conditions guaranteeing exact solvability). Two potentials, $V_1$ and $V_2$, which are supersymmetric partners, are shape invariant if they have the same functional dependence on the relevant variable ($\beta$ in this case), while the values of the parameter are different for each one, and the relative displacement does not depend on the variable [34]. Symbolically, $V_2(\beta; a_1) = V_1(\beta; a_2) + R(a_1)$, where $a_1$, $a_2$ are the parameter values and $R$ the relative displacement, or $A(a_1)A^\dagger(a_1) = A^\dagger(a_2)A(a_2) + R(a_1)$.

Analytical expressions for the wave functions are obtained, being square integrable and ensuring the hermiticity of the Hamiltonian [2, 3]. $B(E2)$ transition rates can then be calculated numerically [2, 3].

Only some specific potentials are shape invariant and can be treated by the above method. These include [30, 31]:

1) the shifted harmonic oscillator,

2) the 3D oscillator and the Coulomb potential, the only ones which are soluble by this method in 3D,

3) the Morse, Rosen–Morse I (trigonometric), Rosen–Morse II (hyperbolic), Scarf I (trigonometric), Scarf II (hyperbolic), Eckart, and Poeschl–Teller potentials.

In what follows, we are going to consider (see Fig. 2) the Davidson potential [35] (a special case of the 3D oscillator) and the Kratzer potential [36] (a special case of the Coulomb problem).

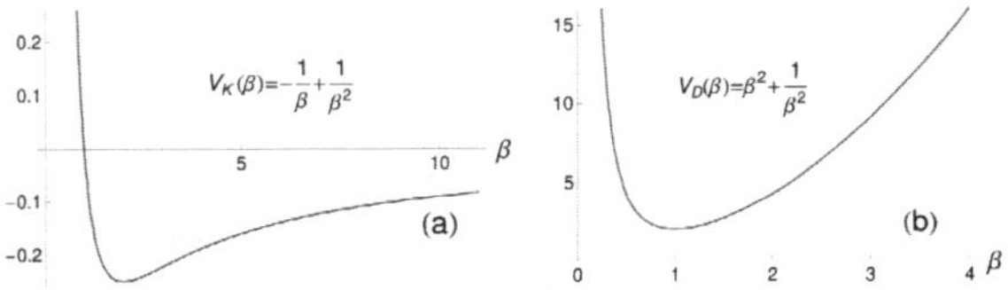


**Figure 2.** The Kratzer (a) and Davidson (b) potentials. The quantities shown are dimensionless, while all free parameters have been set equal to unity, for the sake of simplicity. Figure taken from Ref. [3].

### 4. What is the DDM Bohr Hamiltonian useful for

In both of the DDM Davidson [35] and DDM Kratzer [36] models, the mass depends on the deformation through the form $m(\beta) = m_0/f^2(\beta)$, where $m_0$ is a constant. In the Davidson case, one has $f(\beta) = 1 + a\beta^2$, while in the case of Kratzer one has $f(\beta) = 1 + a\beta$. In both cases, the parameter $a$, expressing the degree of dependence of the mass on the deformation, has to be much less than unity.

In the Davidson case, moments of inertia turn out to be proportional to $\beta^2/(1+a\beta^2)^2$, while in the Kratzer case they increase as $\beta^2/(1 + a\beta)^2$. In both cases, the increase is not following $\beta^2$, as in the plain Bohr case, but is moderated by the denominator, the moderation increasing with increasing $a$. One can easily see that energy ratios $E(L)/E(2)$ increase with increasing $a$, while transition ratios $B(E2; L \to L-2)/B(E2; 2 \to 0)$ decrease with increasing $a$ [2] (see Fig. 1).

Separation of variables can be easily achieved in the following cases.

1) Potentials independent of $\gamma$, referred to as $\gamma$-unstable potentials. They can describe nuclei lying relatively near to closed shells, which exhibit nearly vibrational behaviour. More than 50 nuclei of this kind have been fitted using both the Davidson and the Kratzer potentials, the results given in Refs. [2, 3] (see Fig. 3).

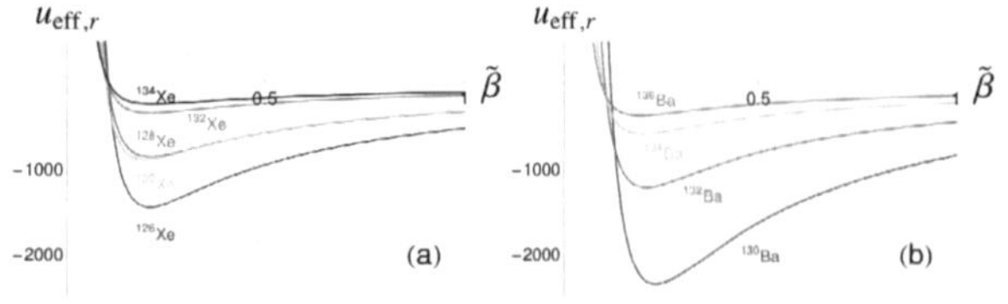

**Figure 3.** Effective DDM Kratzer potentials for $L = 4$ for some Xe (a) and Ba (b) isotopes. Figure taken from Ref. [3].

2) Potentials of the form $v(\beta, \gamma) = u(\beta) + w(\gamma)/\beta^2$, with $w(\gamma) = c\gamma^2$, i.e. a sharp potential centered at $\gamma = 0$. The term $\beta^2$ guarantees exact solvability in this case, since it allows the $w(\gamma)$ term to be absorbed into the centrifugal term. These potentials can describe well deformed nuclei, lying near the middle of the proton and neutron shells. More than 60 nuclei of this kind have been fitted using both the Davidson and the Kratzer potentials, the results given in Refs. [2, 3] (see Fig. 4).

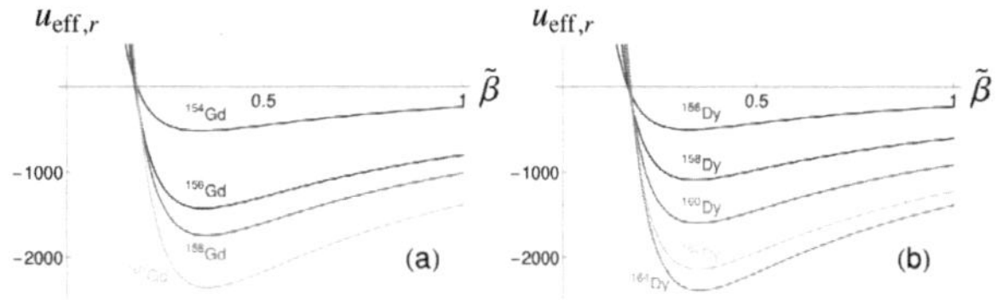

**Figure 4.** Effective DDM Kratzer potentials for $L = 4$ for some Gd (a) and Dy (b) isotopes. Figure taken from Ref. [3].

3) Potentials of the form $v(\beta, \gamma) = u(\beta) + w(\gamma)/\beta^2$, with $w(\gamma) = c(\gamma - 30^o)^2$, i.e. a sharp potential centered at $\gamma = 30^o$. Again the term $\beta^2$ guarantees exact solvability in this case. These potentials can describe nuclei exhibiting features of triaxiality.

It turns out that the Kratzer potential [3] can fit the $N = 90$ nuclei $^{150}$Nd, $^{152}$Sm, $^{154}$Gd, while the Davidson potential [2] cannot fit them. These nuclei are the best experimental manifestations [37] of the X(5) critical point symmetry [38], occurring at the point of the shape/phase transition [39] between spherical and prolate deformed shapes. It is known that the potential at the transition point should be flat [38]. The above result can then be explained by the fact that the Kratzer potential for appropriate parameter values becomes wide and relatively flat, while the Davidson potential always possesses a deep minimum.

## 5. Relation of DDM to other approaches to nuclear collectivity

The extra parameter $a$ introduced by the DDM approach can be connected to quantities appearing in other collective models.

*5.1. Connection of the DDM parameter to the space curvature*

It is known that the Position Dependent Mass (PDM) approach is equivalent to the introduction of curved spaces, or to the employment of deformed canonical commutation relations [40]. Here we briefly describe the way in which the DDM parameter $a$ can be connected to the curvature of space [4].

In the usual Bohr Hamiltonian (where the 5 variables can be enumerated as $q_i$ with $i = 1, \ldots, 5$), it is known that the kinetic energy can be written in the form $T = (m/2)(ds/dt)^2$, where the length element is $ds^2 = g_{ij} dq_i dq_j$, where $g_{ij}$ is a $5 \times 5$ matrix [41].

On the other hand, in the DDM Davidson case it can be proved [2] that the DDM Bohr Hamiltonian can obtain the same form, with matrix elements $g'_{ij} = g_{ij}/f^2$. Therefore, $f$ is a conformal factor, i.e. a factor not affecting the angles (i.e., $\gamma$ and the three Euler angles). One can further see that for the volume element one has $dv' = dv/f^5$, while for the wave functions $\Psi' = \Psi f^{5/2}$, where the primed quantities refer to the DDM Bohr Hamiltonian, while the unprimed ones regard the original Bohr Hamiltonian.

Conformal factors are known to occur during the embedding from $N$ into $N+1$ dimensions [4]. As an example, let us consider the embedding from 2D to 3D. The surface of a sphere represents a 2D space with constant curvature, on which the Pythagorean theorem is not valid. The Pythagorean theorem becomes valid if we embed the 2D sphere into a 3D space, in which $x_1^2 + x_2^2 + x_3^2 = R^2$. The length element in cartesian coordinates reads $dl^2 = dx_1^2 + dx_2^2 + dx_3^2$, while in polar coordinates it reads $dl^2 = dr^2/Q + r^2 d\theta^2$, where $Q = 1 - (r/R)^2$. Performing the conformal transformation $r = r_1/F$, where $F = 1 + (r_1/(2R))^2$, the length element takes the form $dl^2 = (dr_1^2 + r_1^2 d\theta^2)/F^2$, which is identical to the euclidean form up to the conformal factor $F$.

Exactly the same picture appears when embedding the 5D Bohr Hamiltonian into a 6D space [2]. The line element in the 5D euclidean space of the original Bohr Hamiltonian reads $ds^2 = d\beta^2 + \beta^2 d\Omega_4^2$, where $d\Omega_4^2$ is a complicated function of $\gamma$ and the three Euler angles. After the conformal transformation, the line element for the DDM Davidson Hamiltonian [2] reads $(ds')^2 = (d\beta^2 + \beta^2 d\Omega_4^2)/f^2$, with $f = 1 + a\beta^2$. The conformal transformation from the embedding of 5D into 6D reads $\beta' = \beta/F$, where $F = 1 + (\beta/(2R))^2$. Comparing $f$ and $F$ we see that $a = 1/(4R^2)$, i.e. the parameter contained in the dependence of the mass on the deformation is related to the radius of curvature $R$ of a hypersphere in the 5D space [4]. It is also clear that the case of mass independent from the deformation ($a = 0$, as in the original Bohr Hamiltonian) corresponds to a hyperplane in the 5D space [4].

*5.2. Connection of the DDM parameter to IBM parameters*

The procedure inverse to the embedding from $N$ into $N+1$ dimensions is the stereographic projection from $N+1$ to $N$ dimensions. As an example, let us consider the stereographic projection from 3D to 2D. In this procedure, a point $(x_1, x_2, x_3)$ on a sphere is projected onto a point $(X_1, X_2)$ on a plane passing through the south pole of the sphere and being orthogonal to the axis connecting the north and south poles. The projection is carried out through a line starting from the north pole of the sphere, passing through the point $(x_1, x_2, x_3)$ and ending at the point $(X_1, X_2)$. Using cylindrical coordinates $(r, \theta, x_3)$ on the sphere and polar coordinates $(r_1, \Theta)$ on the plane, one can easily prove that $\theta = \Theta$ and $r = r_1/F$, where $F = 1 + (r_1/(2R))^2$. We remark that the conformal factor $F$ appearing in the stereographic projection from 3D to 2D is the same as the conformal factor appearing in the embedding from 2D into 3D. The same conformal factor appears in general in the stereographic projection from $N+1$ to $N$ dimensions and in the embedding from $N$ into $N+1$ dimensions.

The Hamiltonian of the Interacting Boson Model (IBM) [7], possessing an overall U(6) symmetry, is known to live in a 6D space. It is known to possess a classical limit [20, 21, 22, 23], obtained [22] through a stereographic projection from 6D to 5D. It is therefore legitimate

to compare the classical limit of the IBM Hamiltonian to the DDM Davidson Hamiltonian. Preliminary results [42] indicate that the DDM parameter $a$ is related to
a) the strength of the pairing interaction in the U(5) (vibrational) limit,
b) the strength of the quadrupole-quadrupole interaction in the SU(3) (deformed) limit,
c) the difference between the strength of the ($sd$)-pairing and the strength of the ($dd$)-pairing in the O(6) ($\gamma$-unstable) limit.

Thus it turns out that the IBM Hamiltonian has built in the curvature of the 5D space, while the original Bohr Hamiltonian corresponds to a flat 5D space (a 5D hyperplane). To allow agreement with the IBM predictions, the Bohr Hamiltonian has to be enriched by allowing the nuclear mass to depend on the deformation, thus making the DDM Bohr Hamiltonian to correspond to a curved 5D space.

## References

[1] Bonatsos D, Georgoudis P E, Lenis D, Minkov N, and Quesne C 2010 *Phys. Lett. B* **683** 264
[2] Bonatsos D, Georgoudis P E, Lenis D, Minkov N, and Quesne C 2011 *Phys. Rev. C* **83** 044321
[3] Bonatsos D, Georgoudis P E, Minkov N, Petrellis D, and Quesne C 2013 *Phys. Rev. C* **88** 034316
[4] Georgoudis P E 2014 *Phys. Lett. B* **731** 122
[5] Bohr A 1952 *Mat. Fys. Medd. K. Dan. Vidensk. Selsk.* **26** no. 14
[6] Ring P and Schuck P 1980 *The Nuclear Many-Body Problem* (Berlin: Springer)
[7] Iachello F and Arima A 1987 *The Interacting Boson Model* (Cambridge: Cambridge University Press)
[8] BenDaniel D J and Duke C B 1966 *Phys. Rev.* **152** 683
[9] Gora T and Williams F 1969 *Phys. Rev.* **177** 1179
[10] Bastard G 1981 *Phys. Rev. B* **24** 5693
[11] Zhu and Q-G Kroemer H 1983 *Phys. Rev. B* **27** 3519
[12] von Roos O 1983 *Phys. Rev. B* **27** 7547
[13] Morrow R A 1987 *Phys. Rev. B* **35** 8074
[14] Geller M R and Kohn W 1993 *Phys. Rev. Lett.* **70** 3103
[15] Serra L I and Lipparini E 1997 *Europhys. Lett.* **40** 667
[16] Barranco M, Pi M, Gatica S M, Hernández E, and Navarro J 1997 *Phys. Rev. B* **56** 8997
[17] Niksic T, Vretenar D, and Ring P 2011 *Prog. Part. Nucl. Phys.* **55** 519
[18] Arias de Saaverda F, Boronat J, Polls A, and Fabrocini A 1994 *Phys. Rev. B* **50** 4248
[19] Puente A, Serra L I, and Casas M 1994 *Z. Phys. D* **31** 283
[20] Ginocchio J N and Kirson M W 1980 *Phys. Rev. Lett.* **44** 1744
[21] Ginocchio J N and Kirson M W 1980 *Nucl. Phys. A* **350** 41
[22] Dieperink A E L, Scholten O, and Iachello F 1980 *Phys. Rev. Lett.* **44** 1747
[23] van Roosmalen O S 1982 *Ph.D. thesis* (U. Groningen)
[24] Jolos R V and von Brentano P 2007 *Phys. Rev. C* **76** 024309
[25] Jolos R V and von Brentano P 2009 *Phys. Rev. C* **79** 044310
[26] Jolos R V and von Brentano P 2009 *Phys. Rev. C* **80** 034308
[27] Ermamatov M J and Fraser P R 2011 *Phys. Rev. C* **84** 044321
[28] Ermamatov M J, Srivastava P C, Fraser P R, Stránsky P, and Morales I O 2012 *Phys. Rev. C* **85** 034307
[29] Podolsky B 1928 *Phys. Rev.* **32** 812
[30] Cooper F, Khare A, and Sukhatme U 1995 *Phys. Rep.* **251** 267
[31] Cooper F, Khare A, and Sukhatme U 2001 *Supersymmetry in Quantum Mechanics* (Singapore: World Scientific)
[32] Bagchi B, Banerjee A, Quesne C, and Tkachuk V M 2005 *J. Phys. A: Math. Gen.* **38** 2929
[33] Infeld L and Hull T E 1951 *Rev. Mod. Phys.* **23** 21
[34] Balantekin A B 1998 *Phys. Rev. A* **57** 4188
[35] Davidson P M 1932 *Proc. R. Soc. London Ser. A* **135** 459
[36] Kratzer A 1920 *Z. Phys.* **3** 289
[37] Casten R F and McCutchan E A 2007 *J. Phys. G: Nucl. Part. Phys.* **34** R285
[38] Iachello F 2001 *Phys. Rev. Lett.* **87** 052502
[39] Cejnar P, Jolie J, and Casten R F 2010 *Rev. Mod. Phys.* **82** 2155
[40] Quesne C and Tkachuk V M 2004 *J. Phys. A: Math. Gen.* **37** 4267
[41] Sitenko A G and Tartakovskiĭ V K 1975 *Lectures on the Theory of the Nucleus* (Oxford: Pergamon)
[42] Bonatsos D, Minkov N, and Petrellis D 2014 *Bohr Hamiltonian with a deformation-dependent mass term: physical content of the free parameter*, in preparation.

# Future Experiments with Intense Laser Beams and Brilliant Gamma Beams at the ELI-NP Facility

**Dimiter L. Balabanski and the ELI-NP Science Team**

ELI-NP, Horia Hulubei National Institute for Physics and Nuclear Engineering, 30 Reactorului Str., 077125 Mgurele, jud. Ilfov, Romania

E-mail: dimiter.balabanski@eli-np.ro

**Abstract.** The Extreme Light Infrastructure (ELI) Pan-European facility initiative represents a major step forward in quest for extreme electromagnetic fields. Extreme Light Infrastructure Nuclear Physics (ELI-NP) is one of the three pillars of the ELI facility, that aims to use extreme electromagnetic fields for nuclear physics and quantum electrodynamics research. At ELI-NP, high-power lasers together with a very brilliant gamma beam are the main research tools. Their targeted operational parameters are described. The emerging experimental program in the field of nuclear physics of the facility is reported. The different instruments which are considered to operate in the ELI-NP experimental halls are discussed, with an emphasis on the instrumentation which is designed for nuclear structure, reactions and astrophysics research.

## 1. The ELI-NP Research Facility

ELI-NP [1] is one of the three laboratories of the European Extreme Light Infrastructure, the other two being laser-driven secondary beams in Prague, the Czech Republic [2] and the attosecond pulsed lasers in Szeged, Hungary [3]. The mission of the ELI-NP research facility is to promote nuclear physics studies with laser-driven electron, proton or heavy-ion beams and brilliant $\gamma$ beams. The laboratory complex at ELI-NP will host a high-power laser system (HPLS), which will consist of two ultra high-power lasers and a very brilliant gamma-beam system (GBS), which will deliver narrow-bandwidth $\gamma$ beams with parameters beyond the present state-of-the-art. They will be mounted on an anti-vibration slab, damping vibrations to frequencies $\leq$ 10Hz with amplitudes down to $\pm 1\mu$m, thus acting like a large optical table. The HPLS and the GBS will be commissioned in 2017-2018 and the laboratory will be available for experiments as an open-access facility in 2018.

The HPLS will have six output lines – two at 10 PW with a frequency of $\leq 1/60$ Hz, two at 1 PW with a frequency of $\leq 1$ Hz and two at 100 TW with a frequency of $\leq 10$ Hz. Each output will have its optical pulse compressor. The duration of the pulses from each of the six outputs of the HPLS shall be tunable from the best compression level to at least 5 ps pulse duration, with both positive and negative chirp. The HPLS outputs will be synchronized with accuracy below 200 fs [4]. The laser system will deliver pulses synchronously with the GBS electron and $\gamma$ bunches.

The GBS will produce highly polarized ($> 95\%$) tunable $\gamma$ beams of a high spectral density of $10^4$ photons/s/eV in the range from 200 keV to 19.5 MeV with a bandwidth of $> 0.3\%$ [5, 6]. The $\gamma$ beams will be produced through laser Compton backscattering (LCB) off an accelerated electron beam delivered by a linear accelerator. The LBC process can be looked upon as the most efficient frequency amplifier or as a "photon accelerator". Maximum up-shift is achieved in head-on collisions, producing $\gamma$ rays with energies $E_\gamma \sim 4\gamma_e^2 \cdot E_L$, where $\gamma_e$ is the Lorentz factor of the accelerated electron; $\gamma_e = (1-\beta^2)^{-1/2}$, $\beta = v_e/c$, $v_e$ is the electron velocity, $c$ is the speed of light and $E_L$ is the energy of the laser photons. For a green-light laser, $E_L = 2.4$ eV, LCB off 300-MeV electrons results in a $\gamma$ beam of $E_\gamma \leq 3$ MeV, while LCB off 720-MeV electrons yields $\gamma$ rays with $E_\gamma \leq 19.5$ MeV. The relatively low cross section of the process, ($\sigma_{CBS} \approx 10^{-25}$ cm$^2$), will be compensated by high photon and electron densities at the interaction point.

In reality, the production laser shoots at a small angle $\phi$ with respect to the axis of the e$^-$ beam, as presented in Fig. 1. In this case the energy of the $\gamma$ rays is:

$$E_\gamma = \frac{4\gamma^2 E_L}{1+\gamma^2\theta^2 + a_0^2/2}(1-\Delta), \quad (1)$$

where $\theta$ is the observation angle for the $\gamma$ beam, $a_0$ is a laser parameter, $a_0 = 4.3\frac{\lambda_L}{w_0}\sqrt{\frac{U_L}{\sigma_{tL}}}$, which takes into account the laser pulse energy $U_L$, wavelength $\lambda_L$, beam-spot size $w_0$ and FWHM pulse length $\sigma_{tL}$, and the dimensionless parameter $\Delta$, $\Delta \approx \frac{4\gamma}{mc^2}h\nu_L$, which takes into account the red shift due to the electron recoil, $h\nu_L$ is the laser photon energy.

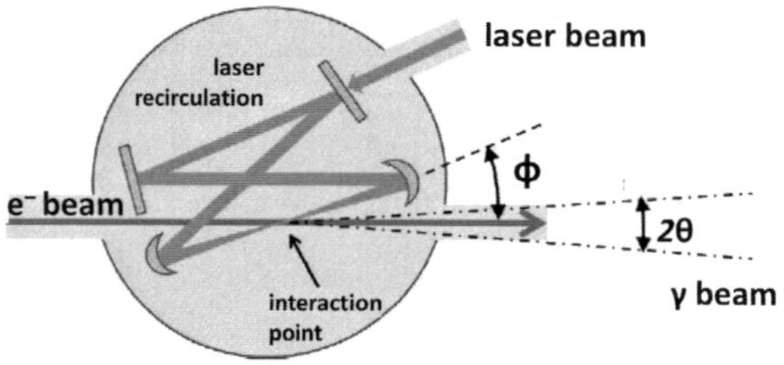


**Figure 1.** Compton backscattering between an electron bunch moving with a relativistic speed and a laser pulse. In the case of a multibunch structure is delivered by the electron accelerator within a macropulse, as in the case of ELI-NP, a laser recirculation system needs to be implemented.

## 2. Nuclear Science at ELI-NP

The nuclear physics research program at ELI-NP will be carried out with both, the ultra high-power lasers of the HPLS and the brilliant $\gamma$ beams of the GBS. A lay-out of the laboratory building is presented in Fig. 2. It covers an area of about 12000 m$^2$ and will host the HPLS, the GBS, experimental areas, support laboratories and workshops.

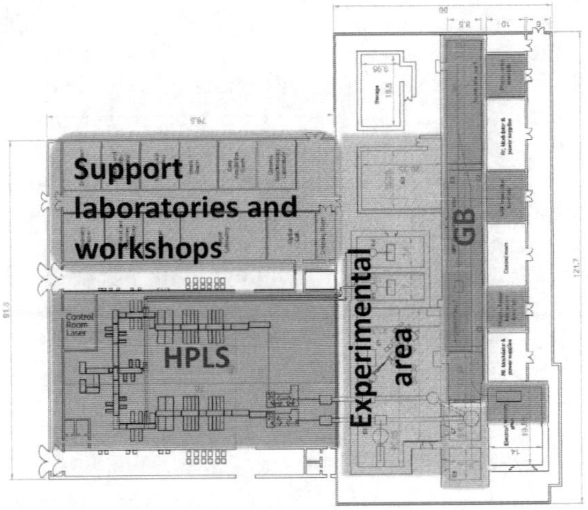


**Figure 2.** Lay-out of the ELI-NP building. The dimensions are in meters.

*2.1. HPLS Nuclear Physics Experiments*

The road towards high-power lasers began with the invention of Chirped Pulse Amplification (CPA) [7], resulted in a dramatic increase of the laser power and catalysed a new field of research, the high-intensity laser interaction with matter. Tajima and Dawson [8] proposed to accelerate electrons with an intense laser pulse. As a result of the laser-matter interaction, a wake of plasma oscillations is produced due to localized volumes of low and high densities of electrons. The wakefield, generated by an intense laser pulse propagating in an underdense plasma, exerts on electrons a ponderomotive force in the longitudinal direction. Experiments with hundred-terawatt lasers demonstrated that it is possible to produce accelerated beams of electrons, protons or heavy ions through such laser-matter interactions. For a recent review see Ref. [9].

At ELI-NP laser-driven ion-beam acceleration will be studied at laser intensities at the order $10^{23}$ W/cm$^2$. The experiments will aim at the production of heavy-ion beams, including actinide beams. The ultimate goal of this program is to explore the suggested fission-fusion mechanism [10]. In this reaction, laser-accelerated actinide ions, *e.g.* $^{232}$Th, impinge on a

$^{232}$Th target, where the target-like and the beam-like Th ions will fission. Due to the very high beam intensity, exceeding existing classical ion beams by many (up to 15) orders of magnitude, high-temperature high-ion density will be created. As a result, subsequent fusion of neutron-rich fragments will occur, resulting in a very neutron-rich fusion product. The most important scientific objective in this case is the study of neutron-rich nuclei in the region of the r-process waiting point $N = 126$, which can be achieved through fusion of two light neutron-rich fragments. Such experiments require huge experimental development in the field of laser-driven ion acceleration. Experiments, aiming the acceleration of heavy ions are in their infancy and, therefore, this experimental program will be implemented in steps, starting with studies aiming at mastering the process of ion-beam acceleration. As a next step, fission of accelerated actinide beams will be realized, and finally, the fission-fusion mechanism will be explored.

For the implementation of this experimental program the construction of a a large-acceptance (gas-filled) separator is suggested. It will allow both, the analysis of the produced heavy-ion beams, as well as the separation of the isotopes of interest, which will be produced in fission-fusion experiments. Behind the separator several measurement stations are considered, such as a $\beta$-decay tape station, combined with a $\gamma$-ray spectrometer and neutron detectors. The development of the instrumentation will follow the stages of the experimental program, starting with beam analysis, characterization of fission products, *e.g.* in a gas catcher combined to a multi-reflection time-of-flight (MRToF) beam-purification trap, etc.

### *2.2. The GBS Experimental Program*

The availability of brilliant narrow-width $\gamma$-ray beams in a large energy range opens an avenue for high-resolution $\gamma$-ray induced studies. As a result, an experimental program, covering different aspects of nuclear structure, reactions and astrophysics research, has been defined.

*2.2.1. Nuclear Resonance Fluorescence Experiments.* The $\gamma$-ray beam brilliance and bandwidth at ELI-NP will increase the sensitivity of nuclear resonance fluorescence (NRF) experiments, compared to the existing present-day facilities. ELI-NP offers the opportunity to perform NRF studies on small target samples. This opens up an entire new area of applicability of the NRF method to materials that may be available only in quantities of a few milligrams. For example, studies the low-energy dipole response in the actinide region will be possible. The low-lying dipole strength, such as the scissors mode, and of the fragmentation of the strength of soft modes, such as the Pigmy Dipole Resonance (PDR), will be studied, too. The brilliance of the $\gamma$-ray beam will enable $\gamma\gamma$ coincidence experiments, which provides an opportunity for detailed investigation of the decay of photo-excited states. Studies of the ground-state transition width $\Gamma_0$ with the self-absorption method [11] will benefit from the narrow bandwidth of the $\gamma$ beam. This will allow measurements of nuclear lifetimes in a model independent way with a very high precision. All these studies will benefit from the high polarization of the $\gamma$-ray beam.

The ELI-NP Array of DEtectors (ELIADE) will be used in these experiments. The ELIADE spectrometer will consist of eight segmented Clover detectors with anti-Compton shields, which will be combined with four 3" x 3" $LaBr_3$ detectors. The Clover detectors will be positioned at the vertical and horizontal positions of 90° and 135° rings. The $LaBr_3$ detectors will be positioned at 90°. The top-performance photopeak efficiency of the array will be $\approx 10\%$.

In addition, different applications of narrow-width $\gamma$ beams will be developed, such as characterization of materials and studies of trace materials. Other area of applications is 3D $\gamma$-ray imaging of bulk samples. These studies are related to non-proliferation of nuclear materials and nuclear-waste management.

*2.2.2. Photofission experiments.* The availability of a brilliant narrow-width $\gamma$-ray beam opens an avenue for photo-fission research, since it makes possible high-resolution studies in $\gamma$-ray induced reactions. This program will address high-resolution photo-fission experiments in the actinides, investigation of the second and third potential minima through studies of transmission resonances [12, 13, 14], angular and mass distribution measurements of fission fragments, measurements of absolute photo-fission cross-sections, studies of rare photo-fission events, such as ternary fission [15], highly asymmetric fission, etc.

These studies call for developments of state-of-the-art fission detectors to exploit the unprecedented properties of the high-flux, LCB $\gamma$ beams having a very small, sub-millimeter beam spot size. A multi-target detector array is under development at MTA Atomki, Debrecen, Hungary, consisting of position sensitive gas detector modules based on the state-of-the-art THGEM technology [16]. The foreseen unprecedented sub-millimeter $\gamma$ beam-spot size allows to develop considerably more compact photofission detectors than those of before. Besides, the well-focused $\gamma$ beam also defines a distinct fission position, such that a remarkably improved angular resolution can be achieved. For the measurement of the mass and atomic number distribution of the fission fragments a highly-efficient, five-folded, Frisch-gridded twin ionization chamber [17], which will be used as Bragg ionization chamber (BIC) [18], is under development at MTA Atomki. The twin ionization chamber will be equipped with double-sided Si strip detectors in order to measure light-particle ($\alpha$) emission probability from the highly-deformed compound state and to detect any ternary particles from fission. An increased $\alpha$ decay probability would also be a conclusive evidence for the HD structure of the fissioning system. Atomic numbers of the fission fragments will be extracted by tracking the Bragg curve of the ions using a desktop digitizer and advanced digital signal processing (DSP) techniques.

In addition, the possibility to produce exotic neutron-rich nuclei in photo-fission and study their structure and decays is investigated. The GBS $\gamma$-ray beam will cover selectively energy region of the giant dipole resonance (GDR) of the fissile target, which makes it an ideal tool to induce photo-fission of the target nuclei. The construction of an IGISOL beam line at ELI-NP is under consideration. The IGISOL technique [19] allows the extraction of the isotopes of refractory elements, which do not come out from standard ISOL targets. So far, photofission has been explored as a mechanism for the production of neutron-rich nuclei at bremsstrahlung facilities, such as ALTO [20] and ARIEL [21]. There mA electron beams of energies $\geq 50$ MeV are sent on a converter, producing bremsstrahlung photons, which cover the energy range of the GDR of the fissile target. The bremsstrahlung hits the target, which is located in a gas cell. The ions, produced in fission, leave the target. They are slowed down in the gas and the isotopes of interest are selected through a combination of a laser ion source and a mass separator. A major bottleneck of this technique is the creation of space charge in the gas cell. Its major source are low-energy $\gamma$ rays, which interact with the gas of the cell. These $\gamma$ rays form the dominant part of the of the bremsstrahlung spectra, and thus set a natural limitation of the application of the technique.

At ELI-NP a $\gamma$ beam, which covers the energy range of the GDR is produced through LCB. The low-energy part of the spectrum can be cut through collimation, due to the space distribution of the LCB $\gamma$ rays.

Yield calculations have been performed in GEANT4. For this purpose, several classes were implemented, inheriting the basic classes of GEANT4, *i.e.* the primary $\gamma$ beam was defined as G4GeneralParticleSource class, and the photofission process, which is not available in GEANT4, has been implemented as a G4PhotoFission class [22].

Benchmark calculations, considering a target which is a stack of thin $^{238}$U foils, with a total mass of 800 mg, were performed. The foils were tilted with respect to the beam, such that the ions, produced in photofission, can be stopped in the gas, not reaching the next foil. About $6 \cdot 10^2$ fission/s per $10^6$ $\gamma$ rays were obtained with this target geometry. It should be noted that less than 30% of the $\gamma$-beam photons interact with the target, the gas-cell window and the gas. Thus, $10^8$ fissions/s can be expected at ELI-NP, considering a $\gamma$ beam of $5 \cdot 10^{10}$ photons/s.

Journal of Physics: Conference Series **590** (2015) 012005 doi:10.1088/1742-6596/590/1/012005

*2.2.3. Experiments above the Neutron Separation Threshold.* The brilliant, narrow-width, highly-polarized $\gamma$-ray beam which will be delivered by the ELI-NP GBS will open up new horizons for the investigation of the nuclear photo-response at and above the particle separation threshold. An example for such studies is the detailed investigation of different soft modes, such as e.g. the pygmy dipole resonance (PDR) at and above the particle threshold, which is essential for nucleosynthesis in astrophysics. PDRs are placed much lower in energy than giant dipole resonances (GDR) and they represent only a small fraction of the total E1 strength (few percent), while GDRs exhaust almost fully the E1 strength. Both, the GDR and the PDR can be covered within the energy range of the ELI-NP beams. The PDR occurs close to the neutron emission threshold and its decay is governed by the coupling to the large number of states around the threshold. These experiments will be complimentary to the PDR NRF studies. At present, high-resolution studies of PDR strength can be done with proton probes at Osaka, at the RCNP high-resolution spectrograph Grand Raiden [23, 24]. Similar or better resolution will be achieved with $\gamma$ probes at ELI-NP.

This experimental program will include also studies of ($\gamma$,n) cross section ,*e.g.* p-process related measurements, such as the $^{138}$La($\gamma$,n)$^{137}$La and the $^{180m}$Ta($\gamma$,n)$^{179}$Ta reactions, which are of key importance for the understanding of the process and provide constraints on the explosive dynamics of massive stars.

Two major instruments are under consideration for the realization of this experimental program, a $4\pi$ neutron spectrometer of $^{3}$He counters and an array of large-volume $LaBr_3$ and liquid scintillator detectors for neutron time-of-flight and $\gamma$-ray correlated measurements.

*2.2.4. Nuclear Astrophysics Studies.* The ELI-NP facility provides unique opportunities for nuclear astrophysics research. For example, the $\gamma$-induced nuclear reactions of astrophysical interest were extensively studied during the years, but still represent a challenge for the experimental and theoretical work. The difficulty arises from the very small cross sections due to the fact that the reactions occur deep below the Coulomb barrier especially for the case of ($\gamma$,$\alpha$) reactions. Therefore a very intense $\gamma$ beam would be of interest for such investigations. All p-nuclei can be synthesized from the destruction of pre-existing nuclei of the s- and r-type by a combination of (p,$\gamma$) captures and ($\gamma$,n), ($\gamma$,p) or ($\gamma$,$\alpha$) photo-reactions. Other studies are related to specific key reactions of astrophysics interest, such as the $^{16}$O($\gamma$,$\alpha$)$^{12}$C reaction. After hydrogen is exhausted in the stellar core, stars leave the main sequence and undergo subsequent nuclear core burning stages involving heavier nuclear species, namely, helium, carbon, neon, oxygen and silicon burning provided that stellar masses are large enough. The outcome of helium burning is the formation of the two elements: carbon and oxygen [25]. The ratio of carbon-to-oxygen (C/O) at the end of helium burning has been identified three decades ago as one of the key open questions in Nuclear Astrophysics [25] and it remains so today. The importance of the C/O ratio for the evolution of heavy stars that evolve to core collapse (type II) supernova has been discussed extensively [26] but more recently it was shown that the C/O ratio is also important for understanding the $^{56}$Ni mass fraction produced by lower mass stars that evolve into Type Ia supernova (SNeIa) [27]. Thus the C/O ratio is also important for understanding the light curve of SNeIa. The principle of detailed balance allows the determination of the cross section of an ($\alpha$,$\gamma$) process from the measurement of the time inverse ($\gamma$,$\alpha$) reaction with $\gamma$-ray beams. Since both the electromagnetic and nuclear interactions are time reversal symmetric the cross sections are related to each other in terms of the spin factors and De Broglie wavelengths by:

$$\omega_A \frac{\sigma_A(\alpha,\gamma)}{\lambda_A} = \omega_B \frac{\sigma_B(\gamma,\alpha)}{\lambda_B}. \qquad (2)$$

One of the advantages of measuring the photo-dissociation of $^{16}$O is a gain due to detailed balance. Such an experiment requires a $\gamma$-ray beam of energies 10 MeV and less (approaching 8 MeV) since the Q value of the $^{12}C(\alpha,\gamma)^{16}O$ reaction is 7.162 MeV.

A number of scientific instruments are under discussion to be used in these experiments. These include a time-projection chamber (TPC) with a GEM read-out, which is currently being developed at the University of Warsaw, Poland, a $4\pi$ Si DSSD array, which is being designed by the INFN LNS Catania, Italy, and a bubble chamber. All of these will allow to achieve the needed sensitivity and approach the above defined problems.

## 3. Summary

The ELI-NP research center will host a HPLS and a GBS with parameters beyond the state-of-the-art. The power of the HPLS lasers exceeds existing lasers by an order of magnitude and will deliver every minute on the target intensities in the range of $10^{23}$ W/cm$^2$. The spectral density of $10^4$ photons/s/eV will provide brilliance and narrow bandwidth of the $\gamma$ beams, which will be delivered by the ELI-NP GBS, that are much beyond the parameters of the beams which are delivered at the existing facilities. This makes it possible to design and perform new classes of nuclear physics experiments, which cannot be done elsewhere.

In short, the outstanding performance of the high-power lasers and the $\gamma$ beam opens the possibility to carry out a versatile research program in nuclear physics and tackle key problems in nuclear structure, astrophysics and reactions.

## Acknowledgments

The Extreme Light Infrastructure Nuclear Physics (ELI-NP) - Phase I project is co-funded by the European Union through the European Regional Development Fund. This paper presents the work of the ELI-NP Science team related to the preparation of the experimental program at ELI-NP and the TDRs for the scientific instruments to be used at in the laboratory. Collaboration with Prof. A. Krasznahorkay and Dr. F. Ibrahim, conveners of the ELI-NP Photofission TDR Working Group, with Dr. P. V. Cuong for the photofission yield estimates is greatfully acknowledged.

## References

[1] Zamfir N V 2014 *EPJ Web of Conferences* **66** 11043
[2] http://www.eli-beams.eu
[3] http://www.eli-hu.hu
[4] Ursescu D *et al.* 2013 *Proc. SPIE* **8780** 87801H
[5] Vaccarezza c *et al.* 2012 *Proc. IPAC 2012 Conference* TUOBB01
[6] Petrillo V *et al.* 2012 *Nucl. Instr. Meth.* A **693** 109
[7] Stickland D and Mourou G 1985 *Opt. Commun.* **56** 219
[8] Tajima T and Dawson J M 1979 *Phys. Rev. Lett.* **43** 267
[9] Zamfir N V 2014 *Eur. Phys. J. Special Topics* **223** 1221
[10] Habs D *et al.* 2011 *Appl. Phys.* B **103** 471
[11] Pietralla N *et al.* 1995 *Phys. Rev.* C **51** 1021
[12] Thirolf P G and Habs D 2002 *Prog. Part. Nucl. Phys.* **49** 352
[13] Krasznahorkay A 2011 *Handbook of Nuclear Chemistry* (Springer, Berlin) 281
[14] Csige L *et al.* 2013 *Phys. Rev.* C **87** 044321
[15] Verboven M, Jacobs E and De Frenne D 1996 *Phys. Rev.* C **49** 991
[16] Shalem C K *et al.* 2006 *Nucl. Instr. Meth.* A **558** 468
[17] Budtz-Jorgensen C *et al.* 1987 *Nucl. Instr. Meth.* A **258** 209
[18] Neubert W *Nucl. Instr. Meth.* A **237** 535
[19] Aysto J *et al.* 1992 *Phys. Rev. Lett.* **69** 1167
[20] Azaiez F *et al.* 2013 *Nucl. Phys. News* **23** issue 2 5
[21] http://www.triumf.ca/ariel
[22] Phan Viet Cuong 2009 PhD Thesis *Development of a new surface ion-source and ion guide in the ALTO Project* (University Paris-Sud, Orsay)

[23] Fujita Y *et al.* 1997 *Nucl. Instr. Meth.* B **126** 274
[24] Fujita H *et al.* 2002 *Nucl. Instr. Meth.* A **484** 17
[25] Fawler W A 1984 Rev. Mod. Phys. **56** 149
[26] Weaver T A and Woosley S E 1993 *Phys. Rep.* **227** 65
[27] Thielemann F K *et al.* 2004 New Astron. Rev. **48** 605

# The Pygmy Dipole Resonance – experimental studies of its structure and new developments

**A Zilges**[1]
[1]Institute for Nuclear Physics, University of Cologne, 50937 Köln, Germany

E-mail: zilges@ikp.uni-koeln.de

**Abstract.** Several experimental studies using electromagnetic probes have found an enhancement of electric dipole strength between about 5 and 10 MeV. This phenomenon is usually denoted as Pygmy Dipole Resonance (PDR). The detailed structure of these excitations is still under debate. This manuscript will concentrate on the results of complementary experiments using hadronic probes to populate the PDR. These studies allow a first insight into the origin of the PDR. Finally, the manuscript will shortly discuss plans for future experiments.

## Introduction

In a macroscopic picture, electric dipole (E1) moments of atomic nuclei are always connected to a breaking of the symmetry between protons and neutrons. The most prominent example is the Giant Dipole Resonance (GDR) which is visualized in a macroscopic picture as an out-of-phase oscillation of the proton and neutron fluids leading to an electric dipole excitation at energies of about $E_x = 31\ A^{-1/3} + 21\ A^{-1/6}$ MeV [1,2]. The GDR exhausts about 100% of the Energy Weighted Sum Rule (EWSR) for E1 transitions, which is approximately given by:

$$\int_0^\infty \sigma(E)dE = 60\frac{NZ}{A}MeV \cdot mb$$

Already in 1971 it has been proposed by Mohan, Danos, and Biedenharn that in a three-fluid hydrodynamical model a second collective E1 mode is generated by the oscillation of excess neutrons against an isospin-saturated proton-neutron core [3]. This mode would be located at lower energies and would carry less E1 strength compared to the GDR. First evidence for the existence of such a Pygmy Dipole Resonance (PDR) came from the observation of an enhancement of $\gamma$-ray emission after neutron capture [4]. The resulting schematic distribution of E1 strength is depicted in Figure 1 which already shows the difficulties of clearly distinguishing the PDR and GDR due to the fragmentation and fine structure. It is currently under debate if the precise knowledge of the properties of the PDR or of the complete distribution of the E1 strength has implications on the symmetry parameter in the equation of state [5-9].

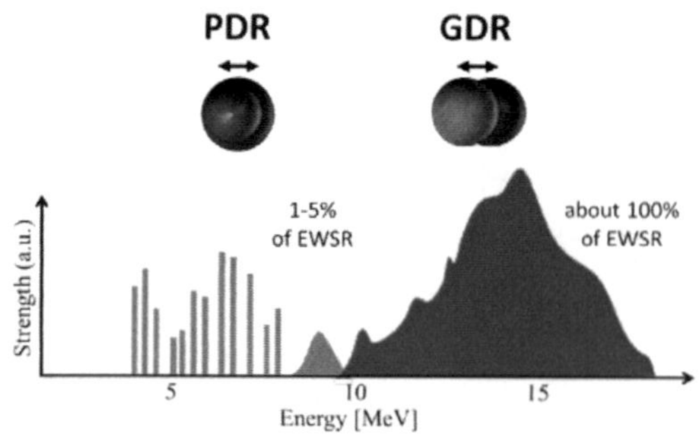


**Figure 1:** Schematic distribution of E1 strength in an atomic nucleus showing the splitting into a Pygmy Dipole Resonance (PDR) and a Giant Dipole Resonance (GDR). Octupole-coupled modes which can generate E1 strength at even lower energies are not shown.

The advent of bremsstrahlung photon sources with endpoint energies up to the particle thresholds from cw electron accelerators and the improvement of $\gamma$ detector setups in the last two decades allowed high precision studies of the photoresponse of stable atomic nuclei in Nuclear Resonance Fluorescence (NRF) ($\gamma$,$\gamma$') experiments [10,11]. In these experiments, mainly dipole transitions are induced from the ground state of the nucleus and the subsequent $\gamma$ decay is observed with high-resolution HPGe detectors. The final result is the detailed dipole strength distribution up to the particle threshold. Quasi-monoenergetic $\gamma$-ray beams from Laser Compton backscattering facilities can yield additional observables like parities and branching ratios of the excited states (see the contribution of Deniz Savran to these proceedings). More recently, the exchange of virtual photons between stable targets and a proton beam at energies of about 300 MeV has been used to study the dipole response below and above the particle threshold in a single experiment. This was achieved by high-resolution spectroscopy of the scattered protons using the Grand Raiden spectrometer at RCNP Osaka [12].

In very neutron-rich, unstable isotopes, one expects an enhancement of PDR-like excitations due to the larger neutron skin. A pioneering study has been performed at a setup at GSI Darmstadt. Radioactive $^{130,132}$Sn nuclei were selected by the Fragment Separator (FRS) and exchanged virtual photons with a high-Z target. A magnet (ALADIN) behind the target chamber selected the ion of interest. Neutrons stemming from the ($\gamma$,xn) reaction were analyzed by the Large Area Neutron Detector LAND in coincidence with $\gamma$ rays detected by scintillators from Crystal Ball surrounding the target chamber. These experiments showed an enhancement of E1 strength in the low energy tail of the GDR at about 10 MeV which exhausts several percent of the EWSR [13]. In a different approach, the radioactive neutron rich nucleus $^{68}$Ni has been studied at the FRS by detecting the $\gamma$ decays with the HECTOR and RISING array [14]. Again an enhancement of the E1 strength at energies of about 11 MeV is observed.

In summary, the studies using electromagnetic interaction observe an enhancement of E1 excitations at energies well below the GDR or on top of the tail of the GDR [15]. In stable nuclei, this strength exhausts about 1% of the isovector EWSR and is found below the neutron separation energy. In the few cases where radioactive nuclei have been investigated, the strength is found at higher energies and exhausts up to 5% of the isovector EWSR. Whether both observations belong to the same excitation mode, i.e., the PDR, is unclear. Obviously, more systematic studies on stable and radioactive species are mandatory to get a better understanding.

Journal of Physics: Conference Series **590** (2015) 012006 doi:10.1088/1742-6596/590/1/012006

## Hadronic probes

One main problem in the interpretation of the data is how to distinguish between excitations belonging to the GDR (having wave functions representing the typical out-of-phase oscillation of protons and neutrons) and excitations belonging to a different type of excitation, i.e., the PDR. One way to answer this question is to use hadronic probes as an alternative excitation mechanism for the dipole modes. Table 1 compares the two experimental approaches. One big advantage of electromagnetic probes is the selective excitation of the lowest spin modes and – if one can use HPGe detectors for the $\gamma$ decay channel – the excellent energy resolution. These two advantages are lost in typical hadron scattering experiments. However, in 1992, Poelhekken, Harakeh *et al.* complemented the QMG/2 particle spectrometer at KVI Groningen by an array of NaI detectors to detect the $\gamma$ rays from the decay in coincidence with the scattered $\alpha$ particles [16]. This coincidence requirement restores the selectivity to low-spin excitations which decay to the groundstate via E1, M1, or E2 transitions. However, due to the limited energy resolution of the NaI detectors, this setup allowed to study only a few selected nuclei with rather low level density.

| | $(\gamma,\gamma')$ or Coulex | $(\alpha,\alpha')$ @ 30 MeV/A |
|---|---|---|
| Interaction | **electromagnetic** | **strong** |
| Location of interaction | **whole** nucleus | **surface** |
| Isospin | **isovector** E1 excitations | dominant **isoscalar** |
| Multipolarity | E1, M1, E2 | E0, E1, E2, E3, ... |
| ΔE | 3-500 keV | 50-200 keV |

**Table 1:** Comparison of electromagnetic and hadronic excitation. At energies of 30 MeV/A, $\alpha$ scattering is strongly dominated by hadronic interaction for the excitations under investigation. Detecting the $\gamma$ decay in coincidence restores the selectivity to low spin excitations.

This limitation was overcome a decade later when we replaced the NaI detectors by an array of HPGe detectors which provided an excellent energy resolution [17]. In these experiments, the scattered $\alpha$ particles were analyzed by the new Big Bite Spectrometer. The coincidence data can be sorted into a matrix where one axis represents the energy loss of the $\alpha$ particle during the scattering process (which yields the excitation energy) and the other axis represents the energy of the $\gamma$ ray measured in coincidence, see Fig. 2.

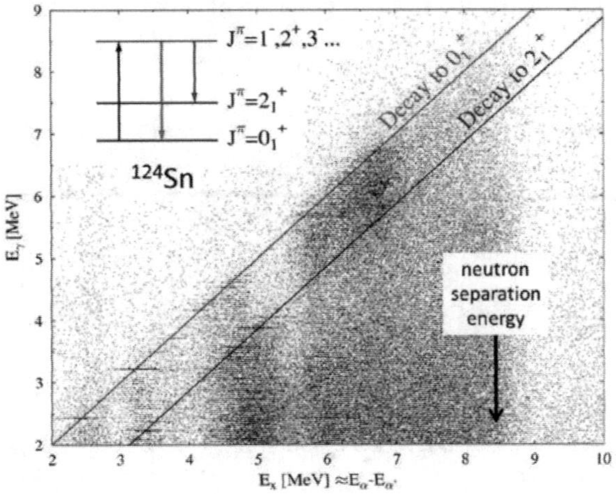


**Figure 2:** $\alpha$-$\gamma$ coincidence matrix of the $(\alpha,\alpha'\gamma)$ experiment on $^{124}Sn$ (adapted from [18]). Thin horizontal lines represent excitations in $^{124}Sn$, mainly $\gamma$ decays to the ground state or first excited state have been observed. The neutron separation energy lies at about 8.5 MeV, above this energy, $\gamma$-ray transitions are strongly suppressed.

Such experiments allow a detailed state by state analysis of the α scattering cross section even in nuclei where the level density is rather high. The result was quite surprising: Whereas below about 7 MeV all E1 excitations seen in (γ,γ') are observed also in (α,α'), this is not the case for the E1 excitations at slightly higher energies [19-22], see Fig. 3. Referring to Table 1, this points to a different underlying nuclear structure: The E1 strength splits into a surface mode with a strong isoscalar component at lower energies and a more isovector mode at higher energies.

Very recently, a group around Angela Bracco at Milano performed $^{17}$O scattering experiments at energies of about 20 MeV/A on $^{208}$Pb and $^{124}$Sn [23,24] at the Tandem-ALPI accelerator complex at INFN Legnaro. Again, the dominantly isocalar hadronic interaction can populate the states of interest. These studies reproduce the observations we made earlier in the (α,α'γ) experiments, i.e., a splitting of the E1 strength into a lower and higher lying part.

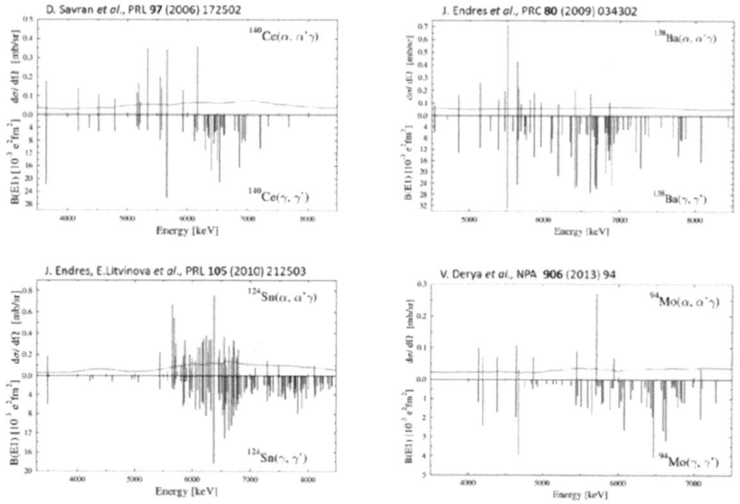


**Figure 3:** Cross sections for E1 excitations observed in the (α,α'γ) experiments (upper parts) in comparison with the E1 strength observed in (γ,γ') (lower parts) for the nuclei $^{140}$Ce, $^{138}$Ba, $^{124}$Sn, and $^{94}$Mo.

These experimental observations are supported by various theoretical calculations, which predict distinctly different transition charge densities for typical $1^-$ states at lower and higher energies: Whereas the former resemble the macroscopic picture of surface neutrons oscillating against an isoscalar core, the latter have charge densities more typical for states belonging to the GDR, see, e.g., Refs [18,25,26].

**Future experiments**

Various theoretical models (also the simplified geometrical picture of a neutron skin oscillating against an isospin saturated core) predict a strong enhancement of the PDR with neutron excess. Therefore, it is interesting to study the isospin character of E1 strength in exotic neutron-rich nuclei as well. One way to do this would be to perform α scattering experiments in inverse kinematics using a radioactive ion beam. A European-Japanese collaboration is currently investigating the Sn isotopic chain from stable proton-magic $^{124}$Sn to the doubly-magic radioactive nucleus $^{132}$Sn in experiments at RIKEN. The large acceptance radioactive ion separator BigRIPS is used to select the isotopes of interest. The beam interacts with a liquid He target, the scattered heavy ions are analyzed by the ZeroDegree spectrometer which is set to collect ions without charge and mass change (i.e., elastic α scattering). The trigger is determined by the detection of coincident γ rays with NaI detectors (DALI2) and additional LaBr detectors in forward direction (see Fig. 4).

**Figure 4**: Setup at RIKEN to study α scattering on neutron rich nuclei. The DALI2 array plus additional LaBr detectors measure the γ decay from the beam after hitting the liquid He target. The scattered beam particles are measured in the ZeroDegree particle spectrometer.

Further experiments using α or proton scattering as an experimental tool to study the PDR in stable nuclei are planned at iThembaLABS Cape Town and at RCNP Osaka, studies on unstable nuclei will be continued or are planned at GSI/FAIR Darmstadt and at RIKEN.

**Acknowledgements**

I want to thank the numerous members of my former and present research groups at TU Darmstadt and the University of Cologne for their outstanding work, especially D. Savran, V. Derya, and J. Endres. For many years of collaboration on the subject I thank M.N. Harakeh and H.J. Wörtche. I thank S.G. Pickstone and M. Spieker for carefully reading the manuscript.

This research was supported by the Deutsche Forschungsgemeinschaft (ZI 510/4-2).

**References**

[1] Migdal A 1944 *J. Phys. (USSR)* **8** 331
[2] Harakeh MN and van der Woude A 2001 in 'Giant Resonances : fundamental high-frequency modes of nuclear excitation', Clarendon Press, Oxford
[3] Mohan R, Danos M, Biedenharn LC 1971 *Phys. Rev.* **3** 1740
[4] Batholemew GA 1961 *Ann. Rev. Nucl. Sci.* **11** 259
[5] Reinhard PG and Nazarewicz W 2010 *Phys. Rev.* **C 81** 051303(R)
[6] Piekarewicz J 2012 *Phys. Rev.* **C 85** 041302(R)
[7] Erler J 2013 *Phys. Rev.* **C 87** 044320
[8] Carbone A et al., 2010 *Phys. Rev.* **C 81** 041301(R)
[9] Brown BA and Schwenk A 2014 *Phys. Rev.* **C 89** 011307(R)
[10] Kneissl U, Pitz HH, Zilges A 1996, *Prog. Part. Nucl. Phys.* **37** 349
[11] Zilges A et al. 2002 *Phys. Lett.* **B 542** 43
[12] Tamii A et al., 2011 *Phys. Rev. Lett.* **107** 062502
[13] Adrich P et al, 2005 *Phys. Rev. Lett.* **95** 132501
[14] Wieland O et al, 2009 *Phys. Rev. Lett.* **102** 024310
[15] Savran D, Aumann T, Zilges A 2013 *Prog. Part. Nucl. Phys.* **C 70** 210
[16] Poelhekken TD at al. 1992 *Phys. Lett.* **B 278** 423
[17] Savran D et al. 2006 Nucl. Instr. and Meth. **A 564** 267
[18] Endres J et al. 2012 Phys. Rev. **C 85** 064331
[19] Savran D et al. 2006 *Phys. Rev. Lett.* **97** 172502
[20] Endres J et al. 2009 Phys. Rev. **C 80** 034302
[21] Endres J et al, 2010 *Phys. Rev.Lett.* **105** 044325

Journal of Physics: Conference Series **590** (2015) 012006 doi:10.1088/1742-6596/590/1/012006

[22] Derya V et al. 2013, Nucl. Phys. **906** 94
[23] Crespi FCL et al., 2014 *Phys. Rev. Lett.* **113** 012501
[24] Pellegri L et al. 2014 *Phys. Lett.* **B 738** 519
[25] Bianco D et al., 2012 *Phys. Rev.* **C 86** 044327
[26] Lanza EG et al. 2014 *Phys. Rev.* **C 89** 041601(R)

# Structure and dynamics of weakly-bound systems: a one-dimensional model [1]

**Andrea Vitturi [1,2] and Laura Moschini [1,2]**

[1] Dipartimento di Fisica e Astronomia, Universitá di Padova, Italy
[2] Istituto Nazionale di Fisica Nucleare, Sezione di Padova, Italy

E-mail: vitturi@pd.infn.it, laura.moschini@pd.infn.it

**Abstract.** A line of research has been developed to describe the structure and the dynamics of weakly-bound systems with one or more valence particles. To simplify the problem we are assuming particles moving in one dimension. Within this model one can describe, for example, one- and two-particle break-up and one- and two-particle transfer processes. In the case of reactions involving two particles one can study in particular the proper reaction mechanism in connection with the role of the pairing interaction. Exact solutions obtained by directly solving the many-body time-dependent Schroedinger equation can be compared with the results obtained with different approximation schemes (first-order, coupled channels, continuum discretization, etc). In this particular contribution we concentrate on one-particle transfer processes.

We address in this contribution the basic problem of the description of both structure and dynamics of weakly-bound systems with one or more valence particles. The problem is relatively easy with one valence particle (one-particle halo systems), but starts to be more complex with two particles (two-particle halo systems), becoming then extremely complicated with more particles. To simplify the problem, we will assume that the particles move just in a one dimension. In spite of this drastic assumption, we expect that the problem will maintain the main features and properties of the full three-dimensional case, for example for the description of pairing correlations and their connection with two-particle transfer or break-up processes.

Different aspects of the one-dimensional problem have been developed and discussed in different papers. In [1] we have discuss the problem of the excitation to the continuum (break-up) of a particle, initially moving in a single-particle orbit of a one-dimensional well, due to the action of an external perturbation (simulating the interaction with a reaction partner). The exact solution has been compared with the results obtained by expanding the wave function in a basis obtained by a discretization of the positive-energy part of the spectrum along the CDCC approach. In [2, 3] we have discussed the situation of a two-particle "borromean" system, bound due to the action of the residual pairing interaction, considering different discretization methods (HO, THO, infinite box) and studying both ground-state properties and the response functions to different operators. In [4] we have studied the break-up of such "borromean" systems due to the action of an external one-body field, pointing out the role of the residual pairing interaction in enhancing the correlated two-particle break-up in comparison with the single one-particle break-up.

[1] Work done in collaboration with Curro Pérez Bernal and Kouichi Hagino

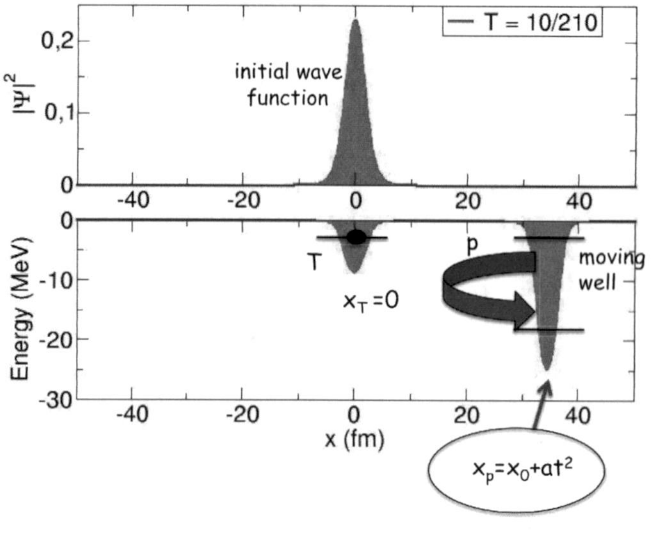


**Figure 1.** Schematic representation of our one-dimentional model. The upper frame gives the square of the one-particle wave funtion at the initial time, the lower frame the two potentials at the same initial time.

In this contribution we focus our attention on one-particle transfer processes. Our model includes a fixed well, with one particle initially sitted on one of the levels of the well. This situation is assumed to be representative of a system characterized by an inert core plus a valence particle. The system is then perturbed (see figure 1) by a second moving well and we follow in time the evolution of the single-particle wave function. For simplicity the motion of the second well is assumed to have a quadratic time dependence ($x_p = x_0 + at^2$, $x_p$ being the position of the center of the well), and by choice of the distance of closest approach and of the acceleration simulates different kinematical conditions of the reaction (bombarding energy, impact parameter, etc). In the same way, the choice of the depths of the two wells (assumed to have a Woods-Saxon shape), as well as the choice of the initial single-particle state, determine the characteristics of the two colliding objects, for example the situation of weak binding in a halo-like system.

In the case depicted in figure 1 the fixed well is chosen in such a way that the single particle is moving in the only bound state (with binding energy $E_b$=−3.10 MeV), while the moving well admits two (initially empty) bound levels. For the parameters defining the trajectory we have taken $x_0$= 10 fm and $a$= 0.3 fm/ps$^2$. The evolution of the wave function with time is illustrated in figure 2. The different frames refer to different times (the total collision time is divided in 210 steps and the corresponding time is quoted in each frame) and in each frame the upper part gives the square of the one-particle wave function while the lower frame gives the actual position of the two potentials at the same time. As apparent from the figure, when the tail of the moving well starts to overlap with the fixed well (frame c) part of the wave function enters in the moving well and then follows its movement (frames d-f). At the end of the process, by taking the overlap of the final wave function with the initial wave function and the wave functions of the single-particle states of the moving well one can determine the probability of elastic scattering and of one-particle transfer process. The presence of a node in the part of wave function inside the moving well already clearly indicates that the transfer takes place to the second single particle

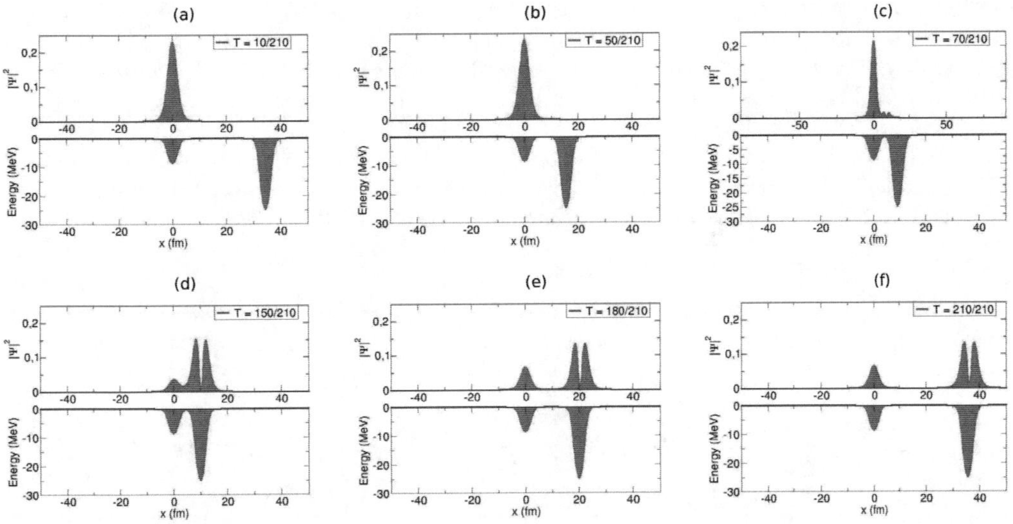


**Figure 2.** In all frames, corresponding to different times, the upper part gives the square of the one-particle wave function at that time, the lower part the actual position of the two potentials at the same time.

state. Note that only a negligible part of the wave function (practically undetectable in the figure) appears outside the two wells, indicating a practicably negligible break-up probability.

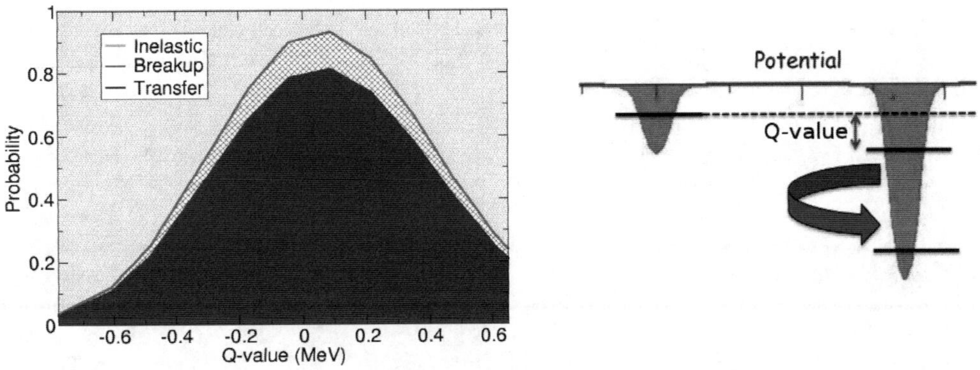


**Figure 3.** Left side: Probabilities for elastic scattering, transfer process and break-up process as a function of the difference between the energies of the single-particle states in the fixed and moving frame ( Q-value). Right side: Schematic picture of the reaction process.

The simplicity of the model allows us to compare easily the results obtained by varying the parameters characterizing the different situations. As an example we can vary the depth of the moving well, so varying the energies of the corresponding bund states. We can therefore study the transfer process as a function of the difference between the energies of the single-particle states in the fixed and moving frame, i.e. the Q-value associated to the transfer process (cf. the right part of figure 3). In the left part of figure 3 we plot the probabilities for elastic scattering, transfer process and (to minor extent) break-up process as a function of the Q-value. As we can see from the figure, the model reproduces the expected bell behavior of the transfer probability around the "optimal Q-value", which in our specific case of transfer of a neutral particle is close to zero. Of course the width of the probability distribution depends on the parameters of the collision (distance of closest approach and acceleration) and can therefore be modified by acting on these parameters.

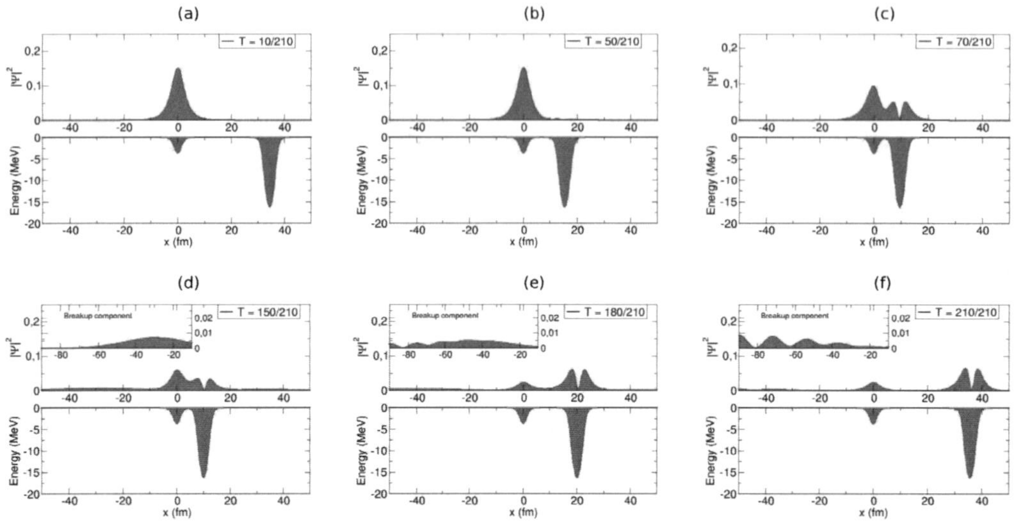


**Figure 4.** Same as in figure 2, but for an initial weakly-bound single-particle state ($E_b$=–0.91 MeV). The wave function is amplified in the insets, to put in evidence the continuum (break-up) part of the wave function.

We move now to the case involving weakly-bound systems. To this end we modify the depth of the potentials in order to obtain weakly-bound single-particle states ($E_b$=–0.91 MeV, with $x_0$=12 fm). As a consequence of the weak binding the initial wave function displays a longer tail, with the result of larger overlaps between the two wells. The corresponding time evolution of the wave function is shown, at different stages of the reaction, in the frames of figure 4. The weak binding situation is also responsible of larger break-up probabilities. To evidence the continuum part of the wave function, which is distributed over a large interval, we also show in the insets an amplification of the wave function in a region far from the two wells.

As in the previous case of well-bound single-particle states, we can study the final probabilities for transfer and break-up as a function of the difference between the energies of the single-particle states in the fixed and moving frame, i.e. the Q-value associated to the transfer process. These probabilities for elastic scattering, transfer process and break-up process as a function of the Q-value are plotted in the left frame of figure 5. As shown by the figure, there is a large contribution from the break-up process, which is strongly favored by the initial weak binding of the particle, and which shows a smoother dependence on the value of the single-particle energies in the moving well. On the converse, the transfer probability displays the expected bell behavior around the optimal Q-value.

Journal of Physics: Conference Series **590** (2015) 012007 doi:10.1088/1742-6596/590/1/012007

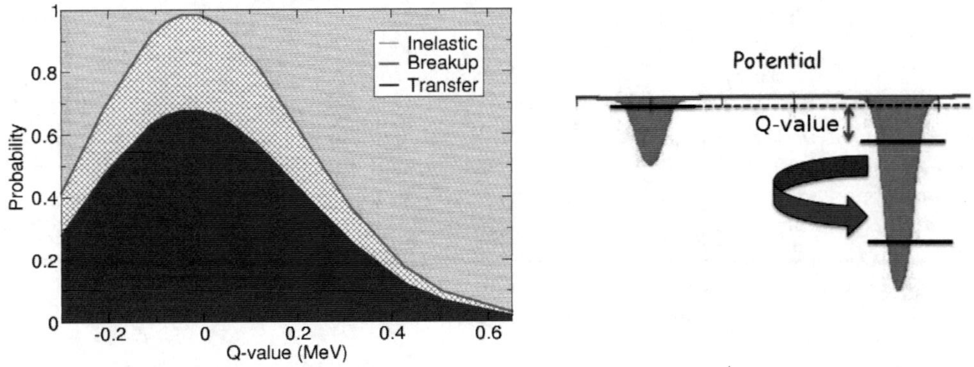


**Figure 5.** Same as in figure 3, but for an inial weakly-bound single-particle state.

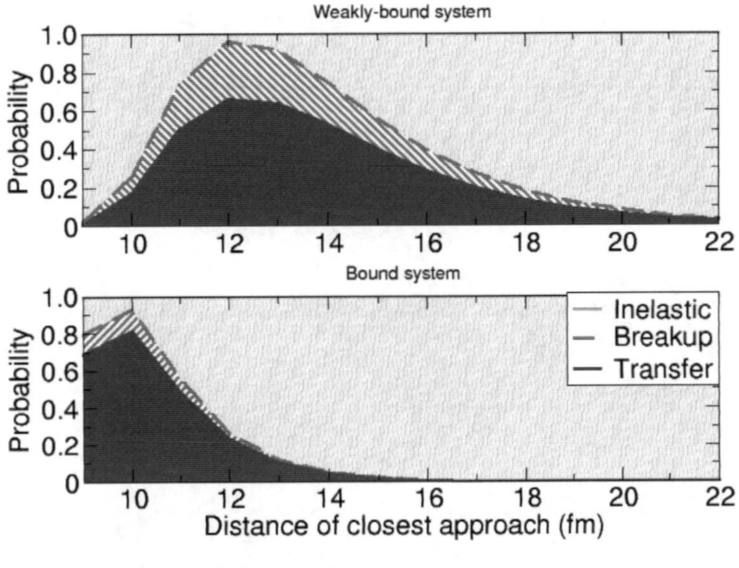


**Figure 6.** Transfer and break-up probabilities as a function of the distance of closest approach $x_0$ between the two colliding wells. The lower frame refers to the case of a well-bound initial single-particle state, the upper frame to the case of a weakly-bound state.

As a final point in this contribution we display in figure 6 the transfer and break-up probabilities as a function of the distance of closest approach $x_0$ between the two colliding wells. The probabilities display the expected exponentially decaying behavior arising from the combination of the tail properties of the two wells and of the single-particle states. It is therefore not surprising that, in the case of the initial weakly-bound system characterized by the long tail, the transfer probability survives to larger distances with a lower decaying rate. Note also that below a certain critical distance (approximately below 10 fm for the well-bound case and 12 fm for the weakly-bound) the transfer probability becomes so large that in a perturbative picture we will have back and forth transfer, with resulting oscillatory behavior. In this regime our model space (characterized by just one active particle) becomes clearly inappropriate.

The results shown in this contribution have been obtained by directly solving the time-dependent one-particle Schroedinger equation. The same equation can be solved within the standard time-dependent coupled-channels formalism by constructing the non-diagonal transfer formfactors and expanding the wave function into the dual basis associated with the two wells. One can in this way test the validity of the first-order approximation and the necessary truncation in the basis. This latter point is particular relevant in the case of weakly-bound systems in connection with the treatment of the positive-energy part of the spectrum and the procedures used for the discretization of the continuum.

Next step will involve the extension to the case of two-particle sytems and two-particle transfer processes. In this case we will have two particles initially sitting in one single-particle state of the fixed well and one will follow the time evolution of the two-particle wave function due to action of the other moving one-body potential plus the residual pairing-like interaction between the two particles. The projections of the final wave function on the asymptotic states corresponding to two-particles in the initial well, one particle in each well, two particles in the moving well and one or two-particles in the continuum will yield the probabilities for elastic, one-partcicle transfer, two-particle transfer and one and two particle break-up, respectively. By switching on and off the residual pairing interaction one will determine its effect in the enhancement of the two-particle transfer probabilities with respect to the one-particle transfer. As in the case of one-particle transfer, the exact results will be compared with those obtained in the conventional coupled-channel scheme or in its lowest-order two-step perturbation limit.

## References

[1] Dasso C H and Vitturi A 2007 *Nucl. Phys.* A **787** 476
Dasso C H and Vitturi A 2009 *Phys. Rev.* C **79** 064620
[2] Pérez Bernal F and Vitturi A 2009 *AIP Conf. Proc.* **1165** 305
Vitturi A and Perez Bérnal F 2010 *Nucl. Phys.* A **834** 428c
Vitturi A and Perez Bérnal F 2012 *AIP Conf. Proc.* **1488** 220
[3] Moschini L 2013, *Master thesis* unpublished
Moschini L 2014 *J. Phys, Conference Series* **566** 012027
[4] Hagino K, Vitturi A, Pérez-Bernal F and Sagawa H 2011 *J. Phys. G: Nucl. Part. Phys.* **38** 015105

# Four-body effects on $^9$Be+$^{208}$Pb scattering and fusion around the Coulomb barrier

**P Descouvemont[1], T Druet[1], L F Canto[2,3] and M S Hussein[4,5,6]**

1 Physique Nucléaire Théorique et Physique Mathématique, C.P. 229, Université Libre de Bruxelles (ULB), B 1050 Brussels, Belgium

2 Instituto de Física, Universidade Federal do Rio de Janeiro, C.P. 68528, 21941-972 Rio de Janeiro, RJ, Brazil

3 Instituto de Fsica, Universidade Federal Fluminense, Av. Gal. Milton Tavares de Souza s/n, Niteri, RJ, Brazil

4 Departamento de Física Matemática, Instituto de Física, Universidade de São Paulo, C.P. 66318, 05314-970, São Paulo, SP, Brazil

5 Instituto de Estudos Avançados, Universidade de São Paulo, C.P. 72012, 05508-970, São Paulo, SP, Brazil

6 Departamento de Física, Instituto Tecnológico de Aeronáutica, CTA, São José dos Campos, São Paulo, SP, Brazil

E-mail: pdesc@ulb.ac.be

**Abstract.** We investigate the $^9$Be+$^{208}$Pb elastic scattering and fusion at energies around the Coulomb barrier. The $^9$Be nucleus is described in a $\alpha + \alpha + n$ three-body model, using the hyperspherical coordinate method. The scattering with $^{208}$Pb is then studied with the Continuum Discretized Coupled Channel (CDCC) method, where the $\alpha + \alpha + n$ continuum is approximated by a discrete number of pseudostates. Optical potentials for the $\alpha$+$^{208}$Pb and $n$+$^{208}$Pb systems are taken from the literature. We present elastic-scattering and fusion cross sections at different energies, and investigate the convergence with respect to the truncation of the $\alpha + \alpha + n$ continuum. A good agreement with experiment is obtained, considering that there is no parameter fitting. We show that continuum effects increase at low energies.

## 1. Introduction

Many experiments have been performed with the $^9$Be nucleus, used as a target or as a projectile [1]. Although $^9$Be is stable, it presents a Borromean structure, as the well known halo nucleus $^6$He. None of the two-body subsystems $\alpha + n$ or $\alpha + \alpha$ is bound in $^9$Be, which has important consequences on the theoretical description of this nucleus. Precise wave functions must include the three-body nature of $^9$Be. The hyperspherical formalism [2] is an ideal tool to describe three-body Borromean systems, as it does not assume a specific two-body structure, and considers the three particles $\alpha + \alpha + n$ on an equal footing.

In the present work, we aim at investigating $^9$Be scattering and fusion on an heavy target. The reaction framework is the Continuum Discretized Coupled Channel (CDCC) method [3–5], which is well adapted to weakly bound projectiles since it allows to include breakup channels. Over the last decades, the CDCC method has been extended in various directions, and in particular to reactions involving three-body projectiles such as $^6$He [6] or $^{11}$Li [7]. Going from two-body projectiles (such as d=p+n or $^7$Li=$\alpha$+t) to three-body projectiles strongly increases the complexity of the calculations, even if both options eventually end up with a standard coupled-channel system.

Many data have been obtained for $^9$Be+$^{208}$Pb elastic scattering [8, 9] and fusion [10, 11]. These experimental data provide a good opportunity to test $^9$Be wave functions. Previous CDCC calculations, using a two-body approximation for $^9$Be, show that breakup effects are important [9, 12, 13]. In a two-body model, $^9$Be is assumed to have a $^8$Be $+ n$ or $^5$He $+ \alpha$ cluster structure. This approximation presents several shortcomings: (*i*) $^8$Be and $^5$He are unbound, and assuming a point-like structure is questionable; (*ii*) in a more rigorous three-body approach, both configurations are strictly equivalent, and the relative importance of the $^8$Be $+ n$ and $^5$He $+ \alpha$ channels is not relevant; (*iii*) a two-body model of $^9$Be requires $^8$Be+$^{208}$Pb and $^5$He+$^{208}$Pb optical potentials, which are not available. These different issues are more precisely addressed in a three-body model of $^9$Be. No assumption should be made about the cluster structure, and $\alpha$+$^{208}$Pb as well as $n$+$^{208}$Pb optical potentials are available in the literature.

## 2. Three-body model of $^9$Be

The determination of the $^9$Be wave functions is the first step for the $^9$Be+$^{208}$Pb CDCC calculation. The spectroscopy of $^9$Be in cluster models has been performed in previous microscopic [14–16] and non-microscopic [17–19] models. For a three-body system, the Hamiltonian is given by

$$H_0 = \sum_{i=1}^{3} \frac{\boldsymbol{p}_i^2}{2m_i} + \sum_{i<j=1}^{3} V_{ij}(\boldsymbol{r}_i - \boldsymbol{r}_j), \tag{1}$$

where $\boldsymbol{r}_i$ and $\boldsymbol{p}_i$ are the space and momentum coordinates of the three particles with masses $m_i$, and $V_{ij}$ a potential between nuclei $i$ and $j$. For the $\alpha + \alpha$ interaction, we use the deep potential of Buck *et al.* [20]. The $\alpha$+n interaction is taken from Kanada *et al.* [21]. Both (real) potentials accurately reproduce the $\alpha + \alpha$ and $\alpha$+nucleon phase shifts over a wide energy range.

With these potentials, the $^9$Be ground state is too bound ($-3.12$ MeV, while the experimental value is $-1.57$ MeV with respect to the $\alpha + \alpha + n$ threshold). We have therefore introduced a phenomenological three-body force as

$$V_{3B}(\rho) = \frac{V_0}{1 + (\rho/\rho_0)^2}, \tag{2}$$

where $\rho_0 = 6$ fm [22]. Using $V_0 = 2.7$ MeV provides the experimental binding energy of the $3/2^-$ ground state. The r.m.s. radius, the quadrupole moment, and the magnetic moment are $\sqrt{r^2} = 2.36$ fm, $Q(3/2^-) = 4.96$ $e$.fm$^2$, and $\mu = -1.33$ $\mu_N$, respectively. These values are in fair agreement with experiment ($2.45 \pm 0.01$ fm, $5.29 \pm 0.04$ $e$.fm$^2$ and $-1.18$ $\mu_N$, respectively)

## 3. The CDCC method

We present here a brief outline of the CDCC method, and we refer to Ref. [23] for specificities of three-body projectiles. The CDCC method is based on approximate solutions of the projectile Hamiltonian (1)

$$H_0 \Phi_k^{jm\pi} = E_{0,k}^{j\pi} \Phi_k^{jm\pi}, \tag{3}$$

where $k$ are the excitation levels in partial wave $j\pi$. Solutions with $E_{0,k}^{j\pi} < 0$ correspond to bound states of the projectile, whereas $E_{0,k}^{j\pi} > 0$ correspond to narrow resonances or to approximations of the three-body continuum. These states cannot be associated with physical states, and are referred to as pseudostates.

Journal of Physics: Conference Series **590** (2015) 012008 doi:10.1088/1742-6596/590/1/012008

The Hamiltonian of the projectile + target system is written as

$$H(\boldsymbol{R},\boldsymbol{x},\boldsymbol{y}) = H_0(\boldsymbol{x},\boldsymbol{y}) - \frac{\hbar^2}{2\mu_{PT}}\Delta_R + V(\boldsymbol{R},\boldsymbol{x},\boldsymbol{y}) \tag{4}$$

where $\mu_{PT}$ is the reduced mass of the system, and $\boldsymbol{R}$ the relative coordinate (see Fig. 1). The Jacobi coordinates $\boldsymbol{x}$ and $\boldsymbol{y}$ are proportional to $\boldsymbol{r}_{\text{Be-n}}$ and $\boldsymbol{r}_{\alpha-\alpha}$, respectively. The potential term reads

$$V(\boldsymbol{R},\boldsymbol{x},\boldsymbol{y}) = V_{1t}(\boldsymbol{R},\boldsymbol{y}) + V_{2t}(\boldsymbol{R},\boldsymbol{x},\boldsymbol{y}) + V_{3t}(\boldsymbol{R},\boldsymbol{x},\boldsymbol{y}) \tag{5}$$

where the three components $V_{it}$ are optical potentials between fragment $i$ and the target.

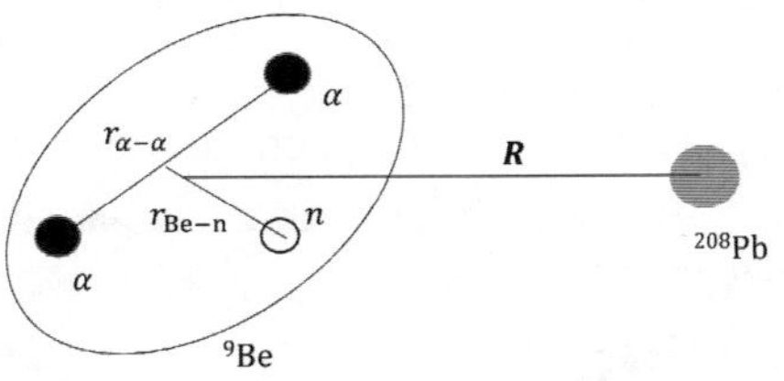


**Figure 1.** Coordinates involved in the present four-body model.

In order to solve the Schrödinger equation associated with (4), the total wave function with angular momentum $J$ and parity $\Pi$ is expanded as

$$\Psi^{JM\Pi}_{(\omega)}(\boldsymbol{R},\boldsymbol{x},\boldsymbol{y}) = \sum_{j\pi kL} \varphi^{JM\Pi}_{j\pi kL}(\Omega_R,\boldsymbol{x},\boldsymbol{y}) g^{J\Pi}_{jkL(\omega)}(R), \tag{6}$$

where $\omega$ is the entrance channel, and where the channel wave functions are given by

$$\varphi^{JM\Pi}_{j\pi kL}(\Omega_R,\boldsymbol{x},\boldsymbol{y}) = i^L \left[\Phi^{j\pi}_k(\boldsymbol{x},\boldsymbol{y}) \otimes Y_L(\Omega_R)\right]^{JM}. \tag{7}$$

The radial functions $g^{J\Pi}_c(R)$ (we use the notation $c = (j,\pi,k,L)$) are obtained from the system

$$\left[-\frac{\hbar^2}{2\mu_{PT}}\left(\frac{d^2}{dR^2} - \frac{L(L+1)}{R^2}\right) + E_c - E\right] g^{J\Pi}_{c(\omega)}(R) + \sum_{c'} V^{J\Pi}_{c,c'}(R) g^{J\Pi}_{c'(\omega)}(R) = 0, \tag{8}$$

where $E$ is the c.m. energy, and where the coupling potentials are given by matrix elements

$$V^{J\Pi}_{c,c'}(R) = \langle \varphi^{JM\pi}_c | V | \varphi^{JM\pi}_{c'} \rangle. \tag{9}$$

As the fragment-target potentials (5) are optical potentials, matrix elements (9) contain a real and an imaginary parts. The integration is performed over $\Omega_R$ and over the internal coordinates $\boldsymbol{x}$ and $\boldsymbol{y}$. The calculation of these coupling potentials is much more complicated for three-body projectiles than for two-body projectiles. Out of the three potentials $V_{it}$ in (5), matrix elements of $V_{1t}$ are the easiest since $V_{1t}$ does not depend on $\boldsymbol{x}$. The calculations of the two remaining terms $V_{2t}$ and $V_{3t}$ are performed by using the Raynal-Revai coefficients (see Refs. [23, 24] for detail). We solve Eq. (8) with the $R$-matrix theory [25].

## 4. Application to the $^9$Be+$^{208}$Pb system

### *4.1. Elastic scattering*

The calculations are performed with the $j^\pi = 3/2^-, 5/2^-, 1/2^+$ partial waves on $^9$Be, and the cutoff energy is 10 MeV. Figure 2 shows $^9$Be+$^{208}$Pb elastic cross sections at various energies (the data are taken from Refs. [8, 26]). In contrast with optical-model calculations [8], no renormalization of the potential is needed. The absorption is simulated by the imaginary parts of the $\alpha-^{208}$Pb and n$-^{208}$Pb interactions, and by the $\alpha + \alpha + n$ discretized continuum. From Fig. 2, it appears that continuum couplings are more important at low energies. At $E = 38$ and 44 MeV, the single-channel calculation, limited to the $^9$Be ground state (dotted lines) is significantly different from the data. This confirms the conclusion of a previous work [27] which suggests that continuum couplings are more important at low energies.

Calculations involving $3/2^-$ pseudostates are shown as dashed lines, and the full calculations as solid lines. The effect of the $1/2^+$ partial wave is minor, and the main differences stem from the $5/2^-$ pseudostates.

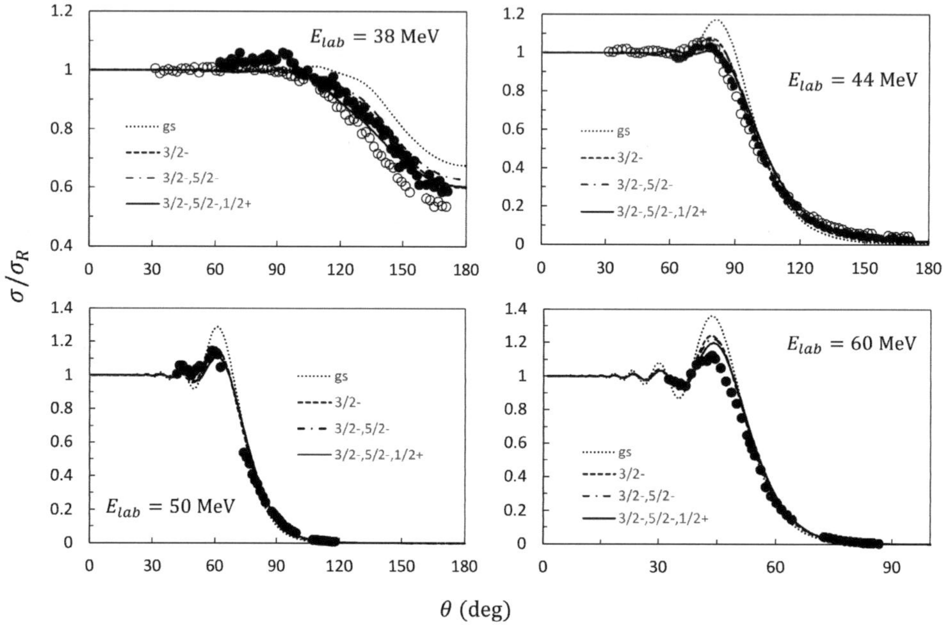


**Figure 2.** $^9$Be+$^{208}$Pb elastic cross sections (divided by the Rutherford cross section) at different $^9$Be laboratory energies, and for different sets of $^9$Be partial waves. The experimental data are taken from Ref. [8] (filled circles) and Ref. [26] (open circles).

### *4.2. Fusion*

The $^9$Be+$^{208}$Pb fusion reaction has been studied theoretically by several groups (see Ref. [28] and references therein). Data about the various mechanisms (total TF, complete CF and incomplete ICF fusion) are also available in the literature [10, 11, 29].

The total fusion cross section $\sigma_{TF}$ is defined from [30–32]

$$\sigma_{TF}(E) = \frac{\pi}{k^2} \sum_{J\Pi} (2J+1) T^{J\Pi}(E), \tag{10}$$

where $k$ is the wave number, and $T^{J\Pi}$ is a transmission coefficient in partial wave $J\Pi$. It can be obtained from the scattering matrices as

$$T^{J\Pi}(E) = \frac{1}{(2I_1+1)(2I_2+1)} \sum_{L_\omega,L} \left(\delta_{L_\omega L} - |U^{J\Pi}_{\omega L_\omega,\omega L}(E)|^2\right), \tag{11}$$

where $I_1$ and $I_2$ are the spins of the colliding nuclei. Using flux conservation properties, the transmission coefficient can be expressed from the imaginary part $W^{J\Pi}_{c,c'}(R)$ of the coupling potentials [32]. It is therefore strictly equivalent to

$$T^{J\Pi}(E) = -\frac{4k}{E}\frac{1}{(2I_1+1)(2I_2+1)} \sum_{L_\omega} \sum_{\alpha,\alpha',L,L'} \int g^{J\Pi}_{\omega L_\omega,\alpha L}(R)\, W^{J\Pi}_{\alpha L,\alpha' L'}(R)\, g^{J\Pi}_{\omega L_\omega,\alpha' L'}(R)\, dR, \tag{12}$$

where index $\alpha$ stands for $\alpha = (j, \pi, k)$, i.e. it does not depend on the angular momentum $L$ (in other words $c = (\alpha, L)$). This derivation is more complicated since it requires the wave functions, but it provides a better accuracy at low energies, where the scattering matrix is close to unity. It also provides an approximate method to distinguish between CF and ICF [33].

The TF cross section is shown in logarithmic (Fig. 3) and linear (Fig. 4) scales. Convergence against $^9$Be states is rather fast, except below the Coulomb barrier where neglecting breakup channels leads to an underestimation of the cross section. Breakup effects, however, introduce a strong enhancement of the cross section at low energies. Similar conclusions have been drawn by Jha *et al.* [28].

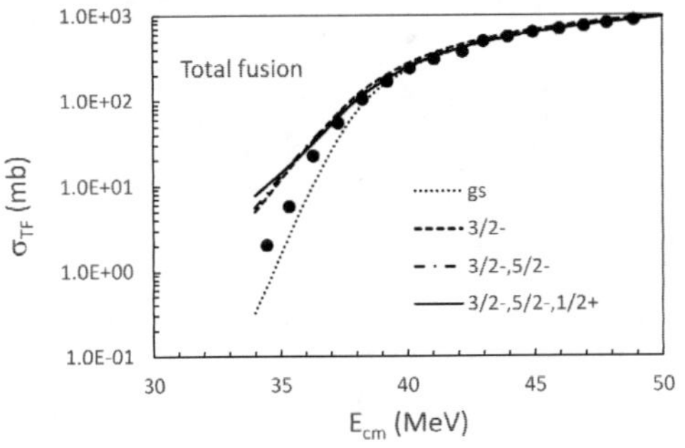


**Figure 3.** $^9$Be+$^{208}$Pb total fusion cross section in a logarithmic scale. The data are taken from [10].

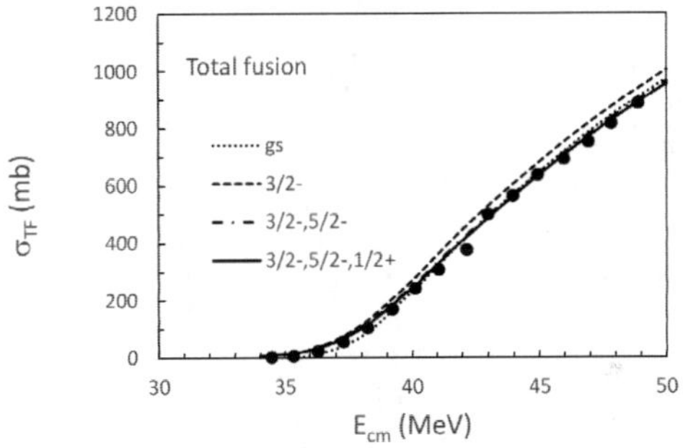


**Figure 4.** See Fig. 3 in a linear scale.

## 5. Conclusion

We have applied the CDCC formalism to the $^9$Be+$^{208}$Pb system. In a first step, we have computed $^9$Be wave functions in a three-body $\alpha + \alpha + n$ model. These calculations were performed for low-lying states, but also for pseudostates, which represent positive-energy approximations of the continuum. The main advantage of the hyperspherical approach is that it treats the three-body continuum without any approximation concerning possible $^8$Be + $n$ or $^5$He + $\alpha$ cluster structures, which are in fact strictly equivalent.

The three-body model of $^9$Be relies on $\alpha + \alpha$ and $\alpha + n$ (real) interactions, which reproduce very well the elastic phase shifts. With these bare interactions, the $^9$Be ground state is slightly too bound. We therefore introduce a phenomenological three-body force to reproduce the experimental ground-state energy. The spectroscopic properties of low-lying states is in fair agreement with experiment.

We used first the $^9$Be wave functions in a CDCC calculation of the $^9$Be+$^{208}$Pb elastic scattering at energies close to the Coulomb barrier. As expected, including continuum channels improves the theoretical cross sections. A fair agreement with the data is obtained.

We also applied the four-body CDCC to the $^9$Be+$^{208}$Pb fusion, by using the same potentials as for elastic scattering. The total fusion cross section is found in good agreement with experiment, except at very low energies, where we overestimate the data.

## References

[1] Keeley N, Alamanos N, Kemper K and Rusek K 2009 *Prog. Part. Nucl. Phys.* **63** 396
[2] Zhukov M V, Danilin B V, Fedorov D V, Bang J M, Thompson I J and Vaagen J S 1993 *Phys. Rep.* **231** 151
[3] Rawitscher G H 1974 *Phys. Rev. C* **9** 2210
[4] Yahiro M, Iseri Y, Kameyama H, Kamimura M and Kawai M 1986 *Prog. Theor. Phys. Suppl.* **89** 32
[5] Yahiro M, Matsumoto T, Minomo K, Sumi T and Watanabe S 2012 *Prog. Theor. Phys. Supp.* **196** 87
[6] Matsumoto T, Hiyama E, Ogata K, Iseri Y, Kamimura M, Chiba S and Yahiro M 2004 *Phys. Rev. C* **70** 061601
[7] Cubero M, Fernández-García J P, Rodríguez-Gallardo M, Acosta L, Alcorta M, Alvarez M A G, Borge M J G, Buchmann L, Diget C A, Falou H A, Fulton B R, Fynbo H O U, Galaviz D, Gómez-Camacho J, Kanungo R, Lay J A, Madurga M, Martel I, Moro A M, Mukha I, Nilsson T, Sánchez-Benítez A M, Shotter A, Tengblad O and Walden P 2012 *Phys. Rev. Lett.* **109** 262701
[8] Woolliscroft R J, Fulton B R, Cowin R L, Dasgupta M, Hinde D J, Morton C R and Berriman A C 2004 *Phys. Rev. C* **69** 044612
[9] Pandit S K, Jha V, Mahata K, Santra S, Palshetkar C S, Ramachandran K, Parkar V V, Shrivastava A, Kumawat H, Roy B J, Chatterjee A and Kailas S 2011 *Phys. Rev. C* **84** 031601
[10] Dasgupta M, Hinde D J, Butt R D, Anjos R M, Berriman A C, Carlin N, Gomes P R S, Morton C R, Newton J O, Szanto de Toledo A and Hagino K 1999 *Phys. Rev. Lett.* **82** 1395
[11] Dasgupta M, Gomes P R S, Hinde D J, Moraes S B, Anjos R M, Berriman A C, Butt R D, Carlin N, Lubian J, Morton C R, Newton J O and Szanto de Toledo A 2004 *Phys. Rev. C* **70** 024606
[12] Keeley N, Kemper K W and Rusek K 2001 *Phys. Rev. C* **64** 031602
[13] Parkar V V, Jha V, Pandit S K, Santra S and Kailas S 2013 *Phys. Rev. C* **87** 034602
[14] Descouvemont P 2001 *Eur. Phys. J. A* **12** 413
[15] Arai K, Descouvemont P, Baye D and Catford W N 2003 *Phys. Rev. C* **68** 014310
[16] Arai K, Ogawa Y, Suzuki Y and Varga K 1996 *Phys. Rev. C* **54** 132
[17] Voronchev V, Kukulin V, Pomerantsev V and Ryzhikh G 1995 *Few-Body Systems* **18** 191
[18] Theeten M, Baye D and Descouvemont P 2006 *Phys. Rev. C* **74** 044304
[19] Álvarez-Rodríguez R, Jensen A S, Garrido E and Fedorov D V 2010 *Phys. Rev. C* **82** 034001
[20] Buck B, Friedrich H and Wheatley C 1977 *Nucl. Phys. A* **275** 246
[21] Kanada H, Kaneko T, Nagata S and Nomoto M 1979 *Prog. Theor. Phys.* **61** 1327
[22] Thompson I J, Danilin B V, Efros V D, Vaagen J S, Bang J M and Zhukov M V 2000 *Phys. Rev. C* **61** 024318
[23] Rodríguez-Gallardo M, Arias J M, Gómez-Camacho J, Johnson R C, Moro A M, Thompson I J and Tostevin J A 2008 *Phys. Rev. C* **77** 064609
[24] Thompson I J, Nunes F M and Danilin B V 2004 *Comp. Phys. Comm.* **161** 87
[25] Descouvemont P and Baye D 2010 *Rep. Prog. Phys.* **73** 036301
[26] Yu N, Zhang H Q, Jia H M, Zhang S T, Ruan M, Yang F, Wu Z D, Xu X X and Bai C L 2010 *J. Phys. G* **37** 075108
[27] Druet T and Descouvemont P 2012 *Eur. Phys. J. A* **48** 147
[28] Jha V, Parkar V V and Kailas S 2014 *Phys. Rev. C* **89** 034605
[29] Dasgupta M, Hinde D J, Sheehy S L and Bouriquet B 2010 *Phys. Rev. C* **81** 024608
[30] Canto L F, Gomes P R S, Donangelo R and Hussein M S 2006 *Phys. Rep.* **424** 1
[31] Hagino K and Takigawa N 2012 *Prog. Theor. Phys.* **128** 1001–1060
[32] Canto L F and Hussein M S 2013 *Scattering Theory of Molecules, Atoms and Nuclei* (World Scientific Publishing)
[33] Diaz-Torres A and Thompson I J 2002 *Phys. Rev. C* **65** 024606

Journal of Physics: Conference Series **590** (2015) 012009 doi:10.1088/1742-6596/590/1/012009

# Fully damped Mott oscillations in sub-barrier elastic scattering of identical heavy ions and the nuclear interaction

**M. S. Hussein**

Instituto de Física and Instituto de Estudos Avançados, Universidade de São Paulo, São Paulo, Brazil
Departamento de Física, Instituto Tecnológico de Aeronáutica, CTA, São José dos Campos, S.P., Brazil

**L. F. Canto**

Instituto de Física, Universidade Federal do Rio de Janeiro, Rio de Janeiro, Brazil
Instituto de Fisica, Universidade Federal Fluminense, Av. Gal. Milton Tavares de Souza s/n, Niterói, R.J., Brazil

**W.Mittig**

National Superconducting Cyclotron Laboratory and Department of Physics and Astronomy, Michigan State University, East Lansing, Michigan 48824, USA

E-mail: hussein@if.usp.br

**Abstract.** We investigate the possible disappearance of Mott oscillations in the scattering of bosonic nuclei at sub-barrier energies. This effect is universal and happens at a critical value of the Sommerfeld parameter. It is also found that the inclusion of the short-range nuclear interaction has a profound influence on this phenomenon. Thus we suggest that the study of this lack of Mott oscillation, which we call, "transverse isotropy" is a potentially useful mean to study the nuclear interaction.

## 1. Introduction

It is quite well known that the scattering amplitude describing the scattering of particles with a long-range interaction, such as the Coulomb one, composed of a slowly varying amplitude and a possibly rapidly varying phase. This results in a quantal cross section which is devoid of oscillations and which exactly coincides with the corresponding classical cross section. The reason for this is understood semi-classically, in the sense that in the evaluation of the quantal amplitude using the stationary phase method, one finds only one dominant contribution. The absence of oscillations in the scattering of non-identical particles with long range interaction, is changed when dealing with identical particles as the scattering amplitude in this case has to be symmetrized (bosons) or anti-symmetrized (fermions). In these cases one encounters rather strong oscillations in the cross section arising from the interference of the amplitude evaluated at an angle $\theta$ and the one at $\pi - \theta$. The resulting, Mott, cross section is oscillatory and can

Journal of Physics: Conference Series **590** (2015) 012009 doi:10.1088/1742-6596/590/1/012009

be used to assess the importance of other weaker interactions which may be present. This idea has been explored by several authors in the case of heavy-ion scattering, where the long range interaction is the Coulomb one and the weaker interactions, can be long-ranged, such as multipole polarizability [1, 2], relativistic effects [5], color van der Waals force [3, 4], or short-ranged, such as the nuclear interaction [5, 6]. In this contribution we demonstrate that the Mott cross section in the absence of the weaker forces (pure Coulomb) can become structureless at a certain value of the Sommerfeld parameter [7]. This surprising effect can be explored to pin down the properties of the short-range nuclear interaction at energies well below the Coulomb barrier. In the next sections we give the details of our recent investigation [8, 9].

## 2. Mott oscillations and their disappearance: Transverse Isotropy

We consider the scattering of two charged particles (nuclei) with charges $Z_p$ and $Z_t$. The interaction is predominantly Coulomb. with a small nuclear contribution. The unsymmetrized scattering amplitude is given by,

$$f_c(\theta) = -\frac{a}{2}\, e^{2i\sigma_0}\, \frac{e^{-i\eta \ln(\sin^2\theta/2)}}{\sin^2(\theta/2)} = A_C(E,\theta)e^{i\Delta_C(E,\theta)}, \quad (1)$$

where the amplitude, $A_C(E,\theta)$ and the phase, $\Delta_C(E,\theta)$ are defined in Eq. (1)

$$A_C(E,\theta) = -\frac{a}{2\sin^2(\theta/2)} \quad (2)$$

$$\Delta_C(E,\theta) = 2\sigma_0 - \eta \ln(\sin^2\theta/2) \quad (3)$$

and where $\eta$ and $a$ are respectively the Sommerfeld parameter, and half the distance of closest approach in a head-on collision (zero impact parameter), given by

$$\eta = \frac{Z_p Z_t e^2}{\hbar v}, \qquad a = \frac{Z_p Z_t e^2}{2E_{\rm c.m.}}. \quad (4)$$

Above, $v$ is the relative velocity, and $\sigma_0$ is the s-wave Coulomb phase shift

$$\sigma_0 = \arg\{\Gamma(1+i\eta)\}, \quad (5)$$

where $\Gamma$ stands for the usual Gamma-function.

An excellent approximation for $\sigma_0$ was obtained in [10],

$$\sigma_0 = \frac{1}{2}\tan^{-1}(\eta) + \eta\Big(\ln\left(\sqrt{1+\eta^2}\right) - 1\Big) - \frac{\eta}{12\,(1+\eta^2)}. \quad (6)$$

The Coulomb amplitude, Eq. (1), must be symmetrized in the case of the scattering of identical bosonic charged particles, which for spin zero, $f(\theta) = f_C(\theta) + f_C(\pi-\theta)$. The resulting Mott cross section, $\sigma_M(\theta)$ contains an incoherent, classical, piece, $\sigma_{\rm inc}(\theta)$, plus an interference one, $\sigma_{\rm int}(\theta)$, viz,

$$\sigma_M(\theta) = \sigma_{\rm inc}(\theta) + \sigma_{\rm int}(\theta), \quad (7)$$

where $\sigma_{\rm inc}(\theta)$ is the incoherent sum of contributions from the two amplitudes, $f_C(\theta)$, and $f_C(180^\circ - \theta)$,

$$\sigma_{\rm inc}(\theta) = |f_C(\theta)|^2 + |f_C\,(180^\circ - \theta)\,|^2, \quad (8)$$

and $\sigma_{\rm int}(\theta)$ is the interference term,

$$\sigma_{\rm int}(\theta) = 2\,{\rm Re}\Big\{f_C^*(\theta)\,\times f_C\,(180^\circ - \theta)\Big\}. \quad (9)$$

Note that the incoherent part of the cross section is positive-definite, whereas the interference term may assume positive or negative values.

It was found in Ref. [7] that the full Mott cross section becomes flat at a critical value, $\eta_c = \sqrt{2}$, of the Sommerfeld parameter. This feature was coined *Transverse Isotropy*, TI.This was accomplished by demanding that the second derivative of the cross section with respect to $\theta$ to be zero at $\theta = 90^o$. Of all the heavy-ion systems which were considered, the best case for studying the TI was found to be the elastic scattering of two $^4$He nuclei, for which the condition $\eta_c = \sqrt{2}$ implies a center of mass energy of $E_{\text{c.m.}}$= 0.397 MeV, well below the height of the Coulomb barrier, $V_B = 0.87$ MeV. This guarantees that, for all practical purposes, the scattering is purely Coulomb. Our calculation also indicated that at an energy below $E_c$ the cross section at 90$^o$ exhibits a maximum, while above the critical energy it exhibits a minimum. In 2006, Abdullah et al. [11] published elastic data for the $\alpha + \alpha$ system at several energies starting from around the Coulomb barrier. In Fig. (1) we show their results,

We see two flat regions followed by the usual oscillatory behavior. The first flat region is at $E_{\text{c.m.}} = 1.0$ MeV, and the second at $E_{\text{c.m.}} = 1.5$ MeV. Our calculation for a pure Coulomb potential Mott scattering predicts one flat region at $E_{\text{c.m.}} \approx 0.4$ MeV. This clearly indicates the influence of the short range nuclear interaction.

## 3. Inclusion of the nuclear interaction

In this section we investigate the influence of the nuclear interaction on the TI. We do this by adding this potential to the Coulomb one and solve the scattering problem. From the analytical point of view the determination of regions of the TI is more difficult here, but the calculation of the second derivative can be easily done numerically. First we give the salient feature of the symmetrized cross section in the presence of the nuclear interaction.

Using as a guide the form of the Rutherford cross section, Eq. (1), we can write the symmetrized cross section as

$$\frac{d\sigma}{d\Omega} = \sigma_{\text{inc}} + \sigma_{int} \tag{10}$$

with the incoherent part $\sigma_{\text{inc}}$, affected only very little by the short range nuclear interaction at sub-barrier energies,

$$\sigma_{\text{inc}} = \frac{a^2}{4}\left[\frac{1}{\sin^4(\theta/2)} + \frac{1}{\cos^4(\theta/2)}\right] \tag{11}$$

and the coherent, interference, part $\sigma_{\text{int}}$,

$$\sigma_{\text{int}} = 2\sqrt{\sigma_{\text{C}}(\theta)\sigma_{\text{C}}(180^o - \theta)}\cos\left([\Delta_{\text{C}}(\theta) + \Delta_{\text{N}}(\theta)] - [\Delta_{\text{C}}(180^o - \theta) + \Delta_{\text{N}}(180^o - \theta)]\right) \tag{12}$$

which gives, with the help of Eq. (1),

$$\sigma_{inc}(\theta) = \frac{a^2}{2}\frac{1}{\sin^2(\theta/2)\cos^2(\theta/2)}\cos\left(2\eta\ln\tan\frac{\theta}{2} + [\Delta_{\text{N}}(\theta) - \Delta_{\text{N}}(180^o - \theta)]\right) \tag{13}$$

The second derivative of the full cross section at $\theta = 90^o$ is evaluated for a nuclear interaction used in Ref. [11]. They employed the following nuclear force,

$$V_{\text{N}}(r) = -V_A e^{-(r/R_A)^2} + V_R e^{-(r/R_R)^2} \tag{14}$$

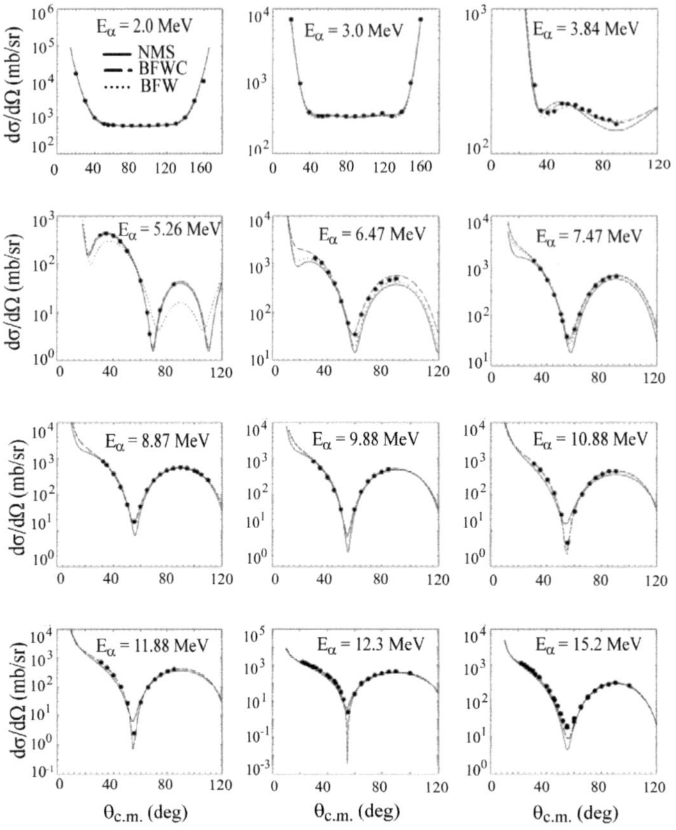


**Figure 1.** (Color online) The experimental angular distributions for the $^4$He+$^4$He system at several collision energies. The figure was taken from Abdullah *et al.* [11]. For details see the text.

where, $V_A = 122.62$ MeV, $R_A = 2.132$ fm, $V_R = 3.0$ MeV, and $R_R = 2.0$ fm, and a Coulomb interaction behaving as $4e^2/r$ for $r > R_C$, with $R_C = 5.8$ fm and for $r < R_C$,

$$V_C(r) = \frac{4e^2}{2R_C}\left[3 - \left(\frac{r}{R_C}\right)^2\right] \tag{15}$$

The above potentials, $V_{\rm N}(r)$, $V_{\rm C}(r)$, and $V_{\rm N}(r) + V_{\rm C}(r)$ are shown in Fig. (2).

The nuclear phase is directly related to the nuclear potential and, to first order, is given by,

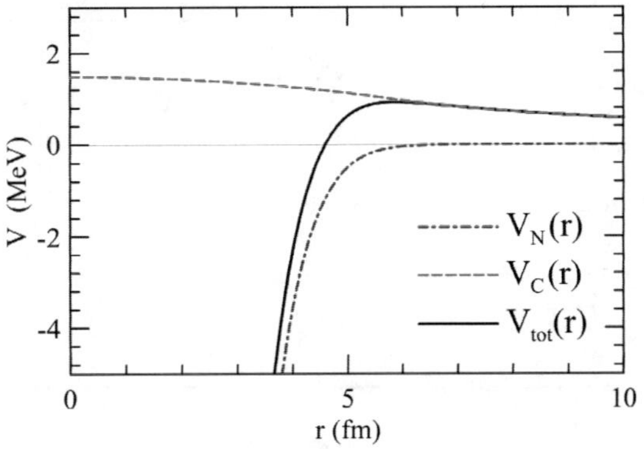


**Figure 2.** (Color online) The $\alpha + \alpha$ potential of Ref. [11]. The nuclear potential is the sum of two Gaussians, Eq.(14). The Coulomb potential is taken as $V_C(r) = \frac{4e^2}{r}\Theta(r - R_C) + \frac{4e^2}{2R_C}[3 - (r/R_C)^2]\Theta(R_C - r)$, where $\Theta(x)$ is the step function.

$$\Delta_{\rm N} = -\sqrt{\frac{2\mu}{\hbar^2}} \int_{r_0}^{\infty} \frac{V_{\rm N}(r)}{\sqrt{E - V_{\rm C}(r) - \frac{\hbar^2 l(l+1)}{2\mu r^2}}} dr \tag{16}$$

where $r_0 = 2a$ is the distance of closest approach for head-on collision.

The above formulae can be used to assess the importance of the nuclear interaction on the Mott scattering. In fact, this procedure, based on perturbation theory, was used in Refs. [3, 4, 5, 6]. It would be instructive to perform the calculation of the second derivative of the cross section and demands that it be zero at $\theta = 0.0$, to find the nuclear modified critical Sommerfeld parameter, which would be $\eta_c = \sqrt{2} + \delta\eta(E)$. We leave this exercise for a future work. In the following, however, we perform the calculation exactly, by solving the scattering Schrödinger equation (optical model calculation)

The result of the optical model calculation of the second derivative of the symmetrized cross section at $\theta = 90^o$ is given in Fig.(3). We see two clear crossings of the horizontal axis at 0.5 MeV and at 2.5 MeV. These crossing points correspond to TI . Between these two points, at $E_{\rm c.m.} \approx 1.4$ MeV, the second derivative attains a very small value, very close to zero. We may take this point as a third TI. It is interesting to note that the TI at the lowest energy, $E_{\rm c.m.} = 0.4$ MeV, is shifted by the nuclear interaction to $E_{\rm c.m.} = 0.5$ MeV, indicating a slightly smaller value of $\eta_c$ than $\sqrt{2}$. The data of Ref. [11] exhibit two TI's, one at $E_{\rm c.m.} = 1.0$ MeV, and the other one at 1.5 MeV. Had the authors of Ref. [11] extended their measurement to energies, below the barrier, they would probably have encountered the TI at around (0.4 - 0.5) MeV. It would therefore be quite interesting to perform measurements of the elastic scattering of $\alpha + \alpha$ at around $E_{\rm c.m.} \approx 0.4$ MeV and verify our findings.

## 4. Conclusions

In this work, we summarized our research on the phenomenon of Transverse Isotropy, predicted to occur in the low energy scattering of identical charged particles. At a critical value of the Sommerfeld parameter $\eta = \sqrt{2}$, the cross section becomes quite flat. The c.m. energy corresponding to this value of $\eta$, is found to be below the height of Coulomb barrier only for a very few systems. We considered the system $\alpha + \alpha$ at sub-barrier energies and compared our findings

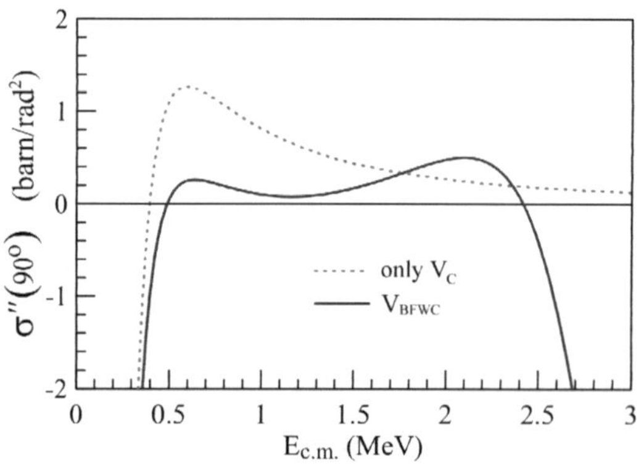


**Figure 3.** (Color online) Second derivative of the Mott cross section at $\theta = 90^{\circ}$ for the BFWC potential of Fig. (2). The dotted line is the usual Mott (only Coulomb) result, while the solid line takes into account both the Coulomb and the Nuclear potentials.

to the experimental data of [11]. We found that whereas our calculation exhibits only one flat region, the data, taken at energies above the barrier, show two flat regions. The region we have predicted at $E_{\text{c.m.}} = 0.4$ MeV was not accessed by the experimental group. The difference in the results of our calculation and the data is attributed to the importance of the nuclear interaction even sub-barrier energies. By including the nuclear interaction in our calculation, we accounted reasonably well for several features of the data. This clearly indicates that the study of TI could shed light on the nuclear interaction at very low energies.

## Acknowledgments

This work was supported by the CNPq, FAPESP, CAPES/ITA, FAPERJ and UFF.

## References

[1] Lynch W G, Tsang M B, Bhang H C, Cramer J G, and Puigh R J 1982 Phys. Rev. Lett. **48** 979
[2] Rodning N L , Knutson L D , Lynch W G, and Tsang M B 1982 Phys. Rev. Lett., **49** 707
[3] Hussein M S, Lima C L, Pato M P and Bertulani C A 1990 Phys. Rev. Lett. **65** 839
[4] Villari A C C et al.1993 Phys. Rev. Lett. **71** 2551
[5] Hussein M S, Franzin V L M, Franzin R, and Baltz A J 1984 Phys. Rev. C **30** 184
[6] Hussein M S, Baltz A J, and Carlson B V 1984 Physics Reports, **113** 133
[7] Canto L F, Donangelo R, and Hussein M S, 2001 Mod. Phys. Lett. A **16** 1027
[8] Canto L F and Hussei M S 2013 *Scattering Theory of Molecules, Atoms and Nuclei* World Scientific
[9] Canto L F, Hussein M S, and Mittig W 2014 Phys. Rev. C **89** 024610
[10] Barata J C A, Canto L F, and M. S. Hussein M S 2011 Braz.J. Phys. **41** 50
[11] Abdullah M N A Sabra M S Rashid M M Shehadeh Z Billah M M Das S K Uddin M A Basak A K Reichstein and I Sen Gupta H M 2006 Nucl. Phys. A **775** 1

# Nuclear Structure Studies with Stable and Radioactive Beams: The SPES radioactive ion beam project

**G. de Angelis[1] for the SPES collaboration,**

**G. Prete[1], A. Andrighetto[1], M.Manzolaro[1], S.Corradetti[1], D.Scarpa[1], M.Rossignoli[1], A.Monetti[1], M.Lollo[1], M.Calderolla[1], J.Vasquez[1], D.Zafiropoulos[1], L.Sarchiapone[1], D.Benini[1], P.Favaron[1], M.Rigato[1], R.Pegoraro[1], D.Maniero[1], L.Calabretta[2], M.Comunian[1], M.Maggiore[1], A.Lombardi[1], L.Piazza[1], A.M.Porcellato[1], C.Roncolato[1], G.Bisoffi[1], A.Pisent[1], A.Galatà[1], M.Giacchini[1], G.Bassato[1], S.Canella[1], F.Gramegna[1], J.Valiente[1], J.Bermudez[1], P.F. Mastinu[1], J.Esposito[1], J.Wyss[1], A. Russo[2], S.Zanella[1]**

[1]INFN Laboratori Nazionali di Legnaro, I-35020, Legnaro (Pd), Italy

[2]INFN Laboratori Nazionali del Sud, I-95123, Catania, Italy

E-mail: giacomo.deangelis@lnl.infn.it

**Abstract**. A new Radioactive Ion Beam (RIB) facility (SPES) is presently under construction at the Legnaro National Laboratories of INFN. The SPES facility is based on the ISOL method using an UCx Direct Target able to sustain a power of 10 kW. The primary proton beam is provided by a high current Cyclotron accelerator with energy of 35-70 MeV and a beam current of 0.2-0.5 mA. Neutron-rich radioactive ions are produced by proton induced fission on an Uranium target at an expected fission rate of the order of $10^{13}$ fissions per second. After ionization and selection the exotic isotopes are re-accelerated by the ALPI superconducting LINAC at energies of 10A MeV for masses in the region A=130 amu. The expected secondary beam rates are of the order of $10^7$ - $10^9$ pps. Aim of the SPES facility is to deliver high intensity radioactive ion beams of neutron rich nuclei for nuclear physics research as well as to be an interdisciplinary research centre for radio-isotopes production for medicine and for neutron beams.

Keywords: Nuclear Physics; Radioactive Ion Beams; Isol facility.

## 1. Introduction

SPES is an acronym for Selective Production of Exotic Species. The aim of the SPES project is to provide high intensity and high-quality beams of neutron-rich nuclei to perform forefront research in nuclear structure, reaction dynamics and interdisciplinary fields like medical, biological and material sciences. SPES is a second generation ISOL radioactive ion beam facility. It represents an intermediate step toward the future generation European ISOL facility EURISOL [1]. The SPES project is part of the INFN Road Map for the Nuclear Physics; it is supported by the Italian national laboratories LNL (Legnaro) and LNS (Catania). It is based on the ISOL method with an UCx Direct

Target able to sustain a power of 10 kW. The primary proton beam is delivered by a Cyclotron accelerator with an energy of more than 40 MeV and a beam current of 200 μA. Neutron-rich radioactive ions will be produced by Uranium fission at an expected fission rate in the target of the order of $10^{13}$ fissions per second. The exotic isotopes will be re-accelerated by the ALPI superconducting LINAC at energies of 10A MeV and higher, for masses in the region of A=130 amu, with an expected rate on the secondary target of $10^7$-$10^9$ pps.

## 2. Challenges of the SPES project

To understand the properties of a nucleus, apart from establishing the interaction between its components, it is necessary to determine the arrangement of the nucleons, i.e. the structure. Presently our knowledge about the structure of nuclei is mostly limited to nuclei close to the valley of stability or nuclei with a deficiency of neutrons. Only recently the availability of beams of unstable ions has given access to unexplored regions of the nuclear chart, especially on the neutron-rich side.

Starting from a nucleus on the stability line and adding successively neutrons, one observes that the binding energy of the last neutron decreases steadily until it vanishes and the nucleus decays by neutron emission. The position in the nuclear chart where this happens defines the neutron drip line. It lies much farther away from the valley of stability than the corresponding drip line associated with protons, owing the absence of electrical repulsion between neutrons. The location of the neutron drip line is largely unknown as experimental data are available only for nuclei with mass up to around 30.

The interest in the study of nuclei with large neutron excess is not only focused on the location of the drip line but also on the investigation of the density dependence of the effective interaction between the nucleons for exotic N/Z ratios. In fact, changes of the nuclear density and size in nuclei with increasing N/Z ratios are expected to lead to different nuclear symmetries and new excitation modes. While in the case of some very light nuclei a halo structure has been identified, for heavier nuclei the formation of a neutron skin has been predicted.

The evolution of nuclear properties towards the neutron drip line depends on how the shell structure changes as a function of neutron excess. This evolution has consequences on the ground state properties of the nuclei and on the single-particle and collective excitations. In particular, studies of neutron-rich nuclei beyond the doubly magic $^{132}$Sn are of key importance to investigate the single-particle structure above the N=82 shell closure and find out how the effective interaction between valence nucleons behaves far from stability.

New modes of collective motion are also expected in connection with the formation of a neutron skin, namely oscillation of the skin against the core, similar to the soft dipole mode already identified in the case of very light halo nuclei. Presently, neither the thickness nor the detailed properties of the neutron skin of exotic nuclei are known. This information is needed to enable a quantitative description of compact systems like neutron stars, where exotic nuclei forming a Coulomb lattice are immersed in a sea of free neutrons, a system which is expected to display the properties of both finite and infinite (nuclear matter) objects. At the energy of SPES, it will be possible to address important questions related to the study of neutron-rich matter such as nuclear forces, level density, viscosity, barrier, neutron pairing and collective modes.

With the high intensity beams delivered by SPES a challenging and broader range of studies in nuclear spectroscopy and reaction mechanism will be performed. Interesting areas where new data will be collected are those in the very neutron rich regions, where shell evolution is an issue. Effects of how the pairing interaction is modified in the nuclear medium will receive significant inputs by measurements of multi-nucleon transfer reactions to specific nuclear states. Effects of rotational damping in the decay of high energy levels, for instance the dynamical dipole emission, will be studied by changing the N/Z of projectile and target. Neutron rich nuclei will be used to investigate the tunneling process in sub-barrier fusion reactions in presence of very positive Q-values, an issue interesting also for astrophysics.

## 3. The SPES radioactive ion beam facility

SPES is an INFN project involving the two national laboratories, LNL and LNS and other INFN sites in Italy. Based on the Isol (Isotope separation on line) technique for radioactive beam production it is presently under construction at the Legnaro National Laboratories. A primary beam provided by a driver accelerator induces nuclear reactions inside a thick target. The reaction products are extracted from the target by thermal process due to the high temperature of the target-ion source system (exceeding 2000° C). Once reached the $1^+$ source, the products are ionized and extracted. After an isotopic selection and a further ionization to $n^+$, the exotic beam is injected into a re-accelerator.

The SPES facility is based on a proton beam to induce fission on a UCx target (Direct Target). The driver is a 70 MeV proton cyclotron with two exit ports providing a total current of 750 μA. Two production Isol targets are planned to be installed and operated alternatively.

As the cyclotron can supply two beams at the same time, a second independent facility can be operated in parallel. In particular the high intensity proton beam will be used to produce radioisotopes for nuclear medicine as well as a neutron source. Concerning the latter possibility, the expected neutron production at SPES will exceed $10^{14}$ n/s with energy up to 70 MeV and a peak between 0.1 and 10 MeV according to target and moderator. The proton beam of the SPES cyclotron is suited for the production of new radioisotopes for medical applications. Taking advantage of the high energy and current, it opens up the possibility to produce $^{82}Sr$, $^{67}Cu$, and $^{47}Sc$ isotopes to be used in therapy and diagnostics. A pictorial sketch of the facility is shown in Figure 1.

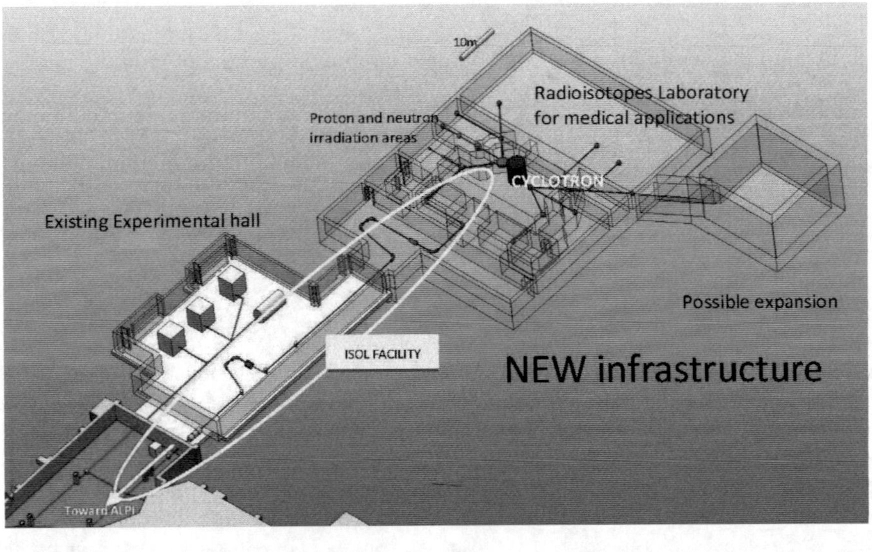


Figure 1. Pictorial sketch of the SPES facility. The ALPI superconductive linac (not shown) is on the left side.

### 3.1. *The Isol front-end*

The Isol front-end is the system which couples the proton beam with the UCx target, the ion source and the first part of the transport line of the exotic beam. The whole system has been completed and is now in operation (with stable beam) at LNL for off-line tests.

The most critical element of the SPES Isol facility is the Direct Target. It follows the basic design of the CERN ISOLDE one but represents an innovation in term of capability to sustain the primary beam power. The design is carefully oriented to optimize the radiative cooling, taking advantage of the target system's high operating temperature, which is of the order of 20000 C.

The SPES target is designed in order to maximize the release efficiency and the heat dissipation. It consists of multiple thin disks housed in a cylindrical graphite box. It is composed of 7 UCx disks

(diameter and thickness of 40 and 1.3 mm, respectively), appropriately spaced in the axial direction in order to dissipate, by thermal radiation, the average power of 8 kW due to the energy release of the proton beam which, passing through them, induces nuclear reactions. Thc graphite box is housed inside a tubular hollow tungsten ohmic heater. In fact, due to the intense heat exchange by radiation, the energy released by the proton beam is not sufficient to make the system reaching the operating temperature. This allows for a better thermal control of the target operation.
An extensive simulation study of the target behavior for thermal and release properties is at the basis of the target-ion-source design. Experimental work to benchmark the simulations was carried out in collaboration with HRIBF [2], the Oak Ridge National Laboratory Isol facility (USA). Partial results on the measured production rates ( and on the expected rates of the reaccelerated SPES beam ) are shown in Figure 2.

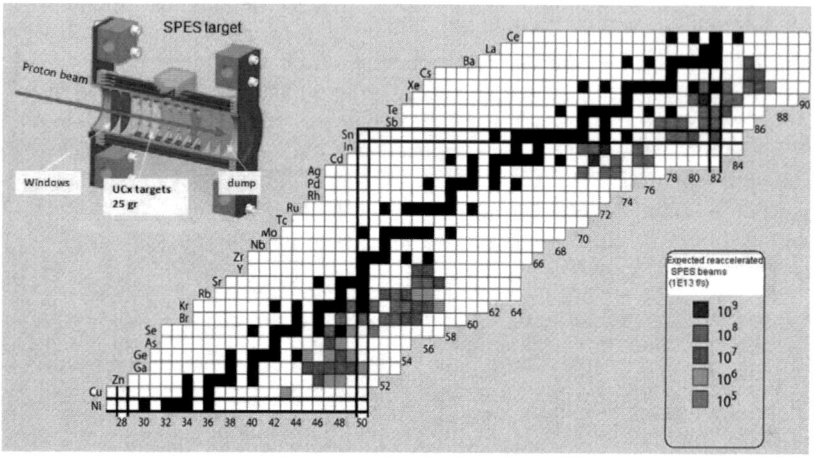


Figure 2. Expected beam on target intensities. Experimental data scaled to: 0.2 mA proton current, 2-5% transport efficiency. ISOL target operated at 2000$^0$C, ionization and extraction with $1^+$ plasma source.

### *3.2. The Ion-source System*

The interaction of the proton beam with the UCx target will produce neutron-rich fission fragments subsequently extracted by thermal motion and ionized at $1^+$ charge state by the Isol source directly connected to the production target. The hot-cavity ion source chosen for the SPES project was designed at CERN (ISOLDE) [3]. The source has the basic structure of the standard high temperature RIB ion sources employed for on-line operation. The ionizer cavity is a tungsten tube (34 mm length, 3 mm inner diameter and 1 mm wall thickness) resistively heated to near 2000C. The isotopes produced in the target diffuse in the target material and then effuse through the transfer tube into the ionizer cavity where they undergo surface or laser ionization. The Surface ionization process can occur when an atom comes into contact with a hot metal surface. In the positive surface ionization, the transfer of a valence electron from the atom to the metal surface is energetically favourable for elements with an ionization potential lower than the work function of the metal. For alkalis and some rare earth elements, high ionization efficiencies can be achieved using the surface ionization technique. Such $1^+$ source has good efficiency and selectivity for the elements as Rb, Cs, Ba. For most of the other elements, the laser resonant photo-ionization, using the same hot cavity cell, is a powerful method to achieve sufficiently selected ionization. This technique, now under development, has been implemented with the aim to produce beams as pure as possible (chemical selectivity) also for metal isotopes. To ionize elements with high ionization potential, as rare gasses, the plasma source is needed. This source ionizes all the elements without any selectivity. At SPES both surface and plasma sources have been developed and are in operation at the off line test stand. The laser source is presently under development.

### *3.3. Beam selection and transport*

The first mass selection is performed by a Wien Filter with 1/100 mass resolution, installed just after the first electrostatic quadrupole triplet inside the production bunker with the aim to confine the larger part of radioactivity inside the high shielded area. The transfer line toward ALPI is equipped with several beam handling systems to purify the beam. A Beam Cooler and a High Resolution Mass Separator (HRMS) with 1/20000 mass resolution will be installed inside the new building. Special efforts have been dedicated to design a mass spectrometer with an effective mass resolution of at least 1/20000. Such design takes advantage of a Beam Cooler, to improve the beam emittance, and of a 260 keV HV platform on which the mass separator is mounted. The mass separator is a scaled-up version of the separator designed for the CARIBU facility [4] in ANL (USA). In the SPES configuration the calculated resolution is of the order of 1/20000 constrained by an emittance of 3 mm mrad and an energy spread of 1.3 eV. Such high selectivity results in an advantage also for the safety issue, reducing the contaminations along the beam transport area, the re-accelerator and the final target location.

Before the injection in the ALPI superconducting linac it is necessary an increase of the charge state from 1+ to n+. This is performed by means of a Charge Breeder. The SPES Charge Breeder is based on ECR method [5] and aims to produce ions with A/q less than 7 at A~130. The design and construction of the Charge-Breeder will be performed in the framework of the SPES-SPIRAL2 collaboration at LPCS (Grenoble, France) following an up-graded version of the Phoenix booster in operation at LPSC.

After the Charge Breeder a second mass separator with 1/1000 mass resolution, will be installed to clean the beam from the contaminants introduced by the Charge Breeder itself. The same basic configuration used for the HRMS is adopted. Finally an RFQ pre-accelerator will increase the beam energy to match the ALPI acceptance. The layout of the radioactive beam line is showed in Figure 3.

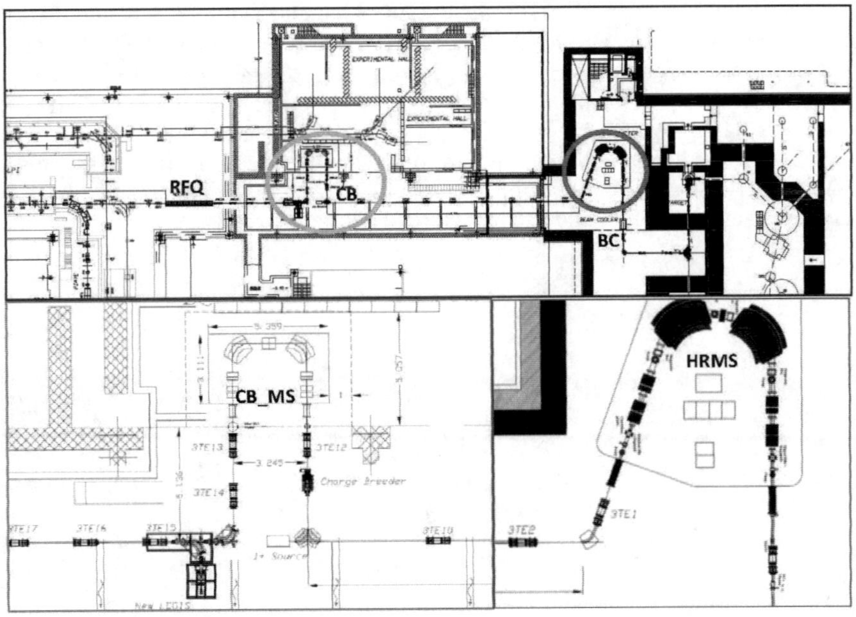


Figure 3. Layout of radioactive beam handling and transfer line. High resolution mass separator and the Charge Breeder mass separator are shown in the lower panels.

### *3.4. Exotic beam reacceleration*

The secondary beams are injected into the ALPI superconducting linac by a new RFQ that will operate in a CW mode (100% duty factor) at a resonant frequency of 80 MHz. This frequency is the same as that of the superconducting structures of the lowest energy section of ALPI. The injection energy of the ions is set to 5.7 keV/u. This choice is a compromise between the desire to reduce the ion energy to simplify the low energy injection and the RFQ bunching section design and the need to increase the beam rigidity in the $1^+$ transport line to reduce space charge effects. The extraction energy of the RFQ is set to 727 keV/u to optimize the beam dynamics of the superconducting linac ALPI. ALPI is the superconducting heavy ion linac in operation at LNL since the early nineties. It is composed by three sections of quarter wave resonators (QWR) for a total of 80 cavities installed and operates routinely at an equivalent acceleration voltage of 50 MV. The linac is constructed in a bended configuration: it is composed by two branches connected by an achromatic and isochronous U-bend. It uses three different kinds of cavities: Low Beta, Medium Beta and High Beta cavities, according to the different velocity along the acceleration path. The ALPI linac has gone through a continuous upgrade in the number and performance of its accelerating cavities, and consequently in the maximum achievable beam energy. In the last years the cavities of the medium energy QWR section have been upgraded using a new Nb sputtered coating in substitution to the original Pb sputtered layer. A further upgrade program is on the way, to improve the accelerating fields of the present QWRs and adding more cavities in the Low Beta section. The final equivalent voltage will exceed 70 MV going well beyond the Coulomb barrier for nuclear reactions involving mid-A projectiles and heavy targets.

*3.5Status of the SPES project*

As already mentioned SPES is presently under construction at LNL. The building is at the excavation phase and will be completed at end of 2014. The cyclotron, provided by the BEST company in Canada, is under construction (the magnet is completed and the magnetic field was positively mapped). It is expected to be transferred to LNL, after the factory commissioning, in the second quarter of 2014. The authorizations to operate the cyclotron have been already obtained allowing the acceleration of the proton beam on non-fissile target at high power and testing the UCx target up to a current of 5 μA. A further authorization request to run the UCx target at design power will be presented to the Italian authorities within this year. The physics design of the radioactive ion beam transfer line ($1^+$ and $n^+$) is completed and the reshaping of the ALPI beam line in the experimental hall is initiated to allow the installation of the Charge Breeder and of the mass separator. The Isol front-end is completed and under operation in the off-line laboratory to test ion sources, assembly and controls. The first reaccelerated exotic beam is expected around 2018 [6].

**References**

[1] www.eurisol.org

[2] Scarpa D, Biasetto L, Corradetti S, Manzolaro M, Andrighetto A, Carturan S, Prete G, Zanonato P and Stracener D W 2011 *Eur. Phys. J. A* **47**

[3] Lindroos M 2004 *CERN-AB-2004-086*

[4] Savard G, Pardo R, Baker S, Davids C, Levand A, Peterson D, Phillips D, Sun T, Vondrasek R, Zabransky B, Zinkann G 2011 *Hyperfine Interactions* **199** 301

[5] Lamy T, Bouly J L, Curdy J C, Sole P, Sortais P, Thuillier T, Vieux-Rochaz J L and Voulot D 2004 *Rev. Sci. Instrum* **75** 1624

[6] https://web.infn.it/spes/

Journal of Physics: Conference Series **590** (2015) 012011 doi:10.1088/1742-6596/590/1/012011

# Recent results on the Pygmy Dipole Resonance

**Deniz Savran**

ExtreMe Matter Institute EMMI and Research Division, GSI Helmholtzzentrum für Schwerionenforschung GmbH, Darmstadt, Germany
and
Frankfurt Institute for Advanced Studies FIAS, Frankfurt am Main, Germany

E-mail: d.savran@gsi.de

**Abstract.** Beside the well-known Giant Dipole Resonance in most nuclei an additional structure has been identified in the low-energy part of the electric dipole strength, which is usually denoted as Pygmy Dipole Resonance (PDR). The PDR has attracted strong interest in the last years in nuclear physics and has been experimentally studied using different approaches in the last decade. Among the other experiments the nuclear resonance fluorescence (NRF) method has played a pioneering role in the research on the PDR.

## 1. Introduction

The Isovector Electric Dipole Giant Resonance (IVGDR) is the dominant feature of the electric dipole (E1) strength in atomic nuclei. This collective excitation mode is understood in a macroscopic interpretation as an out-of-phase oscillation of the protons and neutrons and has been studied in the past in various experiments, see e.g. [1] and references therein.

In many nuclei an additional structure in the low-energy tail of the IVGDR has been observed, which exhausts only a few percent of the total E1 strength and, therefore, is denoted as Pygmy Dipole Resonance (PDR) or Pygmy Dipole Strength. Even though the strength of the PDR is small compared to the IVGDR, its location at low energies around the particle threshold makes it particular interesting from the nuclear structure point of view, since it allows to study details of modern microscopic models which describe the E1 response of atomic nuclei [2]. In addition it has gained interest in the last decade due to its possible connections to the neutron skin of atomic nuclei and properties of nuclear matter [3, 4, 5, 6, 7] as well as reaction rates in the synthesis of heavy elements [8, 9, 10]. However, the microscopic structure of the PDR, its collectivity and the robustness of the connections to other properties mentioned above are a matter of ongoing discussions.

Experimentally no conclusive picture can be drawn so far on the systematic or structure of the PDR. We have recently presented the available experimental data on the PDR in our review [11] describing in detail also the experimental approaches and their limitations. Special attention is made in this review also on the NRF method using different photon beams and its application to investigate low-lying E1 strength below the particle threshold. For a detailed discussion I thus refer to [11].

## 2. Systematics of the PDR

Photon (real or virtual) induced reactions are an ideal tool to investigate the (total) strength of the PDR. The well-known reaction mechanism allows a model independent extraction of absolute transition strengths from the ground state into the PDR from the experimental observables. For stable nuclei the nuclear resonance fluorescence (NRF) method is used most widely to obtain information on the E1 strength distribution below the neutron separation energy. For these experiments photon beams produced via bremsstrahlung [12, 13] and laser Compton backscattering (LCB) [14] have been used. In the analysis of the data different approaches have been used taking into account either only the contribution of isolated peaks resulting in model-independent results but accepting a limited sensitivity (see e.g. [15]) or including the full spectra to extract the full strength but accepting the need of statistical model assumptions in the analysis and thus model dependent results (see e.g. [16]). For a detailed discussion of the different approaches used in NRF experiments including their individual advantages and drawbacks I refer to the review [11] and the references given above.

For excitation energies above the neutron emission threshold the NRF reaction is no longer suitable, since the decay is dominated by neutron evaporation. In this energy region information on the excitation energy thus has to be provided by the incoming photon beam. Consequently, LCB photons have been used to measure the $(\gamma, n)$ cross section as a function of the excitation energy (see e.g. [17, 18]). Just recently a new approach has been proven to result in the extraction of the full E1 response independent of particle thresholds by the means of Coulomb excitation (i.e. virtual photons) in the (p,p') reaction [19, 20], which however requires a separation from the M1 response via a multipole decomposition analysis (see contribution of P. von Neumann-Cosel).

For unstable nuclei the available approaches are much more involved, since experiments have to be performed in inverse (and complete) kinematics including a clean isotope identification on an event-by-event basis, which requires complex experiment setups in addition to the facility producing and providing the exotic ion beam. So far the E1 response in medium-heavy to heavy nuclei has been studied via Coulomb excitation in inverse kinematics in the region of $^{130,132}$Sn [21, 4] and in $^{68}$Ni [22, 23]. The experiments were sensitive to excitation energies above the corresponding particle thresholds only. Thus, deriving a consistent systematic picture on the total strength of the PDR together with the results from stable nuclei is very difficult.

Figure 1 summarizes the fraction of the energy weighted sum rule (EWSR) exhausted by the PDR for a variety of nuclei of different masses as reported in the different corresponding publications (see [11] for details). It has to be stressed, that the EWSR values are not derived in a consistent way along the different experiments, i.e. in some cases contributions of the low energy IVGDR tail are subtracted in others not. Partly different energy regions have been chosen for the integration region. This summary therefore can only provide a first idea on the strength of the PDR.

The strength of the PDR is plotted in Fig. 1 as a function of the parameter $\Delta_{CCF}$, which is defined by

$$\Delta_{CCF} = (S_{2p} - S_{2n})/2 + E_C \tag{1}$$

where $S_{2p}$ and $S_{2n}$ are the two proton and two neutron separation energies, respectively, and $E_C$ is the height of the Coulomb barrier at the nuclear radius (CFF = Coulomb Corrected Fermi energy). The parameter therefore compares the Fermi energies of protons and neutrons corrected for the coulomb contribution, which seems to be a reasonable parameter for the neutron skin thickness (see [11] for details). Although the strength of the PDR seems to be enhanced in the more neutron rich isotopes a clear and conclusive statement on the dependence of the PDR on $\Delta_{CCF}$ cannot be drawn. Additional and especially more complete and consistent data is necessary in order to learn more on the systematics of the PDR. Also, experiments determining the E1 strength distribution alone are not sufficient in order to learn more on the underlying structure of the PDR.

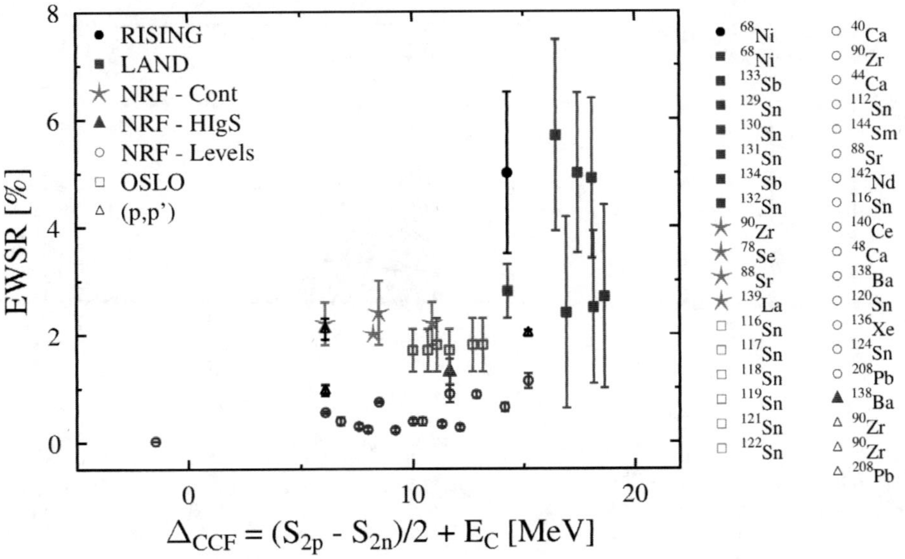


**Figure 1.** Reported exhaustion of the PDR in various nuclei extracted from experiments using different experimental approaches as a function of $\Delta_{CCF}$ (see [11] for details).

## 3. The decay properties of the PDR

A property which can provide further insight into the structure of the E1 strength at low energies is the decay pattern of the PDR. So far, this property could not be directly accessed, since the branching transitions to individual low-lying excited states are small and, therefore, a very sensitive experimental method is needed. On the other hand, information on the decay behavior of the PDR is of high interest since it is also directly connected to the concept of $\gamma$-ray strength functions, which are used within the statistical model to describe the decay of excited states in nuclear synthesis processes. Partly, this concept is also used in the analysis of experimental data to derive information on the total strength of the PDR [16]. An experimental determination of branching ratios and in general the decay pattern of the low-lying E1 strength is therefore of principle interest.

Again, photon-induced reactions can be used in order to address this topic. However, standard NRF experiments are not sensitive enough to investigate the weak branching transitions to low-lying excited states, even when using a mono-energetic LCB photon beam. At the same time these NRF experiments have shown indirectly that the integrated intensity going into these so-called inelastic transitions (i.e. any transition not going directly back to the ground state) cannot be neglected [24, 25, 26, 27]. The method of $\gamma$-$\gamma$ coincidence experiments has been shown to be very sensitive even to very weak transitions. However, so far this method has only been used in particle-induced reactions, which do not sufficiently excite the $J^\pi = 1^-$ states of interest. Therefore, the combination of $\gamma$-$\gamma$ coincidence spectroscopy and the NRF reaction using a mono-energetic photon beam from LCB offers the best possible experimental approach to access the branching pattern of the PDR.

This approach is realized in the newly installed $\gamma^3$ setup at the High Intensity $\gamma$-ray Source (HI$\gamma$S) at TUNL, Duke University [28]. In this setup high-efficiency large LaBr detectors are combined with high-resolution HPGe detectors. The concept is illustrated in the left part of Fig. 2. The high efficiency of the LaBr array allows to gate on the decay of the low-lying state (green)

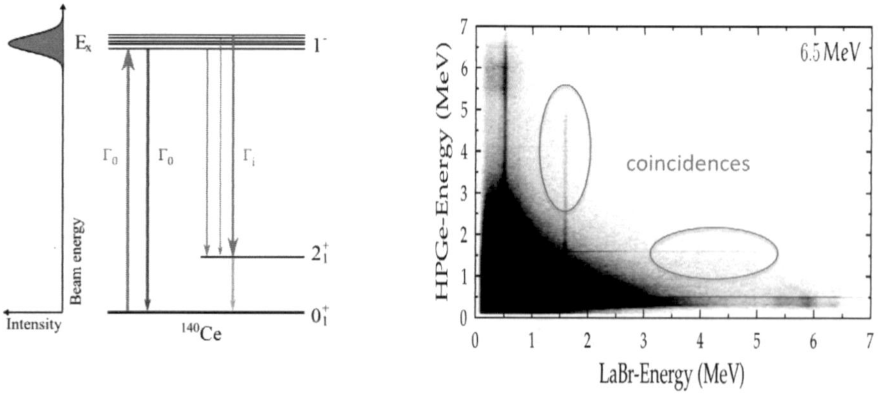


**Figure 2.** Left part: Concept of the $\gamma$-$\gamma$ coincidence technique in combination with the mono-energetic photon beam at HI$\gamma$S. Right part: Measured $\gamma$-$\gamma$ coincidence matrix using the combination of LaBr and HPGe detectors.

and investigate the feeding transitions (red) with high resolution using the HPGe array. Since the photon beam is mono-energetic all transition energies are uniquely defined in a two-step decay cascade . This strongly reduces the background and, therefore, provides the high sensitivity needed to investigate even weak branching transitions. In a first commissioning experiment this has been demonstrated for the case of $^{32}$S, where an improvement of the peak-to-background ratio of about a factor of 50 was achieved [28].

As one of the first cases to investigate the decay pattern of the PDR with the new $\gamma^3$ setup the N=82 isotone $^{140}$Ce has been chosen, since in this nucleus an interesting splitting of the low-lying E1 strength into two parts with different underlying structure has been reported using the $(\alpha, \alpha'\gamma)$ reaction (see contribution of A. Zilges and references [29, 30, 31]). The first results show that the sensitivity is high enough in order to investigate directly the branching ratios into the first excited states. In the right part of Fig 2 one example of a two-dimensional $\gamma$-$\gamma$ coincidence matrix is shown for a beam energy of 6.5 MeV. The coincidence transitions into the first excited $2_1^+$ state at 1596 keV are evident. One first result from this experiment is given in Fig. 3, which shows the determined mean branching ration as a function of the excitation energy (see [32] for details). Calculations within the quasi-particle phonon model (QPM) are in excellent agreement to the data, which shows that the relevant couplings of the PDR to the $2_1^+$ is described in the QPM in an accurate way [33].

These first results show, that the used approach is suitable and sensitive enough to study even very weak decay branchings to excited states of the nucleus, which will allow to investigate the decay pattern of photo-excited states in much greater detail.

**Acknowledgments**
I thank my research group as well as all my collaborators and colleagues who have worked with me in the numerous experiments in the last years, in particular A. Zilges, J. Endres, M.N. Harakeh and W. Tornow. The experiments at the $\gamma^3$ setup would not have been possible without the strong contribution of B. Löher and V. Derya. Also, I acknowledge the support by the Alliance Program of the Helmholtz Association (HA216/EMMI).

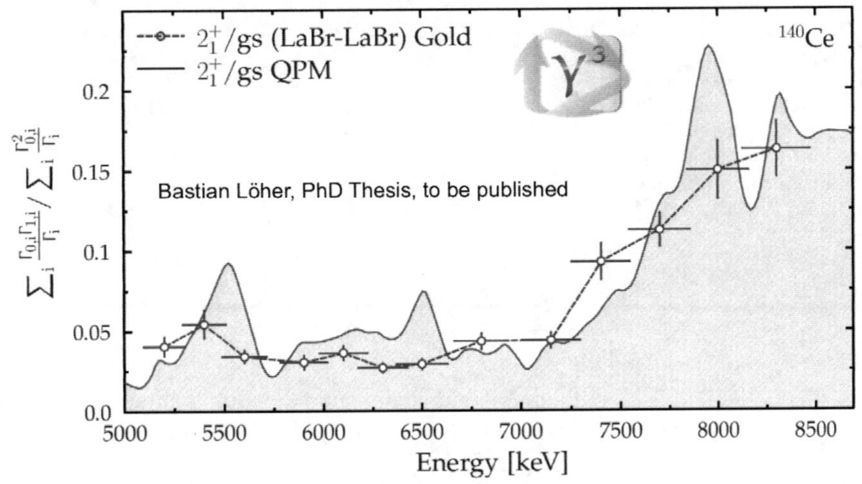


**Figure 3.** First results for the mean branching ratio to the first excited state in the nucleus $^{140}$Ce as determined using the $\gamma^3$ setup and calculated within the QPM model [32].

## References


[1] Harakeh M N and van der Woude A 2001 *Giant Resonances: Fundamental High-Frequency Modes of Nuclear Excitation, Oxford University Press*
[2] Paar N, Vretenar D, Khan E and Colò G 2007 *Rep. Prog. Phys.* **70** 691
[3] Piekarewicz J 2006 *Phys. Rev.* **C73** 044325
[4] Klimkiewicz A, Paar N, Adrich P *et al.* 2007 *Phys. Rev.* **C76** 051603
[5] Tsoneva N and Lenske H 2008 *Phys. Rev.* **C77** 024321
[6] Piekarewicz J 2011 *Phys. Rev.* **C83** 034319
[7] Vretenar D, Niu Y F, Paar N and Meng J 2012 *Phys. Rev.* **C85** 044317
[8] Goriely S 1998 *Phys. Lett.* **B436** 10
[9] Goriely S, Khan E and Samyn M 2004 *Nucl. Phys.* **A739** 331
[10] Litvinova E, Loens H P, Langanke K, Martinez-Pinedo G, Rauscher T, Ring P, Thielemann F K and Tselyaev V 2009 *Nucl. Phys.* **823** 26
[11] Savran D, Aumann T and Zilges A 2013 *Prog. Part. Nucl. Phys.* **70** 210
[12] Schwengner R, Beyer R, Dönau F *et al.* 2005 *Nucl. Instr. and Meth.* **A555**(1-2) 211
[13] Sonnabend K, Savran D, Beller J 2011 *et al. Nucl. Instr. and Meth. Phys. Res.* **A640** 6
[14] Weller H R, Ahmed M W, Gao H, Tornow W, Wu Y K, Gai M and Miskimen R 2009 *Prog. Part. Nucl. Phys.* **62** (1) 257
[15] Savran D, Elvers M, Endres J *et al.* 2011 *Phys. Rev.* **C84** 024326
[16] Schwengner R, Rusev G, Tsoneva N *et al.* 2008 *Phys. Rev.* **C78** 064314
[17] Utsunomiya H, Goriely S, Akimune H *et al.* 2010 *Phys. Rev.* **C81** 035801
[18] Utsunomiya H, Goriely S, Kamata M *et al.* 2011 *Phys. Rev.* **C84** 055805
[19] Tamii A, Poltoratska I, von Neumann-Cosel P *et al.* 2011 *Phys. Rev. Lett.* **107** 062502
[20] Iwamoto C, Utsunomiya H, Tamii A *et al.* 2012 *Phys. Rev. Lett.* **108** 262501
[21] Adrich P, Klimkiewicz A, Fallot M *et al.* 2005 *Phys. Rev. Lett.* **95** 132501
[22] Wieland O, Bracco A, Camera F *et al.* 2009 *Phys. Rev. Lett.* **102** 092502
[23] Rossi D M, Adrich P, Aksouh F *et al.* 2013 *Phys. Rev. Lett.* **111** 242503
[24] Tonchev A P, Hammond S L, Kelley J H, Kwan E, Lenske H, Rusev G, Tornow W and Tsoneva N 2010 *Phys. Rev. Lett.* **104** 072501
[25] Angell C T, Hammond S L, Karwowski H J, Kelley J H, Krtička M, Kwan E, Makinaga A and Rusev G 2012 *Phys. Rev.* **C86** 051302(R)
[26] Isaak J, Savran D, Krticka M *et al.* 2013 *Phys. Lett.* **B727** 361
[27] Scheck M, Ponomarev V Y, Aumann T *et al.* 2013 *Phys. Rev.* **C87** 051304

[28] Löher B, Derya V, Aumann T *et al.* 2013 *Nucl. Instr. and Meth. Phys. Res.* **A723** 136
[29] Savran D, van den Berg A M, Harakeh M N, Hasper J, Matic A, Wörtche H J and Zilges A 2006 *Phys. Rev. Lett.* **97** 172502
[30] Endres J, Savran D, van den Berg A M, Dendooven P, Fritzsche M, Harakeh M N, Hasper J, Wörtche H J and Zilges A 2009 *Phys. Rev.* **C80** 034302
[31] Endres J, Litvinova E, Savran D *et al.* 2010 *Phys. Rev. Lett.* **105** 212503
[32] Löher B 2014 *PhD thesis, University Mainz, to be published*
[33] Ponomarev V 2014 *private communication*

Journal of Physics: Conference Series **590** (2015) 012012 doi:10.1088/1742-6596/590/1/012012

# Recent results on reactions with radioactive beams at RIBRAS (Radioactive Ion Beams in Brazil)

**A Lépine-Szily[1], R Lichtenthäler[1], V Guimarães[1], A Arazi[2], A Barioni[1,3], E A Benjamim[1], P N de Faria[1,4], P Descouvemont[5], L R Gasques[1], E Leistenschneider[1,6], D R Mendes Jr[4], M C Morais[1,4], V Morcelle[1], A M Moro[7], R Pampa Condori[1], K C C Pires[1], M Rodriguez-Gallardo[7], V Scarduelli[1], J M B Shorto[8], J C Zamora[1,9]**

[1] Instituto de Física, Universidade de São Paulo, CP 66318, 05389-970, São Paulo, Brazil
[2] Laboratorio Tandar, Departamento de Física, Comisíon Nacional de Energía Atómica, Av. del Libertador 8250, (1429), Buenos Aires, Argentina
[3] Instituto de Física, Universidade Federal da Bahia, 40210-340, Bahia, Brazil
[4] Instituto de Física - Universidade Federal Fluminense, Niterói,R.J.,24210-340, Brazil
[5] Physique Nucléaire Théorique et Physique Mathématique, C.P. 229, Université Libre de Bruxelles (ULB), B 1050 Brussels, Belgium
[6] TRIUMF, V6T2A3 Vancouver B. C., Canada
[7] Departamento de FAMN, Universidad de Sevilla, Apto. 1065, E-41080 Seville, Spain
[8] Instituto de Pesquisas Energéticas e Nucleares, Av. Prof. Lineu Prestes 2242, São Paulo, Brasil
[9] Technische Universität Darmstadt, Germany

E-mail: alinka@if.usp.br

**Abstract.** We present a quick description of RIBRAS (Radioactive Ion beams in Brazil), which is a superconducting double solenoid system, installed at the Pelletron Laboratory of the University of São Paulo and extends the capabilities of the original Pelletron Tandem Accelerator of 8MV terminal voltage (8UD) by producing secondary beams of unstable nuclei. The experimental program of the RIBRAS covers the study of elastic and inelastic scattering with the objective to study the interaction potential and the reaction mechanisms between weakly bound (RIB) and halo ($^{6}$He and $^{8}$B) projectiles on light, medium and heavy mass targets. With highly purified beams, the study of resonant elastic scattering and resonant transfer reactions, using inverse kinematics and thick targets, have also been included in our recent experimental program.

## 1. Introduction

Great progress in our understanding of nuclear structure is due to experiments performed in many laboratories using radioactive ion beams (RIB). Before the use of RIB, knowledge of nuclear structure and reactions was based on experiments using systems of stable or very long lived nuclei. However, more than 90% of all nuclei, most of them still unknown, are located between the stability line and the drip lines, where many new phenomena can take place.

The use of transfer reactions to produce RIB is possible at quite low energies ($E \leq 10$MeV/n) and, in this way, small university laboratories can also make important contributions in this field. Their main advantage is that they produce low energy RIB. At energies near the Coulomb barrier or above, valuable information on the structure of exotic nuclei and on the dynamics of the nuclear reactions between them can be obtained. Important issues, such as fusion below the Coulomb barrier, and the role of the neutron halo with respect to fusion, can be studied only with low energy beams. Many experiments were performed with light exotic nuclei presenting halo, such as $^6$He, $^8$He, $^{11}$Li, or $^{11}$Be among others [1]. On the other hand, recent experiments with radioactive ion beams have been very successful in nuclear astrophysics investigations, where many stellar scenarios involve short-lived nuclei [2]. In this paper we intend to give a short description of the experimental device [3, 4] and of the scientific results achieved recently.

## 2. Experimental devices and method

The RIBRAS system consists of two superconducting solenoids of 6.5 T maximum central field (5 Tm axial field integral)[5]. The production system (primary target) can be a gas or a solid target, such as a $^9$Be foil, mounted before the first solenoid. The primary beam, but also the secondary beams, in the angular range between zero and 2°, are stopped and collected by a tungsten rod with electron and geometrical suppression (Faraday cup), located downstream after the production target. A collimator at the entrance of the first solenoid limits the maximum angular acceptance to 6°, so, in this set-up, the angular divergence of the secondary beams entering the first solenoid is 4°, between 2° and 6°. The angular divergence of the secondary beam in central chamber is 3.2°, between 1.3° and 4.5°.

**Figure 1.** The RIBRAS facility. The stable beam comes from the left, the production target is located on the left side of the first solenoid. Between the two solenoids one can see the intermediate scattering chamber. At the right extremity, after the second solenoid, the new, large scattering chamber can be seen. Taken from reference [3].

The first solenoid makes an in-flight selection by the magnetic rigidity of the reaction products emerging from the primary target in the forward angle region. As the first magnet focuses all ions with the same magnetic rigidity, i.e., the same $ME/Q^2$ ratio, where $M, E$ and $Q$ stand for mass, energy, and charge state of the ion, the beam of interest can be accompanied by many contaminant beams of the same magnetic rigidity, but with different charges, masses and energies.

Journal of Physics: Conference Series **590** (2015) 012012 doi:10.1088/1742-6596/590/1/012012

**Table 1.** Secondary beams produced at the RIBRAS facility and the nuclear reactions used for their production. The intensities, energy resolution (FWHM) and energies obtained for the beams at the central chamber are also shown. The intensities are reported in *pps* per 1$\mu$A of primary beam.

| Secondary beam | Production reaction | Q-value (MeV) | intensity (pps) | Energy resolution FWHM(keV)/Energy |
|---|---|---|---|---|
| $^{6}$He | $^{9}$Be($^{7}$Li,$^{6}$He)$^{10}$B | -3.390 | $10^5$ - $10^6$ | 1000/22 MeV |
| $^{7}$Be | $^{3}$He($^{6}$Li,$^{7}$Be)d | +0.112 | $10^4$ - $10^5$ | 800/18.8 MeV |
| $^{7}$Be | $^{7}$Li($^{6}$Li,$^{7}$Be)$^{6}$He | -4.369 | $10^4$ - $10^5$ | 1000/22 MeV |
| $^{8}$Li | $^{9}$Be($^{7}$Li,$^{8}$Li)$^{8}$Be | +0.367 | $10^5$ - $10^6$ | 500/25.8 MeV |
| $^{8}$B | $^{3}$He($^{6}$Li,$^{8}$B)n | -1.975 | $10^4$ | 1000/15.6 MeV |
| $^{10}$Be | $^{9}$Be($^{11}$B,$^{10}$Be)$^{10}$B | -4.642 | $10^5$ | 800/23.2 MeV |
| $^{12}$B | $^{9}$Be($^{11}$B,$^{12}$B)$^{8}$Be | +1.705 | $10^5$ | 800/25.0 MeV |

## 3. Elastic scattering of 2-n halo projectile $^{6}$He: Experimental results and analysis

Until recently almost all experiments performed at RIBRAS were on elastic scattering using only the first solenoid. Elastic scattering angular distributions were measured for the available radioactive beams ($^{6}$He, $^{8}$Li, $^{7,10}$Be and $^{8}$B) on light, medium mass and heavy targets at various energies, near and above the Coulomb barrier [4, 6–17]. The $^{6}$He and $^{8}$Li secondary beams were produced respectively by the one-proton and one-neutron transfer reactions, $^{9}$Be($^{7}$Li,$^{6}$He)$^{10}$B (Q= - 3.39 MeV) and $^{9}$Be($^{7}$Li,$^{8}$Li)$^{8}$Be (Q = 0.3677 MeV), using the $^{7}$Li primary beam accelerated by the Pelletron tandem and a $^{9}$Be target. A central scattering chamber, located between the two solenoids, was used and the reaction products were detected using $\Delta$E-E Si telescopes, with detector thicknesses of 20 - 500 (or 1000) $\mu$m respectively.

These measurements are important since they allow the determination of the optical potentials between the radioactive projectiles and the targets. The elastic scattering also gives information on the total reaction cross section and on the size of the nuclei involved. The coupling of the elastic scattering to other important channels, as breakup or transfer, can also be investigated through the elastic scattering experiments. In the case of radioactive projectiles, these effects are strongly enhanced and their study as a function of the target mass or the incident energy was achieved by our measurements.

We have measured the elastic scattering of the 2-neutron halo projectile $^{6}$He, on $^{9}$Be, $^{27}$Al, $^{51}$V, $^{58}$Ni and $^{120}$Sn targets at energies around and above the Coulomb barrier. The data on $^{27}$Al, $^{51}$V were analyzed using optical model and the "São Paulo optical potential" [18, 19]. The normalization of the imaginary part $N_I$ and the diffuseness, $a$, of the projectile were adjusted to reproduce the data [16, 20]. For $^{6}$He + $^{51}$V at 23.0 and 15.4 MeV, both $N_I$ and $a$ had to be increased, when compared to typical, strongly bound, stable nuclei, indicating a larger absorption. This feature is different from the $^{6}$He + $^{27}$Al case [16], where the $N_I$ and $a$ values were similar to the values used for typical, strongly bound, stable nuclei, respectively, $N_I = 0.78$ and $a$=0.56(2)fm.

For the systems with $^{6}$He projectile, on $^{9}$Be, $^{58}$Ni and $^{120}$Sn targets, the elastic scattering angular distributions were compared to Continuum Discretized Coupled Channels (CDCC) calculations [21]. CDCC has been applied to a number of cases in recent years, first for the three-body problem [22–26] where the projectile is considered as a core plus a valence particle. In the three-body model, the $^{6}$He structure is simplified to an alpha-particle core plus a di-neutron which means that the neutron pair is treated as a single particle bound to the alpha core by 0.973 MeV. More recently, full four-body calculations [27–30], applied to the problem of

2-neutron halo projectiles as $^6$He, were also realized.

In Figure 2 we present an example of our data compared to CDCC calculations [6, 9]. For the $^6$He+$^{58}$Ni system (Figure 2) the elastic scattering angular distribution is compared with no-continuum CDCC, three, and four-body CDCC calculations, considering projectile breakup [6]. The agreement between the four-body CDCC and the experimental data is remarkable in the whole angular range.

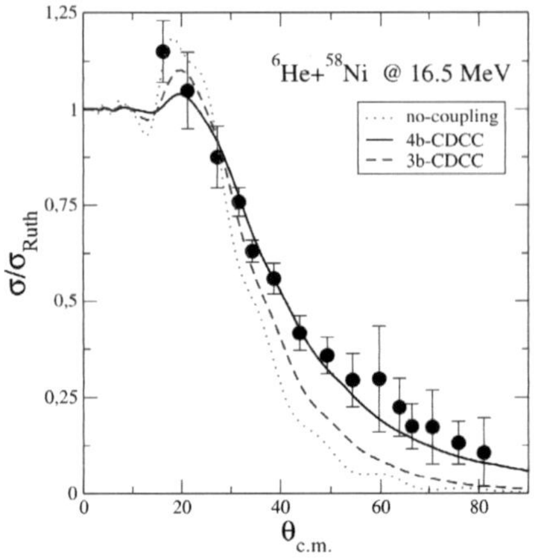


**Figure 2.** Angular distribution of $^6$He+$^{58}$Ni elastic scattering. The experimental angular distribution is compared to calculations: red dotted curve is without coupling to the continuum, blue dashed curve is 3b-CDCC calculation and the black solid line is the 4b-CDCC. Taken from ref. [6].

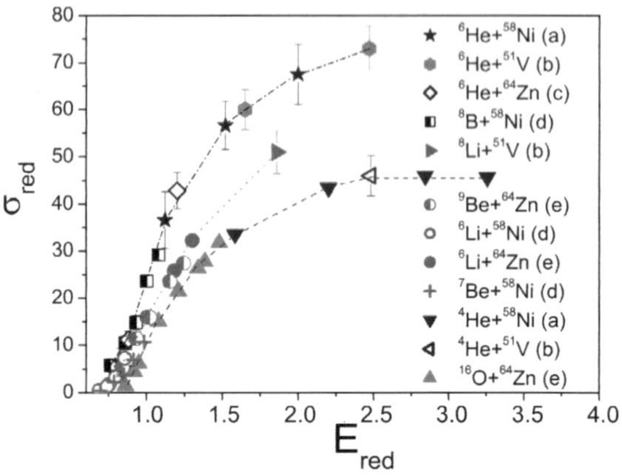


**Figure 3.** Reduced reaction cross sections for different projectiles on intermediate mass ($A \approx 60$) targets: Ref. (a) from [6, 31, 32], (b) from [20], (c) from [33], (d) from [34] and (e) from [35–37]. The units of $\sigma_{red}$ and $E_{red}$ are, respectively, mb and MeV. The lines are to guide the eye. Taken from ref. [4].

The total reaction cross section can also be obtained from the experimental data of elastic scattering or from the optical model or CDCC calculations used to reproduce the elastic data. Comparing total reaction cross sections for different systems can be an interesting way to investigate the influence or competition of reaction mechanisms, such as breakup, transfer, and fusion. This can be particularly revealing, if these systems have weakly bound or exotic projectiles. In order to compare total reaction cross sections for systems of different masses and charges at different energies, it is necessary to use a scaling procedure [35, 38–40] to remove trivial effects due to different sizes and energies with respect to the Coulomb barriers.

We present on Fig.3 the reduced reaction cross sections for different projectiles on intermediate mass ($A \approx 60$) targets. We see three bands, the one with lowest reduced reaction cross section corresponds to tightly bound, $\alpha$-cluster ($^{16}$O and $^4$He) projectiles on $^{64}$Zn and $^{58}$Ni targets, respectively. The one with the highest reduced reaction cross section corresponds to exotic halo nuclei, $^6$He and $^8$B, on $^{64}$Zn, $^{51}$V, and $^{58}$Ni targets. The band between both corresponds to reactions of weakly bound, stable ($^6$Li and $^9$Be) and radioactive ($^8$Li and $^7$Be) projectiles. A strong enhancement is observed in the total reaction cross section of the exotic $^6$He projectile compared with tightly bound and weakly bound stable systems, probably due to the coupling to Coulomb break up.

## 4. Scattering on hydrogen target: $^8$Li $+p$ excitation functions

The use of two magnets in RIBRAS is important to purify the secondary beams. With two solenoids, it is possible to use the differential energy loss in an energy degrader foil, located at the crossover point between the magnets, to select the ion of interest and move the contaminant ions out of the band pass of the second solenoid. The $^8$Li(p,p)$^8$Li measurement was performed in the large scattering chamber located after the second solenoid with a nearly pure (95-99%) $^8$Li beam on the secondary target of a proton-rich $[CH_2]_n$ polyethylene foil of 7.7 $mg/cm^2$ thickness. The $^8$Li(p,p)$^8$Li elastic scattering excitation function and the $^8$Li(p,d)$^7$Li and $^8$Li(p,$\alpha$)$^5$He reactions could also be observed in this experiment [41]. The excitation functions are shown in Figure 4 together with R-matrix calculations [42, 43]. With the R-matrix fits we could determine, or at least constrain, values of spins, parities, total and partial widths ($\Gamma_p$, $\Gamma_\alpha$ and $\Gamma_d$) for three resonances, located at $E_{cm}$ = 1.13, 1.69 and 1.76 MeV.

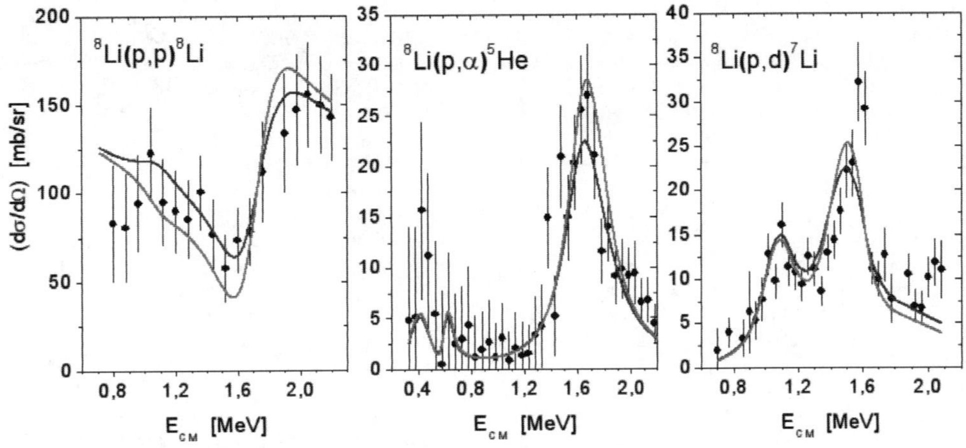


**Figure 4.** Excitation functions for the elastic scattering $^8$Li(p,p)$^8$Li and for the reactions $^8$Li(p,$\alpha$)$^5$He and $^8$Li(p,d)$^7$Li with R-matrix calculations [41]. The blue and red curves are best fits with somewhat different parameters

## 5. Acknowledgements

The authors thank the Fundacão de Amparo à Pesquisa do Estado de São Paulo (FAPESP) and the Conselho Nacional de Desenvolvimento Científico e Tecnológico (CNPq) for financial support.

## References

[1] Keeley N, Alamanos N, Kemper K W and Rusek K 2009 *Progr. Part. Nucl. Phys.* **63** 396 – 447
[2] Langanke K and Wiescher M 2001 *Rep. Prog. Phys.* **64** 1657
[3] Lépine-Szily A, Lichtenthäler R and Guimarães V 2013 *Nucl. Phys. News* **23**(3) 5–11
[4] Lépine-Szily A, Lichtenthäler R and Guimarães V 2014 *Eur. Phys. J. A* **50** 128–163
[5] Lichtenthäler R, Lépine-Szily A, Guimarães V, Perego C, Placco V, Camargo O, Denke R, de Faria P N, Benjamim E A, Added N, Lima G F, Hussein M S, Kolata J J and Arazi A 2005 *Eur. Phys. Jour. A* **25** 733–736

[6] Morcelle V, Pires K C C, Rodríguez-Gallardo M, Lichtenthäler R, Lépine-Szily A, Guimarães V, de Faria P N, Mendes Jr D R, Moro A M, Gasques L R, Leistenschneider E, Pampa Condori R, Scarduelli V, Morais M C, Barioni A, Zamora J C and Shorto J M B 2014 *Phys. Lett. B* **732** 228
[7] Mendes D R, Lépine-Szily A, Descouvemont P, Lichtenthäler R, Guimarães V, de Faria P N, Barioni A, Pires K C C, Morcelle V, Pampa Condori R, Morais M C, Leistenschneider E, Lima C E F, Zamora J C, Alcantara J A, Zagatto V, Assuncão M and Shorto J M B 2012 *Phys. Rev. C* **86**(6) 064321
[8] Zamora J C, Guimarães V, Barioni A, Lépine-Szily A, Lichtenthäler R, de Faria P N, Mendes D R, Gasques L R, Shorto J M B, Scarduelli V, Pires K C C, Morcelle V, Leistenschneider E, Condori R P, Zagatto V A, Morais M C and Crema E 2011 *Phys. Rev. C* **84** 034611
[9] Pires K C C, Lichtenthäler R, Lépine-Szily A, Guimarães V, de Faria P N, Barioni A, Mendes Junior D R, Morcelle V, Pampa Condori R, Morais M C, Zamora J C, Crema E, Moro A M, Rodríguez-Gallardo M, Assuncao M, Shorto J M B and Mukherjee S 2011 *Phys. Rev. C* **83** 064603
[10] de Faria P N, Lichtenthäler R, Pires K C C, Moro A M, Lépine-Szily A, Guimarães V, Mendes D R, Arazi A, Barioni A, Morcelle V and Morais M C 2010 *Phys. Rev. C* **82** 034602
[11] de Faria P N, Lichtenthäler R, Pires K C C, Moro A M, Lépine-Szily A, Guimarães V, Mendes D R, Arazi A, Rodríguez-Gallardo M, Barioni A, Morcelle V, Morais M C, Camargo O, Alcantara Nuñez J and Assuncão M 2010 *Phys. Rev. C* **81** 044605
[12] Mukherjee S, Deshmukh N N, Guimarães V, Lubian J, Gomes P R S, Barioni A, Appannababu S, Lopes C C, Cardozo E N, Pires K C C, Lichtenthäler R, Lépine-Szily A, Monteiro D S, Shorto J M B, de Faria P N, Crema E, Morcelle V, Morais M C and Condori R P 2010 *Eur. Phys. Jour. A* **45** 23–28
[13] Barioni A, Guimarães V, Lépine-Szily A, Lichtenthäler R, Mendes Jr D R, Crema E, Pires K C C, Morais M C, Morcelle V, de Faria P N, Condori R P, Moro A M, Monteiro D S, Shorto J M B, Lubian J and Assuncão M 2009 *Phys. Rev. C* **80** 034617
[14] Lichtenthäler R, de Faria P N, Lépine-Szily A, Guimarães V, Camargo Jr O, Denke R, Benjamim E A, Barioni A, Pires K C C, Mendes D J, Assuncao M, Arazi A, Padron I and Gomes P R S 2007 *Eur. Phys. Jour.-Spec. Top.* **150** 27–30
[15] Lépine-Szily A, Lichtenthäler R and RIBRAS collaboration 2007 *Nucl. Phys. A* **787** 94C–101C
[16] Benjamim E A, Lépine-Szily A, Mendes Jr D R, Lichtenthäler R, Guimarães V, Gomes P R S, Chamon L C, Hussein M S, Moro A M, Arazi A, Padron I, Nunez J A, Assuncao M, Barioni A, Camargo Jr O, Denke R Z, de Faria P N and Pires K C C 2007 *Phys. Lett. B* **647** 30–35
[17] Pires K C C, Lichtenthäler R, Lepine-Szily A and Morcelle V 2014 *Phys. Rev. C* **90** 027605
[18] Chamon L C, Carlson B V, Gasques L R, Pereira Det al., 2002 *Phys. Rev. C* **66** 014610
[19] Alvarez M A G, Chamon L C, Hussein M S, Pereira D, Gasques L R, Rossi E S and Silva C P 2003 *Nucl. Phys. A* **723** 93 – 103
[20] Morcelle V 2007 *Master's Thesis* IFUSP
[21] Austern N, Iseri Y, Kamimura M, Kawai M, Rawitscher G and Yahiro M 1987 *Phys. Rep.* **154** 125 – 204
[22] Aguilera E F, Kolata J J, Nunes F M, Becchetti F D et al, 2000 *Phys. Rev. Lett.* **84** 5058–5061
[23] Aguilera E F, Kolata J J, Becchetti F D et al., 2001 *Phys. Rev. C* **63** 061603
[24] Keeley N, Cook J M, Kemper K W, Roeder B T, Weintraub W D et al. et al.2003 *Phys. Rev. C* **68** 054601
[25] Kakuee O R, Rahighi J, Sánchez-Benitez A M et al., 2003 *Nucl. Phys. A* **728** 339 – 349
[26] Hussein M S, Lichtenthäler R, Nunes F M and Thompson I J 2006 *Phys. Lett. B* **640** 91 – 95
[27] Thompson I J 1998 *Comp. Phys. Rep.* **17** 167
[28] Matsumoto T, Egami T, Ogata K, Iseri Y, Kamimura M and Yahiro M 2006 *Phys. Rev. C* **73** 051602
[29] Rodríguez-Gallardo M, Arias J M, Gómez-Camacho J et al., 2008 *Phys. Rev. C* **77** 064609
[30] Rodríguez-Gallardo M, Arias J M, Gómez-Camacho J, Moro A M et al., 2009 *Phys. Rev. C* **80** 051601
[31] Gasques L R, Chamon L C, Pereira D, Alvarez M A G, Rossi E S et al., 2003 *Phys. Rev. C* **67** 067603
[32] Gasques L R, Chamon L C, Gomes P R S and Lubian J 2006 *Nucl. Phys. A* **764** 135 – 148
[33] Di Pietro A, Figuera P, Amorini F et al., 2003 *Europhys. Lett.* **64** 309
[34] Aguilera E F, Martinez-Quiroz E, Lizcano D et al., 2009 *Phys. Rev. C* **79** 021601
[35] Gomes P R S, Padron I, Rodriguez M D, Marti G V et al., 2004 *Phys. Lett. B* **601** 20 – 26
[36] Moraes S B, Gomes P R S, Lubian J, Alves J J S, Anjos R M, Sant'Anna M M, Padrón I, Muri C, Liguori Neto R and Added N 2000 *Phys. Rev. C* **61** 064608
[37] Gomes P R S, Rodríguez M D, Martí G V et al., 2005 *Phys. Rev. C* **71** 034608
[38] Gomes P R S, Lubian J, Padron I and Anjos R M 2005 *Phys. Rev. C* **71**(1) 017601
[39] Canto L F, Gomes P R S, Lubian J, Chamon L C and Crema E 2009 *Nucl. Phys. A* **821** 51 – 71
[40] Shorto J M B, Gomes P R S, Lubian J, Canto L F, Mukherjee S et al, 2009 *Phys. Lett. B* **678** 77
[41] Leistenschneider E 2014 *Master's thesis* IFUSP
[42] Lane A M and Thomas R G 1958 *Rev. Mod. Phys.* **30** 257
[43] Descouvemont P and Baye D 2010 *Rep. Prog. Phys.* **73** 036301

Journal of Physics: Conference Series **590** (2015) 012013 doi:10.1088/1742-6596/590/1/012013

# Nuclear Astrophysics with the Trojan Horse Method

**Claudio Spitaleri**
Dipartimento di Fisica e Astronomia, Università di Catania, Catania, Italy
INFN, Laboratori Nazionali del Sud, Catania, Italy
E-mail: spitaleri@lns.infn.it

**Abstract.** In stars nuclear reactions take place at physical conditions that make very hard their measurements in terrestrial laboratories. Indeed in astrophysical environments nuclear reactions between charged nuclei occur at energies much lower than the Coulomb barrier and the corresponding cross section values lie in the nano or picobarn regime, that makes their experimental determination extremely difficult. This is due to the very small barrier Coulomb penetration factor, which produces an exponential fall off of the cross section as a function of energy. Additionally, the presence of the electron screening needs to be properly taken into account when dealing with cross section measurements at low-energies. The Trojan Horse Method (THM) represents an independent experimental technique, allowing one to measure astrophysical S(E)-factor bared from both Coulomb penetration and electron screening effects. The main advantages and the most recent results are here shown and discussed.

## 1. Introduction

Because of their crucial role in understanding the first phases of the Universe history and the subsequent stellar evolution. The main bare nucleus cross section measurements of the reaction between chargrd particles at the Ganow energy ($E_G$) should be known with an accuracy better than 10% [1] The effective cross section in the stellar plasma $\sigma_{pl}(E)$, is related to the bare nucleus cross section $\sigma_b(E)$ and to the the stellar electron screening enhancement factor $f_{pl}$ by the relation

$$\sigma_{pl}(E) = \sigma_b(E) f_{pl}(E) = \sigma_b(E) \cdot exp(\pi\eta U_{pl}/E) \quad (1)$$

where $U_{pl}$ is the plasma potential energy and $\eta$ the Sommerfeld parameter. If $\sigma_b(E)$ is measured at the ultralow energies $E_G$ and $f_{pl}$ is estimated within the framework of the Debye-Hückel theory, it is possible estimate the $\sigma_{pl}(E)$f rom equation(1).

In neutron-induced reactions, the lak of a Coulomb barrier and the typical $\sigma_b(E) \sim 1/E^{1/2}$ energy dependance lead to a precise knowledge of $\sigma_b(E_G)$ for many cases. Whereas the Coulomb barrier of height $E_C$ in charged-particle induced reactions cause an exponential decrease of the cross section $\sigma_b(E)$ at $E > E_C$, $\sigma_b(E) \sim exp(-2\pi\eta)$, leading to a low-energy limit of direct measurements of $\sigma_b(E)$, which are typically much larger than $E_G$.

Owing to the strong Coulomb suppression, the behavior of the cross section at $E_G$ is usually extrapolated from the higher energies by using the definition of the smoother astrophysical factor S(E):

$$S_b(E) = E\sigma_b(E) exp(2\pi\eta) \quad (2)$$

where $exp(2\pi\eta)$ is the inverse of the Gamow factor, which removes the dominant energy dependence of $\sigma(E)$ due to the barrier penetrability.

Recently the availability of hight-current low-energy accelerators, such as that at the underground Laboratori Nazionali del Gran Sasso [2], together with the improved target production and detection techniques have allowed us to perform $\sigma_b(E)$ measurements down to (or at least close to) $E_G$. Therefore no $\sigma_b(E)$ extrapolation would be needed anymore for these reactions. However, measurements in laboratory are affected by the electron screening effects, which leads to an exponensial increase of the measured [3, 4] value of thecross section $\sigma_s(E)$ [or equivalently of the astrophysical factor $S_s(E)$] with decreasing energy relative to the case of bare nuclei. This can be described by an enhancement factor defined by the relation

$$f_{lab}(E) = \sigma_s(E)/\sigma_b(E) = exp(\pi\eta U_e/E) \tag{3}$$

Therefore even if it is possible to measure cross sections in the Gamow energy range, the bare cross section $\sigma_b$ is extracted by extrapolating the direct data from higher energies where a negligible electron screening contribution is expected.

Thus, alternative methods for determining bare nucleus cross sections of astrophysical interest are needed. In this contest a number of indirect methods, e.g. the Coulomb dissociation [5, 6], the ANC (asymptotic normalization coefficient) method ([7] and references therein), and the Trojan-horse method (THM) were developed [8, 9, 10, 11, 12, 13]. The THM has already been applied several times to reactions connected with fundamental astrophysical problems. In this paper we will show the result of the measurement of the astrhysical S(E)-factor in the case of application of THM to the measurement of the two-body reactions $^{10}$B(p,$\alpha_o$)$^{7}$Be [14] which are a reaction involved in the depletion of light nuclei.

Indeeed the THM is currently an alternative technique for measuring the cross section of reactions induced by neutrons [15].

## 2. The Trojan Horse method

The basic idea underlying indirect methods is the use of nuclear reaction theory to link the cross section of astrophysical importance to the one of a different process, easier to be studied with present-day facilities. Indirect techniques have been developed over the past several decades to determine reaction rates that cannot be measured in the laboratory.

The THM is a powerful indirect technique that allows one to determine the bare nucleus astrophysical S-factor for rearrangement reactions. The THM involves obtaining the cross section of the binary process x + B $\rightarrow$ C + D at astrophysical energies by measuring the Trojan Horse (TH) reaction (the two-body to three-body process ( 2 $\rightarrow$ 3 particles)) A+B $\rightarrow$ C + D+S in quasi-free (QF) kinematics regime, where the "Trojan Horse" particle, A = (Sx), which has a dominant cluster structure, is accelerated at energies above the Coulomb barrier.

After penetrating the barrier, the TH nucleus A undergoes a breakup leaving particle x(participant) to interact with target B while projectile S(spectator) flies away. From the measured cross section of TH reaction, the energy dependence of the binary sub-process is determined [1]

The advantage of the THM is that the extracted cross section of the binary sub-process does not contain the Coulomb barrier factor and electron screening does not play a role in the cross section. Consequently, the TH cross section can be used to determine the energy dependence of the astrophysical factor, $S_(E_{xB})$, of the binary process down to zero value of relative kinetic energy $E_{xB}$ of particles x and B, without distortion due to the electron screening [9, 10].

At low energies, where electron screening becomes important, the comparison between the astrophysical factor determined through the THM and the direct result allows to estimate of the screening potential.

The THM has been successfully employed to determine the bare nucleus cross section of reactions between charged particles at sub-Coulomb energies. Furthermore, the equations

Journal of Physics: Conference Series **590** (2015) 012013 doi:10.1088/1742-6596/590/1/012013

presented in the literature are obtained for an s-wave bound state of A =(Sx), and therefore, a limited number of "virtual target or projectile" might be used for the application of THM. In addition, the THM cannot be applied to radiative capture reactions (p,$\gamma$), (n,$\gamma$), ($\alpha$,$\gamma$). The THM has already been applied several time [10, 12, 16, 17, 18] to the rate determination of nuclear reactions of the type (p,$\alpha$) and sometimes of the type ($\alpha$,n). The description of the quasi-free reaction ($QFR$) is very simple in the Impulse Approximation ($IA$) and can be represented by the Feynman diagram in figure 1. In the $IA$ approximation the three body

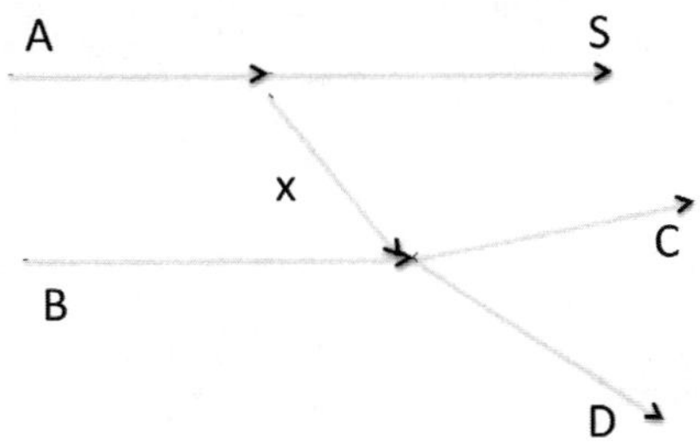


**Figure 1.** Feynman diagram representing the quasi-free$B(A, CD)S$reaction.

reaction cross section is proportional to the cross section of the binary reaction [19]. Following the Plane Wave Impulse Approximation (PWIA) the three body reaction can be factorized into two terms corresponding to the vertices of figure 1 and is given by:

$$\frac{d^3\sigma}{d\Omega_C d\Omega_D dE_C} = (KF) \cdot |\phi(p_{xS})|^2 \cdot \left[\frac{d\sigma}{d\Omega}\right]^{HOES} \tag{4}$$

where:

(i) $\left[\frac{d\sigma}{d\Omega}\right]^{HOES}$ is the half-off-energy-shell differential cross section for the binary $B(x, C)D$ reaction at the center of mass energy $E_{CM}$ given in post-collision prescription ($PCP$).

(ii) $KF$ is a kinematical factor containing the final state phase space factor that is a function of the masses, momenta, and angles of the outgoing particles;

(iii) $\phi^2(k_{xS}$ is the Fourier transform of the radial wave function for the $\chi(r_{xS})$ inter-cluster motion usually described in terms of Hankel, Eckart and Hulthen function depending on the $xS$ system properties.

The applicability of the pole approximation is limited to small momentum $p_{xS}$. Namely, the region where the pole diagram is expected to be predominant in the reaction mechanism was suggested to be:

$$0 \leq p_{xS} \leq k_{xS} \tag{5}$$

being $p_{xS}$ the momentum, in the alf off energy shell ($HOES$), of the cluster $x$ when it interacts with the particle $B$ and $k_{xS}$ the momentum in the half off energy shell ($HOES$) defined by the relation

$$k_{xS} = \sqrt{2\mu_{xS} B_{xS}} \tag{6}$$

where $\mu_{xS}$ is the reduced mass and $B_{xS}$ is the binding energy of the system $xS$, and $Q_{xS}$ is the $Q - value$ for the virtual decay. In order to select the data to be analysed, which correspond to

Journal of Physics: Conference Series **590** (2015) 012013 doi:10.1088/1742-6596/590/1/012013

the region where $QF$ mechanism is dominant, coincidence events for momentum $S$- spectator ( defined by the equation (5) )were considered in the standard analysis. The shape of $\phi^2(k_{xS})$ can be related to angular momentum and the radial wave function of the $xS$ component of the wave function. The equation (4) can be applied only to specific cases in which involved reactions with the presence of related contributions to a single $l_t h$ partial wave (for example $l_0$).

*2.1. Incident energy prescriptions.*
The incident energy $E_{AB}$ of the projectile must be chosen in such a way that the bombarding energy $E_{AB}$ overcomes the Coulomb barrier in the entry channe:l $E_{AB}>(E_{AB})^{Coul.Barr.}$: Thus, particle $x$ is brought into the nuclear interaction zone to induce the reaction of interest x + B → C + D. The $QF$ kinematical conditions must be chosen in such a way that the $E_{xB}$ relative energy can span the astrophysical region of interest below the Coulomb barrier: $E_{xB} < (E_{xB})^{Coul.Barr.}$ [8]

In post collision prescription $E_{xB}= E_{CD}- Q_{2b} = E_{cm}$ where the symbol $E_{CM}$ is often used instead of $E_{x-B}$, $Q_{2b}$ is the $Q-value$ of the $B+x \rightarrow C+D$ two-body reaction and $E_{CD}$ is the relative energy between the outgoing particles $C$ and $D$. The approach based on the idea that the initial projectile velocity is compensated for by the binding energy of particle $x$ inside $A$: $E_{cm}= E_{x-B} - B_{xS}$ with $E_{x-B}$ center of mass system.Thus the relative energy of the fragments in the initial channel $x+B$ of the binary reaction can be very low and even negative. In contrast, this condition is difficult or impossible to achieve in direct measurements due to the Coulomb barrier.

## 3. Data analysis

*3.1. Extraction of two-body cross section from the measured three-body reaction*
In the analysis the two body cross section is derived by dividing the selected three-body coincidence yield (Yd) by the result of the Monte Carlo calculation of the product of kinematical factor time the interval of experimental momentum distribution

$$\left[\frac{d\sigma}{d\Omega}\right]^{HOES} \propto \left[\frac{Y_l}{(KF)\phi^2(p_{xS})}\right] \tag{7}$$

where $\phi^2(p_{xS})$ is refered to the range of momentum distribution corresponding to the $p_{xS}$ range that can be considered equation (5). We stress that this relation is a limit of the pole approximation applicability. This cross section can be normalized to the binary reaction obtaining the absolute value in the energy range $\Delta(E_{CM})$ of the excitation function above the Coulomb barrier.

.In the THM application the reaction x+ B → C+D is induced inside short-range nuclear field the penetration probability of the Coulomb barrier has to be introduced to compare the THM cross section with the direct data (taken from literature) from literature in the energy region below the Coulomb barrier

$$P_l(k_{xB}r_{xB}) = \frac{k_{xB}r_{xB}}{F_l^2(k_{xB}r_{xB})G_l^2(k_{xB}r_{xB})} \tag{8}$$

where $F_l$ and $G_l$ are the regular and irregular Coulomb wave functions, $k_{xB}$ and $r_{xB}$ the x-B relative wave number and interaction radius, respectively. In this case the two body cross section correspond to the nuclear contribution to the cross section and the TH two body cross section is defined by the relation

$$\left[\frac{d\sigma}{d\Omega}\right]^{THM} \propto \left[\frac{Y_l}{(KF)\phi^2(p_{xS}}\right][P_l(k_{xB}r_{xB})] \tag{9}$$

and the quantities to be compared for the validity test are

$$\left[\frac{d\sigma}{d\Omega}\right]^{THM} \propto \left[\frac{d\sigma}{d\Omega}\right]^{OES} \tag{10}$$

## 4. Astrophysical S-factor

The THM provides an indipendent bare nucleus cross section measurement, $\sigma_b(E)$ (or equivalently the bare nucleus astrophysical factor $S_b(E)$). Thus, it is possible to extract information on the electron screening potential [Spitaleri2011] by comparing $S_b(E)$ with the one obtained from shielded nuclei. Once the two-body cross section has been extracted , the THM astrophysical S(E)-factor can be obtained according to:

$$[S_b(E)]^{THM} = E[\sigma_b(E)]^{THM} exp(2\pi\eta) \tag{11}$$

where $\sigma_b(E)^{THM}$ is the bare nucleus cross section obtained via the THM by the Eq. (12). Thus, the energy dependence of $[S_b(E)]^{THM}$ should show the same trend of the one derived by the direct measurements, excepted in ultra low energy range $\Delta E_{ES}$ where the two data sets should differ due to the effects of electron screening:

$$[S_b(E]^{THM}] \propto [S_b(E]^{OES}] \tag{12}$$

where E>$\Delta E_{ES}$. If needed, the value for the electron screening potential $U_e$ can then obtained by comparing the two data sets. Clearly, this procedure does not allow us to extract the absolute value of the astrophysical S-factor. However, the absolute scale for $S_b(E)^{THM}$ can be obtained by normalization of the THM data to the direct $S(E)^{OES}$ ones obtained at energies E* where the electron screening effects are negligible.

$$N_{abs.value} = \frac{[S_b(E^*)]^{OES}}{[S_b(E^*)]^{THM}} \tag{13}$$

## 5. Example of THM application: $^{10}$B(p,$\alpha_0$)$^7$Be reaction E= 24.5 MeV

Fig. 1 shows the TH $^{10}$B(p,$\alpha_0$)$^7$Be S(E) factor obtained in an experiment performad at Tandem of the Laboratori Nazionali del Sud- INFN, Catania [14].

The data are reported at infinite resolution, together with their upper and lower limits as given by the corresponding uncertainties. The THM data are compared with the low-energy direct data [21]. While at energies lower than 30 keV direct data are strongly influenced by electron screening effects. The TH S(E) factor describes the typical bare-nucleus behavior.

## 6. Conclusions

In summary, the THM has proved to be an extremely versatile and powerful techique to investigate both reactions of astrophysical interest at the extreme of sub-Coulomb energies and purely nuclear phenomena otherwise difficult to study by a direct approach. Important results have already been obtained in both areas and new applications of the method to further cases of interest (e.g., reactions involving neutrons and/or reactions with Radioactive Ion Beams), are currently underway.

## 7. Acknowledgments

This work has been partially supported by the Italian Ministry of University MIUR under the grant LNS-Astrofisica Nucleare (fondi premiali).

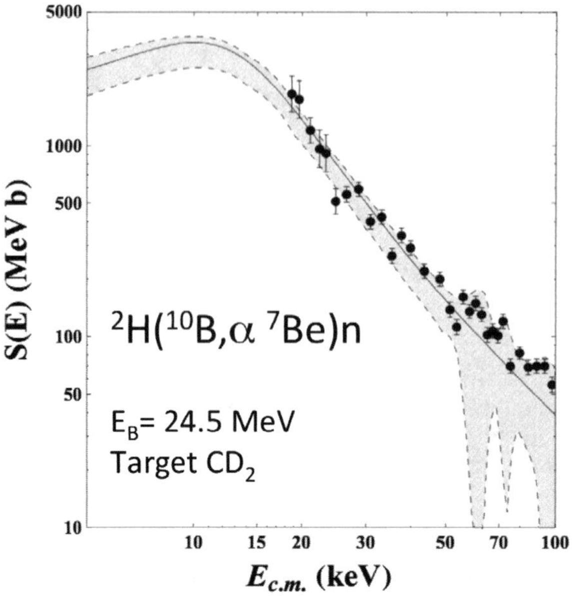


**Figure 2.** The TH $^{10}B(p,\alpha_0)^7Be$ astrophysical S(E) factor compared with the low-energy direct data [14].

## 8. References

[1] Rolfs C e and Rodney W S 1988 *Cauldrons in the Cosmos*, University of Chicago Press, Chicago
[2] Bonetti R et al.1999 *Phys. Rev. Lett.* **82** 5205.
[3] Assenbaum H J, Laganke K and Rolfs C 1987 *Z.Phys.*A **327** 461.
[4] Strieder F, Rolfs C, Spitaleri C and Corvisiero P 2001*Naturrwissenschaften* **88** 461.
[5] Baur G et al.1994 100.
[6] Baur G et al.1996 100.
[7] Mukhamedzhanov A M, et al. 2008 *Phys. Rev.* C **78** 015804.
[8] Baur G,1986 *Phys. Lett.* B **178** 135.
[9] Tribble R E, Bertulani C, La Cognata M, Mukhamedzhanov A M and Spitaleri C 2014 *Rep. Progr. Phys.* **77** 106901.
[10] Spitaleri C et al. 2011 *Phys. Atom. Nucl.* **74** 1763.
[11] Spitaleri C et al. 2003 Nucl. Phys. A **719** 99c.
[12] Spitaleri C 1991 *Problem of fundamental modern physics II* Editors Cherubini, Dal Piaz, Minetti, ISBN 981-02-0295-4 (World Scientific).
[13] Typel S and Baur G 2003 *Ann.Phys.(NY)* **305** 228.
[14] Spitaleri c et al 2014 *Phys. Rev.* C **90** 035801.
[15] Gulino M et al. 2010 *Journ. Phys.G -Nuclear and Particle Physics* **37** 1205105.
[16] Spitaleri C et al 1999 *Phys.Rev.* C **60** 055802.
[17] Spitaleri C et al 2001 *Phys.Rev.* C **00** 000000.
[18] Spitaleri C et al 2004 *Phys.Rev.* C **00** 000000
[19] Chew G F and Wick G C 1952 *Phys. Rev.* C **85** 636.
[20] Shapiro I S 1967 *Interaction of High-Energy Particles with Nuclei, International School of Physics Enrico Fermi, Course 38* edited by E.O. Ericson (Academic, New York, 1967) pp. 210-251.
[21] Angulo C et al. (NACRE) 1999 *Nucl. Phys.* A **656** 3.

Journal of Physics: Conference Series **590** (2015) 012014 doi:10.1088/1742-6596/590/1/012014

# The structure of $^{12}$C and $^{16}$O

**Martin Freer**

School of Physics and Astronomy, University of Birmingham, Birmingham, B15 2TT, UK

M.Freer@bham.ac.uk

**Abstract**. The recent advances in the understanding of the two nuclei $^{12}$C and $^{16}$O are discussed in the light of recent measurements and developments in nuclear theory. The focus is the collective excitations, which may be related to dynamical symmetries which in turn reveal the nature of the underlying nuclear structure.

## 1. Nuclear Correlations

The complexity revealed in the binding of light nuclei is revealing. It exposes in sharp relief details of the strong interaction in a manner that is akin to the grazing incidence light on the moon's surface exposes detail of historical impacts. Figure 1 illustrates, for nuclei up to Neon isotopes, bound and unbound nuclei. At the neutron drip-line the effects of neutron-neutron correlations are manifest in the Borromean nature of $^{6}$He, $^{8}$He, $^{11}$Li and $^{14}$Be, with their $A$-1n counterparts $^{5}$He, $^{7}$He, $^{10}$Li and $^{13}$Be all being unbound to neutron decay. Similarly, on the proton-rich side $^{10}$C may be thought of as super-Borromean (or perhaps more accurately Brunnian [1]), composed of the constituents $^{4}$He+$^{4}$He+p+p. The lack of the existence of bound $^{9}$B and $^{6}$Be means that the removal of any of the 4 constituents, results in the dissociation of $^{10}$C; and indeed, the lack of bound $^{8}$Be and $^{5}$Li nuclei also implies that removal of any two components results in the spontaneous decay of the remaining ones. This rather delicate balance of bound and unbound substructures again may be linked to correlations. In this particular instance it is the very high binding energy of the alpha-particle which is playing a driving role.

**Figure 1.** Bound light nuclei, with their correlations, or clusterization, illustrated.

This high binding owes its origins to the maximal correlations between pairs of protons and neutrons and indeed between the n-p pairs. This yields one of the highest binding energies of light nuclei and is the origin of the unbound nature of $^{8}$Be, which decays into two alpha-particles with a Q-value of 92 keV. Moreover, the correlations are manifest in the very high lying (>20 MeV) first excited state. It is these

Journal of Physics: Conference Series **590** (2015) 012014 doi:10.1088/1742-6596/590/1/012014

properties that make the nucleus $^4$He such a good nuclear cluster. The evidence for the influence of the alpha-cluster on the properties of light nuclei has deep historical roots, extending back to the 1930s with the work of Hafstad and Teller [2] through the 1960s with the contribution of Ikeda [3] to more recent developments involving molecular type structures where neutrons are exchanged between alpha-particle clusters [4].

The treatment of the phenomenon of clustering from a theoretical perspective had, until recently, been rather rudimentary, where the cluster and core were considered to be pre-existing entities and collective excitations are generated through orbital motion of the cluster about the core. Great advances have been made in theory in which $A$-body calculations may be performed that reveal the single-particle and cluster like states within the same framework [5]. Similarly, chiral effective field theory (cEFT) calculations performed on the lattice are beginning to provide a real insight into the structure of light nuclear systems from a perspective in which the interaction is grown from the QCD degrees of freedom. In particular, the focus for such calculations have been the structure of $^{12}$C and in particular the Hoyle-state [6,7].

## 2. Carbon-12

Given the underlying influence of the correlations described above, the structure of $^{12}$C naturally may be thought of, in zeroth order, as an assembly of three alpha-particles. The Ikeda picture [3] reveals that the full realization of clustering does not occur until the appropriate decay threshold is reached. In the case of $^{12}$C the 3 alpha-particle decay threshold is just above 7 MeV, and the Hoyle-state at 7.65 MeV famously has not only been associated with a well-developed alpha-cluster structure, but also plays a key role in nucleosynthesis [8]. Below this threshold, it is expected that the cluster-structure is reduced and eventually the shell-model like states are arrived at.

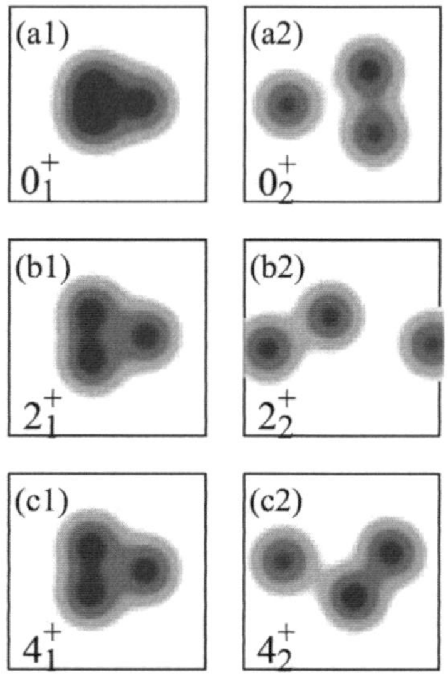


**Figure 2.** AMD calculations from [5]. The ground-state is depicted in (a1), with rotational states (b1) and (c1). The second $0^+$ state, which would be linked to the 7.65 MeV Hoyle-state, is shown in (a2) and the excitations are in (b2) and (c2).

Figure 2 shows the Antisymmetrised Molecular Dynamics (AMD) calculations [5] for the structure of the ground-state (a1) and Hoyle-state (a2). The ground-state is compact, though still reveals the 3alpha structure, which is also strongly influenced by the $^8$Be+alpha substructure. The "collective" excitations of this state (b1) and (c1) – $2^+$ and $4^+$ states - reveal a very similar intrinsic structure. The Hoyle-state is a loose assembly of alpha-particles, again influenced by the $^8$Be+alpha partition. However, the excitations do not strongly resemble the intrinsic structure of the Hoyle-state and hence in this model a rotational description might not be appropriate.

The condensate (Bose Einstein Condensate (BEC)) description of the 3alpha system was also presented at the present meeting by Funaki-san. This reveals that the Hoyle-state may be described in terms of a wave-function that captures the bosonic properties of the alpha-particles, whilst recognizing their internal fermionic structure which is revealed as the alpha-particles overlap through an antisymmetrisation operator. The description of the Hoyle-state in this model lies outside the low energy shell model basis, and can reproduce many of the experimental characteristics; most notably the angular distribution for electron inelastic scattering from the ground state to the Hoyle-state [9]. Here, the underpinning wave-function is one which has the three alpha-particles with a common $s$-state wave function, in a potential formed from their mutual interaction. The antisymmetrisation operator then serves to suppress the amplitudes at small distances and the cluster structure then emerges. Collective excitations then, in large part, correspond to rotations. The contribution to the present meeting from Funaki-san indicates that the $2^+$ excitation is dominated by an alpha-particle rotating around an $^8Be_{gs}$ with two units of angular momentum, with a smaller component of $^8Be(2^+)$+alpha. A $4^+$ excitation has dominant components from the rotation of the alpha-particle around the $^8Be_{gs}$ core with $L$=4 and $^8Be(2^+)$ core with $L$=2 and only a small contribution from $^8Be(4^+)$.

Understanding these collective excitations is rather important, inasmuch that they are fundamentally linked to the nature of the Hoyle-state itself. If, as indicated by the AMD calculations, the dominant component is governed by $^8$Be+alpha, then the collective modes would be linked to either rotations of the $^8$Be component, the alpha-particle motion around the $^8$Be, or combinations of these collective degrees of freedom. This is a similar description to that of the condensate picture. Guidance may be derived from the collective excitations of the $^8$Be nucleus; the excitation energies of the $2^+$ and $4^+$ excitations are 3.03 and 11.35 MeV. If these are mirrored in $^{12}$C then the $2^+$ and $4^+$ excitations of the Hoyle-state should lie at 7.65+3.03 = 10.68 MeV and 7.65+11.35 = 19.00 MeV.

The experimental situation has over recent years become clearer. A $2^+$ excitation at ~9.8 MeV [10-13] with a width of 700-800 keV (perhaps even larger [13]) has been firmly identified. There is also an indication of a broad $4^+$ state close to 13.3 MeV [14]. The experimental $2^+$ state lies in close proximity to that extrapolated from $^8$Be, but the $4^+$ lies somewhat lower in energy by about 6.5 MeV. The lowering in energy with respect to the collective excitations of $^8$Be may be interpreted as either the 2 correlated alpha-particles possess a larger separation than that in the $^8Be_{gs}$ band, or that the states are generated in a different manner. If as indicated by both the AMD and BEC approaches the motion of the alpha-particle contributes strongly to the collectivity, then it is possible that the energy of the collective excitations may be lowered such that they lie closer to the experimental equivalents. The BEC approach indeed does predict lower energy $2^+$ and $4^+$ states, but actually lower in energy than the experimental counterparts. This, thus, remains an open question, which could perhaps be resolved if it might be possible to measure the B(E2) transition rates between these states.

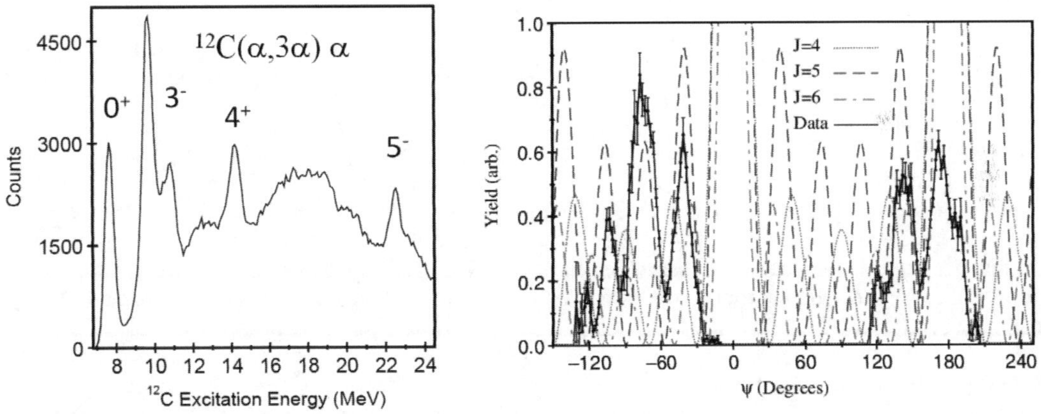


**Figure 3.** (left) The $^{12}$C excitation energy spectrum from the $^{12}$C(alpha,3alpha)alpha reaction, for decays to the $^8$Be ground-state. The $5^-$ state is observed at 22.5 MeV. (right) The angular distribution for the $5^-$ state showing the close agreement with $J^\pi=5^-$ [15].

The nature of the ground state of $^{12}$C as revealed in the AMD calculations needs to be verified. The equation describing the rotations of a spinning top, predicted by Hafstad and Teller, and applied to $^{12}$C is

$$E_{J,K} = \frac{\hbar^2}{2I_{Be}}J(J+1) - \frac{\hbar^2 K^2}{4I_{Be}}$$

Here $J$ and $K$ are the total angular momentum and the projection of the total angular momentum onto the symmetry axis, respectively, of the three centered object, where in a simple model the alpha-particles reside at the vertices of an equilateral triangle. For $K = 0$ then the rotations are about a symmetry axis which passes through the plane of the triangle – passing through one vertex and bisecting the other two. The gives rise to $0^+$, $2^+$ and $4^+$ states. For the next possible value of $K$, $K = 3$, then the rotation is about an axis which is perpendicular to, and passes through the center of, the triangle. In this instance each alpha-particle carries one unit of angular momentum and would produce $3^-$, $4^-$, $5^-$.... excitations.
The ground state rotational band is known for the $2^+$ and $4^+$ members at 4.44 and 14.08 MeV, as is the $3^-$ excitation at 9.64 MeV. Recent measurements of the width of the $3^-$ state have indicated that it has a well-developed alpha-cluster structure [16] and the $4^-$ member of the band has been identified at 13.2 MeV [17,18]. A key measurement has been the determination of the $5^-$ member of the $K$ =3 band at 22.5 MeV – as shown in Fig. 3. The observation of this state would appear to confirm the dynamical symmetry associated with the ground state structure.

Returning to the structure of the Hoyle-state; an alternative approach to establishing its nature would be to examine its' decay properties. The possible charged particle decay modes are either sequential via $^8Be_{gs}$+alpha, followed by the decay of $^8Be$ into two alpha-particles, or direct 3alpha decay. The Fermi breakup model allows the two and three-body decay phase space to be calculated. Integrating over the energies of the decay products gives

$$W_2 = \frac{1}{2}K_2\left(\frac{1}{\sum_{b=1}^{2} m_b}\prod_{b=1}^{2} m_b\right)^{3/2}\frac{(2\pi)^{1.5}}{\Gamma(3/2)}Q_2^{0.5}$$

$$W_3 = \frac{1}{6}K_3\left(\frac{1}{\sum_{b=1}^{3} m_b}\prod_{b=1}^{3} m_b\right)^{3/2}\frac{(2\pi)^{3}}{\Gamma(3)}Q_3^{2}$$

where

$K_n = \left(\frac{V}{(2\pi\hbar)^3}\right)^{n-1}$ and $V = \frac{4}{3}\pi\left(\left(1.4 \times 12^{1/3}\right) \times 10^{-15}\right)^3$. Also $Q_2$ and $Q_3$ are the two and three body Q-values, respectively, and $m_b$ are the masses of the decay products.
For low decay energies the three-body phase-space is highly constrained, and the precise calculation depends on the magnitude of the of volume term $V$. Nevertheless, ratio of $W_3/W_2$ is close to 0.2 to 0.3%. An enhancement of the three-body decay probability over that predicted by phase space alone would be interesting and might signal a BEC type structure. The recent measurements of the ratio of the three- to two-body decay give an upper limit of 0.2%. The experimental precision is thus very close to the region where probing the structure of the Hoyle-state through this approach becomes interesting [19].

## 3. Oxygen-16

The fingerprint of the $D_{3h}$ dynamical symmetries found in the spectroscopy of $^{12}C$ are remarkable and although have been explored recently, may be traced back to the work of Hafstad and Teller [2]. Similarly, the structure of $^{16}O$ may be interpreted in terms of a tetrahedral arrangement of 4 alpha-particles. The recent work of Bijker and Iachello [20] shows the spectrum that would be expected for such a tetrahedral structure; the sequence $0^+$, $3^-$, $4^+$, $6^{+/-}$ would be expected. The $0^+$ state corresponds to the ground state, the $3^-$ state would be the observed 6.13 MeV excitation, there is a $4^+$ state at 10.36 MeV and several $6^+$ state at 16.23, 21.05 and 21.65 MeV. The $3^-$ state would be the obvious analogue of the 9.64 MeV $3^-$ state in $^{12}C$, with the tetrahedron rotating such that the axis of rotation passes through the canter of the base and the vertex of the upper alpha-particle.

This is a compelling picture, however, would require some reinterpretation of the nature of the rotational bands which have been "established" for $^{16}O$. In order to distinguish between the competing interpretations, precision measurements are required. These involve both a study of electromagnetic decay strengths coupled with precision measurements of the charged particle decay widths. For example, the measurements presented in [21] provide characterization of the high spin members of the $K^{\pi} = 0^{+/-}$ cluster bands in $^{16}O$ which appear to confirm a common structure. The spectroscopy of $^{16}O$ is much more complex than $^{12}C$ and hence a concerted experimental effort is required to pin down the exact details of the underlying structure.

The analogue of the Hoyle-state in $^{16}O$ has yet to be confirmed, however, as indicated by Funaki-san it is most likely the 15.2 MeV $0^+$ state. The 4alpha decay threshold is at 14.4 MeV and consequently the 15.2 MeV state cannot decay strongly into this channel. There are, however, a number of resonances that decay to the $^8Be$+$^8Be$ or $^{12}C$(Hoyle)+alpha final-states. The pioneering measurements of Chevalier et al., [22] revealed both the energy and dominant angular momenta of the $^{16}O$ resonances that decayed to $^8Be$+$^8Be$ as populated in the $^{12}C(alpha,^8Be)^8Be$ reaction. The energy-spin systematics of selected, narrow, resonances remarkably fell onto a $J(J+1)$ trajectory with a moment of inertia commensurate with a structure in which the a alpha-particles were arranged in a linear fashion; an alpha-particle chain. This work was published in 1967 and until the present has been held up as an example of extreme alpha-clustering.

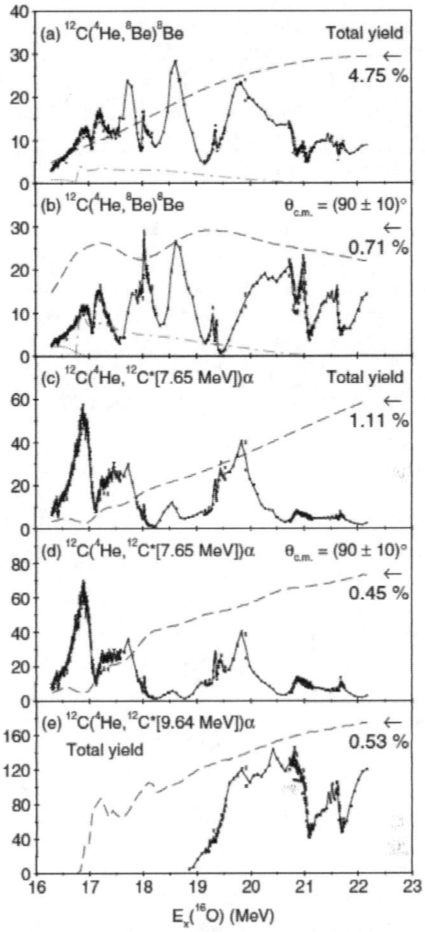


**Figure 4.** Excitation functions for reactions populating resonances in $^{16}O$ decay to 4 alpha-particle final states, from [23].

Confirmation of such an exotic structure is clearly vital. There are a number of possible approaches; one of which is to confirm the details of the excitation function, the second is to search for higher spin members of the 4 alpha-particle chain band – the band was only observed up to spin 6. The highly detailed excitation functions for the $^{12}C(alpha,^8Be)^8Be$ and $^{12}C(alpha,^{12}C[7.65$ MeV])alpha reactions presented in [23] show that both the original structure that was interpreted as resonances in the earlier work [22] was more complex and no evidence for an $8^+$ state could be identified (Figure 4). This most recent study contained over 400 measurements at different energies, with significant coverage of the angular distributions which should permit the components from resonances and transfer-like processes to be disentangled.

As with $^{12}C$ the field is considerably closer to being able to unpick some of the questions that surround the spectroscopy of $^{16}O$ and address the underlying structure. However, comprehensive, precision, data are the key.

## 4. Summary

Understanding the structure of light nuclei lies on the critical path to comprehending the nature of the strong interaction on a nuclear scale. The development of the ability to apply realistic interactions to nuclei, with the number of constituents beyond a few, transforms the potential to better understand how the correlations which a manifest in systems such as $^{9}$Be, $^{12}$C and $^{16}$O. Advances in experimental techniques are permitting a clearer understanding of the spectroscopy of light nuclei to constrain better the state-of-the-art models. The next steps in experimental characterization is to measure electromagnetic transition rates for states above the cluster decay threshold where branching ratios are $10^{-6}$ or less! Such studies have the potential to discriminate to the extent that it is possible to arrive at a detailed understanding of the role of cluster correlations in nuclei.

## Acknowledgements

The author would like to acknowledge the contribution from a very large number of collaborators to the development of these ideas, particularly within the Birmingham group. This work has been supported by the UK STFC research council.

## References

[1] Curtis N, et al. 2008 *Phys. Rev. C* **77** 021301(R)
[2] Hafstad L R, Teller E, 1938 *Phys. Rev.* **54** 681
[3] Ikeda K, et al. 1968 *Prog. Theor. Phys. Suppl., Extra Numbers* 464
[4] von Oertzen W, Freer M, Kanada-En'yo Y 2006 *Physics Reports* **432** 43
[5] Kanada En'yo Y 2007 *Prog. Theor. Phys.* **117** 655
[6] Epelbaum E, et al. 2012 *Phys. Rev. Lett.* **109** 252501
[7] Epelbaum E, et al. 2011 *Phys. Rev. Lett.* **106** 192501
[8] Freer M, Fynbo H O U 2014 *Progress in Particle and Nuclear Physics* **78**, 1
[9] Funaki Y, et al. 2006 *Eur. Phys. J. A* **28** 259
[10] Freer M, et al. 2012 *Phys. Rev. C* **86** 034320
[11] Itoh M, et al. 2011 *Phys. Rev. C* **84** 054308
[12] Freer M, et al. 2009 *Phys. Rev. C* **80** 041303
[13] Zimmerman W R, et al. 2013 *Phys. Rev. Lett.* **110** 152502
[14] Freer M, et al. 2011 *Phys. Rev. C* **83** 034314
[15] Marín-Lámbarri D J, Bijker R, Freer M, Gai M, Kokalova Tz, Parker D J and Wheldon C 2014 *Phys. Rev. Lett.* **113** 012502
[16] Kokalova Tz, Freer M, Buthelezi Z, Carter J, Fearick R W, Förtsch S V, Fujita H, Neveling R, Papka P , Smit F D, Swartz J A and Usman I 2013 *Phys. Rev. C* **87** 057307
[17] Freer M, et al. 2007 *Phys. Rev. C* **76** 034320
[18] Kirsebom O S et al. 2010 *Phys. Rev. C* **81** 064313
[19] Itoh M, et al. 2014 *Phys. Rev Lett.* **113** 102501
[20] Bijker R and Iachello F 2014 *Phys. Rev. Lett.* **112** 152501
[21] Wheldon C, et al. 2011 *Phys. Rev. C* **83** 064324
[22] Chevalier P, et al. 1967 *Phys. Rev.* **160** 827
[23] Curtis N, et al. 2013 *Phys. Rev. C* **88** 064309

Journal of Physics: Conference Series **590** (2015) 012015 doi:10.1088/1742-6596/590/1/012015

# Nuclear Structure and Setting Constraints on the r-process?

**Ani Aprahamian**
Nuclear Science Laboratory, University of Notre Dame, Notre Dame, IN 46556
E-mail: aapraham@nd.edu

**Abstract.** One of the open questions in all of physics today has to do with the site of the r-process. There are uncertainties associated with the astrophysics, the observations, and the physics of nuclei far from stability. This paper uses existing nuclear physics models and their implied properties for nuclei far from stability to determine the nuclei that have the greatest impact on a given astrophysical scenario. The properties considered are nuclear masses, beta-decay rates, and neutron capture rates individually as well as in a correlated way to identify the most important nuclei in a given r-process trajectory. The ultimate goal is to set some constraints on potential sites of the r-process based on the nuclear properties.

## 1. Introduction

The NRC report "Connecting Quarks to the Cosmos" identified eleven of the most challenging open questions in all of Physics, Astrophysics, Astronomy, and Cosmology for the 21st Century. One of these questions is the identification of the site for the synthesis of the heavy elements via the r-process. Nuclei are made in stars from the primordial building blocks that resulted from the big bang and observations of early stars show abundances for the elements up to Fe. Beyond Fe, charged particle reactions are no longer effective and the heavier elements have to be made in processes that involve slow (s-process) and rapid (r-process) neutron capture reactions. The s-process and its path lie close to stability where the nuclear physics is well understood and the associated astrophysical sites have been identified. The r-process is thought to occur far from stability where the nuclear physics is unknown or challenging to measure and no astrophysical sites have yet been identified. While we have limited knowledge of the properties of specific nuclei far from stability, the nuclear processes that make up the r-process are well known: successive neutron captures and subsequent $\beta$-decays. The nuclear properties most relevant to the r-process include nuclear masses, $\beta$-decay rates, n-capture rates, and $\beta$-delayed neutron emission probabilities.

Experimental access to the very rich nuclei far from stability is a worldwide quest as evidenced by the developments of radioactive ion beam facilities across the globe. New facilities are being developed in the USA (CARIBU at ANL, NSCL and FRIB at MSU), in Canada (ISAC at TRIUMF), in Europe (ISOLDE at CERN), in France (SPIRAL II at GANIL), in Finland (Jyvaskyla), in Germany (FAIR at GSI Darmstadt), in Italy (Legnaro-SPES collaboration), in Japan (RIKEN), and in China (BRIF,CARIF in CIAE Beijing) to study the properties of exotic nuclei far from stability thought to be involved in the r-process. A specific r-process path may involve over three thousand nuclei and a large number of these remain a challenge to experimental reach. The question is "which nuclei are the most important or most impactful to the r-process?" and "which ones should be given priority for measurements at the new facilities as they come online?".

The approach we have chosen is one of simulation taking advantage of nuclear models for the properties of the most neutron rich nuclei, and comparing with the observed abundances of the r-process elements. The determination of nuclei whose changes in the relevant nuclear properties have the largest impact on the resulting abundances when compared to the observed r-process elemental abundances.

In the sensitivity studies described below, we have compared nuclear mass models in a fixed astrophysical scenario and then investigated the impact of a fixed nuclear model with variations in the r-process scenarios. Candidates for the site of the r-process are many, we have chosen two of the most likely scenarios which include two-neutron star mergers and supernova and determined the most important or most impactful nuclei using sensitivity studies. The elemental abundances resulting from the conditions in the cosmos are the results of complex processes that depend on the uncertainties in nuclear physics as well as the specifics of an astrophysical scenario. Our longer term goal is to disentangle the nuclear uncertainties from the astrophysical ones and to attempt to set some constraints on potential astrophysical scenarios for the r-process.

## 2. Sensitivity Studies

The r-process proceeds via a sequence of neutron captures, photo-dissociations and $\beta$ decays. Simulations of the r-process therefore require tabulations of $\beta$-decay lifetimes, neutron capture rates and neutron separation energies; photodissociation rates are determined from the capture rates and separation energies by detailed balance [**?**]:

$$\lambda_\gamma(Z, A) \propto T^{3/2} \exp\left[-\frac{S_n(Z, A)}{kT}\right] \langle\sigma v\rangle_{(Z,A-1)} \tag{1}$$

In this equation, $T$ is the temperature, $\langle\sigma v\rangle_{(Z,A-1)}$ is the thermally-averaged value of the neutron capture cross section for the neighboring nucleus with one less neutron, and $S_n(Z, A)$ is the neutron separation energy—the difference in binding between the nuclei $(Z, A)$ and $(Z, A - 1)$. We have examined the impact of individual nuclear properties such as masses, $\beta$-decay rates, neutron capture rates, and $\beta$-delayed neutron emission probabilities on the r-process.

For example, in the sensitivity studies exploring the impact of nuclear masses on the r-process, we initially chose a specific astrophysical trajectory and examined examined the individual neutron separation energies [?] and eventually $\pm1$ MeV in binding energy [?] within a given mass model, as they appear in Eqn. **??**, in an attempt to determine the nuclei that have the greatest impact on the overall r-process abundances. All the calculations were done for the same initial astrophysical conditions. This comparison can be quantified by summing the differences in the final mass fractions:

$$F_\pm = 100 \sum_A |X_{baseline}(A) - X_{\pm\Delta S_n}(A))|, \tag{2}$$

where $X(A) = AY(A)$ is the mass fraction of nuclei with mass number $A$ (such that $\sum_A X(A) = 1$), and the sum of $A$ ranges over the entire abundance curve. The values of $F = (F_+ + F_-)/2$ are calculated for 3010 nuclei from $^{58}$Fe to $^{294}$Fm. Nuclei that have the greatest impact on the r-process are those neutron rich nuclei near the closed shells at Z=28 and 50, and N=50, 82, and 126. Neutron separation energies include differences in mass of two nuclei. We followed with a complementary study with variations in the binding energy of an individual nucleus by $\pm1$ MeV[?]. The results of the binding energy sensitivity study for the three nuclear masses are presented in Fig. 1. This figure shows the same bulk features as the separation energy

Journal of Physics: Conference Series **590** (2015) 012015 doi:10.1088/1742-6596/590/1/012015

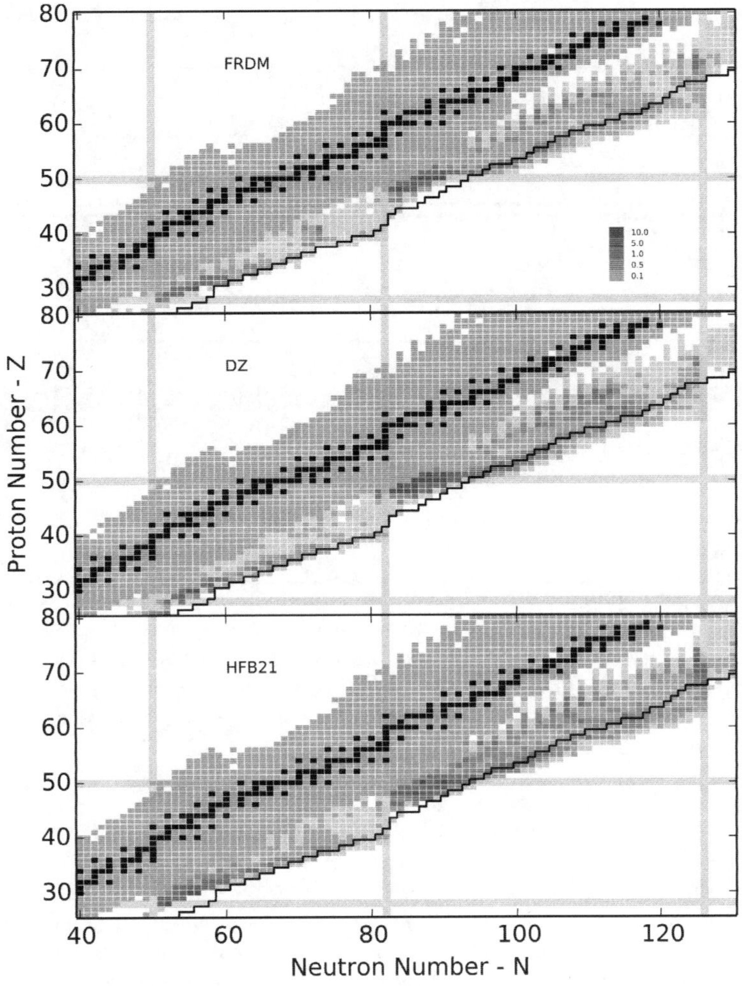


**Figure 1.** Sensitivity measures $F$ for each nucleus $(Z, N)$ in the network, for three binding energy sensitivity studies using FRDM[?] (top panel), DZ[?] (middle panel), and HFB-21[?] (bottom panel) masses. All three studies use binding energy variations of $\pm 1$MeV and astrophysical conditions as used in Brett et al.[?], similar to the 'H' trajectory from Qian[?]. Stable nuclei are represented by solid black boxes. Overlaid in gray is the region of nuclear masses that have been reported as measured in AME2012[?] and the solid black line represents the predicted limits of accessibility for the production rates from FRIB[?]. It is clear from the concentration of $F$ intensities that FRIB will allow the measurements of a significant portion of the most impactful nuclei.

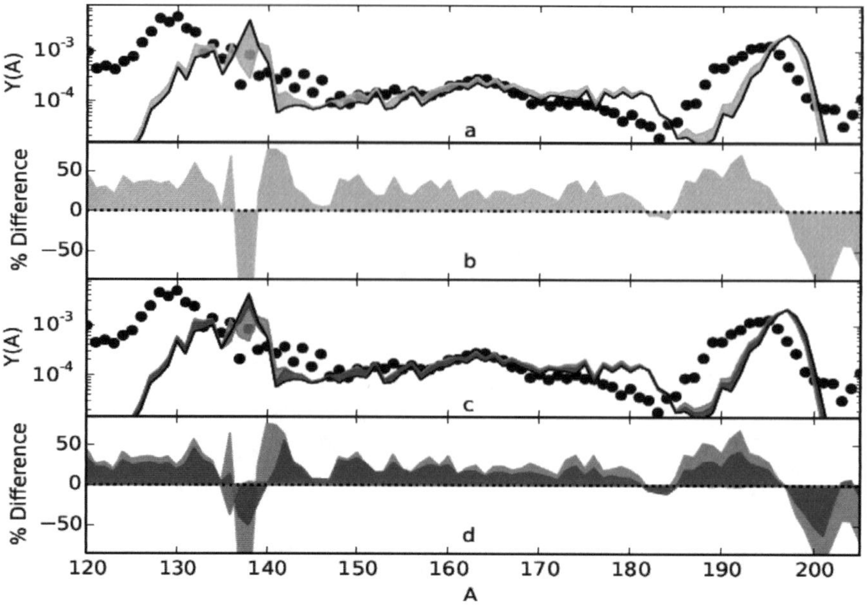


**Figure 2.** (a) Shows the change in final $r$-process abundances when the mass of $^{140}$Sn is increased by 500 keV and all dependencies have been considered compared to baseline simulation (black) and solar data (black circles) from [?]. (b) Percent difference of the abundances in (a) to baseline. (c) Separate simulations show the contribution of the mass variation when only propagating changes to neutron captures (red), $\beta$-decays (blue) and separation energies (green). (d) Percent difference of the abundances in (c) to baseline.

sensitivity studies shown in Fig. 3 of Brett et al.[?]. That is, the largest concentration of nuclei with great impact on the $r$-process as determined from the binding energy sensitivity studies and the separation energy sensitivity studies are clustered around the closed shells. The largest sensitivity measures $F$ are produced by variations in the masses of nuclei along and near the $r$-process path, particularly at the closed shells but the effect is somewhat more smeared out over the shell closures for the binding energy sensitivity studies. There are some minor differences between three mass models shown in Fig.1 for the binding energy sensitivity studies. However, there is a significant overlap in the nuclei that rise to the surface. Similarly, we examined the impact of individual $\beta$-decay rates[**?**] and $\beta-$delayed neutron emission probabilities[**?**] across the chart of nuclides in the context of a main $r$ process which produces nuclear flow out to the third ($A$ = 195) abundance peak. All of these studies have provided insight into how the r-process proceeds, particularly at late times in the process once (n,$\gamma$)-($\gamma$,n) equilibrium has failed, and have pointed out the masses, decay rates, or capture rates with the greatest leverage on the final abundance pattern.

These studies, however, have looked at each piece of nuclear data in isolation. More recently[?], we have looked at the impact of these nuclear properties in a correlated way. That is we have made changes in mass value of an individual nucleus and then propagated the changes that such a variation causes on the relevant quantities of nearby nuclei. A change in one mass results in 12 changes for nuclei nearby. An example of the impact on the r-process abundances for

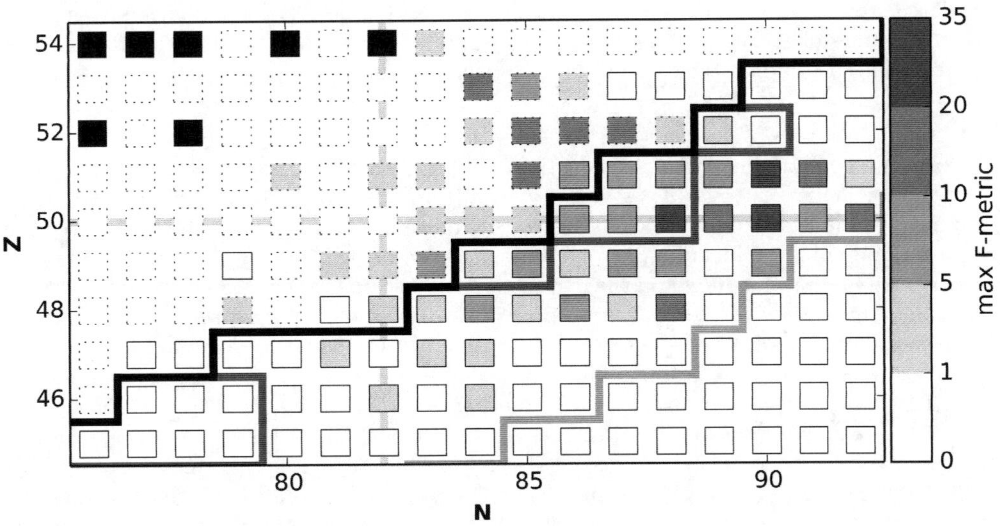


**Figure 3.** Nuclei from this study that significantly impact final $r$-process abundances in the $N = 82$ region with an uncertainty of $\pm 500$ keV. The black line shows the extent of measured masses from the AME 2012. Nuclei with masses measured within 100 keV are denoted by a dotted border and those greater denoted by solid border. Accessibility limits shown for CARIBU (dark gray) and FRIB (light gray).

each of the nuclear properties (masses, n-capture rates, $\beta$-decay rates, and $\beta$-delayed neutron emission probabilities) for the case of $^{140}$Sn, is shown in Fig. ??. In panel (a) and (b) we added 500 keV to the mass of $^{140}$Sn ($Z = 50$) and propagated changes to all 12 nuclear physics properties of nearby nuclei.

Simulations shown in the Fig. ?? break down the impact of a this single mass change to the various processes. For example, the change in final abundances seen in panel (a) is dominated by the shift in separation energies, A smaller change comes from the $\beta$-decay rates as can be seen from the blue line in panel (c) of Fig. ??. A mass increase of 500 keV to $^{140}$Sn lowers the half-life of this nucleus by 30%, however this is not the dominant effect. Instead, the prevailing effect comes from the change in neutron emission probabilities. These quantities are extremely dependent on the two separation energies, as the Q-values control the distribution of $\beta$-strength that goes into each channel. In this situation, the increase of the $\beta$-decay rate of $^{140}$Sn coupled with the changes to $P_{jn}$ values ensures a speedup of the nuclear flow in this region. panels(c) and (d). Breaking down by the color used in Fig. ?? we propagated changes to neutron capture rates (red), $\beta$-decay rates and associated $P_{jn}$ values (blue) and one-neutron separation energies (green). The green simulation is the same type of calculation used in [?], except here we use different $\beta$-decay rates and $r$-process conditions. The effects of individual variations in 170 nuclear masses on r-process abundances are summarized in Fig. ??. Here we have taken the maximum F-value between the case where 500 keV was added to or subtracted from the FRDM mass. Qualitatively, the distribution of important nuclei agrees with previous work from [?] and lies within experimental reach of CARIBU and FRIB.

Mass uncertainties of $\pm 0.5$ MeV can produce large local changes, on the order of 100% difference, and smaller global changes, on the order of 20-40% difference, to $r$-process abundances. In terms of our metric, a value of $F \sim 30$ for a mass variation in a single nucleus can yield over an order of magnitude local change in the $r$-process abundance predictions. The correlated studies were done with mass uncertainties as low as $\pm 0.1$ MeV, which are smaller than the RMS values of all the existing mass models. Yet the results suggest that nuclear masses in this region need to be known more accurately than 100 keV.

## 3. Conclusions

The broader goal of this work is to use nuclear properties to disentangle the complexities of the heavy element nucleosynthesis and potentially to gain some insight on how the r-process proceeds. We have done this by studying via simulations, the impact of all the relevant nuclear properties individually (including masses, beta decay rates, neutron capture rates, and $\beta$-delayed neutron emission probabilities) on the r-process. We have further examined the correlated effects of a change in mass of 500 keV and the propagated effect on twelve associated properties that one mass change impacts. As a result, we have shown the nuclei that are the most impactful to the r-process elemental abundances in a number of astrophysical scenarios. Even mass changes of $\pm 100keV$ have a significant impact on the r-process. This leads us to the conclusion that most of the available mass models with rms values of more than 500 keV on average do not have the precision needed to resolve between different astrophysical scenarios in a significant way.

Our results lead us to identify the most important nuclei to measure, to put direct constraints on the precision needed for measurements of nuclear properties and theoretical models, and perhaps in the longterm constrain $r$-process models.

### *3.1. Acknowledgments*

The work presented here was carried out in collaboration with Rebecca Surman, Matthew Mumpower, Mary Beard, Ian Bentley, Dong-Liang Fang, N. Paul, and S. Brett. This research is funded by the NSF through the Joint Institute of Nuclear Astrophysics under contract numbers PHY0822648 and PHY1419765.

## 4. References

[1] Fowler W, Caughlan G, Zimmerman B 1967 Ann. Rev. Astron. Astrophys. **5**, 525
[2] Brett S, Bentley I, Paul N, Surman R, and Aprahamian A, 2012 European Journal of Physics A, **48**, 184
[3] Aprahamian A, Bentley I, Mumpower M, and Surman R 2014 AIP Advances, **4**(4):041101
[4] Möller P, Nix J R , Myers W D, Swiatecki W J 1995 Atomic Data and Nuclear Data Tables **59**, 185
[5] Duflo J, Zuker A P 1995 Phys. Rev. C **52**, R23
[6] Goriely S, Chamel N, Pearson J M 2010 Phys. Rev. C **82**, 035804
[7] Qian Y-Z, Vogel P, Wasserburg G J 1998 Astrophys. J **494**, 285
[8] Wang M, Audi G, Wapstra A H , Kondev F G, MacCormick M, Xu X, and Pfeiffer B 2012 Chinese Physics C **36**, (12) 1603
[9] Tarasov O B , Hausmann M 2005 LISE++ development: Abrasion-Fission, Technical report, NSCL, MSUCL1300
[10] Mumpower M R, Cass J, Passucci G, Surman R, and Aprahamian A, 2014 AIP Advances **4**, 041009
[11] Surman R, Mumpower M and Aprahamian A, 2014 subm. to proceedings of ARIS2014
[12] Arlandini C, Kaeppeler F, Wisshak K, Gallino R, Lugaro M, Busso M, and Straniero O 1999 Astrophys. J. **525**, 886
[13] Mumpower M R, Fang D-L , Surman R, Beard M, and Aprahamian A 2014 subm. Phys. Rev. Lett. arXiv:1410.7275v1

Journal of Physics: Conference Series **590** (2015) 012016 doi:10.1088/1742-6596/590/1/012016

# Studies of the Shapes of Heavy Nuclei at ISOLDE

**Peter A. Butler**

University of Liverpool, Oliver Lodge Laboratory, Oxford Street, Liverpool
L69 7ZE, UK

E-mail: peter.butler@liverpool.ac.uk

In this talk I will describe how recent experiments carried out at REX-ISOLDE have furthered our understanding of shape co-existence and reflection-asymmetric shapes in atomic nuclei. I will discuss how the experimental results have constrained nuclear theory and how the latter measurements might contribute to tests of extensions of the Standard Model. I will also discuss future prospects for measuring nuclear shapes from inelastic scattering.

## 1. Quadrupole shape coexistence

The evolution and microscopic origin of quadrupole collectivity and shape coexistence at low excitation energies in neutron mid-shell nuclei near the $Z = 82$ shell closure are still not fully understood [1,2]. The shape coexistence phenomenon in neutron-deficient nuclei close to $Z = 82$ was first observed when isotope shift measurements at ISOLDE for Hg nuclides revealed a sharp transition between $^{187}$Hg and $^{185}$Hg [3]. This change was interpreted as a transition from a weakly oblate shape to a more deformed prolate structure [4, 5]. Nowadays there exists a large body of experimental information supporting the coexistence of different shapes at low excitation energies in mercury isotopes. Similarly, in neutron deficient even-mass Pb isotopes intruder states were first identified in $\alpha$- and $\beta$-decay studies using on-line isotope separators [6]. The low-lying excited $0^+$ states associated with weakly deformed oblate proton 2p–2h intruder structures have been observed in $\alpha$-decay in several Pb isotopes with $N \geq 106$ [7, 8], and prolate deformed bands built upon 4p-4h excitations have also been identified in separate experiments (e.g. [9]). The lowest excited states in $^{186}$Pb were found to have spin and parity $0^+$ and are associated with oblate and (predicted) more deformed prolate shape respectively [10].

## 2. Recent experiments on shape coexistence

In order to determine the structure of the coexisting states in this mass region, Coulomb excitation experiments have been carried out [11] at REX-ISOLDE, CERN. Here, beams of neutron deficient mercury isotopes were accelerated to energies of ~ 3MeV/u and allowed to impinge upon various targets in the MINIBALL [12] $\gamma$-ray spectrometer. In order to determine $E2$ matrix elements in $^{182-188}$Hg, the Coulomb excitation least-squares fitting code GOSIA [13] was used. The code fits a set of reduced matrix elements to reproduce the measured yield of $\gamma$-ray transitions depopulating the Coulomb-excited states of $^{182-188}$Hg, taking into account known spectroscopic data related to electromagnetic matrix elements: branching ratios and conversion coefficients [e.g. 14], and lifetimes of the yrast and non-yrast states [15,16]. Crucial for this analysis was the knowledge of the conversion coefficient of the $2_2^+ \rightarrow 2_1^+$ transitions, as they contain a large $E0$ component [14].

The extracted matrix elements can be analyzed in terms of the quadrupole deformation parameters, Q and δ, that describe the charge distribution of a nucleus in a given nuclear state. Using the quadrupole sum rules approach [17] the quadrupole invariants, $\langle Q^2 \rangle$ and $\langle Q^3 \cos(3\delta) \rangle$, can then be obtained. The sums of products of the relevant matrix elements between $0^+$ and $2^+$ states are shown in Fig. 1: the sum of squared E2 matrix elements (SSM) related to $\langle Q^2 \rangle$ and the sum of triple products of E2 matrix elements (STM) related to $\langle Q^3 \cos(3\delta) \rangle$. The quadrupole invariants can be further related to the collective model variables $\beta$ (overall deformation parameter) and $\gamma$ (triaxiality parameter). It can be concluded that the ground states of the even-even mercury isotopes are weakly deformed with a $\beta$ value close to 0.15 and are consistent with an oblate like deformation [$\langle \cos(3\delta) \rangle \simeq -1$], while the excited $0^+$ states are more deformed.

These data are compared to the equivalent sums calculated from beyond-mean field

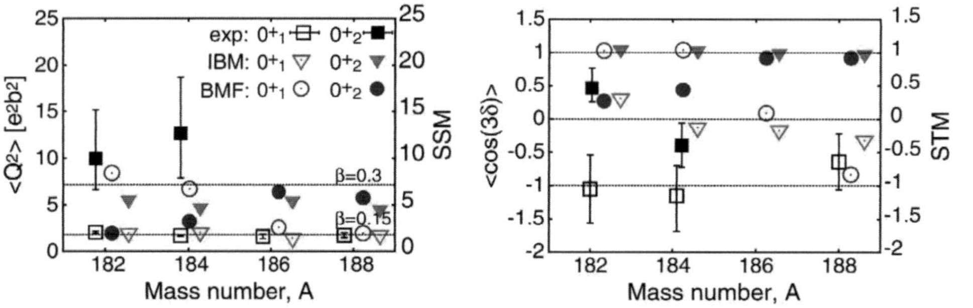


**Fig. 1.** The SSM and STM values of the $0_1^+$ (open symbols) and $0_2^+$ (full symbols) extracted from the experimental matrix elements [11] (black squares) are compared to the equivalent values from the BMF [18] (blue circles) and IBM [19] (red triangles) calculations.

(BMF) [18] and interacting-boson based models (IBM) [19] (Fig. 1). According to the BMF calculation, for N > 104 the Hg ground states are predicted to be predominantly oblate and the first excited $0^+$ state to be prolate, whereas for $100 \leq N \leq 104$, the ground state is predominantly prolate and the excited $0^+$ state is an almost equal mixture of prolate and oblate configurations [18]. For the latter, this is inverted with respect to experiment. In the IBM approach [19], whereby particle-hole pair excitations across the $Z$=82 closed shell are explicitly included, only partial agreement between experiment and theory is observed. The disagreements for the two models are not understood and point to missing ingredients in the calculations.

## 3. Octupole shapes and EDMs

Strong octupole correlations leading to pear shapes can arise when nucleons near the Fermi surface occupy states of opposite parity with orbital and total angular momentum differing by 3. This condition is met for proton number $Z \sim 34$, 56 and 88 and neutron number $N \sim 34$, 56, 88 and 134. The largest array of evidence for reflection asymmetry is seen at the values of $Z \sim 88$ and $N \sim 134$, where phenomena such as interleaved positive-

and negative-parity rotational bands in even-even nuclei, parity doublets in odd-mass nuclei, and enhanced $E1$ transition moments have been observed [20]. Many theoretical approaches have been developed to describe the observed experimental features: shell-corrected liquid-drop models, mean-field approaches using various interactions, models that assume $\alpha$-particle clustering in the nucleus, algebraic models and other semi-phenomenological approaches.

Atoms with octupole-deformed nuclei are very important in the search for permanent atomic electric-dipole moments (EDMs). The observation of a non-zero EDM at the level of contemporary experimental sensitivity would indicate time-reversal ($T$) or equivalently charge–parity ($CP$) violation due to physics beyond the standard model. Experimental limits on EDMs provide important constraints on many proposed extensions to the standard model. For a neutral atom in its ground state, the Schiff moment (the electric-dipole distribution weighted by radius squared) is the lowest-order observable nuclear moment. Octupole-deformed nuclei with odd nucleon number $A$ will have enhanced nuclear Schiff moments owing to the presence of the large octupole collectivity and the occurrence of nearly degenerate parity doublets that naturally arise if the deformation is static [21,22].

## 4. Electric octupole moments and recent experiments

In order to determine the shape of nuclei, the rotational model can be used to connect the intrinsic deformation, which is not directly observable, to the electric charge moments that arise from the non-spherical charge distribution. The $E3$ transition moment is collective in behaviour (> 10 single particle units) and is largely insensitive to single-particle effects, as it is generated by coherent contributions arising from the quadrupole-octupole shape. The $E3$ moment is therefore an observable that should provide direct evidence for enhanced octupole correlations and, for deformed nuclei, can be related to the intrinsic octupole deformation parameters. Twenty years ago Coulomb excitation was applied to the detailed measurements of the octupole shape of $^{148}$Nd in the $Z \sim 56$, $N \sim 88$ region [23] and of $^{226}$Ra in the $Z \sim 88$ and $N \sim 134$ region [24], but measurements of other nuclei in the latter mass region have had to wait for the development of accelerated radioactive beams.

Experiments have been carried out that used REX-ISOLDE to accelerate beams of radon and radium to bombard various targets in the MINIBALL $\gamma$-ray spectrometer. The measured values [25] of the $E2$ and $E3$ matrix elements in $^{220}$Rn and $^{224}$Ra are all consistent with the geometric predictions expected from a rotating, deformed distribution of electric charge, although these data do not distinguish whether the negative-parity states arise from the projection of a quadrupole-octupole deformed shape or from an octupole oscillation of a quadrupole shape. Figure 2 compares the experimental values of $Q_\lambda$ derived from the matrix elements connecting the lowest states for nuclei near $Z = 88$ and $N = 134$ measured by Coulomb excitation. It is striking that while the $E2$ moment increases by a factor of 6 between $^{208}$Pb and $^{234}$U, the $E3$ moment changes by only 50% in the entire mass region. Nevertheless, the larger $Q_3$ values for $^{224}$Ra and $^{226}$Ra indicate an enhancement in octupole collectivity that is consistent with an onset of octupole deformation in this mass region. On the other hand, $^{220}$Rn has similar octupole strength to $^{208}$Pb, $^{230,232}$Th and $^{234}$U, consistent with it being an octupole vibrator. In the case of a vibrator, the coupling of an octupole phonon to the ground state rotational band will give

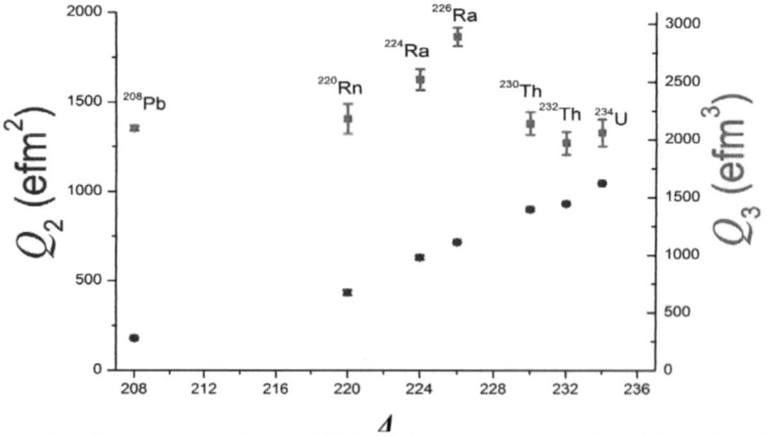


**Fig. 2.** The systematics of measured $E2$ and $E3$ intrinsic moments $Q_\lambda$ for $\lambda \rightarrow 0$ transitions. See table 2 in reference [25] for details.

zero values for matrix elements such as $<1^-||E3||4^+>$, because an aligned octupole phonon would couple the $4^+$ state to a $7^-$ state. Although the radioactive beam experiments do not have sensitivity to this quantity, this effect has been observed for $^{148}$Nd [23], consistent with behaviour expected for an octupole vibrator. In contrast the intrinsic moment derived from the measured $<1^-||E3||4^+>$ in $^{226}$Ra [24] is similar to that derived from the value of $<0^+||E3||3^->$, as expected for a static deformed system.

The values of $Q_3$, deduced from the measured transition matrix elements, are plotted in figure 3 as a function of $N$ and compared to various theoretical calculations. The trend of the experimental data is that the values decrease from a peak near $^{226}$Ra with decreasing $N$ (or $A$). This is in marked contrast to the predictions of the cluster model calculations [26]. It is also at variance with the Gogny HFB [27], Woods-Saxon macroscopic-microscopic [28] and relativistic Hartree-Bogoliubov (mapped onto an IBM Hamiltonian) [29] mean-field predictions of a maximum for $^{224}$Ra, although the agreement with the measurements for $^{220}$Rn and $^{224}$Ra by themselves is quite good. As can be seen, the relativistic mean field calculations [30] predict that the maximum value of $Q_3$ occurs for radium isotopes between $A$ = 226 and 230, depending on the parameterization, and Skyrme Hartree-Fock calculations [31] predict that $^{226}$Ra has the largest octupole deformation, consistent with the data.

The recent measurements of $Q_3$ values in $^{220}$Rn and $^{224}$Ra are also consistent with suggestions from the systematic studies of energy levels [32] that the even-even isotopes $^{218-222}$Rn and $^{220}$Ra have vibrational behaviour while $^{222-228}$Ra have octupole-deformed character. The parity doubling condition that leads to enhancement of the Schiff moment is therefore unlikely to be met in $^{219,221}$Rn. On the other hand $^{223,225}$Ra, having parity doublets separated by ~50 keV, will have large enhancement of their Schiff moments, provided that the states of opposite parity have the same intrinsic structure.

## 5. Future developments

Future experiments at the HIE-ISOLDE facility [33] at CERN will study $^{182,184}$Hg at higher energies, in order to carry out stringent tests of beyond-mean field and algebraic-model based calculations, and carry out new measurements of $B(E3)$s in $^{222,228}$Ra and $^{222-226}$Rn, which should greatly enhance our knowledge of the systematic behaviour of $E3$ moments in this mass region and help us to assess the likelihood of parity doubling in the odd-$A$ nuclei. There are also plans to measure directly the low-lying structure of odd-mass Rn isotopes, from in-beam conversion-electron and $\gamma$-ray measurements following Coulomb excitation or from the decay of the astatine parent. These measurements are vital in order to identify the best candidates for the radon EDM programme. More challenging is the measurement of the inter-band $E3$ matrix elements

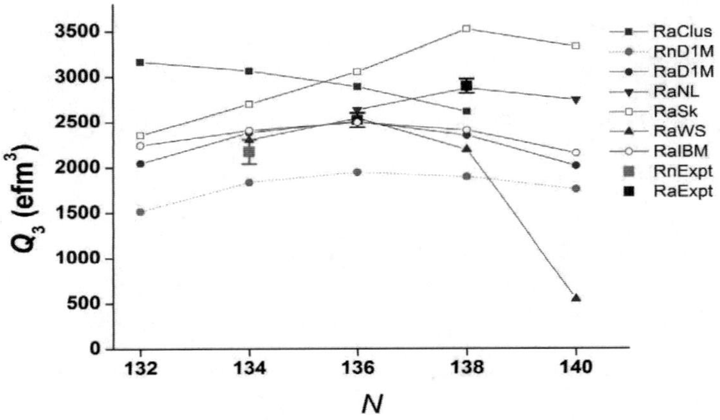


**Fig. 3.** Values of for low-lying transitions as a function of *N*. Measured [24,25] values for $3^- \rightarrow 0^+$ transitions are compared to various theoretical models: cluster model [26], Gogny HFB with D1M parameterization [27], shell-corrected liquid drop models WS [28], relativistic Hartree-Bogoliubov (mapped onto an IBM Hamiltonian) [29], relativistic mean field NL [30], and Skyrme HF SkO' [31].

between the low-lying states in odd-mass nuclei, required for both radon and radium EDM programmes. This will require advances in precision spectroscopy. An approach advocated here is to employ (d,d') scattering [34] to populate low-lying states in odd-*A* nuclei via single-step excitation. We have carried out simulations that indicate that ~ 20 keV energy resolution for the final states can be reached by using a helical orbit spectrometer (HELIOS) such as that developed at the Argonne National Laboratory [35], provided that the radioactive heavy beams are cooled in a storage ring. The cooling should reduce the transverse emittance of the beam to 0.01 mm mrad or better and the FWHM energy spread should be 0.025% or better. Such beam characteristics can be achieved using the Heidelberg Test Storage Ring (TSR) [36] that can be coupled to HIE-ISOLDE. In order to reduce the contribution to the final energy spread from multiple scattering and energy loss in the target, several targets of thickness 20 μg/cm$^2$ will be employed with separations along the beam axis sufficient to be identified by time-of-flight. The other requirement is that the position resolution of the linear silicon array should be at least 1 mm along and perpendicular to the beam direction, and the intrinsic energy resolution should be 10 keV or better. The required homogeneity in the magnetic field in HELIOS ($10^{-3}$) is easily achieved in MRI-type magnets.

## 6. Conclusion

We have demonstrated that radioactive beams of heavy nuclei with $A \approx 200$ can be successfully accelerated with sufficient intensity to extract precise values of electric-quadrupole and electric-octupole moments. Comparing our data with predictions of *E*2 and *E*3 strengths from recent models, we find that there are detailed differences with most calculations. We have confirmed the suitability of Ra isotopes with $N$~136 for EDM searches. Further studies of Hg, Rn and Ra isotopes will be extended at the HIE-ISOLDE facility.

**Acknowledgments**

I gratefully acknowledge the help of N Bree, K Wrzosek-Lipska and A Petts who analysed the $^{182\text{-}188}$Hg Coulex data, L Gaffney who analysed the $^{220}$Rn and $^{224}$Ra Coulex data, and who with M Hackstein analysed the $^{184,186}$Hg lifetime data, M Scheck and N Warr who played key roles in preparing the Coulex experiments, T Stora and F Wenander who led the teams that developed radioactive ion beams, D Cline, M Huyse, D Jenkins and P Van Duppen who are my long-time collaborators, and the other members of the collaboration (see references 11 and 25) who assisted in these experiments. This work was supported by the UK STFC.

## References

[1] Julin R Helariutta K and Muikku M 2001 *J. Phys. G* **27** R109
[2] Heyde K and Wood JL 2001 *Rev. Mod. Phys.* **83** 1467
[3] Bonn J et al. 1972 *Phys. Lett. B* **38** 308
[4] Frauendorf S and Pashkevich V V 1975 *Phys. Lett. B* **55** 365
[5] Bonn J, Huber G, Kluge H-J and Otten E W 1976 *Z. Phys. A* **276** 203
[6] Van Duppen P *et al.* 1984 *Phys. Rev. Lett.* **52** 1974
[7] Wood J L *et al.* 1992 *Phys. Rep.* **215** 101
[8] Bijnens N *et al.* 1996 *Z. Phys. A* **356** 3
[9] Jenkins D G *et al.* 2000 *Phys. Rev. C* **62** 021302(R)
[10] Andreyev A N *et al.* 2000 *Nature* **405** 430
[11] Bree N *et al.* 2014 *Phys. Rev. Lett.* **112** 162701
[12] Warr N *et al.* 2013 Eur. Phys. J. A **49** 40
[13] Czosnyka T Cline D and Wu CY 1982 *Bull. Am. Phys. Soc.* **28** 745
[14] Rapisarda E *et al.* (to be published)
[15] Scheck M *et al.* 2010 *Phys. Rev. C* **81** 014310
[16] Gaffney LP *et al.* 2014 *Phys. Rev. C* **89** 024307
[17] Kumar K 1972 *Phys. Rev. Lett.* **28** 249
[18] Yao JM Bender M and Heenen PH 2013 *Phys. Rev. C* **87** 034322
[19] García-Ramos JE and Heyde K 2014 *Phys. Rev. C* **89** 014306
[20] Butler PA and Nazarewicz W 1996 *Rev. Mod. Phys.* **68** 349
[21] Auerbach N Flambaum VV and Spevak V 1996 *Phys. Lett. Lett.* **76** 4316
[22] Dobaczewski J and Engel J 2005 *Phys. Rev. Lett.* **94** 232502
[23] Ibbotson RW *et al.* 1997 *Nucl. Phys. A* **619** 213
[24] Wollersheim HJ *et al.* 1993 *Nucl. Phys. A* **556** 261
[25] Gaffney LP *et al.* 2013 Nature **497** 199
[26] Shneidman TM *et al.* 2003 *Phys. Rev. C* **67** 014313
[27] Robledo LM and Butler PA 2013 *Phys. Rev. C* **88** 051302(R)
[28] Nazarewicz W *et al.* 1984 *Nucl. Phys. A* **429** 269
[29] Nomura K *et al.* 2014 *Phys. Rev. C* **89** 024312
[30] Rutz K *et al.* 1995 *Nucl. Phys. A* **590** 680
[31] Engel J *et al.* 2003 *Phys. Rev. C* **68** 025501
[32] Cocks JFC *et al.* 1999 *Nucl. Phys. A* **645** 61
[33] Lindroos M Butler PA Huyse M and Riisager K 2008 *Nucl. Instrum. Meth. B* **266** 4687
[34] Thorsteinsen TF Nybø K and Løvhøiden G 1990 *Physica Scripta* **42** 141
[35] Lighthall JC *et al.* 2010 *Nucl. Instrum Meth. A* **622** 97
[36] M Gieser M *et al.* 2012 *Eur. Phys. J. Special Topics* **207** 1

Journal of Physics: Conference Series **590** (2015) 012017 doi:10.1088/1742-6596/590/1/012017

# Complete electric dipole response in $^{120}$Sn and $^{208}$Pb and implications for neutron skin and symmetry energy

**Peter von Neumann-Cosel**

Institut für Kernphysik, Technische Universität Darmstadt, 64289 Darmstadt, Germany

E-mail: `vnc@ikp.tu-darmstadt.de`

**Abstract.** Polarized proton scattering at energies of a few 100 MeV and very forward angles including 0° has been established as a new tool to extract the complete E1 strength distribution in nuclei for excitation energies between about 5 and 20 MeV. A case study of $^{208}$Pb demonstrates excellent agreement with other electromagnetic probes. From the information on the B(E1) strength one can derive the electric dipole dipole polarizability, which is strongly correlated to the neutron skin and to parameters of the symmetry energy. Recently, we have extracted the polarizability of $^{120}$Sn with a comparable precision. The combination of both results further constrains the symmetry energy parameters and presents a challenge for mean-field models, since relativistic and many Skyrme parameterizations cannot reproduce both experimental results simultaneously.

## 1. Introduction

The electric dipole (E1) response of nuclei is dominated by the giant dipole resonance (GDR), a collective mode at high excitation energy which exhausts most of the energy-weighted sum rule [1]. Properties of the GDR are fairly well understood but recent interest focuses on evidence for a soft mode in neutron-rich nuclei at lower excitation energies termed pygmy dipole resonance (PDR), for a recent review see Ref. [2]. Because of the saturation of nuclear density, excess neutrons are expected to concentrate on the surface forming a skin whose oscillations against an $N \approx Z$ core should give rise to a low-energy E1 mode [3]. Therefore, the PDR may shed light onto the formation of neutron skins in nuclei [4]. Another quantity related to nuclear E1 modes is the symmetry energy acting as restoring force. The E1 strength distribution carries information on its magnitude and density dependence [5]. The neutron matter equation of state in turn is a key ingredience for the modeling of the equilibrium properties of neutron stars [6].

Studies of energy density functionals (EDFs) using Skyrme forces or a relativistic framework suggest the nuclear dipole polarizability $\alpha_D$ as an observable carrying information on both neutron skin and symmetry energy [7]. The polarizability is related to the photoabsorption cross section $\sigma_{abs}$ by [8]

$$\alpha_D = \frac{\hbar c}{2\pi^2 e^2} \int \frac{\sigma_{abs}}{\omega^2} d\omega, \tag{1}$$

where $\omega$ denotes the photon energy. Because of the inverse energy weighting knowledge of the E1 strength at low energies – despite its smallness – contributes substantially to the polarizability.

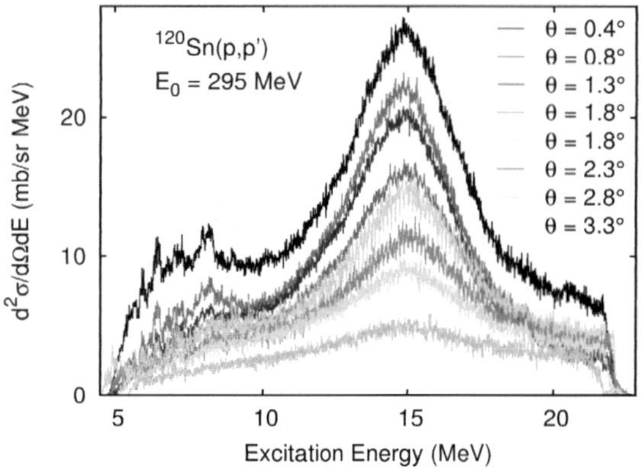


**Figure 1.** Differential cross sections of the $^{120}$Sn(p,p) reaction at $E_0 = 295$ MeV and very forward angles

Data on the PDR in very neutron-rich nuclei are still scarce [9, 10, 11, 12]. Stable nuclei at different shell closures have been explored with the $(\gamma, \gamma')$ reaction [2]. However, the centroid of the PDR in heavy nuclei typically lies in the vicinity of the neutron emission threshold $S_n$ and most of the experimental techniques sensitive to the E1 strength provide data either below or above $S_n$ only. This contribution discusses a novel approach, viz. polarized proton scattering at angles close to and including 0°, allowing to extract the *complete E1 response in nuclei* covering the ecitation region of PDR and GDR.. Here, besides published data on $^{208}$Pb first results for $^{120}$Sn are shown and the impact of the combined results on EDFs is explored.

## 2. Experiments

The experimental techniques are described in Ref. [13] and details of the $^{208}$Pb measurement in Ref. [14]. Spectra of the $^{120}$Sn(p,p) reaction at $E_0 = 295$ MeV and forward angles are displayed in Fig. 1. They are dominated by the broad GDR peak centered at about 15 MeV. Towards larger angles the differential cross sections exhibit a fast fall-off characteristic for Coulomb excitation. Between 5 and 9 MeV a structured bump is visible, whose cross sections also decrease with angle. It is expected to contain the PDR but also contributions from spin-flip M1 excitations. For proton energies of 200 − 400 MeV the nuclear cross sections at small momentum transfers are dominated by isovector spinflip-M1 transitions [15], whose cross sections also peak at 0° because of the $\Delta L = 0$ angular momentum transfer.

## 3. Decomposition of E1/M1 cross sections

An extraction of the cross sections due to M1 and E1 excitations was performed in two independent ways based on a multipole decomposition of cross section angular distributions (MDA) and on the measurement of polarization transfer observables (PTA). The MDA is explained in detail in Refs. [16, 17]. It is based on model calculations using the code DWBA07, microscopic transition densities from the quasiparticle-phonon model (QPM), and the Love-Franey proton-nucleus effective interaction [18]. In contrast to the case of $^{208}$Pb, in $^{120}$Sn due to the higher level density no individual transitions could be resolved at low excitation energy despite the good energy resolution of about 25 keV (FWHM). Therefore the MDA was performed in energy bins of 200 keV (400 keV above 12 MeV). Under such conditions no sizable contributions to the cross sections from multipoles other than E1 or M1 are indicated by the experimental angular distributions except the excitation of the isoscalar giant quadrupole resonance (GQR). Its contributions to the cross sections were determined by normalizing the QPM predictions for the GQR to the experimental isocalar E2 strength distribution taken from

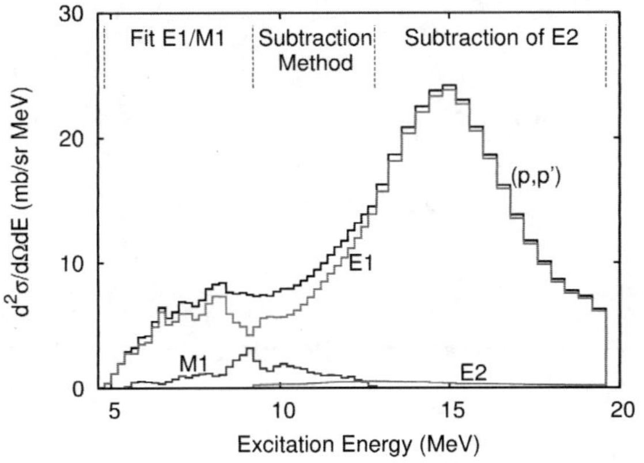


**Figure 2.** Result of the multipole decomposition for $^{120}Sn$ into E1 and M1 cross sections. E2 cross sections have been calculated with the code DWBA07 using the experimental strength distribution of Ref. [19] and the Love-Franey effective interaction [18] and subtracted prior to the MDA.

Ref. [19] and subtracting the resulting cross sections from the spectra at all angles prior to the MDA. The decomposition into E1 and M1 cross sections is shown in Fig. 2 together with the E2 cross sections resulting from the above described procedure.

Alternatively, the measurement of the polarization transfer coefficients $D_{SS}$, $D_{NN}$, and $D_{LL}$ can be used for the decomposition, where $S$, $N$, and $L$ denote sidewards, normal, and longitudinal polarization spanning a coordinate system defined with respect to the moving incident proton. The combination of these polarization transfers observables takes specific values for spinflip ($-1$) and non-spinflip ($+3$) transitions [20]. At $0°$, $D_{SS}$ and $D_{NN}$ are indistinguishable and it suffices to measure on of them ($D_{SS}$ in the present case). The relation of Ref. [20] can then be rearranged to introduce a total spin transfer $\Sigma$

$$\Sigma = \frac{3 - (2D_{SS} + D_{LL})}{4}, \tag{2}$$

which takes a value of 0 for $\Delta S = 0$ (non-spinflip) and 1 for $\Delta S = 1$ (spinflip) transitions. The different reaction mechanisms allow an idenfication of spinflip with M1 and non-spinflip with E1 transitions, respectively.

Figure 3 presents a comparison of the results from MDA and PTA in the excitation energy region up to 12.5 MeV (at higher $E_x$ the M1 contribution is negligible). Good agreement of the two independent methods is obtained except for the energy region between 8 and 9 MeV, where the PTA yields somewhat larger $\Delta S = 1$ cross sections although both methods still agree within error bars.

## 4. Polarizability

The E1 cross sections of Fig. 2 can be converted into reduced B(E1) transition strengths with the aid of the semiclassical model of Coulomb excitation [21]. The electric dipole polarizability of $^{120}Sn$ can then be determined using Eq. (1). Additional data from excitation energy regions not accessible in the present experiment were also considered. Below 5 MeV, the E1 strength was taken from the $(\gamma, \gamma)$ data [22]. Above 29 MeV, photoabsorption cross sections from a natural Sn target [23] were taken into account assuming that the absolute cross sections and energy dependence are equal for all stable tin isotopes. In the giant resonance region, photoabsorption cross sections deduced from the present work were averaged with data from ($\gamma$,n) [24] and ($\gamma$,xn) [25, 26] experiments.

The resulting polarizability of $^{120}Sn$ is plotted in Fig. 4 versus the polarizability of $^{208}Pb$. The yellow bands indicate the respective experimental uncertainties. The predictions of a variety of

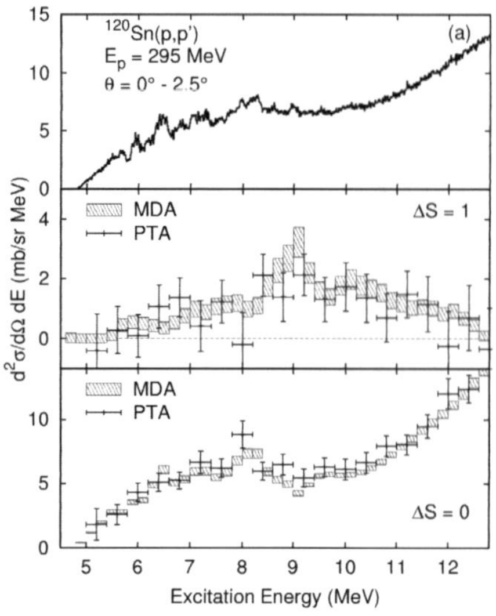


**Figure 3.** Comparison of the decomposition of E1/M1 cross sections for $^{120}$Sn using MDA and PTA.

EDFs based on Skyrme forces for these two quantities a shown in comparison. A description of SV-min and SV-Bas can be found in Ref. [27]. The open circle shows the results from latest attempt of the UNEDEF collaboration aiming at the determination of an optimal EDF [28]. References to the other forces are summarized in Ref. [29].

While some models (SKT6, SG-II) generally predict much too large values of $\alpha_D$ for both nuclei, most Skyrme forces are capable to reproduce the experimental result for $^{208}$Pb but find a somewhat too large value for $^{120}$Sn. Exceptions are SLy6 and BSK4 which predict polarizabilities compatible with both values. There is no estimate of the theoretical uncertainties except for SV-min and SV-bas, where a corresponding procedure is decribed in Ref. [27]. By way of construction the errors are much larger for SV-min than for SV-bas. However, for both models there is some overlap with the experimental data point for $^{120}$Sn taking these error bars into account.

Figure 5 depicts the same correlation as Fig. 4 but in comparison to relativistic mean-field models. Specifically, results for PC (full symbols) [30], DD-ME2 (open symbols) [31] and FSU (cross) [4] are shown. In the relativistic models one can study the dependence on the symmetry

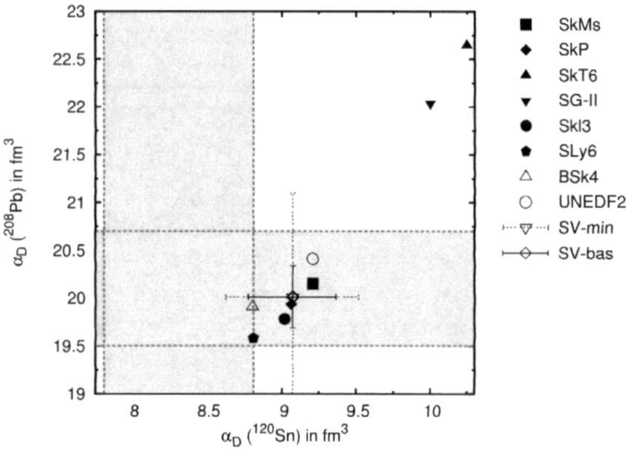


**Figure 4.** Polarizabilites of $^{208}$Pb and $^{120}$Sn in comparison to EDFs based on Skyrme forces.

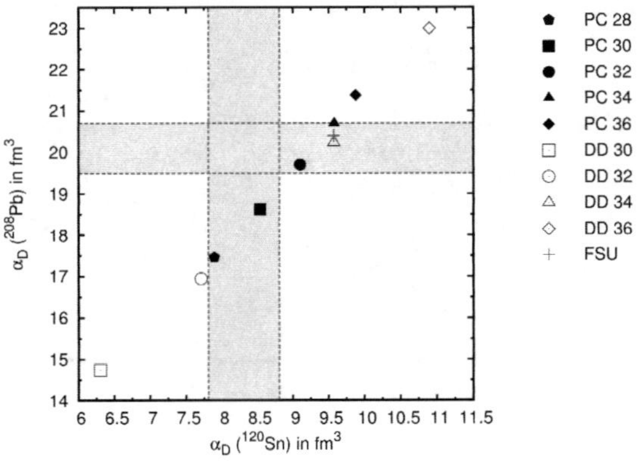


**Figure 5.** Polarizabilites of $^{208}$Pb and $^{120}$Sn in comparison to EDFs based on relativistic models.

energy by varying its magnitude described by the parameter $J$ (also callled $a_0$) [5]. The summary of presently available results favor a value $J \simeq 30-35$ MeV [32]. The effect of a variation between $J = 28$ and 36 MeV is demonstrated in Fig. 5. All results approximately fall on a straight line, i.e., a proportionality of both $\alpha_D$ values with $J$ is observed. For $^{208}$Pb, such a correlation was already established by Roca-Maza et al. [33]. However, the absolute value of $J$ needed to describe $\alpha_D(^{208}\mathrm{Pb})$ differs for the PC and DD-ME2 forces. Furthermore, all three models fail to describe both polarizabilities simultaneously. This implies that the density dependence of the symmetry energy predicted by the relativistic models is not correct.

## 5. Concluding Remarks

Polarized proton scattering at very forward angles allows to study, with high energy resolution, the complete electric dipole response of nuclei from low excitation energies up to the GDR. The E1 strength distribution deduced in a benchmark experiment on $^{208}$Pb is in excellent agreement with all available data. A precise value for the E1 polarizability can be extracted with important consequences for a determination of the neutron skin and the symmetry energy in neutron-rich nuclei. Since the correlation between polarizability, neutron skin thickness and symmetry energy ($\Delta r_{np} \propto \alpha_D \cdot J$ [34],) is model-dependent, experimental tests in other nuclei are important as a constraint of the poorly determined isovector interaction in EDFs.

The present work presents a first result for the semimagic nucleus $^{120}$Sn. The correlation between both polarizabilites reveals a sytematic problem of RMFs, which cannot reproduce both values simultaneously irrespective of the choice of the symmetry energy strength. This points to a problem of RMF models to provide the correct density dependence of the symmetry energy, which is crucial for an extrpolation into the high-density regime of neutron stars. EDFs based on Skyrme forces tend to give a somewhat too large value for $^{120}$Sn but the agreement is probably still acceptable taking theoretical uncertainties into account. In any case, in view of these new results a systematic study of the dipole polarizability over a wider range of nuclei is of utmost importance and work along these lines is underway.

The $(p, p')$ data also provide information on other interesting questions not dicussed here because of lack of space. Utilizing isospin symmetry, the M1 cross sections resulting from the MDA and PTA decomposition can be converted into reduced B(M1) transition strengths [35]. This promises systematic information on the hardly explored spin-M1 resonance in heavy nuclei [15, 36]. Excellent agreement with results from eletromagnetic probes [37] is found for $^{208}$Pb. Furthermore, because of the high energy resolution fine structure is systematically observed in the excitation region of the GDR. With a wavelet analysis [38, 39] one can extract information on important decay mechanisms of the GDR and the level density of $J^{\pi} = 1^{-}$ states [40].

## Acknowledgements

I am grateful to all collaborators of experiment E316 and, in particular, T. Hashimoto, A.M. Krumbholz and A. Tamii for their important contributions to the present work. P.-G. Reinhard is thanked for providing the thoeretical results on the polarizabilites of $^{208}$Pb and $^{120}$Sn. This work was supported by the DFG under contracts SFB 634 and NE 679/3-1.

## References

[1] Berman B L and Fultz S C 1975 *Rev. Mod. Phys.* **47** 713
[2] Savran D, Aumann T and Zilges A, 2013 *Prog. Part. Nucl. Phys.* **70** 210
[3] Paar N, Vretenar D, Khan E and Colò G 2007 *Rep. Prog. Phys.* **70** 691
[4] Piekarewicz J 2006 *Phys. Rev. C* **73** 044325
[5] Tsang M B *et al.* 2012 *Phys. Rev. C* **86** 015803
[6] Horowitz C J and Piekarewicz J 2001 *Phys. Rev. Lett.* **86** 5647
[7] Piekarewicz J *et al.* 2012 *Phys. Rev. C* **85** 041302(R)
[8] Bohigas O, Van Giai N and Vautherin B 1981 *Phys. Lett. B* **102** 105
[9] Adrich P *et al.* 2005 *Phys. Rev. Lett.* **95** 132501
[10] Klimkiewicz A *et al.* 2007 *Phys. Rev. C* **76** 051603(R)
[11] Wieland O *et al.* 2009 *Phys. Rev. Lett.* **102** 092502
[12] Rossi D M *et al.* 2013 *Phys. Rev. Lett.* **111** 242503
[13] Tamii A *et al.*, *Nucl. Instrum. Meth. A* **605** 3
[14] Poltoratska I 2011 Doctoral thesis D17, Technische Universität Darmstadt
[15] Heyde K, von Neumann-Cosel P and Richter A 2010 *Rev. Mod. Phys.* **82** 2365
[16] Tamii A *et al.* 2011 *Phys. Rev. Lett.* **107** 062502
[17] Poltoratska I *et al.* 2012 *Phys. Rev. C* **85** 041304(R)
[18] Love W G and Franey M A 1981 *Phys. Rev. C* **24** 1073; Franey M A and Love W G 1985 *Phys. Rev. C* **31** 488
[19] Li T *et al.* 2010 *Phys. Rev. C* **81** 034309
[20] Suzuki T 2000 *Prog. Theo. Phys.* **103** 859
[21] Bertulani C A and Baur G 1988 *Phys. Rep.* **163** 299
[22] Özel-Tashenov B *et al.* 2014 *Phys. Rev. C* **90** 024304
[23] Leprêtre A *et al.* 1981 *Nucl. Phys. A* **367** 237
[24] Utsunomiya H *et al.* 2011 *Phys. Rev. C* **84** 055805
[25] Fultz S C *et al.* 1969 *Phys. Rev.* **186** 1255
[26] Leprêtre A *et al.* 1974 *Nucl. Phys. A* **219** 39.
[27] Reinhard P-G and Nazarewicz W 2013 *Phys. Rev. C* **87** 014324
[28] Kortelainen M *et al.* 2014 *Phys. Rev. C* **89** 054314.
[29] Bender M, Heenen P-H and Reinhard P-G 2003 *Rev. Mod. Phys.* **75** 121
[30] Zhao P W, Li Z P, Yao J M and Meng J 2010 *Phys. Rev. C* **82** 054319
[31] Lalazissis G A, Nikšić T, Vretenar D and Ring P 2005 *Phys. Rev. C* **71** 024312
[32] Tamii A, von Neumann-Cosel P and Poltoratska I 2014 *Eur. Phys. J. A* **50** 28
[33] Roca-Maza X *et al.* 2013 *Phys. Rev. C* **88** 024316
[34] Satuła W, Wyss R A and Rafalski M 2006 *Phys. Rev. C* **74** 011301(R)
[35] Birkhan J *et al.* to be published.
[36] Frekers D *et al.* 1990 *Phys. Lett. B* **244** 178
[37] Laszewski R M, Alarcon A, Dale D S and Hoblit S D 1988 *Phys. Rev. Lett.* **61** 1710
[38] Shevchenko A *et al.* 2004 *Phys. Rev. Lett.* **93** 122501
[39] Shevchenko A *et al.* 2008 *Phys. Rev. C* **77** 024302
[40] Poltoratska I *et al.* 2014 *Phys. Rev. C* **89** 054322

Journal of Physics: Conference Series **590** (2015) 012018 doi:10.1088/1742-6596/590/1/012018

# Known and exotic light hypernuclei studied with the $^{6}$Li+$^{12}$C reaction at 2 $A$ GeV

**Take R. Saito**
GSI Helmholtz Centre for Heavy Ion Research, Planckstrasse 1, 64291 Darmstadt, Germany
Johannes Gutenberg-Universität Mainz, J.J.Becherweg 40, 55099 Mainz, Germany
The Helmholtz Institute Mainz (HIM), J.J.Becherweg 40, 55099 Mainz, Germany
Iwate University, Morioka, Iwate 020-8550, Japan

E-mail: t.saito@gsi.de

**Vakkas Bozkurt**
Nigde University, 51100 Nigde, Turkey
GSI Helmholtz Centre for Heavy Ion Research, Planckstrasse 1, 64291 Darmstadt, Germany

**Christophe Rappold**
Justus-Liebig-Universität Giessen, Heinrich-Buff-Ring 16, 35392 Giessen, Germany
GSI Helmholtz Centre for Heavy Ion Research, Planckstrasse 1, 64291 Darmstadt, Germany

**and the HypHI collaboration**

**Abstract.** Light hypernuclei have been studied by using induced reactions of $^{6}$Li projectiles bombarded on the natural carbon target by means of the invariant mass method, and $^{3}_{\Lambda}$H and $^{4}_{\Lambda}$H hypernuclei as well as Λ-hyperons are produced and identified. Significance values of $^{3}_{\Lambda}$H, $^{4}_{\Lambda}$H and Λ are respectively $4.7\sigma$, $4.9\sigma$ and $6.7\sigma$. By analyzing decay vertices of $^{3}_{\Lambda}$H, $^{4}_{\Lambda}$H and Λ by means of unbinned maximum likelihood fitting their lifetime values have been deduced to be $183^{+42}_{-32}$ ps, $140^{+48}_{-33}$ ps and $262^{+56}_{-43}$ ps, respectively. In addition, signals appeared for the d+$\pi^-$ and t+$\pi^-$ final states in the analyses on invariant mass and lifetime. Those signals have suggested a possible bound state of two neutrons with Λ, $^{3}_{\Lambda}$n. It has also been revealed for the first time that the lifetime of $^{3}_{\Lambda}$H is significantly shorter than that of Λ, which is consistent to the statistical analyses including all the lifetime data reported in the past. However, there is no theoretical model to explain the short lifetime of $^{3}_{\Lambda}$H as well as the existence of the $^{3}_{\Lambda}$n bound state. These should be further experimentally and theoretically studied.

## 1. Introduction

One of the important issues in nuclear and hadron physics is the comprehensive understanding of baryon-baryon interactions under the flavor SU(3) symmetry with up, down and strange quarks. A baryon involving at least one strange quark is called a hyperon, and the lightest hyperon is Λ. Because the Λ-hyperon decays via weak interaction with a lifetime of 263.2 ps [1], it has not been practical to study the nucleon-Λ and Λ-Λ interactions by direct reaction experiments with projectiles and targets involving the Λ-hyperon, thus a Λ-hypernucleus, a bound subatomic system containing a $\Lambda$, was studied to extract information on the nucleon-$\Lambda$ and $\Lambda$-$\Lambda$ interactions. About 40 kinds of $\Lambda$-hypernuclei have been so far investigated experimentally mainly with emulsion techniques [2], secondary meson beams [3] and primary electron beams [4]. In those experiments, the isospin of the produced hypernuclei was limited by the reaction mechanism since a nucleus in the stable target material is converted to a $\Lambda$-hypernucleus by production or exchange of strangeness in a single nucleon. While, a different approach to studying hypernuclei by using projectile fragmentation reactions of heavy ion beams was employed for the present work. In such reactions, a projectile fragment can capture a hyperon produced in the hot participant region to produce a hypernucleus. In this reaction, the produced hypernucleus has a large Lorentz factor, and the decay of the hypernucleus takes place well behind the production target. This makes it possible to study hypernuclei in flight. Since a hypernucleus is produced from a projectile fragment, isospin and mass values of the produced hypernuclei, unlike in other hypernuclear experiments, can be widely distributed in similar fashion.

The HypHI collaboration has proposed a series of experiments at the GSI Helmholtz Centre for Heavy Ion Research, using induced reactions of stable heavy ion beams and rare-isotope beams, with the aim of producing and measuring hypernuclei with an invariant mass method [5]. In the proposed experiments, charged particles and neutrons from the mesonic or non-mesonic weak decay of hypernuclei are tracked and identified in order to reconstruct the hypernuclear mass values. The lifetime of produced hypernuclei can be extracted by measuring of the proper time in the rest frame of the hypernuclear decay. This methodology also allows investigating neutron- and proton-rich hypernuclei as well as hypernuclei with more than two units of strangeness, and several hypernuclei can be studied in a single data taking.

The first experiment aiming to study ${}^{3}_{\Lambda}$H and ${}^{4}_{\Lambda}$H hypernuclei was performed by mean of projectile fragmentation reactions of ${}^{6}$Li projectiles at 2 $A$ GeV delivered on a carbon target. In the experiment, we have succeeded to observe signals of ${}^{3}_{\Lambda}$H and ${}^{4}_{\Lambda}$H as well as those in the d+$\pi^-$ and t+$\pi^-$ as the final states. The later two signals may indicate the existence of a possible bound state of two neutrons and a $\Lambda$-hyperon (${}^{3}_{\Lambda}$n). In this article, we discuss on the observation of these signals. The discussions below are based on our already published articles [6, 7, 8].

## 2. Experiment

The first HypHI experiment, Phase 0, took place at the GSI Helmholtz Centre for Heavy Ion Research. Details of the experimental setup and analyses can be found in [6]. Projectiles of ${}^{6}$Li at 2 $A$ GeV with an average intensity of $3\times10^6$ beam particle per second were propelled at a carbon graphite target 8.84 g/cm$^2$ thick. As discussed in [6], the experimental apparatus consisted of three tracking stations of scintillating fiber detector arrays, (*TR0, TR1, TR2*), and two drift chambers (*BDC, SDC*) for the displaced vertex measurement. As well, three scintillating hodoscope walls (*TOF+, TFW, ALADiN TOF*) were appended to the tracking systems for tracking, energy loss and time-of-flight measurements of charged particles across a large acceptance dipole magnet. The tracking system for vertexing was placed in front of the dipole magnet around the expected decay volume of hypernuclei, while the *SDC* drift chamber was set just behind the magnet. Two detached detection branches consisting of *TOF+* and *TFW & ALADiN TOF* hodoscope walls were situated behind the magnet in such a way to measure separately positively and negatively charged particles respectively.

At the trigger level, the event topology was selected in order to acquire and record potential events combining a displaced vertex, a Helium isotope and a $\pi^-$ meson as decay particles. The displaced vertex trigger used the information of *TR0*, *TR1* and *TR2* fiber detector arrays around the decay volume of hypernuclei and $\Lambda$ hyperons. The Helium isotope trigger used the time-over-threshold measurement of the energy loss from the *TOF+* hodoscope wall dedicated solely for the positively-charge particle and fragments. The $\pi^-$ trigger involved the hit detection on the *TFW* hodoscope wall dedicated exclusively for $\pi^-$ detection behind the dipole magnet.

This topological combination at hardware level was exploited in order to cope with the large difference of magnitude between the hypernuclear production cross section and the total reaction cross section.

The particle identification was based on the track reconstruction across the magnet, the measurements of the time-of-flight, and the energy deposit with the hodoscope walls. The four-vectors of the detected particles and fragments were then deduced. The invariant mass of final states of interest was calculated and a lifetime estimation was inferred from the observed decay vertex position after the decay vertex finding. Details of the analyses are discussed in [6].

## 3. Results

After the particle identification and the reconstruction of the decay vertexes, the invariant mass of the final states of interest has been deduced. Details of the analyses can be found in [6, 7]. The top panels of Figure 1 show the invariant mass distributions for p + $\pi^-$, $^3$He + $\pi^-$ and $^4$He + $\pi^-$ candidates. In each distribution, a peak at around the mass of $\Lambda$, $^3_\Lambda$H and $^4_\Lambda$H can be seen. The background contributions to the invariant mass distributions initially were estimated with the event mixing method [9, 10]. The background contribution estimated via mixed event analysis is shown in Figure 1 by the open triangles. The separated signal and background contributions have been estimated via the binned maximum likelihood fitting method. A signal in the invariant mass distributions is represented by a Gaussian probability density function. The background contribution was then modeled by a Chebychev polynomial probability density function of the first kind. An extended maximum likelihood estimator allowed us to obtain the share of signal and background in each invariant mass spectrum. Significance values of the signals are 6.7$\sigma$, 4.7$\sigma$ and 4.9$\sigma$ for $\Lambda$ hyperon, $^3_\Lambda$H and $^4_\Lambda$H, respectively.

After reconstructing the invariant mass, the lifetime of $\Lambda$, $^3_\Lambda$H and $^4_\Lambda$H was extracted. The signal contribution was determined by subtracting the background contribution from the signal-plus-background contribution. For this purpose, two data sets are built. The signal-plus-background contribution was deduced in the fitted peak region within the fitted values of $\bar{m} \pm 2\sigma_m$, and the background-only data set was extracted from the adjacent sideband region within the intervals of $[\bar{m} - 4\sigma_m, \bar{m} - 2\sigma_m]$ and $[\bar{m} + 2\sigma_m, \bar{m} + 4\sigma_m]$ with proper normalization. In each data set, the proper decay time $t = l/(\beta\gamma\, c)$ is then calculated in the rest frame of the mother state of interest from the measured decay length $l$, where $\beta\gamma\, c = p/m$, $p$ and $m$ the momentum and the mass of the mother state of interest. The yield at different decay lengths was also corrected by the deduced acceptance and reconstruction efficiency based on full Monte Carlo simulations. The lifetime values were then extracted utilizing an *unbinned* maximum likelihood fitting method. The profiled likelihood ratios for estimation of the mean decay length $c\tau$ of $\Lambda$ hyperon, $^3_\Lambda$H and $^4_\Lambda$H hypernuclei are shown in the middle panels of Figure 1. Deduced lifetime values for $\Lambda$, $^3_\Lambda$H and $^4_\Lambda$H are respectively $262^{+56}_{-43}$ ps, $183^{+42}_{-32}$ ps and $140^{+48}_{-33}$ ps. The goodness of the fitting was represented by the $\chi^2$ test by using the modeled proper decay time function with the background [6], and they are shown in the bottom panels of Figure 1.

Analyses with the same manner were applied for all possible final states, and all the final states were investigated. Surprisingly. signals involving the d+$\pi^-$ and t+$\pi^-$ final states have been appeared [7], as shown in panels (a1) and (a2) of Figure 2. Significance values of these signals are respectively 5.0 $\sigma$ and 5.3 $\sigma$. The lifetime values of the state finalizing to d+$\pi^-$ and t+$\pi^-$ have been deduced in the same manner for $\Lambda$, $^3_\Lambda$H and $^4_\Lambda$H, and they are respectively $181^{+30}_{-24}$ ps and $190^{+47}_{-35}$ ps [7]. The results of the lifetime analyses are also shown in Figure 2.

After the careful investigations of possibility of creating artificial peaks and the mis-reconstruction of multi-body decays of the other hypernuclei [7], we have suggested that those signals are from the two- and three-body decays of an unknown neutral hypernuclear bound state with two neutrons and one $\Lambda$-hyperon, $^3_\Lambda$n [7].

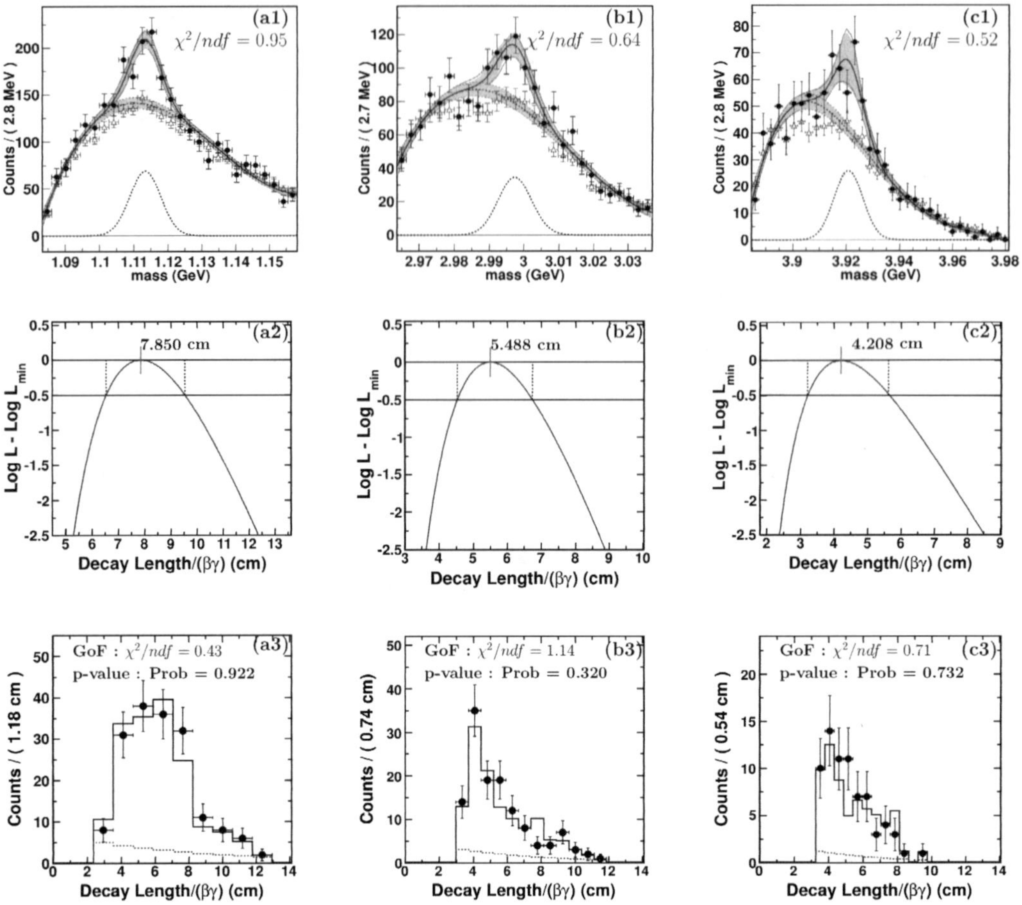


**Figure 1.** (Color online) Invariant mass distribution for candidates of $\Lambda$, ${}^3_\Lambda$H and ${}^4_\Lambda$H, are represented by the filled circles in panels (a1), (b1), and (c1), respectively. The shaded orange region represents one standard deviation of the fitted model centered at the solid blue line. The dotted lines show the separate contributions of the signal and the background with, respectively, black and colored lines. The data represented by open triangles correspond to invariant mass distributions of the mixed event analysis. Profiled likelihood ratio for interval estimation of $\Lambda$ hyperon (a2), ${}^3_\Lambda$H (b2), and ${}^4_\Lambda$H hypernuclei (c2) is shown. Interval estimation for 1 standard deviation is shown on each profiled likelihood ratio. *Binned* decay length distributions of the signal region of $\Lambda$ hyperon (a3), ${}^3_\Lambda$H (b3), and ${}^4_\Lambda$H hypernuclei (c3) with the fitted model, which included the exponential function resulted from the *unbinned* maximum likelihood fit and the background contribution estimated by the sidebands. The black line represents the fitted model, while the blue dotted line represents the contribution of the exponential function. Figures are taken from [6] and rearranged.

## 4. Discussion

There are two striking results presented by the current work. First, as discussed above, the lifetime of ${}^3_\Lambda$H has been observed to be $183^{+42}_{-32}$ ps, which is significantly shorter than that of the

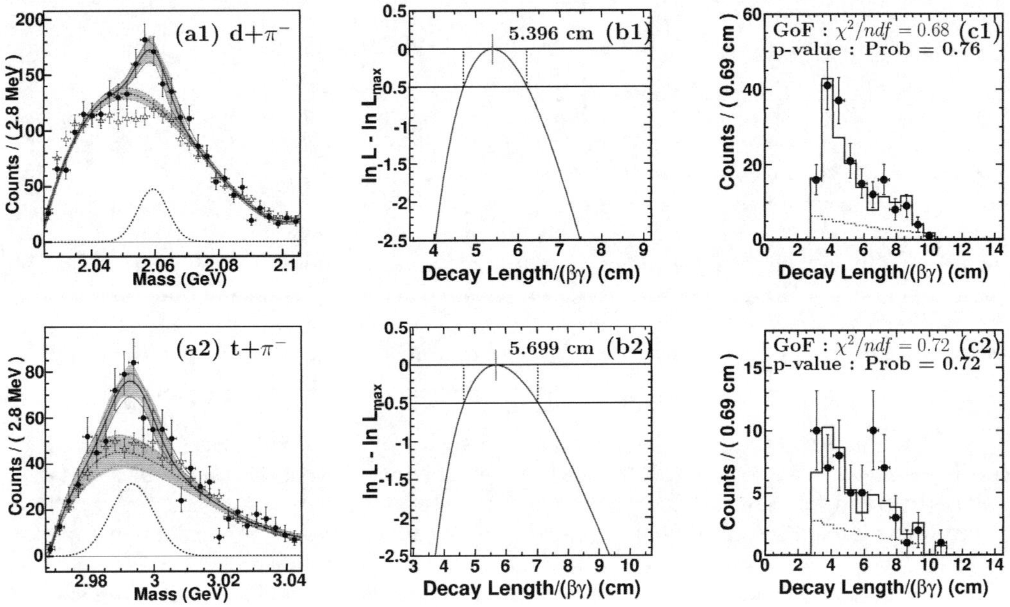

**Figure 2.** (Color online) Invariant mass distributions of d+$\pi^-$ final state candidate in panels (a1) and of t+$\pi^-$ in panels (a2). They are for -10 cm $< Z <$ 30 cm. Profiled likelihood ratio for interval estimation of d+$\pi^-$ (b1) and t+$\pi^-$ (b2) as well as *Binned* decay length distributions of the signal region of d+$\pi^-$ (c1) and t+$\pi^-$ (c2) are shown. Same convention of the representation of figures in Fig. 1 is used. Figures are taken from [7] and rearranged.

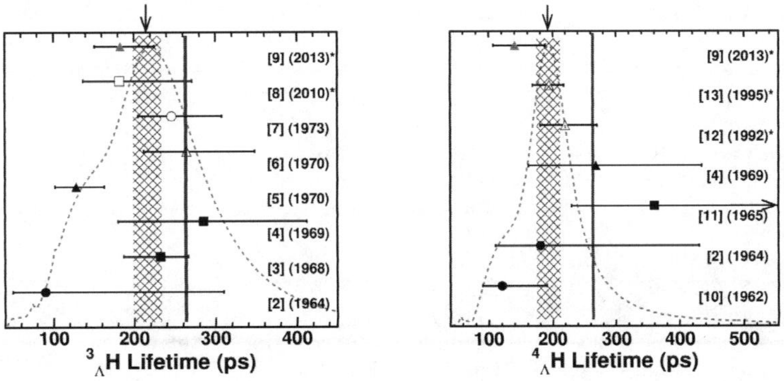

**Figure 3.** (Color online) World data comparison of ${}^3_\Lambda$H and ${}^4_\Lambda$H lifetimes. The combined average is represented by the arrow at the top, while the width of the hatched band corresponds to the one standard deviation of the average. The vertical line at 263.2 ps with width of $\pm$ 2 ps shows the known lifetime of $\Lambda$ hyperon. References to counter experiments is marked by an asterisk. Figures are taken from [8].

$\Lambda$-hyperon (263.2 ps). Since the binding energy of the $\Lambda$-hyperon to the deuteron core in ${}^3_\Lambda$H is known to be only 130 keV, it has been expected that the lifetime of ${}^3_\Lambda$H should not be deviated from that of the Λ-hyperon. However, the lifetime of ${}^{3}_{\Lambda}$H has not been discussed extensively since the experimental data of the ${}^{3}_{\Lambda}$H lifetime was rather scarce and scattered with large errors. We have analyzed all the existing data for the lifetime of ${}^{3}_{\Lambda}$H and ${}^{4}_{\Lambda}$H with statistical methods [8], and the combined analyses have shown that the possible range of the ${}^{3}_{\Lambda}$H lifetime is also significantly lower that the Λ lifetime, as shown in Figure 3. It has to be noted that the short lifetime of ${}^{3}_{\Lambda}$H has also recently been observed by the STAR and ALICE collaborations [11, 12], however, there is no theory to reproduce the observed short lifetime of ${}^{3}_{\Lambda}$H.

After our observation of the signals indicating the possible existence of ${}^{3}_{\Lambda}$n by observing the d+$\pi^-$ and t+$\pi^-$ final states, there have been several theoretical calculations to study the bound state of ${}^{3}_{\Lambda}$n [13, 14, 15]. They have shown that it can not be bound by basing on the current theoretical framework. However, it has to be emphasized hereby once again that the current theories can not explain the short lifetime of ${}^{3}_{\Lambda}$H which is a mirror hypernucleus of ${}^{3}_{\Lambda}$n, and therefore, the conclusion of the existence of ${}^{3}_{\Lambda}$n can not yet be given. Both the existence of ${}^{3}_{\Lambda}$n and the short lifetime of ${}^{3}_{\Lambda}$H have to be further experimentally and theoretically studied.

The authors would like to thank the GSI Departments of Accelerator, of Experimental Electronics, of the Detector Laboratory and of the Target Laboratory and the Electronics Department of the Institute for Nuclear Physics of Mainz University for supporting the project. The HypHI project is funded by the Helmholtz association as Helmholtz-University Young Investigators Group VH-NG-239 at GSI, and the German Research Foundation (DFG) under contract number SA 1696/1-1. The authors acknowledge the financial support provided by the Ministry of Education, Science and Culture of Japan, Grant-in-Aid for Scientific Research on Priority Areas 449, and Grant-in-Aid for promotion of Cooperative Research in Osaka Electro-Communication University (2004-2006). This work is also supported by the Ministry of Education, Science and Culture of Japan, Grants-in-Aid for Scientific Research 18042008 and EU FP7 Hadron-Physics-2 SPHERE. A part of this work was carried out on the HIMSTER high performance computing infrastructure provided by the Helmholtz-Institute Mainz We would also like to thank to A. Botavina, A. Gal, K.D. Gross, N. Hermann, E. Hiyama, R. Klanner, K. Langanke, I. Mishustin, H. Nemura, J. Schaffner-Bielich, C. Scheidenberger, H. Stöcker and J. Stroth for the involved discussions.

[1] J. Beringer, et al., Phys. Rev. D 86 (2012) 010001.
[2] D. Davis, Nucl. Phys. A 754 (2005) 3.
[3] O. Hashimoto, H. Tamura, Prog. Part. Nucl. Phys. 57 (2006) 564.
[4] O. Hashimoto, et al., Nucl. Phys. A 835 (2010) 121.
[5] T. Saito, et al., Letter of intent, http://www.gsi-schwerionenforschung.org/documents/DOC-2005-Feb-432-1.ps, 2006.
[6] C. Rappold, et al., Nucl. Phys. A 913 (2013) 170.
[7] C. Rappold, et al., Phys. Rev. C 88 (2013) 041001(R) .
[8] C. Rappold, et al., Phys. Lett. B 728 (2013) 543 .
[9] D. Drijard, H. Fischer, T. Nakada, Nucl. Instrum. Methods A 225 (1984) 367.
[10] P. Crochet, P. Braun-Munzinger, Nucl. Instrum. Methods A 484 (2002) 564.
[11] STAR-collaboration presented in the conference NUFRA2013 in Turkey, 2013.
[12] ALICE-collaboration presented in the conference NUFRA2013 in Turkey, 2013.
[13] E. Hiyama, et al., Phys. Rev. C89 (2014) 061302(R).
[14] A. Gal, et al., Phys. Lett. B736 (2014) 93.
[15] H. Garcilazo, et al., Phys. Rev. C89 (2014) 057001.

# Spin Responses in Nuclei and Nuclear Weak Processes in Stars

**Toshio Suzuki**

Department of Physics, College of Humanities and Sciences, Nihon University
Sakurajosui 3-25-40, Setagaya-ku, Tokyo 156-8550, Japan

E-mail: suzuki@phys.chs.nihon-u.ac.jp

**Abstract.** New shell-model Hamiltonians which can successfuly describe spin-dependent transition rates in nuclei are applied to study nuclear weak processes in stars. New $\nu$-induced reaction cross sections in $^{12}$C and $^{56}$Fe evaluated by the new Hamiltonians are shown to reproduce well the experimental data. Nucleosynthesis of light elements in supernova explosions (SNe) as well as $\nu$ oscillation effects are discussed with the new cross sections. Electron capture and $\beta$-decay rates in stellar environments are re-evaluated in $fp$- and $sd$-shell nuclei. Nucleosynthesis in Type-Ia SNe, rp-process and X-ray burst are discussed with the new reaction rates in Ni isotopes. Important roles of accurate e-capture and $\beta$-decay rates in $sd$-shell nuclei on the cooling of stars with 8-10 solar masses by nuclear URCA processes and the fate of the stars are demonstrated. $\beta$-decay half-lives of waiting-point nuclei at $N$ =126 are evaluated by shell-model calculations, and r-process nucleosynthesis up to Th and U region in both core-collapse SNe and binary neutron star mergers are studied.

## 1. Introduction

Now several new shell-model Hamiltonians, for example, SFO[1] in $p$-shell and GXPF1[2] in $fp$-shell, are available. They can reproduce Gamow-Teller (GT) strengths in C, Fe and Ni isotopes. The monopole-based universal interaction, VMU[3], is also introduced to take into account important roles of tensor forces in shell evolutions and spin responses in nuclei. Monopole terms of tensor forces obey general sign rule; they are attractive between $j_> = \ell+1/2$ and $j_< = \ell-1/2$ orbits while they are repulsive between $j_>$ ($j_<$) and $j_>$ ($j_<$) orbits[4]. This feature is embedded both in good phenomenological interactions such as GXPF1 and SFO as well as SDPF-M[5] and microscopic G-matrices derived from realistic nucleon-nucleon interactions[3]. The sign rule leads to proper shell evolutions and change of magic numbers toward drip-lines[4, 3].

We apply these new Hamiltonians to evaluate spin-dependent transition strengths and nuclear weak rates at stellar environments. In Sect. 2, $\nu$-nucleus reactions and nucleosynthesis of light elements by $\nu$-processes in SNe are studied. In Sect. 3, new GT strengths in $fp$-shell nuclei are used to obtain e-capture rates in Ni isotopes at high densities and high temperatures. Possible consequences of the new rates in nucleosynthesis in Type-Ia SNe, rp-process and X-ray burst are discussed. In Sect. 4, e-capture and $\beta$-decay rates in $sd$-shell nuclei are used study nuclear URCA processes and cooling of stars with O-Ne-Mg cores. In Sect. 5, $\beta$-decays of isotones with $N$=126 are studied and r-process nucleosynthesis in core-collapse SNe and neutron star mergers (NSM) are discussed.

## 2. Neutrino-nucleus reactions and light-element nucleosynthesis in SNe

### *2.1. $\nu$-induced reactions on $^{12}C$ and $^{56}Fe$*

The GT strength in $^{12}$C is well described by SFO with a small quenching for the axial-vector coupling constant; $g_A^{eff}/g_A$ =0.95. The exclusive reaction cross section for $^{12}$C($\nu_e$, $e^-$)$^{12}$N ($1^+_{g.s.}$) is, therefore, well reproduced by SFO as shown in Fig. 1. SFO also reproduces neutral-current and inclusive charged-current reaction cross sections on $^{12}$C for DAR $\nu$[6]. The GT strength in $^{56}$Fe obtained by GXPF1J[7], together with contributions from other multipoles evaluated by RPA, can reproduce reaction cross section for $^{56}$Fe($\nu_e$, $e^-$)$^{56}$Co induced by DAR $\nu$[8, 9]. Calculated value is $259\times10^{-42}$cm$^2$ while the experimental one is $(256\pm108\pm43)\times10^{-42}$cm$^2$. Thus, it becomes possible now to evaluate $\nu$-nucleus reaction cross sections on $^{12}$C and $^{56}$Fe, the only targets with available experimental data, accurately with the use of new shell-model Hamiltonians[1, 7].

### *2.2. Synthesis of $^{11}B$ and $^7Li$ and effects of $\nu$ oscillations*

New $\nu$-induced cross sections for $^{12}$C, enhanced compared with those by conventional Hamiltonians, are used to study light element synthesis in core-collapse SNe[6, 10]. $^{11}$B is produced mainly by $^{12}$C($\nu$, $\nu$'p)$^{11}$B raection as well as $\alpha$ capture on $^7$Li, while $^7$Li is produced mostly by $^4$He($\nu$, $\nu$'p)$^3$H. The enhancement of $^{11}$B and $^7$Li abundances compared with previous calculations[11] are found.

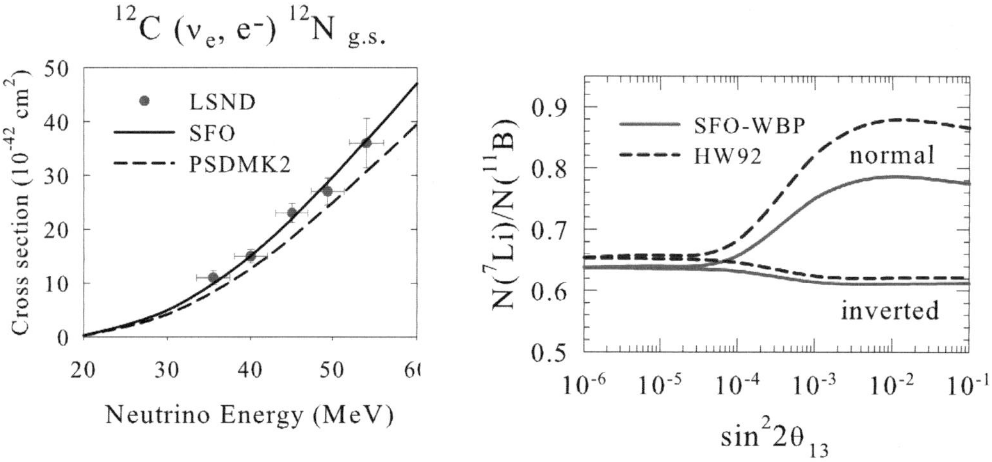


**Figure 1.** (Left figure) Exclusive charge-exchange reaction cross sections for $^{12}$C induced by DAR neutrinos. The calculated values obtained by shell model Hamiltonians, SFO[1] and PSDMK2[12, 13], as well as the experimental values, LSND[14], are shown. (Right figure) Dependence of the abundance ratio $^7$Li/$^{11}$B on the mixing angle $\theta_{13}$ and the $\nu$ mass hierarchies. Figures taken from Ref. [10].

Effects of $\nu$ matter oscillations (MSW effects) on the abundances of $^{11}$B and $^7$Li are also examined. Increase in the rates of charged-current reactions induced by $\nu_e$ converted from $\nu_\mu$ and $\nu_\tau$ with higher energies occur in the He layer. The abundance ratio for $^7$Li/$^{11}$B is found to be sensitive to the $\nu$ mass hierarchy[10]. In case of normal hierarchy, because of MSW

Journal of Physics: Conference Series **590** (2015) 012019 doi:10.1088/1742-6596/590/1/012019

effect at high-density resonance, charged-current reactions induced by $\nu_e$ become more effective to enhance the abundance ratio, while in case of inverted hierarchy there is no high-density resonance and no enhancement of the ratio. The dependence of the ratio on the mixing angle $\theta_{13}$ and the mass hierarchy is shown in Fig. 1 for the new Hamiltonians (SFO-WBP[15], where WBP is used to evaluate $\nu$-$^4$He cross sections) and previous one (HW92[11]). As the value of $\sin^2 2\theta_{13}$ is now known to be about 0.1[16], it is in principle possible to determine the $\nu$ mass hierarchy from the abundance ratio. Recent analysis of supernova X-grains[17] lead to a result that inverted mass hierarchy is statistically more favored[18].

### *2.3. $\nu$-$^{13}$C and $\nu$-$^{16}$O reactions*
The SFO is applied to evaluate new cross sections for $\nu$-$^{13}$C reactions leading to low-lying states in $^{13}$C and $^{13}$N[19]. Since the threshold energy for $\nu$-$^{12}$C reactions is about 13 MeV, $^{13}$C is a good candidate for detection of low energy $\nu$ with $E_\nu < 10$ MeV such as reactor neutrinos. Detailed discussion is given in Ref. [19].

New reaction cross sections for $\nu$-$^{16}$O are obtained with the use of SFO-tls Hamiltonian[20], which has been modified from SFO by inclusion of tensor components of $\pi+\rho$-meson exchanges in the p-sd cross-shell matrix elements. Energies of spin-dipole states in $^{16}$O are found to be well described by SFO-tls. The cross section for $\alpha$p emission channel, which produces $^{11}$B, is found to be as large as about 20% of $^{12}$C($\nu$, $\nu$'p)$^{11}$B reaction cross section. A certain amount of $^{11}$B is also produced from $\nu$-$^{16}$O reactions in the O layer[21].

## 3. The electron capture rates in $fp$-shell nuclei at stellar environments
### *3.1. GT strength in Ni isotopes and e-capture rates*
GT strength in $fp$-shell nuclei are studied with GXPF1J[7] and applied to e-capture reactions at stellar environments[25]. Calculated GT strengths in $^{56}$Ni and $^{55}$Co obtained with GXPF1J with the quenching factor of $f_q = g_A^{eff}/g_A = 0.74$ are shown in Fig. 2, and compared with those obtained by KB3G[22]. The distribution of the GT strengths by GXPF1J are found to be generally more spread compared to those by KB3G in Fe and Ni isotopes. In particular, the characteristic two-peak structure is found in the GT strength in $^{56}$Ni for GXPF1J as shown in Fig. 2(a). This two-peak structure has been recently confirmed by (p, n) reaction experiment [24]. The experimental GT strength in $^{55}$Co[24] is also well reproduced by GXPF1J.

The evaluation of the electron capture reaction rates in type-Ia SNe is important for the synthesis of nuclei around Ni and Fe region. As more strength is found at higher energy region for GXPF1J than for KB3G, the capture rates for GXPF1J become smaller compared to KB3G at high temperatures and densities (see Ref. [25] for $^{56}$Ni for the details). A systematic difference is found for the calculated capture rates between GXPF1J and KB3G

### *3.2. Applications to astrophysical processes*
Accretion of matter to white-dwarfs from their binary stars ignites type-Ia supernova explosions when the white-dwarf mass exceeds the Chandrasekhar limit. A large amount of $^{56}$Ni is produced in type-Ia SNe. As e-capture process on $^{56}$Ni proceeds, neutron-rich nuclei are produced and the lepton-to-baryon ratio (or proton fraction) $Y_e$ gets smaller. The smaler the e-capture rates on $^{56}$Ni, the less production yields of neutron-rich isotopes such as $^{58}$Ni and the higher value for $Y_e$ in statistical equilibrium calculations. The production yield ratio, $^{58}$Ni/$^{56}$Ni, can be reduced nearly by half for GXPF1J compared to KB3G[26]. The problem of over-production of $^{58}$Ni, $^{54}$Cr and $^{54}$Fe compared to the solar abundance[27] could be solved by using smaller e-capture rates of GXPF1J.

Though main processes of rp-process in stars are (p, $\gamma$), ($\alpha$, $\gamma$) and ($\alpha$, p) reactions, nuclear weak processes such as $\beta^+$-decay and e-capture reactions on $^{50}$Fe and e-capture reactions on

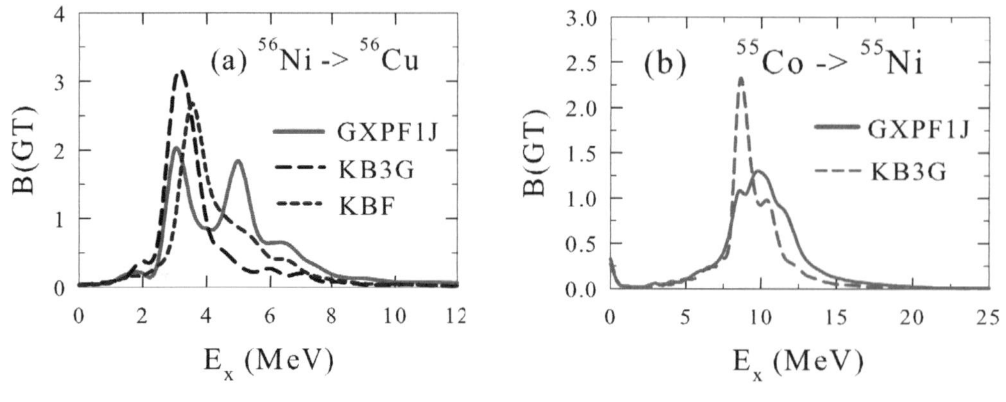


**Figure 2.** GT strengths in (a) $^{56}$Ni and (b) $^{55}$Co obtained by shell-model calculations with GXPF1J and KB3G (and KBF for $^{56}$Ni).

$^{55}$Ni, which is a mirror nucleus of $^{55}$Co, are important and give sizable contributions to the rp-process nucleosynthesis. In particular, e-capture reaction on $^{55}$Ni succeeded by (p, $\gamma$) reaction produces $^{56}$Ni.

At the envelopes of accreting neutron stars, X-ray bursts can occur through nucleosynthesis by rp-process. The temperature at the surface of the star rises up with the synthesis of proton-rich nuclei, and comes to the peak value when $^{56}$Ni is produced. Then, the temperature decreases rapidly, and there appears a bump in the temperature curve when waiting point nuclei such as $^{60}$Zn, $^{64}$Ge and $^{68}$Se are formed. It continues to decrease gradually till the Sn-Sb-Te cycle, where the temperature rises up a little. Then, the temperature decreases rapidly. The peak value of the temperature depends on the e-capture rates on $^{55}$Ni. As the rates are smaller for GXPF1J compared to KB3G and KBF[22], the maximum temperature for GXPF1J becomes a bit smaller. Thus the e-capture rates in Ni and Fe region at densities around $\rho Y_e = 10^6$ g/cm$^3$ and at temperatures $T_9$ =1-10 ($T = T_9 \times 10^9$ K) play important roles in the rp-process and X-ray bursts.

## 4. Electron capture and $\beta$-decay rates in *sd*-shell and nuclear URCA processes

Evolutions of stars with 8-10 solar masses and nuclear URCA processes are discussed. The fate of the stars is sensitive to its mass and nuclear e-capture and $\beta$-decay rates. The stars with 8-10 $M_\odot$ can end with O-Ne-Mg white dwarfs, or e-capture SNe with neutron star (NS) remnant, or core-collapse SNe with NS. Cooling of the O-Ne-Mg core by nuclear URCA processes determines whether the star ends with e-capture SNe or core-collapse SNe. Important URCA processes in *sd*-shell occur at A=23, 25 and 27, where Q-values for the capture reactions become quite small. Detailed e-capture rates on $^{23}$Na, $^{25}$Mg and $^{27}$Al and $\beta$-decay rates on $^{23}$Ne, $^{25}$Na and $^{27}$Mg are obtained at high densities and temperatures. At URCA densities, where e-capture and $\beta$-decay rates coincide each other and both the processes occur simultaneously emitting $\nu_e$ and $\bar{\nu}_e$, thus resulting in remarkable cooling of the O-Ne-Mg core of the star.

The GT strengths for *sd*-shell nuclei are evaluated by USDB interaction with a quenching factor of q=0.764[28]. The URCA densities for A=23 and A=25 are obtained at $\log_{10}(\rho Y_e)$ =

8.92 and and 8.78, respectively, while no clear URCA density is found for A=27[29]. The cooling of O-Ne-Mg core is found to occur at URCA densities for A=23 and 25 in $8.8M_\odot$ star[30]. The $8.8M_\odot$ star collapses triggered by subsequent e-captures on $^{24}$Mg and $^{20}$Ne, thus leading to the e-capture SNe. For $9.5M_\odot$ star, neon-burning shell propagates to the center of the core before the core density reaches the URCA densities and end with a core-collapse SNe[30]. Screening effects on the capture rates shift the URCA densities toward a higher density region; the URCA density is increased to $\log_{10}(\rho Y_e) = 8.81$ for A=25.

## 5. $\beta$ half-lives of waiting-point nuclei at $N$=126 and r-process nucleosynthesis

$\beta$-decay rates for exotic nuclei at $N$=126 relevant to r-process nucleosynthesis are studied by shell-model calculations. The half-lives for $N$=126 isotones are evaluated in a range of $Z$=64-78, which is an extension of our previous work for the isotones with $Z$=64-73[31]. Half-lives obtained by including both the Gamow-Teller (GT) and first-forbidden (FF) transitions are found to be short compared with standard values by FRDM[32], while they are close to another shell-model evaluation[33] as shown in Fig. 3. Here, a quenching factor of $g_A^{eff}/g_A$ =0.70 is adopted for the GT transitions, while quenching factors for $g_A$ and $g_V$ for FF transitions are taken to be 0.34 and 0.67, respectively, which are determined to explain FF beta-decay rates in nuclei near $^{208}$Pb[34, 35]. The FF contributions are important for larger $Z$ and dominant at $Z > 72$. Calculated half-life for $Z$ =78 is found to be consistent with the experimental data[36].

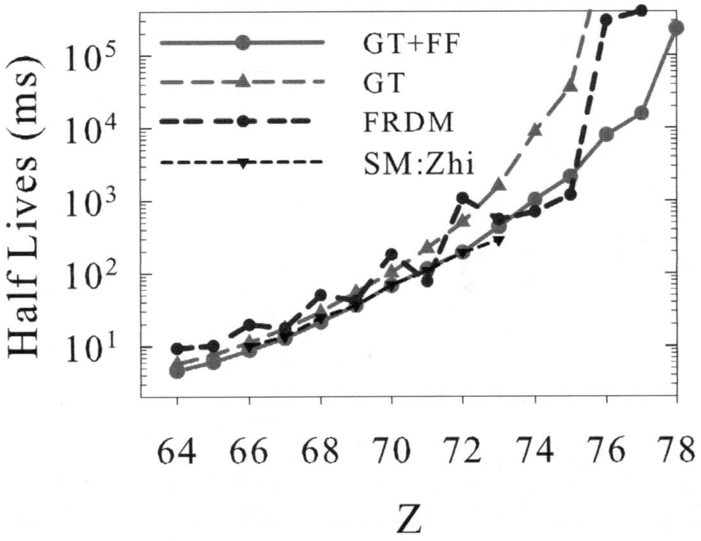


**Figure 3.** Half-lives of $N$=126 isotones obtained by shell-model calculations and FRDM[32]. The present shell-model results with GT only and GT+FF are shown by dashed and solid curves, respectively. Results from Refs. [32] (FRDM) and [33] are denoted by dash-dotted and short-dashed curves, respectively.

The short half-lives obtained here are used to study r-process nucleosynthesis in core-collapse SNe and binary neutron star mergers (NSM). The element abundances are obtained up to Th and U region in both SNe and NSM. The abundances of Th and U are found to be enhanced about by three times for the present half-lives at $N$ =126 in the SNe model[37]. In case of binary NS-NS mergers, based on fully general relativistic simulation of a NS-NS merger that allows mixing of various values of $Y_e$ = 0.09-0.45[38], calculated element abundances are found to be in reasonable agreement with the solar r-process abundance in full-mass range, A=90-240[37]. It is expected that the r-process site can be determined by extending our study on nucleosynthesis in both the astrophysical sites.

The author would like to thank T. Kajino, T. Otsuka, M. Honma, T. Yoshida, S. Chiba, M. Famiano, A. B. Balantekin, S. Wanajo, K. Nomoto, H. Toki and S. Jones for collaboration of the works presented here. This work has been supported in part by Grants-in-Aid for Scientific Research (C) 22540290.

[1] Suzuki T, Fujimoto R and Otsuka T 2003 *Phys. Rev.* C **67** 044302
[2] Honma M, Otsuka T, Brown B A and Mizusaki T 2002 *Phys. Rev.* C **65** 061301(R); **69** 034335
[3] Otsuka T, Suzuki T, Honma M, Utsuno Y, Tsunoda N, Tsukiyama K and Hjorth-Jensen M 2010 *Phys. Rev. Lett.* **104** 012501
[4] Otsuka T, Suzuki T, Fujimoto R, Grawe H and Akaishi Y 2005 *Phys. Rev. Lett.* **95** 232502
[5] Utsuno Y, Otsuka T, Mizusaki T and Honma M 1999 *Phys. Rev.* C **60** 054315
[6] Suzuki T, Chiba S, Yoshida T, Kajino T and Otsuka T 2006 *Phys. Rev.* C **74** 034307.
[7] Honma M, Otsuka T, Mizusaki T, Hjorth-Jensen M and Brown B A 2005 *J. Phys. Conf. Ser.* **20** 7
[8] Suzuki T et al. 2009 *Phys. Rev.* C **79** 061603(R)
[9] Parr N, Suzuki T, Honma M, Marketin T and Vretenar D 2011 *Phys. Rev.* C **84** 047305
[10] Suzuki T and Kajino T 2013 *J. Phys. G: Nucl. Part. Phys.* **40** 08101
[11] Woosley S E, Hartmann D H, Hoffman R D and Haxton W C 1990 *Astrophys. J.* **356** 272
[12] Cohen S and KurathD 1965 *Nucl. Phys.* **73** 1; Millener D J and Kurath D 1975 *Nucl. Phys.* **A 255** 315
[13] OXBASH, The Oxford, Buenos-Aires, Michigan State, Shell Model Program, Brown B A, Etchegoyen A and Rae W D M 1986 MSU Cyclotron Laboratory Report No. 524
[14] Athanassopoulos C et al. (LSND Collaboration) 1997 *Phys. Rev.* C **55** 2078
Auerbach L B et al. (LSND Collaboration) 2001 *Phys. Rev.* C **64** 065501
[15] Warburton E K and Brown B A 1992 *Phys. Rev.* C **46** 923
[16] Abe K et al. (T2K Collaboration) 2011 *Phys. Rev. Lett.* **107** 041801
An F et al. (Daya Bay Collaboration) 2012 *Phys. Rev. Lett.* **108** 171803
[17] Fujiya W, Hoppe P and Ott U 2011 *Astrophys. J.* **730** L7
[18] Mathews G J, Kajino T, Aoki W, Fujiya W and Pitts J B *Phys. Rev.* D **85** 105023
[19] Suzuki T, Balantekin A B and Kajino T 2012 *Phys. Rev.* C **86** 015502
[20] Suzuki T and Otsuka T 2008 *Phys. Rev.* C **78** 061301(R)
[21] Suzuki T, Yoshida T and Chiba S 2014 in preparation
[22] E. Caurier, G. Martinez-Pinedo, F. Nowacki, A. Poves, and A. P. Zuker 2005 *Rev. Mod. Phys.* **77** 427
[23] Poves A, Sanchez-Solano J, Caurier E and Nowacki F 2001 *Nucl. Phys.* **A 694** 157
[24] Sasano M et al. 2011 *Phys. Rev. Lett.* **107** 202501; Sasano M et al 2012 *Phys. Rev.* C **86** 034324
[25] Suzuki T, Honma M, Mao H, Otsuka T and Kajino T 2011 *Phys. Rev.* C **83** 044619
[26] Famiano M 2013 private communication
[27] Iwamoto K et al 1999 *Astrophys. J. Suppl.* **125** 439
[28] Brown B A and Richter W A 2006 *Phys. Rev.* C **74** 034315
Richter W A, Mkhize S and Brown B A 2008 ibid. **78** 064302
[29] Toki H, Suzuki T, Nomoto K, Jones S and Hirschi R 2013 *Phys. Rev.* C **88** 015806
[30] Jones S et al 2013 *Astrophys. J.* **772** 150
[31] Suzuki T, Yoshida T, Kajino T and Otsuka T 2012 *Phys. Rev.* C **85** 015802
[32] Moller P, Nix J R and Kratz K-L 1997 *At. Data Nucl. Data Tables* **66** 131
Moller P, Pfeiffer B and Kratz K-L 2003 *Phys. Rev.* C **67** 055802
[33] Zhi Q, Caurier E, Cuenca-Garcia J J, Langanke K, Martinez-Pinedo G and Sieja K 2013 *Phys. Rev.* C **87** 025803
[34] Steer S J et al. 2008 *Phys. Rev.* C **78** 061302
Rydstrom L, BlomqvistJ , Liotta R J and Pomar C 1990 *Nucl. Phys.* **A 512** 217
[35] Warburton E K *Phys. Rev.* C **42** 2479
[36] Morales A I et al. 2014 *Phys. Rev. Lett.* **113** 022702
[37] Suzuki T, Yoshida T, Wanajo S, Shibagaki S, Kajino T and Otsuka T 2014 in preparation
[38] Wanajo S et al. 2014 *Astrophys. J.* **789** L39

Journal of Physics: Conference Series **590** (2015) 012020 doi:10.1088/1742-6596/590/1/012020

# Subbarrier fusion of carbon isotopes: from resonance structure to fusion oscillations

**K. Hagino**[1,2] **and N. Rowley**[3]

[1] Department of Physics, Tohoku University, Sendai 980-8578, Japan

[2] Research Center for Electron Photon Science, Tohoku University, 1-2-1 Mikamine, Sendai 982-0826, Japan

[3] Institut de Physique Nucléaire, UMR 8608, CNRS-IN2P3 et Université de Paris Sud, 91406 Orsay Cedex, France

**Abstract.** At energies below the Coulomb barrier, the fusion excitation function for the $^{12}$C+$^{12}$C system shows prominent fine structures, whereas that for the $^{12}$C+$^{13}$C system behaves more smoothly as a function of energy. We demonstrate that these different behaviors can be simultaneously reproduced using an optical potential in which the strength of the imaginary part is proportional to the level density of each compound nucleus. We also discuss the oscillatory behavior of fusion excitation function for these systems observed at energies above the Coulomb barrier from a view point of quantum mechanical systems with identical particles.

## 1. Introduction

The $^{12}$C+$^{12}$C fusion reaction plays an important role in several astrophysical phenomena, such as the carbon burning in stellar evolution, type Ia supernovae, and the X-ray superburst of an accreting neutron star. Even though there is a long history of research on this reaction, the reaction still attracts lots of attention and new experimental and theoretical works have continuously been carried out [1, 2, 3, 4, 5]. A characteristic feature of the $^{12}$C+$^{12}$C fusion reaction is that the cross sections have many fine structures, at energies both below and above the Coulomb barrier, while the cross sections for the neighboring systems, $^{12}$C+$^{13}$C and $^{13}$C+$^{13}$C, are much less structured. In this contribution, we simultaneously analyze fusion reactions for the $^{12}$C+$^{12}$C and $^{12}$C+$^{13}$C systems and discuss possible origins for the different behavior of fusion excitation functions for these systems. Notice that most of the previous studies, except for Ref. [4], have concentrated only on the $^{12}$C+$^{12}$C system. In contrast, we shall analyze both the $^{12}$C+$^{12}$C and $^{12}$C+$^{13}$C systems and clarify the dynamics of subbarrier fusion of two carbon isotopes.

## 2. Subbarrier molecular resonances

We first discuss the resonance behavior of fusion cross sections for the $^{12}$C+$^{12}$C system at energies below the Coulomb barrier. Figure 1(a) shows the experimental data for the modified astrophysical $S$-factor, defined as $S^*(E) = \sigma(E)E\,\exp(87.21/\sqrt{E} + 0.46E)$, where $\sigma(E)$ is the fusion cross section and the energy $E$ is in units of MeV, for the $^{12}$C+$^{12}$C and the $^{12}$C+$^{13}$C systems [2]. As has been well known [6], the fusion cross sections for the $^{12}$C+$^{12}$C system exhibit a few resonance peaks. Those relatively narrow resonances are known as molecular resonances, for which the excitation of $^{12}$C to the first $2^+$ state plays an important role [7, 8, 9].

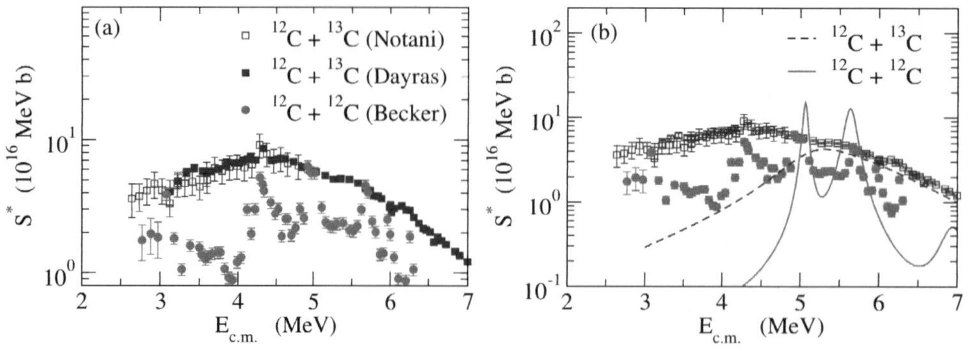


**Figure 1.** (The left panel) The experimental data for the modified astrophysical $S$-factor for the $^{12}$C+$^{12}$C system (the filled circles) and for the $^{12}$C+$^{13}$C system (the filled and open squares). The experimental data are taken from Refs. [2, 10, 11]. (The right panel) Same as the left panel, but with theoretical curves for the $^{12}$C+$^{12}$C system (the solid line) and for the $^{12}$C+$^{13}$C system (the dashed line) obtained with the level density dependent optical potentials.

As has been pointed out in Ref. [2], the fusion cross sections for the $^{12}$C+$^{12}$C system show inhibitions as compared to those for the $^{12}$C+$^{13}$C system except at a few resonance energies, at which the fusion cross sections for the two systems somehow match with each other. Recently, Jiang *et al.* have argued that the fusion inhibition in the $^{12}$C+$^{12}$C system is attributed to i) the smaller fusion $Q$-value in the entrance channel compared to that for the $^{12}$C+$^{13}$C and $^{13}$C+$^{13}$C systems, ii) a smaller level density of the compound nucleus $^{24}$Mg than $^{25}$Mg and $^{26}$Mg at a given excitation energy, and iii) the fact that only states with positive parity and even spin of the compound nucleus are populated in fusion because the entrance channel consists of identical spin-zero bosons [3]. Jiang *et al.* have succeeded to explain the *average* behavior of fusion cross sections for the $^{12}$C+$^{12}$C system based on this idea.

Our first aim in this contribution is to implement the idea of Jiang *et al.* into coupled-channels calculations and discuss the difference between the $^{12}$C+$^{12}$C and $^{12}$C+$^{13}$C systems. To this end, we carry out coupled-channels calculations with an optical potential, in which the strength of the imaginary part is proportional to the level density of the compound nucleus. That is, the imaginary part of the optical potential is assumed to be

$$W(r) = -w_0 \rho_J(E^*) f(r), \tag{1}$$

where $w_0$ is an overall strength, $\rho_J(E^*)$ is the level density of the compound nucleus at the excitation energy of $E^*$ with the angular momentum $J$, and $f(r)$ determines the radial dependence of the imaginary potential, which we assume to be a Woods-Saxon form. The level-density dependent imaginary potential has been employed in Refs. [9, 12], which has further been investigated in Refs. [13, 14]. Eq. (1) may be justified in terms of the Fermi's golden rule for a transition from the entrance channel to compound nucleus states [9, 12, 13, 14]. In this approach, the energy, the angular momentum, and the system dependences of the imaginary potential are taken into account through the level density of the compound nucleus.

In order to perform the coupled-channels calculations for the $^{12}$C+$^{12}$C and $^{12}$C+$^{13}$C systems, we closely follow the calculations presented in Ref. [8]. That is, we use the same geometry for the optical potential as in Ref. [8], and we include the excitations of both the projectile and the target nuclei up to the first member of the ground state rotational band (including also the mutual excitation channels). We take the deformation parameters from Ref. [4], while we

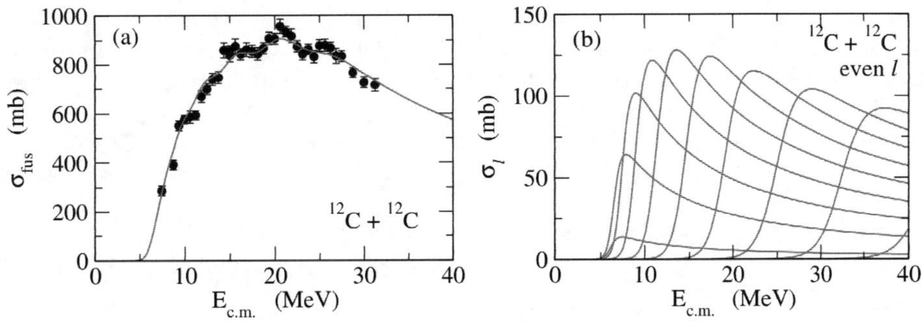


**Figure 2.** (The left panel) Comparison of the experimental fusion cross sections to the result of optical model calculation for the $^{12}C+^{12}C$ system. The experimental data are taken from Ref. [20]. (The right panel) The partial fusion cross sections corresponding to the theoretical result shown in Fig. 2(a).

take the empirical level density parameters from Ref. [3]. The overall strength $w_0$ in Eq. (1) is determined so as to reproduce the results of Ref. [8] if all the parameters are set identical to those in Ref. [8]. We use the same value of $w_0$ both for the $^{12}C+^{12}C$ and $^{12}C+^{13}C$ systems. Most of the channels are closed, and the coupled-channels equations are solved with the variational method [8, 15, 16] in order to avoid a numerical instability.

Figure 1(b) shows the results so obtained. The solid and the dashed lines show the modified astrophysical $S$-factors for the $^{12}C+^{12}C$ and the $^{12}C+^{13}C$ systems, respectively. One can see that the system dependence of fusion cross sections is qualitatively well reproduced. That is, the calculated $S$-factors for the $^{12}C+^{12}C$ system show resonance peaks, while those for the the $^{12}C+^{13}C$ system behave rather smoothly. This indicates that a promising origin for the resonance behavior in the $^{12}C+^{12}C$ system is attributed to the properties of the compound nucleus $^{24}Mg$, as has been suggested by Jiang *et al.* [3].

The calculations still underestimate the fusion cross sections below 5 MeV for both the systems. This would be due to those channels which are not included in the present calculations. Possible candidates are the first $3^-$ state and the second $0^+$ state (that is, the Hoyle state) [4, 5]. The $\alpha$-transfer channel may also play an important role [17, 18, 19]. It would be an interesting future work to repeat the present calculation by including those channels.

## 3. Fusion oscillations above the Coulomb barrier

We next discuss fusion cross sections at energies above the Coulomb barrier. The experimental data for the $^{12}C+^{12}C$ system are shown in Fig. 2 (a). One can see that the fusion cross sections significantly oscillate as a function of energy. The origin for the fusion oscillations may be different from the origin for the resonance behavior at subbarrier energies shown in Fig. 1, however.

At energies above the barrier, the level density is large enough so that resonances in the compound nucleus appreciably overlap with each other. This results in the picture of strong absorption, with which fusion cross sections are given by,

$$\sigma(E) = \frac{\pi}{k^2}\sum_l (2l+1)P_l(E) = \sum_l \sigma_l(E), \tag{2}$$

where $k$ is the wave number related to the incident energy in the center of mass frame, $P_l$ is the penetrability of the Coulomb barrier with the partial wave $l$, and $\sigma_l(E)$ is a partial fusion cross

Journal of Physics: Conference Series **590** (2015) 012020 doi:10.1088/1742-6596/590/1/012020

section. For each partial wave, the penetrability $P_l(E)$ is close to unity and does not change significantly at energies well above the barrier. The partial cross section $\sigma_l(E)$ then decreases as $1/k^2$ as a function of energy. Since the partial waves are discrete, one would then obtain a structure in fusion cross sections, which are associated with the energy dependence of partial fusion cross sections. This is the origin for the fusion oscillations advocated in Refs. [21, 22] (see also Ref. [23]). That is, the fusion oscillations are due to the addition of successive individual partial waves as the energy increases.

For the $^{12}C+^{12}C$ system, since the wave function of the whole system has to be symmetric with respect to the interchange of two identical spin-zero bosons, only even partial waves contribute to fusion cross sections. That is,

$$\sigma(E) = \frac{\pi}{k^2}\sum_l (1+(-1)^l)(2l+1)P_l(E) = \sum_l (1+(-1)^1)\sigma_l(E). \tag{3}$$

This leads to an enhancement of oscillations, since the energy spacing between successive contributing partial waves then increases.

The solid line in Fig. 2(a) shows a potential model fit to the experimental fusion cross sections. Since the effect of channel couplings are small for this system at energies above the Coulomb barrier, we perform single-channel calculations for simplicity. In order to account for the observed decrease of fusion cross sections at energies higher than 25 MeV, we have reduced the penetrability for $l = 14$ by a factor of 2 and set the penetrabilities for all the higher partial waves to be zero. Such an assumption may be justified from a consideration based on the excitation energy of the compound nucleus relative to the yrast energy [24, 25]. This calculation fairly well reproduces the experimental data, including the fusion oscillations. The structure of fusion cross sections is evident if the corresponding partial fusion cross sections are plotted individually, as is done in Fig. 2(b).

Within the parabolic approximation to the Coulomb barrier, one can derive a compact formula for the oscillatory part of fusion cross sections [21, 24, 25, 26],

$$\sigma_{\rm osc}(E) = 2\pi R_E^2 \frac{\hbar\Omega_E}{E} \exp\left(-\frac{\pi\mu R_E^2\hbar\Omega_E}{(2l_g+1)\hbar^2}\right)\sin(\pi l_g), \tag{4}$$

which is added to the smooth part of fusion cross sections given by the well known Wong formula [27]. In this equation, $\mu$ is the reduced mass of the system, and $R_E$ and $\hbar\Omega_E$ are the barrier position and the curvature of the Coulomb barrier, respectively, evaluated at the grazing angular momentum, $l_g$. For a spatially anti-symmetric configuration, the negative sign has to be multiplied to this formula [24, 25]. This formula indicates that the amplitude of fusion oscillations exponentially decrease for heavy systems (with large values of $\mu$), and one has a better chance to see the fusion oscillations in light systems, such as $^{12}C+^{12}C$ and $^{16}O+^{16}O$.

The experimental fusion cross sections for the $^{13}C+^{13}C$ are shown in Fig. 3(a). One can see that the amplitude of fusion oscillation is much smaller than that for the $^{12}C+^{12}C$ system shown in Fig. 2(a). The $^{13}C+^{13}C$ system is a system of two identical spin-1/2 fermions, and both the $S$=0 (which gives even $l$ with weight 1/4) the $S$=1 (which gives odd $l$ with weight 3/4) configurations contribute. Since the oscillatory cross section for odd partial waves has the opposite sign to that for even partial waves, the oscillation is reduced by a factor 2 and has the opposite phase from the symmetric system. The solid line in Fig. 3(a) is obtained in this way.

For the $^{12}C+^{13}C$ system, since this is not a system with identical particles, the oscillations from even $l$ and odd $l$ would cancel out without any further effect. However, the presence of the elastic neutron transfer channel will introduce a parity dependence into the problem. This is most easily seen by considering the total elastic scattering, where one must add an exchange term $f_{\rm trans}(\pi-\theta)$ for elastic transfer to the amplitude $f_{\rm el}(\theta)$ for direct elastic scattering. This yields a

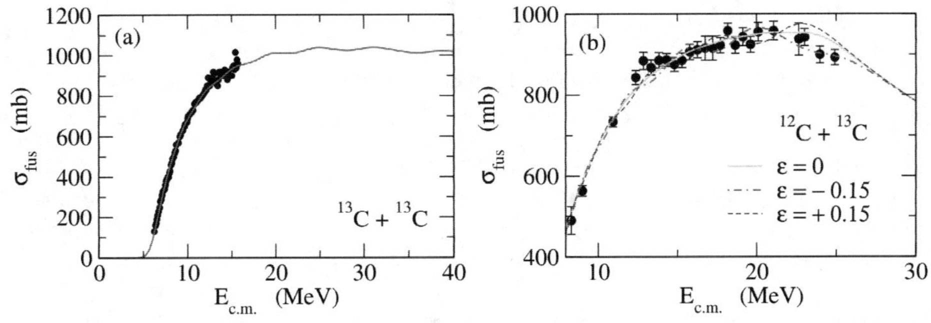


**Figure 3.** (The left panel) Same as Fig. 2(a), but for the $^{13}$C+$^{13}$C system. The experimental data are taken from Ref. [28]. (The right panel) The experimental fusion cross sections for the $^{12}$C+$^{13}$C system taken from Ref. [20]. The three curves are obtained with a parity dependent potential with the depth parameter of $V = (1 + (-1)^l\epsilon)V_0$, where $V_0$ is a negative value, with different values of $\epsilon$ as indicated in the figure.

total scattering amplitude, $f_{\rm total}(\theta) = f_{\rm el}(\theta)+f_{\rm trans}(\pi-\theta)$. Using $P_l(\cos(\pi-\theta)) = (-1)^l P_l(\cos\theta)$, one obtains $S_l^{\rm eff} = S_l^{\rm el} + (-1)^l S_l^{\rm trans}$, that is, different effective $S$-matrix elements for the odd and even partial waves. The same mechanism with $\alpha$ transfer has been discussed in Ref. [22] for the fusion oscillations observed in the $^{12}$C+$^{16}$O system. As is well known, the parity dependence due to elastic transfer can be well mocked up with a parity dependent potential [22, 29, 30, 31]. The three different lines in Fig. 3(b) are obtained with a parity dependent potential, whose depth parameter is defined as $V = (1 + (-1)^l\epsilon)V_0$ (with a negative value of $V_0$). Treating $\epsilon$ as a free parameter, we obtain a good fit to the experimental data with $\epsilon = -0.15$, that is, a shallower potential (and thus a higher Coulomb barrier) for even partial waves. We mention that the sign of $\epsilon$ found for the $^{12}$C+$^{13}$C system is consistent with a simple rule proposed by Baye based on the resonating group method (RGM) with a two-center harmonic oscillator shell model [32, 33].

Notice that, around $E \sim$13 MeV, the calculation with $\epsilon = +0.15$ appears more consistent with the experimental data compared with the result with $\epsilon = -0.15$. It would be interesting to remeasure fusion cross sections in this energy region with higher precision to clarify whether there indeed exists a possible shift in phase of the oscillations as a function of energy.

## 4. Summary

The fusion reaction of carbon isotopes is important from the astrophysical point of view, and at the same time it makes also an interesting quantum mechanical problem. We have discussed this reaction from the subbarrier to the above barrier regions by comparing the $^{12}$C+$^{12}$C system to the neighboring $^{12}$C+$^{13}$C and $^{13}$C+$^{13}$C systems.

At subbarrier energies, the fusion cross sections for the $^{12}$C+$^{12}$C system show many resonance peaks whereas those for the other systems behave smoothly. We have demonstrated that this fact can be naturally explained if one considers properties of the compound nuclei, as had been conjectured by Jiang *et al.*. To this end, we have carried out coupled-channels calculations by including excitations to the first excited state in the ground state rotational band. In these calculations, we have employed an optical potential whose strength is directly proportional to the level density of each compound nucleus.

At energies above the Coulomb barrier, the fusion cross sections for the $^{12}$C+$^{12}$C system oscillate as a function of energy. We have demonstrated that this oscillation can be interpreted as due to the addition of successive individual partial waves. The oscillation is stronger in the $^{12}$C+$^{12}$C system than in the $^{13}$C+$^{13}$C system, because only even partial waves contribute to fusion cross sections for the former system while both even and odd partial waves contribute (with a statistical weight of 1:3) for the latter system. We have also shown that the fusion oscillation observed in the $^{12}$C+$^{13}$C system is due to elastic transfer of neutron, whose effect is well mocked up in terms of a parity dependent potential.

It still remains as a challenging problem to quantitatively explain observed fusion cross sections for the C+C systems at deep subbarrier energies. In order to achieve this goal, one would have to take into account the effects of several collective excitations as well as the alpha transfer channel.

**Acknowledgments**
We thank X.D. Tang and S. Shimoura for useful discussions. This work was supported by JSPS KAKENHI Grant Number 25105503.

## References

[1] Spillane T *et al.* 2007 *Phys. Rev. Lett.* **98** 122501
[2] Notani M *et al.* 2012 *Phys. Rev.* C **85** 014607
[3] Jiang C L *et al.* 2013 *Phys. Rev. Lett.* **110** 072701
[4] Esbensen H, Tang X and Liang C L 2011 *Phys. Rev.* C **84** 064613
[5] Assunção M and Descouvemont P 2013 *Phys. Lett.* B **723** 355
[6] Almqvist E, Kuehner J A and Bromley D A 1960 *Phys. Rev. Lett.* **4** 515
[7] Imanishi B 1968 *Phys. Lett.* **27B** 267; *ibid* 1969 *Nucl. Phys.* A **125** 33
[8] Kondo Y, Matsuse T and Abe Y 1978 *Prog. Theo. Phys.* **59** 465
[9] Fink H -J, Scheid W and Greiner W 1972 *Nucl. Phys.* A **188** 259
[10] Becker H W *et al.* 1981 *Z. Phys.* A **303** 305
[11] Dayras R A *et al.* 1976 *Nucl. Phys.* A **265** 153
[12] Helling G, Scheid W and Greiner W 1971 *Phys. Lett.* **36B** 64
[13] Quesada J M, Lozano M and Madurga G 1983 *Phys. Lett.* B **125** 14
[14] Andres M V, Quesada J M, Lozano M and Madurga G 1985 *Nucl. Phys.* A **443** 380
[15] Matsuse T, Abe Y and Kondo Y 1978 *Prog. Theo. Phys.* **59** 1904
[16] Kamimura M 1977 *Prog. Theo. Phys. Suppl.* **62** 236.
[17] Čujec B, Hunyadi I and Szöghy I M 1989 *Phys. Rev.* C **39** 1326
[18] Wada R, Schimizu J and Takimoto K 1977 *Phys. Rev. Lett.* **38** 1341
[19] Wada R *et al.* 1980 *Phys. Rev.* C **22** 557
[20] Kovar D G 1979 *Phys. Rev.* C **20** 1305
[21] Poffe N, Rowley N and Lindsay R 1983 *Nucl. Phys.* A **410** 498
[22] Kabir A, Kermode M W and Rowley N 1988 *Nucl. Phys.* A **481** 94
[23] Esbensen H 2012 *Phys. Rev.* C **85** 064611
[24] Hagino K and Rowley N 2014 *Preprint* arXiv:1403.1384 [nucl-th]
[25] Hagino K and Rowley N 2014 *to be published*
[26] Hagino K and Takigawa N 2012 *Prog. Theo. Phys.* **128** 1061
[27] Wong C Y 1973 *Phys. Rev. Lett.* **31** 766
[28] Charvet J L 1982 *Nucl. Phys.* A **376** 292
[29] von Oertzen W and Bohlen H G 1975 *Phys. Rep.* **19C** 1
[30] Vitturi A and Dasso C H 1986 *Nucl. Phys.* A **458** 157
[31] Christley J A, Nagarajan M A and Vitturi A 1995 *Nucl. Phys.* A **591** 341
[32] Baye D, Deenen J and Salmon Y 1977 *Nucl. Phys.* A **289** 511
[33] Baye D 1986 *Nucl. Phys.* A **460** 581

Journal of Physics: Conference Series **590** (2015) 012021 doi:10.1088/1742-6596/590/1/012021

# Single-shell states for multi-shell description of nuclei

**Erdal Dikmen**

Department of Physics, Süleyman Demirel University, 32260 Isparta, Turkey

E-mail: erdaldikmen@sdu.edu.tr

**Abstract.** The DUSM method is described for constructing the single-shell states and the corresponding matrix elements needed for describing the isotopes of a given nucleus within a major shell in the framework of the nuclear shell model. The advantage of this approach is that a streamline set of states are constructed, thereby reducing the calculation requirements. The idea is to truncate the full space of permutational irreducible representations with two columns to subspaces where the second column is restricted in its number of boxes - depending on the isospin value. The usefulness of this method for large-scale shell model calculations in the *fp*- and *sdgh*-shell is emphasized. An application to Sn isotopes in the *sdgh*-shell is presented.

## 1. Introduction

The nuclear shell model provides systematic descriptions for some series of isotopes within major shells. These studies are particularly important to recognize the systematics trends in the nuclear properties as function of proton or neutron numbers. It became the standard approach for describing the systematics found in the spectra and transition matrix elements of the *p*-shell nuclei ( $1p_{\frac{3}{2}}$ and $1p_{\frac{1}{2}}$) - [1], the *sd*-shell nuclei ( $1d_{\frac{5}{2}}$, $1d_{\frac{3}{2}}$, and $2s_{\frac{1}{2}}$ )- [2, 3], the *fp*-shell nuclei $0f_{\frac{7}{2}}$, $0f_{\frac{5}{2}}$, $1p_{\frac{3}{2}}$, and $1p_{\frac{1}{2}}$ ) [4, 5, 6], and the *sdgh*-shell nuclei ( $2s_{\frac{1}{2}}$, $1d_{\frac{5}{2}}$, $1d_{\frac{3}{2}}$, $0g_{\frac{7}{2}}$, and $0h_{\frac{11}{2}}$ ) [7, 8, 9, 10, 11, 12].

The shell model approach applied to the description of nuclei is notoriously difficult to perform numerically. In particular the study of neutron (proton) rich nuclei demand systematic description of many nucleons systems in large multi-shell spaces. These nuclei are at the focus of the studies performed with the radioactive beam facilities and crucial for astrophysical applications. A major difficulty for the shell model description of these nuclei is the multi-shell basis dimensions which might be enormous. In the $m$ scheme formalism, all the states with angular momentum $J > J_z$ and isospin $T > T_z$ are included in the calculations. This leads to the main difficulty of the $m$ scheme formalism in that the number of basis states becomes very large even for few valence nucleons. The projection onto states of good total angular momentum $J$ and isospin $T$ is achieved automatically due to the Hamiltonian being a scalar. The number of basis states for states of good $J$ and $T$ in the $J-T$ scheme formalism is much smaller than the number of states in the $m$ scheme. The $J-T$ description is clearly advantageous here over the $m$ scheme.

In this contribution, a great simplification in building the model space for the nuclei with large isospin states (stretched or near stretched states) is offered in the Drexel University Shell Model (DUSM) approach developed in $J-T$ scheme [13, 14, 15]. It is based on the limits on the permissible Young partitions describing the states. These limits are exact; yet they are very

Journal of Physics: Conference Series **590** (2015) 012021 doi:10.1088/1742-6596/590/1/012021

difficult to implement in conventional approaches to the shell model. However, it will be shown below a natural framework to implement such a truncation.

## 2. The DUSM Formalism

A different approach for performing multi-shell nuclear shell model calculations was introduced almost 20 years ago [13, 14, 15]. This algorithm is based on extensive use of permutation group concepts. In this method the orbital angular momentum states and isospin states are treated separately. A state with good isospin belongs to an *arbitrary* irreducible representation (irrep) of the symmetric group and appropriate orbital angular momentum states belong to the conjugate irrep. The restriction of the isospin irreps to two rows ($t = \frac{1}{2}$) implies a restriction of the orbital angular momentum irreps to two columns. By using the inner-product isoscalar factor we obtain globally antisymmetric states from any two states that belong to two conjugate irreps. This procedure enables us to originally and efficiently construct a full $J - T$ basis.

The global orbital states for multi-shell configuration are constructed, by using outer-product isoscalar factors, from single-shell states that also belong to *arbitrary* irreps of the symmetric group. However, since the global orbital states are restricted to two columns irreps, the single shell states are also restricted to these irreps. (In this contribution it is shown that we can put an additional restriction on the number of boxes in the second column, depending on the isospin value.) The one- and two-body matrix elements must then be calculated between states which belong to these conjugate orbital and isospin irreps. The concept of a summation over paths in Young diagram space is used to take into account the contributions from all possible irreps in both orbital and isospin spaces.

The Drexel University Shell Model (DUSM) computer code [15] implemented this new approach to perform shell model calculations in the isospin scheme where the nucleons are distributed in several single-shells. A parallel version of the DUSM code - DUPSM was written [6, 9] using PVM [16], a dynamic environment for parallel programming, supporting the message passing paradigm.

The parallel calculation of the Hamiltonian matrix in the DUPSM code is very efficient since every processor calculates small independent submatrices in which the single-shell states belong to given permutational irreps. In addition, mainly because of the permutational symmetry selection rules, the matrices are sparse (for matrices of order of hundreds of thousands only few percent of the matrix elements are non-zero). With the DUPSM code we can calculate matrices of order of few millions on few tens of parallel machines in a relatively short time.

The DUSM algorithm recursively builds the single-shell many-body states based on the permutation group labels as illustrated in Figure 1. The states are built with a two rows restriction in isospin space and a related two columns restriction in orbital space. The matrix elements of any combination of elementary creation and annihilation operators, from $a^+$ to $a^+a^+a\ a$, are then calculated between these states, taking care of all the different paths. These form the building blocks from which to calculate the matrix elements of any one- and two-body operator (Hamiltonian and transition operators) in the multi-shell calculations describing nuclei with *any given value of isospin.*

## 3. Exact Truncation for Single-Shell Irreps

The DUSM approach offers originally an advantage in Ref. [13, 14, 15] in that the orbital and isospin sub-spaces are separated leading to an enhanced sparsity of the Hamiltonian matrix and to an economy of space for the single-shell one- and two-body operators. This is achieved by restriction of Young frames into two-column irreps including all irreps in Figure 1. Even then extra truncation might be required to render the calculations feasible for the very large Hamiltonian matrix.

Journal of Physics: Conference Series **590** (2015) 012021 doi:10.1088/1742-6596/590/1/012021

An extra exact truncation is presented in this contribution: the restriction of the Young frames into the number of boxes in the second-column irreps, e.g., no-box, one-box, two-boxes, etc depending on the isospin value of the nuclei.

Let's start with all the isotopes with $T_{max}$ i.e., stretched (maximum) isospin. For example, the Ca isotopes in the $fp$-shell and the Sn isotopes in the $sdgh$-shell. For any given number $n$ of valence nucleons (actually, only neutrons - no valence protons in these cases) the appropriate isospin state belongs to the permutational irrep $[n]$, i.e., one row with $n$ boxes. Therefore, its adjoint states should belong to the permutational irrep [111...11], i.e., one column irreps with $n$ boxes - below the blue line in Figure 1. Consequently, all the single-shell irreps must also have only one column. It is trivial to justify that we can take just these one column irreps since by removing one box from any given one column irrep we can get only another one column irrep with one less box - see Figure 1. This truncation is *exact*.

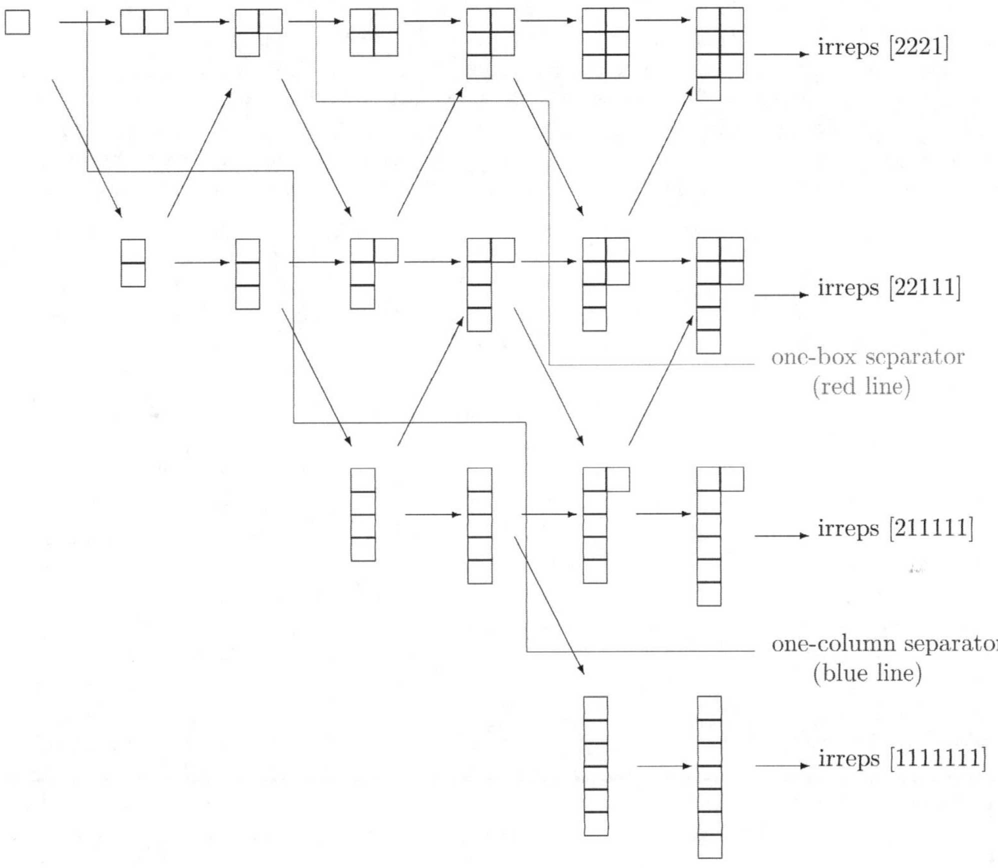


**Figure 1.** Two columns irreps for 1-7 particles. The arrows indicate irreps connected by the addition of one-box. The red line splits between the irreps with at most one-box in the second column and those with more than one-box. The blue line splits between the irreps with one-column and those with more than one-column

Journal of Physics: Conference Series **590** (2015) 012021 doi:10.1088/1742-6596/590/1/012021

Consider *all* the Ca isotopes in the $fp$-shell; in this case it is sufficient to truncate the $f_{\frac{7}{2}}$ states to those belonging to the irreps [1], [11], [111], ...[11111111], and the other shells to the [1], [11], [111], ..., [111111] $f_{\frac{5}{2}}$ states, the [1], [11], [111], and [1111] $p_{\frac{3}{2}}$ states, and the [1] and [11] $p_{\frac{1}{2}}$ states (see irreps below the blue line in Figure 1). The memory required for the single-shell states and the matrix elements related to these irreps is negligible - less than 1MB.

Let's assume that we want to calculate *all* the isotopes with $T_{max} - 1$ in a given major shell, i.e., the Sc isotopes in the $fp$-shell and the Sb isotopes in the $sdgh$-shell. In this case the isospin state for a nucleus with $n$ valence nucleons belongs to the irrep $[n-1,1]$, i.e., a two-rows irrep with only one-box in the second row. The adjoint irrep is [2111...111], i.e., a two columns irrep with only one-box in the second column. Therefore, *all* the orbital single-shell irreps should also belong to the irreps with one-box in the second column - see Figure 1. The full space of two-columns irreps is truncated to a subspace which includes only those irreps with at most one-box in the second column, i.e., the irreps between the red and blue lines in Figure 1. This truncation is possible since deleting one-box from any given irrep of the type [2111...1111] leaves either an irrep of the same type, if we remove the last box from the first column, or a one-column irrep if we remove the box in the second column. Both should be included in the case of $n-1$ valence nucleons as pointed out in Figure 1.

The Sc isotopes have $T_{max} - 1$ isospin. To calculate an arbitrary nucleus in the middle of the $fp$-shell would require to include all the single-shell states of $f_{\frac{7}{2}}$ which belong to the 44 two-columns irreps (16 valence nucleons with maximum 8 rows). To model the Sc isotopes it is sufficient to consider only the eight one-column irreps, and, in addition, the eight irreps with two-boxes in the first row i.e., [2], [21], [211], ..., [21111111] (see irreps between the red and blue lines in Figure 1). For $f_{\frac{5}{2}}$ we have to add only six such irreps, i.e., till [211111], for $p_{\frac{3}{2}}$ we have to add four irreps till [2111], and for $p_{\frac{1}{2}}$ we have to add only the irreps [2] and $[2,1]$. This truncation is obviously exact.

It is important to note that, actually, for the last case, we are reducing the number of single-shell matrix elements by a factor much larger than $\frac{24}{90}$. The reason is that for more than two particles the number of states included in the one-column irrep is less than those included in the two-columns irrep with one-box in the second column. As the single-particle angular momentum $j$ is becoming larger and more nucleons occupy the shell - much more states are included in the latter irrep. The reason was actually explained above - more irreps contribute to the build-up of the two-columns irreps than the one-column irrep - see Figure 1 (Whereas, only one irrep with one-column contributes to the next one-column irrep, two irreps contribute to the two-columns irrep with one-box in the second column.) Similarly, the number of states in the two-columns irrep with one-box in the second row is less than the number of states included in the irrep with two-boxes in the second row (except for the 4-particle irreps).

A similar truncation can apply to the isotopes with $T_{max} - 2$. In this case we have to add the irreps with two-boxes in the second column to each single-shell. For example, we have to add 7 more irreps to the $f_{\frac{7}{2}}$ single-shell, [22], [221], [2211], ..., [22111111], for a full multi-shell calculation. Similarly, for the $h_{\frac{11}{2}}$ single-shell we have to add 11 more irreps - starting with [22] and ending with [2211111111111]. This amounts to keeping only 35 irreps all together and leaving out 55 irreps. This is a major truncation which will enable us to run easily the Te isotopes (as far as the dimension of the Hamiltonian matrix remains manageable.)

This truncation can also be applied to the isotopes with $T_{max} - 3$; this requires adding the permutational irreps with 3 boxes in the second column, and, consequently to add to each sub-shell list of irreps. For example, for the $f_{\frac{5}{2}}$ shell we need to add the irrep [222], which will close the list of all possible irreps for this single-shell. On the other hand, for the $h_{\frac{11}{2}}$ shell we will need to add 10 more irreps - [222] till [222111111111] - and yet still be left with a truncation of

45 irreps - half of all possible irreps.

We can continue with this procedure for $T_{max} - 4$, $T_{max} - 5$, etc., and add in each case an extra box in the second column, until the minimum value of the isospin - $T_{min}$ is reached. It is clear that for this last case we must use all of the two-columns irreps for each sub-shell, with no possible truncation.

## 4. Application to Sn Isotopes

Sn isotopes in the *sdgh*-shell, including single-particle orbits $2s_{\frac{1}{2}}$, $1d_{\frac{5}{2}}$, $1d_{\frac{3}{2}}$, $0g_{\frac{7}{2}}$, and $0h_{\frac{11}{2}}$, have the strecched isospin with $T_{max}$. In this case it is sufficient to truncate the $h_{\frac{11}{2}}$ states to those belonging to the 12 one-column irreps [1], [11], [111], ... [111111111111] compared to the 90 two-column irreps. The other shells belong to the $2j+1$ one-column irreps (where $j$ is the spin of each single-shell); 2 irreps [1], [11] for $s_{\frac{1}{2}}$, 4 irreps [1], [11], [111], [1111] for $d_{\frac{3}{2}}$, 6 irreps [1], [11], [111], ...,[111111] for $d_{\frac{5}{2}}$, 8 irreps [1], [11], [111], ..., [11111111] for $g_{\frac{7}{2}}$.

This great saving of unnecassay calculation of single-shell irreps allows us to calculate the coefficient of fractional parentage (CFP) and to build the many-body basis states for Sn isotopes. The Hamiltonian matrix for a specific J and T of an Sn isotope are then calculated by using the matrix elements of the elementary operators, the coupling coefficients for outer and inner products, and the given single particle and potential matrix elements in the basis. Finally, the Hamiltonian matrix are diagonalized exactly or via the Lanczos algorithm [17] depending on size.

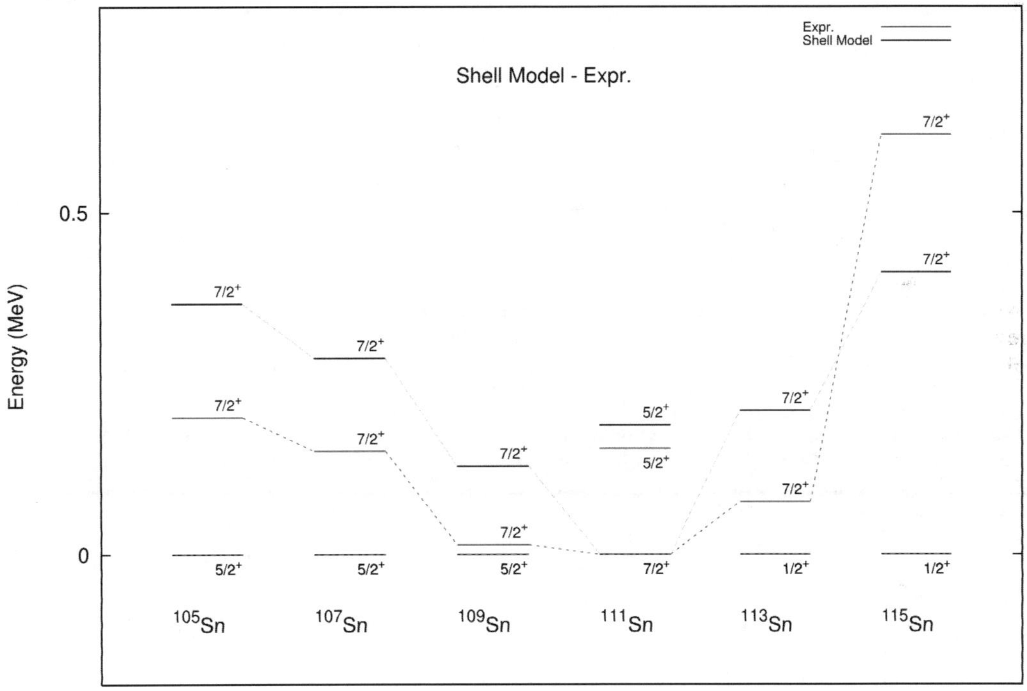


**Figure 2.** Experimental and theoretical ground state and low-lying state $\frac{7}{2}^{+}$ of odd-mass Sn isotopes with $A = 105 - 115$.

In our calculations for Sn isotopes we assume that $^{100}$Sn is a closed core and let the valence neutrons occupy the single particle orbits $2s_{1/2}$, $1d_{5/2}$, $1d_{3/2}$, $0g_{7/2}$, and $0h_{11/2}$. We have used the CD-Bonn effective interactions as two-body matrix elements given to us by M. Hjorth-Jensen *et al.* [18]. We choose the single particle energies for neutrons as $\varepsilon_{1d_{5/2}} = 0.00$ MeV, $\varepsilon_{0g_{7/2}} = 0.172$ MeV, $\varepsilon_{2s_{1/2}} = 2.45$ MeV, $\varepsilon_{1d_{3/2}} = 2.55$ MeV, and $\varepsilon_{0h_{11/2}} = 3.20$ MeV. The evalution of the ground states and low-lying state $\frac{7}{2}^+$ of odd-mass Sn isotopes with $A = 105 - 115$ are shown in Figure 2.

## 5. Conclusion

In conclusion, we presented an algorithm based on the DUSM approach for constructing single-shell states which implements an exact truncation based on the occurrence of only certain Young diagrams in building the multi-shell model states. The rule is that if the isospin is less by $n$ from its maximum value we have to keep the irreps with at most $n$ boxes in their second column. This truncation scheme is extremely important for isotopes with isospin near stretched value. It reduces drastically the necessary number of single-shell one- and two-body matrix elements. This truncation is naturally implemented in the DUSM algorithm to perform shell model calculations since this algorithm is based on the use of the permutation group to build the many-body basis states. This truncation will allow the description of isotopes in the middle of the $fp$-shell and also of isotopes in the $sdgh$-shell relatively far from the beginning of the $gdhs$ shell.

The exact truncation described in this contribution are applied to the odd-mass Sn isotopes with $A = 105 - 115$ and found to be very good agreement with experiments. It is a challenging problem to extent these studies to the Sn isotopes with $A = 117 - 120$ including the most stable naturally found $^{120}$Sn isotope.

## References

[1] Cohen S and Kurath D 1965 *Nucl. Phys.* **73** 1
[2] Wildenthal B H 1984 *Prog. Part. Nucl. Phys.* **11** 5
[3] Brown B A, Richter W A, Julies R E and Wildenthal B H 1988 *Ann. Phys.* **182** 191
[4] Caurier E, Zuker A P, Poves A and Martinez-Pinedo G 1994 *Phys. Rev.* C **50** 225
[5] Martinez-Pinedo G, Zuker A P, Poves A and Caurier E 1997 *Phys. Rev.* C **55** 187
[6] Novoselsky A, Vallières M and La'adan O 1997 *Phys. Rev. Lett.* **79** 4341
[7] Hjorth-Jensen M, Kuo T T S and Osnes E 1995 *Phys. Rep.* **261** 125
[8] Covello A*et al.* 1997 *Prog. Part. Nucl. Phys.* **38** 165
[9] Dikmen E, Novoselsky A and Vallieres M 2001 *Phys. Rev.* C **64** 067305
[10] Dikmen E, Novoselsky A and Vallieres M 2002 *Phys. Rev.* C **66** 057302
[11] Dikmen E, Novoselsky A and Vallieres M 2007 *J. Phys. G: Nucl. Part.* **34** 529
[12] Dikmen E, Ozturk O and Vallieres M 2009 *J. Phys. G: Nucl. Part.* **36** 045102
[13] Novoselsky A, Vallieres M and Gilmore R 1988 *Phys. Rev.* C **38** 1440
[14] Chen J Q, Novoselsky A, Vallieres M and Gilmore R 1989 *Phys. Rev.* C **39** 1088
[15] Vallieres M and Novoselsky A 1993 *Nucl. Phys.* A **570** 345c
[16] Geist A, Beguelin A, Dongarra J, Jiang W, Manchek R and Sunderam V 1994 *PVM - Parallel Virtual Machine* (Cambridge, MA, MIT Press)
[17] Lanczos C 1950 *J. Res. Nat. Bur. Stand.* **45** 55
[18] Hjorth-Jensen M *et al.* 1998 (private communication)

Journal of Physics: Conference Series **590** (2015) 012022 doi:10.1088/1742-6596/590/1/012022

# Coulomb and nuclear components of the breakup, their interference and effect on the fusion process

**P R S Gomes[1], J Lubian[1], L F Canto[1,2], D R Otomar[3], M S Hussein[4,5]**
1-Instituto de Física, Universidade Federal Fluminense, Av. Gal. Milton Tavares de Souza s/n, Niterói, R.J., Brazil
2-Instituto de Física, Universidade Federal do Rio de Janeiro, Rio de Janeiro, Brazil
3- Departamento de Matemática, Universidade Estadual do Centro-Oeste, campus de Irati, Paraná, Brazil
4- Instituto de Física and Instituto de Estudos Avançados, Universidade de São Paulo, São Paulo, Brazil
5- Departamento de Física, Instituto Tecnológico de Aeronáutica, CTA, São José dos Campos, S.P., Brazil

E-mail: paulogom@if.uff.br

**Abstract.** We discuss reaction mechanisms involving weakly bound nuclei, at near barrier energies, and the couplings between different reaction channels. This paper may be thought as a brief description of state of the art of this field, particularly on breakup reactions and their influence on the fusion cross section. Recent experimental and theoretical results are presented, including the interference between Coulomb and nuclear components of the breakup and the systematics so far reached on the static effects due to the characteristic of weakly bound nuclei, especially halo-nuclei and the dynamic effects of the breakup coupling on the fusion cross section.

## 1- Introduction

Several reaction mechanisms may occur when weakly bound nuclei are involved, at energies near the Coulomb barrier, apart from the usual processes which are present when tightly bound nuclei interact (inelastic excitations, direct transfer of nucleons or clusters of nucleons, fusion). If at least one of the colliding nuclei (usually the projectile) has small breakup threshold energy, typically smaller than 3 MeV, this nucleus may breakup in the field of the partner nucleus and different characteristic processes may occur, such as sequential complete fusion, when all fragments fuse, incomplete fusion (ICF) - when some but not all fragments fuse and non-capture breakup (NCBU) - when neither fragment fuses. Complete fusion (CF) is the sum of sequential and direct complete fusion. Total fusion (TF) is the sum of CF and ICF.

It has been observed that the optical potential in the elastic scattering of weakly bound nuclei does not have the usual energy dependence of tightly bound nuclei, namely the threshold anomaly, corresponding to the decrease of the imaginary potential when the bombarding energy decreases towards the barrier energy, but rather a different behaviour that was named breakup threshold anomaly (BTA) [1]. The BTA occurs because the breakup cross section is still important at sub-barrier energies and it is possible to observe that the imaginary potential may even increase at energies close to the barrier, because breakup produces repulsive polarization potentials [2-4]. Theoretically, the most suitable calculations involving breakup are the so-called CDCC (continuum discretized coupled channel) calculations, since the breakup feeds states in the continuum. It has been shown that if one wants to describe the behaviour of elastic scattering angular distributions of weakly bound nuclei, it is essential that continuum-continuum couplings are included in the CDCC calculations [3,5].

These and other special features of reactions involving weakly bound nuclei, both stable and radioactive, and particularly the very specific characteristics of halo-nuclei, made this field a subject of great interest, both theoretically and experimentally, in the last years. A comprehensive review report has been published on this field [6]. Recently it has been observed experimentally [7] that at sub-barrier energies the breakup following direct transfer of nucleons of stable weakly bound nuclei ($^6$Li, $^7$Li, $^9$Be) predominates over the direct breakup of those nuclei.

Among the most important questions on this subject, one finds: (i) Does the breakup enhance or suppress the CF and or the TF cross sections? (ii) What is the relative importance of the Coulomb and nuclear components of the breakup? Do they interfere? (iii) How large are the NCBU cross sections, compared with the fusion cross sections? The answers to these questions depend on the energy regime (above or below the barrier), the target mass or charge and if the projectile has halo-characteristics.

## 2- The non-capture breakup

The measurement of NCBU cross section is a very difficult task. It requires very accurate exclusive experiments with coincidences between the fragments and then to convert the events in integrated cross sections. A clear identification of the processes, including sequential breakup (breakup following transfer) is possible through the Q-values of the reactions. [7]. However, if one is interested in the investigation of the effect of breakup on the fusion cross section, it is of fundamental importance to have indications on the time scale of the breakup. If the breakup occurs when the projectile approaches the target, what is called prompt breakup, it may affect fusion. Otherwise, if the breakup occurs when the projectile is already far from the target and moving away from it, what is called delayed breakup, the process cannot affect fusion. There are different forms of delayed breakup, those corresponding to the population of a long-lived resonance before the breakup and the breakup following a direct transfer of nucleons. In the prompt breakup, the relative energy between the fragments usually is large, whereas for delayed it is small. The relative energy may be determined in some experiments where the trajectory of the fragments can be determined [7].

One may predict the whole direct breakup (prompt + delayed) by performing reliable CDCC calculations. This was done [8, 9] for the $^6$Li projectile on four different targets, from $^{59}$Co to $^{208}$Pb. By reliable calculation we mean that no free parameter was used, only predictions using potentials which agree with available elastic scattering angular distributions. In what follows we give some details of our CDCC calculations.

The wave function with total angular momentum $J$ and z-projection $M$ can be schematically written as

$$\Psi^{JM}\left(\vec{R},\vec{r}\right)=\sum_{i}\frac{F_i^J(R)}{R}Y_i^{JM}\left(\hat{R},\vec{r}\right), \qquad (1)$$

where $i$ stands for the set o f the quantum numbers $\{\varepsilon_l, l_i, j_i, L\}$, being the centroid bin energy, the $\alpha$ -d angular momentum, $j = l + s$ (where $s$ is the spin of the deuteron) and the $^6$Li – target angular momentum, respectively. $\vec{r}$ represents the internal coordinate of the $^6$Li projectile; $\vec{R}$ is the projectile-target separation vector; $\hat{R}$ stands for its angular degrees of freedom. $Y_i^{JM}\left(\hat{R},\vec{r}\right)$ is the tensor product of the internal wave function of the projectile and the angular part of the projectile-target relative motion wave function. Inserting this wave function in the usual Shrödinger equation, and after some manipulations one obtains the following set of coupled channels equations

$$\left[T_L+U_{ii}^J(R)-E+\varepsilon_i\right]F_i^J(R)=-\sum_{j\neq i}U_{ij}^J(R)F_j^J(R)\cdot \qquad (2)$$

The index $i = 0$ stands for the elastic channel, while $i > 0$ are associated with the continuum bin states. Within the cluster model approximation, the interaction of the $^6$Li with the targets is split into two parts

$$V(\vec{R},\vec{r}) = V_{\alpha T}(\vec{R},\vec{r}) + V_{dT}(\vec{R},\vec{r}), \tag{3}$$

corresponding to the $\alpha$- target and deuteron-target interactions, respectively. The matrix-elements in expression (2) are given by

$$U_{ij}^{J}(R) = \int d^2\hat{R}d^3\vec{r}\,Y_i^{JM*}(\hat{R},\vec{r})V(\hat{R},\vec{r})Y_j^{JM}(\hat{R},\vec{r}). \tag{4}$$

The calculations were performed by switching on and off the Coulomb and nuclear components of the potential of the equation (3), to investigate separately the nuclear and Coulomb components of the breakup. The calculations have shown that for large scattering angles, corresponding to short range scattering, the nuclear breakup component may predominate over the Coulomb component, and that at energies above the barrier of light systems, the integrated nuclear component may also predominate. Very interesting results [8, 9] are the observation that there is a strong destructive interference between the two breakup components, and that the Coulomb component is larger than the total breakup at energies below the barrier and, for heavy targets, even for some energies slightly above the barrier. Otomar *et al.* and Hussein *et al.* [8, 9] were able to show that the nuclear breakup component cross section increases linearly with $A_t^{1/3}$, for the same $E_{c.m.}/V_B$ energy, where $A_t$ and $V_B$ are the target mass and height of the Coulomb barrier, respectively. For the Coulomb component of the breakup, the cross section increases linearly with the $Z_t$. The calculated integrated total breakup cross section was found to be larger than the experimental fusion cross section at sub-barrier energies for medium mass and heavy targets, and smaller than fusion cross section at energies above the barrier. This may be understood because at sub-barrier energies it is required the tunneling of the barrier to occur fusion, whereas breakup is a direct process and no tunneling is required.

### 3- Effect of breakup on the fusion cross section

Now we turn to the discussion of fusion cross section and the effect of the breakup on the fusion mechanism. The first point that has to be clear when one talks about enhancement or suppression of fusion of weakly bound nuclei owing to the breakup is in relation to what one is talking about. One may compare data with theoretical predictions or compare data between weakly bound and tightly bound systems. However, another very important point to make is to disentangle between two different effects, the static and the dynamic. The former is due to the longer tail of the optical potential owing to the weakly bound nucleons, especially if the nucleus is of halo-type. The latter is due to the strong coupling between the elastic channel and the continuum states representing the breakup channel. When one compares experimental fusion cross sections with theoretical predictions, the difference between them should be the "ingredients" missing in the theoretical calculations. So, if the calculations consider only a single channel, without any couplings and furthermore, standard densities for the weakly bound nuclei, the difference between fusion data and theory corresponds to all static plus dynamic effects. However, standard densities for weakly bound nuclei, especially halo-nuclei, are not realistic, since those nuclei have much more diffuse densities than the tightly bound ones. Figure 1 shows fusion cross section data for the $^6$He + $^{209}$Bi system [10] and two theoretical curves. In both calculations there are no couplings and the double folding Sao Paulo potential [11] was used. The dashed curve is the result if one uses as the neutron density of the neutron-halo $^6$He a "standard" density corresponding to the density of $^4$He scaled to consider the ratio of the number of neutrons (4 and 2). Therefore, the difference between fusion data and the dashed curve corresponds to all static plus dynamic effects on the fusion cross section. The full curve is the result when one considers the realistic proton and neutron densities of $^6$He, a neutron-halo nucleus, which produces a smaller Coulomb barrier height [12]. The difference between the fusion data and the full curve corresponds to all dynamic effects or all coupling effects not considered in the theory. So, the difference between the full and dashed curves represents the static effects of the neutron halo-characteristics of $^6$He. One can

Journal of Physics: Conference Series **590** (2015) 012022 doi:10.1088/1742-6596/590/1/012022

observe from figure 1 that this effect is of one magnitude order at sub-barrier energies, which can be considered as a large effect, although smaller than the low lying coupling effects of highly deformed nuclei, like $^{154}$Sm, at sub-barrier energies [13-15]. We believe that the static effect of halo-nuclei, enhancing the sub-barrier fusion cross section is well understood at the present. Figure 1 also shows that the static effects also enhance the fusion cross section at energies above the Coulomb barrier.

## 4- Fusion cross section reduction method

If one wants to compare fusion cross section data of different systems, including tightly and weakly bound ones, in the same figure, of course one can not compare cross sections versus $E_{c.m.}$, since the gross dependences of sizes and charges, as well as diffused densities, leading to different barriers, should be removed before the comparison is made. The most widely "reduction" methods are to divide, in the vertical axis, the cross section by $R_B^2$, where $R_B$ is the position of the Coulomb barrier (related to the size of the system) and to divide $E_{c.m.}$ by $V_B$ or to use $E_{c.m.} - V_B$, in the horizontal axis. Gomes et al [16] suggested an improved and alternative method to deal with weakly bound systems. The conclusions of the comparisons between fusion cross sections of different systems were shown to change depending on the reduction procedure used [17], even for very similar systems. Furthermore, it was shown that none of the above mentioned methods fully eliminates the static effects [17, 18]. Canto et al. [17, 18] proposed the use of dimensionless quantities, which appropriately eliminates static effects, as a procedure to investigate dynamic effects on the fusion cross sections due to the breakup couplings. Furthermore, the proposed method allows reaching a systematic understanding of this subject, since it allows the comparison of any kind of system in the same graphic. This method uses a benchmark curve, called the Universal Fusion Function (UFF), given by $F_0(x) = \ln[1 + \exp(2\pi x)]$, where $x = (E-V_B) / \hbar\omega$ and $F(x) = (2 E_{c.m.} / \pi R^2_B \hbar\omega)\, \sigma_{fus}$. $\hbar\omega$ is related to the barrier curvature, $\sigma_{fus}$ is the fusion cross section and $F(x)$ is called fusion function. Since this method is inspired by the Wong formula [19], which is not valid for light systems at sub-barrier energies and one is not usually interested in the well known dynamic effects on fusion owing to inelastic excitation couplings, the experimental fusion function has to be renormalized by using appropriated coupled channel calculations, in order to be compared with the UFF curve. Details of this procedure may be found in refs. [17, 18]. The differences between the renormalized experimental fusion functions and the UFF curve are dynamic effects due to the channels left out of the coupled channel calculations, in this case, breakup and transfer reactions. In our calculations, the bare potential used was the double folding Sao Paulo potential [11]. The method proposed by Canto et al. [17] was later extended for the analysis of total reaction cross section [20].

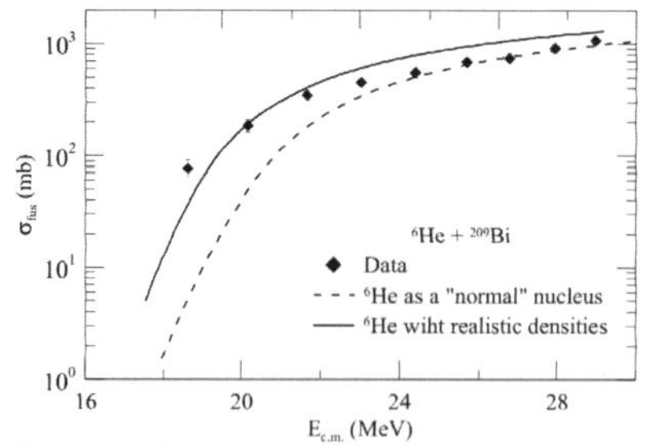


**Figure 1:** Comparison of experimental fusion cross section of the neutron-halo $^6$He nucleus with two theoretical predictions. See text for details.

Figure 2 illustrates, with four systems, the systematics obtained for the dynamic effects of breakup plus transfer channels on the total fusion of stable weakly bound nuclei, complete fusion of stable weakly bound nuclei, total fusion of neutron-halo nuclei and total fusion of proton-halo nuclei. The figure on the left is in logarithmic scale, more suitable to analyze the sub-barrier energy region and the figure on the right, in linear scale, is more appropriate to analyze the effects at energies above the barrier. Total fusion of several stable tightly and weakly bound systems, represented in figure 2 by the $^{9}$Be + $^{208}$Pb system [21], from very light to heavy targets, coincide with the UFF at energies above the barrier. This means that there is no dynamic effect of breakup plus transfer on the total fusion cross section at this energy regime. At energies below the barrier, some small enhancement may be observed. For the complete fusion of stable weakly bound nuclei, represented in figure 2 by the $^{9}$Be + $^{208}$Pb system [21], some suppression of the order of 30% is found, at energies above the barrier, for several systems, with target masses larger than around 90, since there are no CF cross section data available for light systems. It is very interesting to observe that for each weakly bound projectile, the CF suppression factor is independent of the target mass [22]. An analytical relation of the suppression factor with the breakup threshold energy was derived by Wang et al. [22]. However, a physical explanation for that expression is still missing. For the fusion of neutron-halo nuclei, represented in figure 2 by the $^{6}$He + $^{209}$Bi system [10], suppression above the barrier and some enhancement below the barrier is also observed for most of the available data. However, a quantitative estimation of the suppression factor has not been achieved, due to the lack of a larger number of systems investigated and the large error bars in the fusion cross sections of radioactive projectiles. Finally, for fusion with proton-halo nuclei, there are only two reported systems, for $^{8}$Be + $^{58}$Ni [23] and $^{8}$Be + $^{28}$Si [24], with contradictory results. Fusion data by Pakou *et al.* [24] follow the same systematics as for neutron-halo systems, whereas Aguilera`s data [23], shown in figure 2, presents a very unusual enhancement at energies above the barrier. Rangel et al. [25] suggested that the fusion cross section for the $^{8}$B + $^{58}$Ni system may be overestimated, because the protons detected and considered by the authors to originate from fusion evaporation have also contributions from protons from breakup.

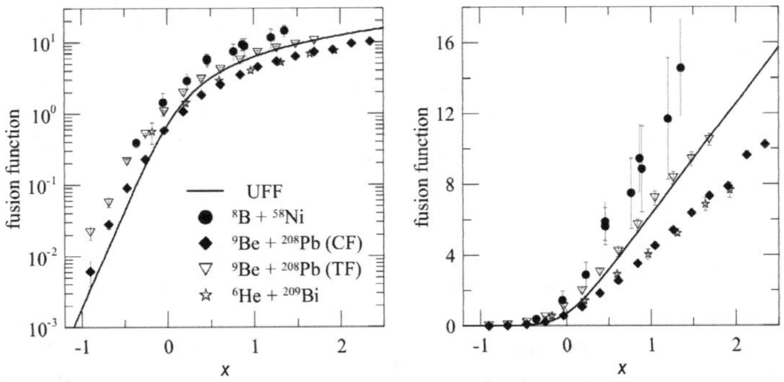


**Figure 2:** Renormalized experimental fusion function for some selected systems. The curve is the benchmark UFF curve. See text for details.

An interesting question is why the dynamic effect of breakup plus transfer on the fusion cross section is to suppress fusion at energies above the barrier and to enhance fusion below the barrier. Gomes et al. [26] suggested an explanation based on the energy dependent optical model and dynamic polarization potentials (DPP). Calculations for direct breakup [2-4] and for quasi-elastic barrier distributions [27, 28] show that the direct breakup produces repulsive DPP, owing to the couplings among continuum breakup states (continuum-continuum couplings), which increases the barrier height and suppress fusion. On the other hand, recent experimental evidences mentioned at the beginning of this paper [7] show that breakup of stable weakly bound nuclei triggered by nucleon transfer predominates over the direct breakup, at sub-barrier energies. So, the polarization potentials for each one should be evaluated separately and the results summed. Thus, the suppression of CF above the Coulomb barrier should result from the predominance of the DPP associated with direct breakup, whereas transfer and transfer followed by breakup, both producing attractive DPP, predominates at sub-barrier energies, especially for neutron-halo nuclei. The suppression above the barrier can also be explained by the BTA already mentioned.

Finally, we would like to mention the apparent contradiction between the observed facts that the effect of the breakup on the suppression of the complete fusion cross section does not depend on the target mass (or charge), but the breakup cross section increases with the target mass and charge. As mentioned at the beginning of this paper, only prompt breakup may affect fusion, and the calculated breakup contains both prompt and delayed breakups. So, the conclusion is that the prompt breakup cross section may not depend on the target mass or charge.

**Acknowledgements**
The authors would like to thank FAPERJ, CNPq, CAPES and FAPESP for the financial support.

**References**

[1] Hussein M S, Gomes P R S , Lubian J and Chamon L C 2006 *Phys. Rev. C* **73** 044610
[2] Lubian J et al. 2007 *Nucl. Phys. A* **791** 24.
[3] Canto L F, Lubian J, Gomes P R S and Hussein M S 2009 *Phys. Rev. C* **80** 047601
[4] Santra S et al. 2011 *Phys. Rev. C* **83** 034616
[5] Lubian J et al. 2009 *Phys. Rev. C* **79** 064615
[6] Canto L F, Gomes P R S, Donangelo R and Hussein M S 2006, *Phys. Rep.* **424** 1
[7] Luong D H et al. 2011 *Phys. Lett. B* **695** 105; 2013 *Phys. Rev. C* **88** 034609
[8] Otomar D R, Gomes P R S, Lubian J, Canto L F and Hussein M S 2013 *Phys.Rev. C* **87** 014615
[9] Hussein M S, Gomes P R S, Lubian J, Otomar D R and Canto L F 2013 *Phys.Rev. C* **88** 047601
[10] Kolata J J et al. 1998 *Phys. Rev. Lett.* **81** 4580
[11] Chamon L C et al. 1997 *Phys. Rev.Lett.* **791** 5218; 2002 *Phys. Rev. C* **66** 014610
[12] Gomes P R S, Canto L F, Lubian J and Hussein M S 2011 *Phys. Lett. B* **695** 320
[13] Stokstad R G et al. 1978 *Phys. Rev. Lett.* **41** 465
[14] Gil S et al. 1990 *Phys.Rev.Lett.* **65** 3100
[15] Gomes P R S et al. 1994 *Phys.Rev. C* **49** 1218
[16] Gomes P R S, Lubian J, Padron I and Anjos R M 2005 *Phys. Rev. C* **71** 017601
[17] Canto L F, Gomes P R S, Chamon L C and Crema E 2009 *Nucl. Phys A* **821** 51
[18] Canto L F, Gomes P R S, Chamon L C and Crema E 2009 *J. Phys. G* **36** 015109
[19] Wong C Y 1973 *Phys.Rev.Lett.* **31** 766
[20] Shorto J M B et al. 2009 *Phys.Lett. B* **678** 77
[21] Dasgupta M et al. 1999 *Phys.Rev.Lett.* **82** 1395; 2004 *Phys.Rev. C* **70** 024606
[22] Wang B, Zhao J, Gomes P R S and Zhou Shan-Gui 2014 *Phys.Rev. C* **90** 034612
[23] Aguilera E F et al. 2011 *Phys.Rev.Lett.* **107** 092701
[24] Pakou A et al. 2013 *Phys.Rev. C* **87** 014619
[25] Rangel J et al. 2013 *Eur. Phys. J A* **49** 57
[26] P R S et al. 2012 *J. Phys. G* **39** 115103
[27] Monteiro D S et al. 2009 *Phys.Rev. C* **79** 014601
[28] Lubian J, Correa T, Gomes P R S and Canto L F 2008 *Phys.Rev. C* **78** 064615

Journal of Physics: Conference Series **590** (2015) 012023 doi:10.1088/1742-6596/590/1/012023

# Photoneutron Reactions in Nuclear Astrophysics

**H. Utsunomiya[1,2], S. Goriely[3], D.M. Filipescu[4], O. Tesileanu[4], I. Gheorghe[5], T. Glodariu[5], H.-T. Nyhus[6], T. Renstrøm[6], Y.-W. Lui[7], T. Shima[8], K. Takahisa[8], S. Miyamoto[9]**

[1] Department of Physics, Konan University, Kobe, Japan
[2] Center of Nuclear Science, University of Tokyo, Saitama, Japan
[3] Institut d'Astronomie et d'Astrophysique, Université Libre de Bruxelles, Brussels, Belgium
[4] Extreme Light Infrastructure Nuclear Physics, Bucharest-Magurele, Romania
[5] National Institute for Physics and Nuclear Engineering Horia Hulubei, Bucharest-Magurele, Romania
[6] Department of Physics, University of Oslo, Oslo, Norway
[7] Texas A&M University, College Station, Texas, USA
[8] Research Center for Nuclear Physics, Osaka University, Osaka, Japan
[9] Laboratory of Advanced Science and Technology for Industry, University of Hyogo, Hyogo, Japan

E-mail: hiro@konan-u.ac.jp

**Abstract**. Highly-monochromatic γ-ray beams are produced at the NewSUBARU synchrotron radiation facility by the inverse Compton scattering of laser photons from relativistic electrons. The latest s-process study in nuclear astrophysics with the γ-ray beam is presented.

## 1. Laser Compton scattering - photon accelerator at NewSUBARU

Highly-monochromatic γ-ray beams are produced at the synchrotron radiation facility NewSUBARU in the inverse Compton scattering of laser photons from relativistic electrons circulating in a storage ring [1].

The laser inverse Compton scattering plays a role of *accelerating* the mass-less particle, photon. The energy of scattered photons is expressed by

$$E_\gamma = \frac{4\gamma^2 \varepsilon_L}{1+(\gamma\theta)^2 + 4\gamma\varepsilon_L/(mc^2)}, \qquad (1)$$

where $\varepsilon_L$ is the energy of a laser photon, $mc^2$ is the rest mass energy of an electron, $\theta$ is the scattering angle of a laser photon with respect to the electron incident direction, and $\gamma$ is the Lorentz factor for electron, $\gamma = E_e/mc^2$, defined by the total electron energy $E_e$ and the rest mass energy. The energy amplification factor in nearly head-on collisions ($\theta \approx 0$), $E_\gamma/\varepsilon_L \approx 4\gamma^2$, is very large on the order of $10^6$ - $10^7$ for several hundred MeV to a few GeV electrons so that an eV laser beam can be converted to an MeV γ-ray beam in the laser inverse Compton scattering. At NewSUBARU, one can produce low-energy γ-ray beams at a few MeV using a $CO_2$ laser (wave length $\lambda$=10.59 μm) and high-energy γ-ray beams at a few tens of MeV using a Nd:$YVO_4$ laser ($\lambda$ =1064nm) in collisions with 0.5 - 1.5 GeV electrons.

Recently, the electron beam energy of the NewSUBARU storage ring was calibrated with the accuracy of the order of $10^{-5}$ by using the low-energy laser Compton scattered (LCS) γ-ray beam produced with a grating-fixed $CO_2$ laser oscillated at a single line of the strongest master transition P(20) [2]. The central wave length of the P(20) transition is known ($\lambda$=10.5915μ m ± 3 Å) [3] with the band width 1.3 Å in the full width at half maximum FWHM [4]. Including the band width, the accuracy of the wave length of $CO_2$ laser is $4.1\times10^{-5}$. A coaxial HPGe detector (64 mm in diameter × 60 mm in length), which was calibrated with the standard γ-ray sources, $^{60}$Co including the sum peak, $^{133}$Ba, $^{137}$Cs, and $^{152}$Eu and a natural radioactivity $^{40}$K, was used to measure the low energy LCS γ rays.

The full energy peaks of the measured γ-ray spectra were reproduced with the Monte Carlo code EGS4/PRESTA [5], where the end-point energy of the full energy peak is sensitive to the electron beam energy. Figure 1 shows the result of the calibration, the difference of the calibrated energy $E_e^c$ from the nominal energy $E_e^n$, $\Delta E = E_e^c - E_e^n$ as a function of $E_e^n$. The absolute energy calibration of the electron beams offers a standard for the energy calibration of high-energy LCS γ-ray beams produced with a Nd:$YVO_4$ laser.

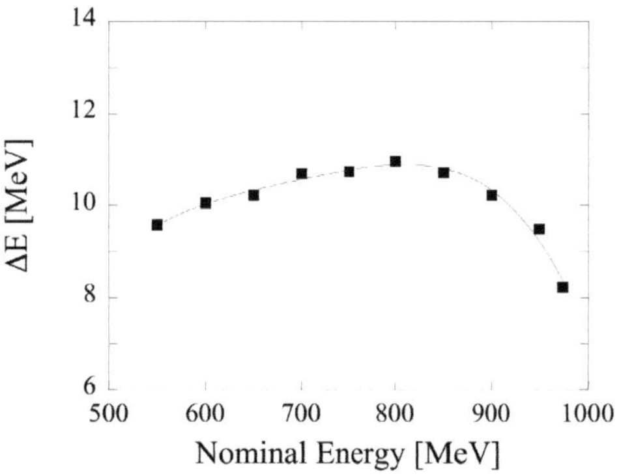


Figure 1. The difference of the calibrated energy $E_e^c$ from the nominal energy $E_e^n$ $\Delta E = E_e^c - E_e^n$ at the NewSUBARU storage ring.

Figure 2 shows typical spectra of the high-energy LCS γ-ray beams recorded with a 3.5" × 4.0" $LaBr_3$:Ce detector (solid lines) along with the GEANT4 simulations of the detector response function (dotted lines) and the incident γ-ray beam (gray lines). The experimental response functions are excellently reproduced by the GEANT4 simulation. Energy spreads of 1.2%, 1.4% and 1.6% in FWHM were obtained for the three incident γ-ray beams of 6.5, 10.0 and 13.0 MeV maximum energy, respectively.

## 2. s-process

Nuclear astrophysics links the microscopic system of the atomic nucleus with large galactic processes. Among many unknowns related to the synthesis of elements, a great challenge is to determine the radiative neutron capture cross section for unstable nuclei. In particular, radioactive nuclei along the line of β stability in the medium- to heavy-mass region has gained increasing interest. The cross sections are important to determine the s-process path at branching points where neutron capture and β decay compete [6]. Although experimental data of radiative neutron capture cross sections for stable nuclei are well documented [7, 8], those for radioactive nuclei are scarce due to the difficulty of direct measurements that requires both an intense neutron beam and radioactive samples. While some of radioactive nuclei with half-lives ($T_{1/2}$) of the order of years have become objectives for direct measurements at the European Organization of Nuclear Research, the Los Alamos Neutron Science Center, and the Japan Proton Accelerator Research Complex, those with $T_{1/2}$ of the order of days and shorter are far beyond the experimental scope at present.

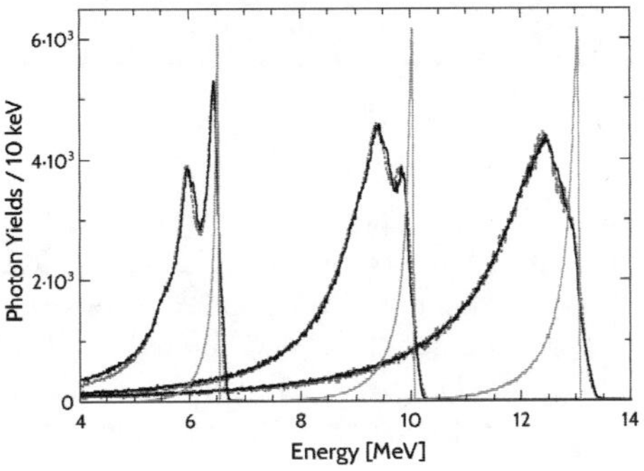


Figure 2. Typical spectra of the γ-ray beams recorded with a $LaBr_3$:Ce detector (solid lines), the simulations of the response function (dotted lines) and of the incident γ-ray beam (gray lines).

The well-established γ-ray strength function (γSF) method can be used for constraining the (n, γ) cross sections for radioactive nuclei [9, 10]. Relying on the Brink hypothesis [11] linking the photo-deexcitation process to photoabsorption, the method determines the γSF, which commonly quantifies radiative neutron capture and photoneutron cross sections. Photoneutron cross sections provide a stringent experimental constraint in absolute scale on the γSF around $S_n$. When auxiliary (γ, γ') and particle-γ coincidence data that help to construct the γSF below $S_n$ are unavailable, the method requires a systematic measurement of photoneutron cross sections for neighboring stable isotopes of a radioactive nucleus of interest, in addition to the existing (n, γ) data which serve as experimental constraints on the γSF below $S_n$. Thus, a unified understanding of (n, γ) and (γ, n) cross sections throughout an isotopic chain offers a detailed information on the γSF for a given nucleus formed by neutron capture on the radioactive nucleus. Such a systematic approach with the γSF method was applied to zirconium [12], tin [13] and molybdenum [14] isotopes.

Very recently, the γSF method has been applied to the neodymium [15] and samarium [16] isotope chains. Figure 3 shows photoneutron cross sections for neodymium isotopes in comparison with the Saclay data [17]. While a good agreement is obtained in the $^{146,148}$Nd cases, significant discrepancies are observed for the light Nd isotopes. Our experiment leads to cross section lowered by typically 20%. Such an overestimate by the Saclay photodata was also reported in previous comparisons for $^{142}$Nd with a renormalization by 0.86 [18], for $^{144}$Sm by 0.80 [19], and for $^{nat}$Rb, $^{nat}$Sr, $^{89}$Y, $^{90}$Zr, $^{93}$Nb, $^{127}$I, $^{197}$Au, and $^{208}$Pb by a factor 0. 80– 0. 93 [20].

Our new photoneutron cross sections are compared to theoretical calculations obtained with the TALYS nuclear reaction code [21, 22] and two different models of the γSF, namely the Skyrme HFB plus QRPA model [23] based on the BSk7 interaction, and the axially-symmetric-deformed Gogny HFB plus QRPA model based on the D1M interaction [24, 25, 26, 27]. As seen in Fig. 3, cross sections around the neutron threshold are better described by the D1M+QRPA model, although some deviations can be seen.

Journal of Physics: Conference Series **590** (2015) 012023 doi:10.1088/1742-6596/590/1/012023

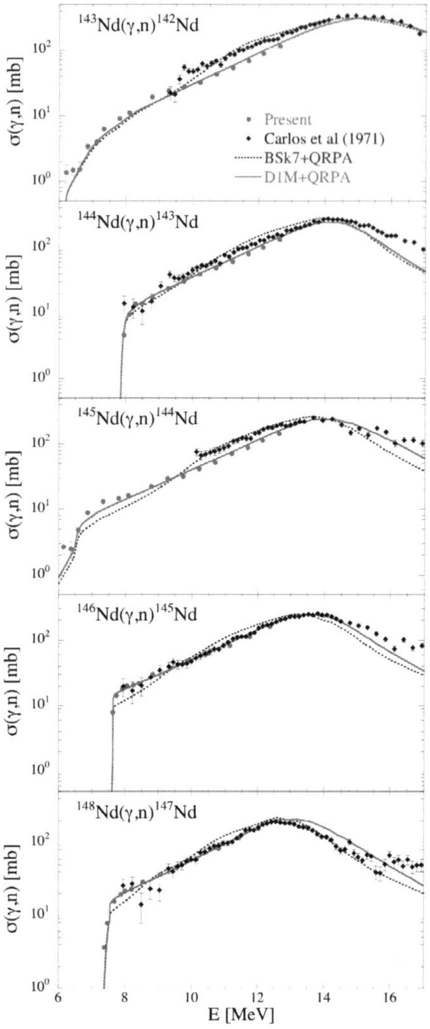


Figure 3. Comparison between the present photoneutronemission cross sections and previously measured ones [17] for $^{143-146,148}$Nd. Also included are the predictions from Skyrme HFB+QRPA (based on the BSk7 interaction) [23] and axially deformed Gogny HFB+QRPA models (based on the D1M interaction) [24].

We now turn to the reverse radiative neutron capture channel. It should be kept in mind that the corresponding cross section for incident keV neutrons depends sensitively on the γSF, but in a rather lower energy range below the neutron threshold, typically around 6 MeV of γ-ray energy. The predicted tail of the γSF at low energies therefore plays a fundamental role. On the basis of the Gogny HFB plus QRPA γ-ray strength [24], the reverse radiative neutron capture cross sections are now estimated with the TALYS reaction code [21, 22] for the stable and experimentally known $^{142-146,148}$Nd isotopes and compared with experimental cross sections in Fig. 4. In addition to the E1 strength function, the cross section calculation also depends on the adopted nuclear level density. We have used here the temperature-dependent HFB plus combinatorial model [28] normalized to the experimental s-wave spacing $D_0$ values [29].

As can be seen in Fig. 4, the TALYS calculation agrees well with experimental data for all the 6 Nd isotopes, which shows that within the uncertainties affecting the experimental γSF and $D_0$ value, all γSF data are compatible with both the photoabsorption above the threshold and the radiative capture channels below the threshold.

The γSF method can now be applied to the experimentally unknown cross section $^{147}$Nd(n, γ)$^{148}$Nd making use of the same nuclear inputs. In the case of $^{148}$Nd, experimental information exists on the resonance spacing at the neutron binding energy, namely $D_0 = 4.0 \pm 1.5$ eV [29]. The 40% error on $D_0$ corresponds to the major uncertainty still affecting the prediction of the $^{147}$Nd(n, γ)$^{148}$Nd cross section, giving rise to a range of predicted cross section within a typically 20–30%, as shown in Fig. 5. The American ENDF/B-VII.1 evaluations [41] is seen to be in relatively good agreement with our estimate, but the Russian ROSFOND-2010 evaluation gives rather lower cross sections above typically a few keVs and the Japanese JENDL-4.0 evaluation larger values above typically 100 keV.

The resulting Maxwellian-averaged cross section of astrophysical interest amounts, at 30 keV, to 880 ± 170 mb. Our Maxwellian-averaged cross section (and consequently, also the ENDF/B-VII.1 and JENDL-4.0 ones) is found to be significantly larger than the value of 544 ± 90 mb recommended in Ref. [7] as well as all the theoretical values compiled in the KADONIS library [8] ranging between 387 and 663 mb.

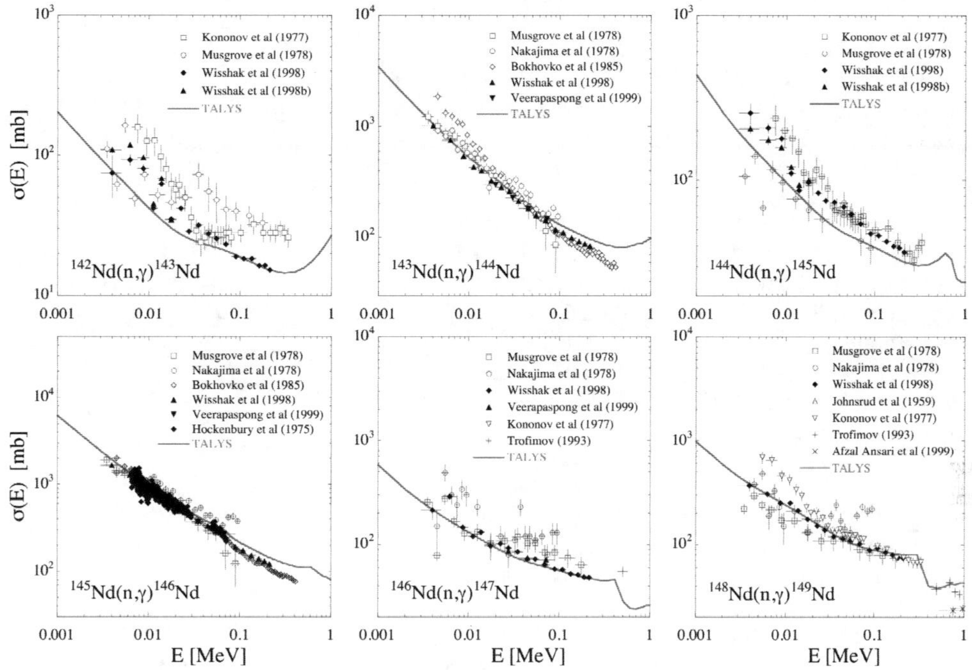


Figure 4. Comparison between the measured radiative neutron capture cross sections [30, 31, 32, 33, 34, 35, 36, 37, 38, 39, 40] with TALYS calculations making use of the D1M+QRPA E1 strength.

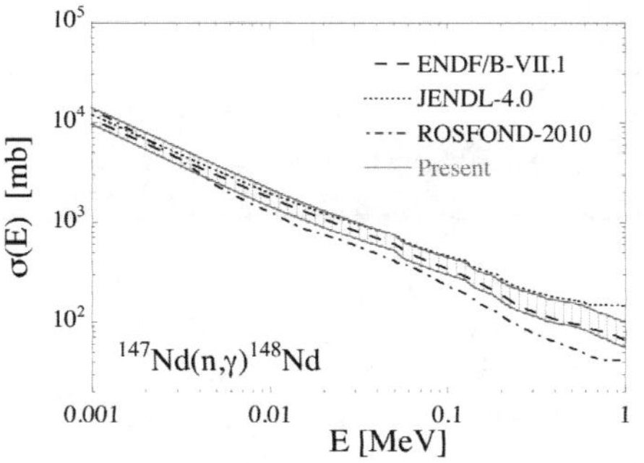


Figure 5. Prediction of the $^{147}$Nd(n,γ)$^{148}$Nd cross section. The dotted, dashed, and dash-dot curves correspond to the Japanese JENDL-4.0, American ENDF/B-VII.1, and Russian ROSFOND-2010 evaluations [41].

## Acknowledgments

This work was supported by the Japan Private School Promotion Foundation and partly by the JSPS-FNRS bilateral program. We are grateful to M. Igashira of the Tokyo Institute of Technology for lending the $^{143,145,146}$Nd samples. We acknowledge PRACE for awarding us access to the resource CURIE based in France at TGCC-CEA. D.M.F. and O.T. acknowledge financial support from the Extreme Light Infrastructure Nuclear Physics (ELI-NP) Phase I, a project co-financed by the European Union through the European Regional Development Fund. H.T.N. and T.R. acknowledge financial support from the Norwegian Research Council (NFR), project number 210007. S.G. acknowledges the financial support from the F.N.R.S.

**References**

[1] Amano S et al. 2009 *Nucl. Instrum. Methods in Phy. Res. A* **602** 337
[2] Utsunomiya H et al. 2014 *IEEE Transactions on Nuclear Science* **61** 1252
[3] Patel C K N 1964 *Phys. Rev. Lett.* **13** 617
[4] Abrams R L 1974 *Applied Physics Letters* **25** 609
[5] Ohgaki H et al. 2000 Nucl. Instrum. Methods in Phys. Res. A **455** 54
[6] Käppeler F, Gallino, Bisterzo R S, and Aoki W 2011 *Rev. Mod. Phys.* **83** 157
[7] Bao Z et al. 2000 *At. Data Nucl. Data Tables* **76** 70
[8] Dillmann I et al. 2006 *AIP Conf. Proc.* **819** 123
[9] Utsunomiya H et al. 2009 *Phys. Rev. C* **80** 055806
[10] Utsunomiya H et al. *Phys. Rev. C* **82** 064610
[11] Brink, D M 1955 *Ph.D Thesis*, Oxford University
[12] Utsunomiya H et al. 2010 *Phys. Rev. C* 81 035801
[13] Utsunomiya H et al. 2011 *Phys. Rev. C* 84 055805
[14] Utsunomiya H et al. 2013 *Phys. Rev. C* 88 015805
[15] Nyhus H T et al. 2014 *Phys. Rev. Lett.* submitted
[16] Filipescu D M et al. 2014 *Phys. Rev. C* submitted
[17] Carlos P et al. 1971 *Nucl. Phys. A* **172** 437
[18] Angell C T et al. 2012 *Phys. Rev. C* **86** 051302(R)
[19] Nair C et al. 2010 *Phys. Rev. C* **81** 055806
[20] Berman B L et al. 1987 *Phys. Rev. C* **36** 1286
[21] Koning A J et al. M 2008 Nucl. Data Sci. and Tech. (EDP Sciences eds Bersillon O et al.) p. 211
[22] Koning A J and Rochman D 2012 *Nuclear Data Sheets* **113** 2841
[23] Goriely S, Khan E, and Samyn M 2004 *Nucl. Phys. A* **739** 331
[24] Péru S and Goutte H 2008 *Phys. Rev. C* **77** 044313
[25] Goriely S, Hilaire S, Girod M, and Péru S 2009 *Phys. Rev. Lett.* **102** 242501
[26] Martini M, Hilaire S, Goriely S, Koning A J, and Péru S *Nuclear Data Sheets* **118** 273
[27] Péru S, and Martini M 2014 *Eur. Phys. J.* A **50** 88
[28] Hilaire S, Girod M, Goriely S, and Koning A J 2012 *Phys. Rev. C* **86** 064317
[29] Capote R, Herman M, Oblozinsky P et al. 2009 *Nuclear Data Sheets* **110** 3107
[30] Kononov V N et al. 1977 *Yadernye Konstanty* **22** 29
[31] Musgrove A R DeL 1978 *Int. Conf. on Neutr. Phys. and Nucl. Data* (Harwell), p. 449
[32] Wisshak K, Voss F, Käppeler F, Kazakov, and F L, Reffo G 1998 *Phys. Rev. C* **57** 391
[33] Wisshak K, Voss F, and Käppeler F 1998 *Phys. Rev. C* **57** 3452
[34] Nakajima Y et al. 1978 *Int. Conf. on Neutr. Phys. and Nucl. Data* (Harwell) p. 438
[35] Bokhovko M V et al. 1985 *Vop. At. Nauki i Tekhn. Ser.Yadernye Konstanty* Issue 3 p. 12
[36] Veerapaspong T et al. 1999 *Jour. of Nuclear Science and Technology* **36** 855
[37] Hockenbury R W, Koste W R, and Shaw R A 1975 *Bull. of the American Phys. Soc.* **20** 560
[38] Trofimov Yu N 1993 *Vop. At. Nauki i Tekhn. Ser. Yadernye Konstanty* Issue 1 p. 17
[39] Johnsrud A E, Silbert M G, and Barschall H H 1959 *Phys. Rev.* **116** 927
[40] Afzal Ansari M, Singh R K Y, Gautam R P, and Kailas S 1999 *Ann. Nucl. Energy* **26** 553
[41] *Evaluated Nuclear Data Files* 2011 http://wwwnds.iaea.or.at/exfor.htm

# Photonuclear reactions induced by a clinical linac

**I. Boztosun[1,2],H. Đapo[1,2],M. Karakoc[1,2],S.F. Ozmen[1,2],Y. Cecen[1,3], A. Coban[1,2],A. Cesur[1,2],T. Caner[1,2],E. Bayram[1,4], T.R. Saito[5,6,7],T. Akdogan[8],V. Bozkurt[5,9],D. Kaya[1,2],Y. Kucuk[1,2]**

[1] Department of Physics, Akdeniz University, TR-07058, Antalya, Turkey
[2] Nuclear Research and Application Center, Akdeniz University, TR-07058, Antalya, Turkey
[3]Department of Radiation Oncology, Akdeniz University,TR-07058, Antalya, Turkey
[4] Department of Chemistry, Akdeniz University, TR-07058, Antalya, Turkey
[5] GSI Helmholtz Center for Heavy Ion Research, Planckstrasse 1, 64291 Darmstadt, Germany
[6] Johannes Gutenberg-Universität Mainz, J.J.Becherweg 40, 55099 Mainz, Germany
[7] The Helmholtz Institute Mainz (HIM), J.J.Becherweg 40, 55099 Mainz, Germany
[8] Department of Physics, Bogazici University, 34342
[9] Nigde University, 51100 Nigde, Turkey

E-mail: boztosun@akdeniz.edu.tr, haris@akdeniz.edu.tr

**Abstract.** The use of a clinical linac in photonuclear experiments is a novel concept. At Akdeniz University a program has been started focused on inducing photonuclear reactions with a Philips SLI-25 clinical linac. A bremsstrahlung photon beam produced by the linac is used to create radioactive nuclei. The decay of said nuclei is observed with a high-purity germanium detector. During the course of the experiment the spectrum of the irradiated sample is recorded. By analyzing the spectrum the energies of gamma-ray transitions in the daughter nuclei as well as the half-life of the parent nucleus are obtained. The results obtained thus far, starting with Zinc nuclei will be presented.

The study of the photon-induced nuclear reactions has been a focus of interest over the years. The motivation for studying the interaction of photons with nucleus are many, but primarily photonuclear reactions experiments are aimed at and motivated by astrophysical references to nucleosynthesis, see Ref. [1, 2, 3, 4] and references there-in. In addition applications of the photonuclear reaction in elemental analysis are many. The appeal for studying photonuclear reaction also comes from availability of a good radiation source.

Photoactivation experiments involving linear accelerators have been performed at several institutions around the world Ref. [5, 6, 7, 8, 9]. In Ref. [10] a new idea of using clinical linac for photonuclear studies was presented. Locally we have had some initial success with this idea Ref. [11] and now wish to expand on it. The aim of the present work is to further expand the concept of using clinical linacs in photonuclear studies by setting up an experiment centered around a clinical linac in order to study nuclear reactions and nuclear structure and to investigate how such setups can contribute to the expansion of the nuclear physics knowledge base. For this purpose, we have chosen to study transition energies and half-lifes of isotopes created by photonuclear reactions with Zinc.

The photonuclear reactions require a source of photons to excite the target nuclei. For us this was an clinical electron linac SLI-25 manufactured by Philips Medical Systems (currently part

Journal of Physics: Conference Series **590** (2015) 012024 doi:10.1088/1742-6596/590/1/012024

of Elekta TM Synergy TM). The accelerator technical documentation can be found in Ref. [12]. Its properties compare well with dedicated linacs Ref. [10].

The cLINAC primary electron beam is generated by an electron gun with an energy of about 50 keV. The electron gun in SLI-25 is a diode design with a 400 Hz pulse repetition frequency. After injection into linac's copper cavity, the electrons are accelerated by a radio-frequency wave with 3 GHz(2856 MhZ), S-band. SLI-25 copper cavity is a traveling wave design. The power is provided by the magnetron with nominal power of 2.5 MW at 4 MeV (low energies) and 5 MW at 25 MeV (high energies).

After exciting the copper cavity, the beam falls onto the a high-Z element target, in our case, Tungsten. The Tungsten target is 0.3 mm thick and serves as an electron stopper and a bremsstrahlung photons source. The photons are collimated and flattened with several filters, as illustrated in Fig. 1 left, resulting in a spatially uniform photon beam with no position dependance. The focusing and collimation is a standard feature of all cLINAC, as it is paramount to maintain excellent spatial dose profile control when irradiating a patient. In fact, it is common to require dose knowledge to better than 3% accuracy.

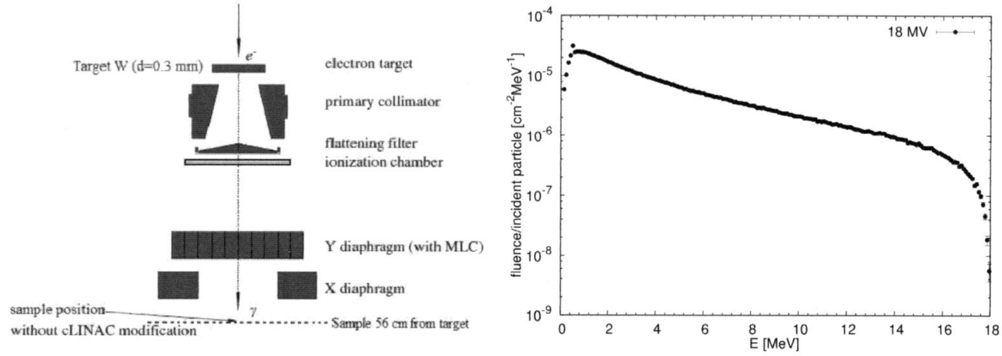


**Figure 1.** Left: Schematic view of the linac. Right: EGS simulation of bremsstrahlung energy distribution from an electron beam accelerated over an 18 MV potential difference impacting on thick Tungsten target in Philips SLI-25 clinical linac. The distribution is normalized per incoming particle.

In our experiment, we have collimated and flattened the beam such that it covers a $40 \times 40$ cm$^2$ area at a 100 cm distance. This is a maximum area at this distance and it was chosen primarily for ease of use. Other, smaller or irregularly-shaped openings of collimators are easily achievable.

Under all of these conditions, the question of bremsstrahlung energy distribution, which comes from the linac, becomes a quite complicated one due to presence of all the filters and collimators. A simulation of the photon energy distribution coming from the SLI-25 was performed using the BEAMnrc package. The resulting distribution is shown in Fig. 1 right. For the estimate of the flux we used the delivered dose rate 5 Gy/min Ref. [12] leading to $5 \times 10^5/(\mathrm{MeVcm}^2\mathrm{s}$ photons at $\bar{E} = 6$ MeV.

In the experiment, one Zinc sample was placed immediately outside the cLINAC head at about 56 cm source to sample distance, see Fig. 1. The sample in question was a Zinc disk weighing 5 g with 5 mm diameter and 1 mm thickness. The sample was irradiated for about 35 min. The sample was made primarily of zinc (67% by mass fraction). Since no special effort was made to isotopically enrich the target, the presence of different Zn isotopes can be considered as those of their natural abundances.

Journal of Physics: Conference Series **590** (2015) 012024 doi:10.1088/1742-6596/590/1/012024

After the irradiation, the sample was transported to the Physics Department of Akdeniz University, where the high-purity Germanium detector(HPGe) detector was located. It is a p-type, coaxial, electrically cooled HPGe, placed in a well-shielded cavity. The shield is 10 cm thick lead with an inner surface covered by a 2 mm copper foil to reduce the X-rays originating in lead. The HPGe used is a gamma-ray spectrometer from AMATEK-ORTEC (GEM40P4-83) with 40% relative efficiency and 768 eV FWHM at 122 keV for $^{57}$Co and 1.85 keV FWHM at 1332 keV for $^{60}$Co Ref. [13]. The HPGe is connected to a standard NIM equipment. The sample was placed into the detector about 10 min after the irradiation and counted for three days.

Immediately before counting the sample, a set of calibration sources were counted. After the sample was counted, an equivalently long natural background spectrum was recorded. Once the background spectrum was recorded, the experiment was concluded with a second measurement of the calibration sources.

The process of photactivation applies to all the nuclei for which the bremsstrahlung radiation exceeds the reaction threshold. Given that the end-point energy of our beam was 18 MeV, it is expected that all of the stable Zn isotopes will get activated. However, our experimental setup was not designed for direct observation of the stable Zinc isotopes, so we have to rely on the radioactive decay of the created nuclei. In this respect, we list the decay reactions which we have observed in the experiment presented

$$^{63}\mathrm{Zn} \rightarrow ^{63}\mathrm{Cu}^* + e^+ + \nu, \tag{1}$$
$$^{65}\mathrm{Zn} \rightarrow ^{65}\mathrm{Cu}^* + e^+ + \nu, \tag{2}$$
$$^{69}\mathrm{Zn}^* \rightarrow ^{69}\mathrm{Zn} + \gamma, \tag{3}$$
$$^{67}\mathrm{Cu} \rightarrow ^{67}\mathrm{Zn}^* + e + \bar{\nu}. \tag{4}$$

Here too, we have yet another limitation to what we can observe. As is noticeable the beta-decay of $^{69}$Zn is missing even though it most certainly did happen. The limitation here is the low branching ratio of the decay into the excited states of $^{69}$Ga which goes, almost exclusively (99.998%), to the ground state

After the data acquisition by the MAESTRO software, the peak analysis needed to be performed. However due to the limitation of the MAESTRO software, a decision was made to seek out other analysis programs. After some testing, we realized that the best option for the data analysis was to combine two different tools. One was the standard RadWare code developed for analysis of gamma-ray coincidence data, by David Radford of the Physics Division at Oak Ridge National Laboratory. The other was the ROOT, which is a package with extensive library structure, developed by CERN. The motivation for using two different programs was to strike a balance between the desired accuracy and time consumption of the data analysis.

The results obtained after analysis the sample eliminating the background peaks and performing both calibration before and after are show in Table 1 left. For calibration we used quadratic fit. And propagated the error from calibration according to:

$$\sigma_E^2 = \sum_i^n \left(\frac{\partial E}{\partial a_i}\right)^2 \sigma_{a_i}^2 + 2\sum_i^n \sum_{j>i}^n cov_{ij} \left(\frac{\partial E}{\partial a_i}\right)\left(\frac{\partial E}{\partial a_j}\right) + \left(\frac{\partial E}{\partial ch}\right)^2 \sigma_{ch}^2 \tag{5}$$

where $E = \sum_i^n a_i ch^i$ is the calibration polynomial, with $a_i$ calibration parameters, $\sigma_{a_i}$ are the errors of the fit parameters, $\sigma_{ch}$ is the errors of the centroid and $cov_{ij}$ is the covariance matrix.

Next we combined the two calibrations into a final result via Ref. [14]:

$$\sigma_E^2 = \frac{\sigma_{E_{bef}}^2 + \sigma_{E_{aft}}^2 + (\bar{E}_{bef} - \bar{E})^2 + (\bar{E}_{aft} - \bar{E})^2}{2} \tag{6}$$

$$\bar{E} = \frac{\bar{E}_{bef} + \bar{E}_{aft}}{2} \tag{7}$$

where $\bar{E}_{bef}$ and $\bar{E}_{aft}$ are the energies from before and after calibration and $\sigma_{E_{bef}}$ and $\sigma_{E_{aft}}$ are the corresponding errors, respectively.

If we compare these results with the literature values, it is possible to assign these peaks to specific isotopes and their transitions. In figure Fig. 2, we show the irradiated sample spectrum without any background subtractions. The assignment of peaks is evident in the figure. All the unsigned peaks are either background peaks or sum and escape peaks.

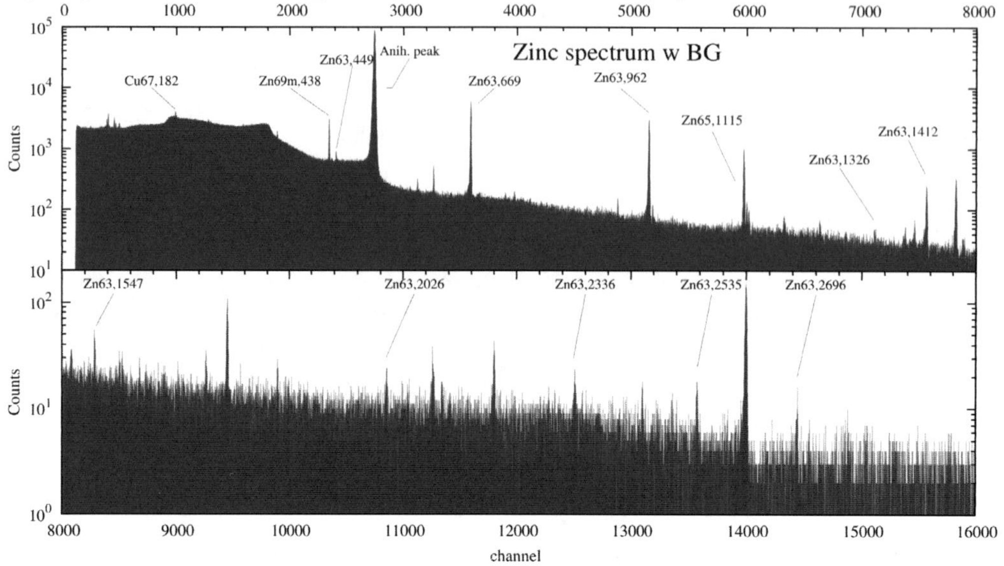


**Figure 2.** Full Zn spectrum with no subtractions, for the irradiated sample after 3 days of counting. Assigned peaks based on energy calibration.

Table 1 on the left shows the average energy and the combined variance in comparison to the literature values. The results quoted are taken from NUDAT and come from Nuclear data sheets publications, in these cases those are Ref. [15, 16, 17, 18]. In turn Nuclear data sheets results are taken from Ref. [19] for $^{63}$Cu, Ref. [20, 21] for $^{67}$Zn Ref. [22, 23] for $^{69}$Zn and Ref. [24] and others for $^{65}$Cu. It may be observed that the agreement is good for the most part. Outside of few exceptions majority of the results have overlapping error bars.

The primary motivation for this study has been the study of proton rich nuclei. The case of $^{63}$Cu, as the decay product of of $^{63}$Zn a proton rich nucleus, is where we see the best results. We draw the reader's attention to the especially good results for 1327.03 keV, 1392.55 keV, 1412.08 keV and 2026.8 keV, where our results fit almost completely within the NUDAT error bars, while having a better accuracy. In general, this range of values, between 1000 keV and 2000 keV, is where we expect that these kind of renewed studies on intermediate mass proton rich nuclei could contribute to the improvement of data on other nuclei as well. It is also our intent to continue these efforts in the future.

In addition to the transition energies, the analysis performed provided information about counts and their time evolution. This information can be used to determine the half-life of the parent nucleus as the decay of the daughter levels are in secular equilibrium with the $\beta$-decay of parent nucleus. Secular equilibrium is a very good assumption here since the half-life of the states in question is orders of magnitude smaller than the parents. The only exception is the $^{69}$Zn, which is an isomeric transition and is thus observed directly. The summary of the half-life fits for $^{63}$Zn is given in Table 1 right.

| Ele. | NUDAT E(keV) | $\sigma$ | $E$(keV) | $\sigma_E$ |
|---|---|---|---|---|
| $^{63}$Cu | 449.93 | 0.05 | 449.98 | 0.04 |
| $^{63}$Cu | 669.62 | 0.05 | 669.72 | 0.02 |
| $^{63}$Cu | 742.25 | 0.10 | 742.21 | 0.07 |
| $^{63}$Cu | 962.06 | 0.04 | 962.09 | 0.02 |
| $^{63}$Cu | 1123.72 | 0.07 | 1123.77 | 0.08 |
| $^{63}$Cu | 1327.03 | 0.08 | 1327.06 | 0.06 |
| $^{63}$Cu | 1374.47 | 0.13 | 1374.36 | 0.04 |
| $^{63}$Cu | 1392.55 | 0.08 | 1392.51 | 0.09 |
| $^{63}$Cu | 1412.08 | 0.05 | 1412.11 | 0.02 |
| $^{63}$Cu | 1547.04 | 0.06 | 1547.11 | 0.06 |
| $^{63}$Cu | 2026.8 | 0.3 | 2026.79 | 0.13 |
| $^{63}$Cu | 2336.5 | 0.3 | 2336.79 | 0.12 |
| $^{63}$Cu | 2536.0 | 0.3 | 2536.47 | 0.14 |
| $^{63}$Cu | 2696.6 | 0.3 | 2697.39 | 0.17 |
| $^{67}$Zn | 184.577 | 0.010 | 184.63 | 0.04 |
| $^{69}$Zn | 438.634 | 0.018 | 438.64 | 0.02 |
| $^{65}$Cu | 1115.539 | 0.002 | 1115.48 | 0.02 |

| Transition | $T_{1/2}$ [min] | $\sigma$ | $\chi^2$/n.d.f |
|---|---|---|---|
| Combined | 38.84 | 0.15 | 0.782 |
| 449 | 42.1 | 3.3 | 0.700 |
| 669 | 38.70 | 0.19 | 1.616 |
| 962 | 39.19 | 0.26 | 1.003 |
| 1412 | 39.2 | 1.2 | 0.977 |

**Table 1.** Left: Sample peak energies from energy calibration before and after measurement of the sample. Right: $^{63}$Zn half lifes, NUDAT value $38.47 \pm 0.05$ min.

In addition to $^{63}$Zn, we also fitted the decay of $^{69}$Zn isomeric transition and obtained 13.76±0.18 h while the literature value is 13.76±0.02 h in good agreement with each other. Of the other two observed decays $^{65}$Zn and $^{67}$Cu, we were not able to fit them. The $^{65}$Zn was not fittable due to unfavorable ratio of its long half-life (244 d) and observation time (3 d). The $^{67}$Cu was not fittable due to the weakness of the signal. Like in the case of $^{63}$Cu, peaks not listed in the Table 1 right had too big errors to allow for a reasonable fit.

In the experiment presented, we have investigated the spectra and half-lifes created by photonuclear reactions on Zn isotopes. The photonuclear reactions were induced by a bremsstrahlung photon beam generated by a linac. The particularly interesting point at which we differed from previous such experiments is that this was a medical linac.

For the photonuclear reactions, we used a beam of 18 MeV end-point energy, which is well above of proton and neutron separation energies of all Zinc isotopes. However, our experiment was not designed to observe Zn levels since it was offline. Instead we observed the energy levels of the daughters of produced nuclei. These transitions, as well as the half-lifes of the parent nuclei, were our intended goal.

The experiment was separated into several parts, the crucial ones being the sample spectrum measurement, analysis and the calibrations. We have payed close attention to the calibration quality and we were confident that the end result had a good robustness and was reliable.

We have demonstrated quite confidently that the experiments such as ours can shed new light on transition energies. Indeed, we have obtained results, which are in good agreement with the literature values. But most importantly, we have obtained results better than those found in the literature in several instances. Our main contribution can be seen in the case of $^{63}$Cu, whose levels we have consistently determined to an accuracy level that is same or better than found in the literature. Thus, we believe that linac-based studies on proton rich nuclei can offer improvements of data in many intermediate mass nuclei. On the half-life side, our data is not as nearly impressive as for the transition energies. However, given the limitation of this study, it is still a nice check of consistency. Obviously, increased experience in these kind of experiments

Journal of Physics: Conference Series **590** (2015) 012024 doi:10.1088/1742-6596/590/1/012024

will also contribute to improvement of results in the future.

In this paper, we have shown the results of an ongoing effort on part of the Physics Department atAkdeniz University to establish an experimental nuclear physics group. The campaign started last year with our first photonuclear reactions experiment, Ref. [11], marking a milestone in experimental nuclear physics in Turkey. Since then, as this paper illustrates, we have made significant progress in our effort to contribute to nuclear physics at large. The quality of the results shown in this paper clearly illustrates that the approach we have adopted can contribute significantly to photonuclear reaction research at large. In addition, this effort has lead to the expansion of experimental nuclear physics at Akdeniz University in terms of both experimental know-how and new possibilities of applied nuclear physics.

[1] Lambert D 1992 *The Astronomy and Astrophysics Review* **3** 201–256 ISSN 0935-4956 URL `http://dx.doi.org/10.1007/BF00872527`
[2] Rauscher T, Heger A, Hoffman R and Woosley S 2002 *Astrophys.J.* **576** 323 (*Preprint* `astro-ph/0112478`)
[3] Arnould M and Goriely S 2003 *Physics Reports* **384** 1 – 84 ISSN 0370-1573 URL `http://www.sciencedirect.com/science/article/pii/S0370157303002424`
[4] Utsunomiya H, Mohr P, Zilges A and Rayet M 2006 *Nucl.Phys.* **A777** 459–478 (*Preprint* `nucl-ex/0502011`)
[5] Guildford 2007 *J. Radioanal. Nucl. Chem.* **271** Proceedings of the 11th International Conference on Modern Trends in Activation Analysis (MTAA-11), Surrey
[6] Masumoto K and Segebade C 2006 *Photon Activation Analysis* (John Wiley & Sons, Ltd) ISBN 9780470027318 URL `http://dx.doi.org/10.1002/9780470027318.a6211`
[7] Mohr P, *et al* 1999 *Nuclear Instruments and Methods in Physics Research Section A: Accelerators, Spectrometers, Detectors and Associated Equipment* **423** 480 – 488 ISSN 0168-9002 URL `http://www.sciencedirect.com/science/article/pii/S0168900298013485`
[8] Sonnabend K, *et al* 2011 *Nuclear Instruments and Methods in Physics Research Section A: Accelerators, Spectrometers, Detectors and Associated Equipment* **640** 6 – 12 ISSN 0168-9002 URL `http://www.sciencedirect.com/science/article/pii/S0168900211005535`
[9] Schwengner R, *et al* 2005 *Nuclear Instruments and Methods in Physics Research Section A: Accelerators, Spectrometers, Detectors and Associated Equipment* **555** 211 – 219 ISSN 0168-9002 URL `http://www.sciencedirect.com/science/article/pii/S0168900205018164`
[10] Mohr P, Brieger S, Witucki G and Maetz M 2007 *Nucl.Instrum.Meth.* **A580** 1201–1208 (*Preprint* `0707.2933`)
[11] Boztosun I, *et al* 2014 *Turk.J.Phys.* **38** 1–9
[12] 2003 *Elekta Digital Accelerator, Technical Training Guide* **(unpublished)**
[13] MAESTRO 2012 URL `http://www.ortec-online.com/download/A65-B32-MAESTRO-32-Emulation-Software.pdf`
[14] Gun A, Gupta K and Dasgupta B 2008 *Fundamentals Of Statistics* (World Press) ISBN 9788187567813 URL `http://books.google.com.tr/books?id=PIkRPwAACAAJ`
[15] ERJUN B and JUNDE H 2001 *Nuclear Data Sheets* **92** 147 – 251 ISSN 0090-3752 URL `http://www.sciencedirect.com/science/article/pii/S009037520190002X`
[16] Browne E and Tuli J 2010 *Nuclear Data Sheets* **111** 2425 – 2553 ISSN 0090-3752 URL `http://www.sciencedirect.com/science/article/pii/S0090375210000864`
[17] Junde H, Xiaolong H and Tuli J 2005 *Nuclear Data Sheets* **106** 159 – 250 ISSN 0090-3752 URL `http://www.sciencedirect.com/science/article/pii/S0090375205000761`
[18] Nesaraja C 2014 *Nuclear Data Sheets* **115** 1 – 134 ISSN 0090-3752 URL `http://www.sciencedirect.com/science/article/pii/S0090375213000938`
[19] Klaasse A and Goudsmit P 1974 *Zeitschrift für Physik* **266** 75–82 ISSN 0044-3328 URL `http://dx.doi.org/10.1007/BF01668668`
[20] Meyer R A, Prindle A L, Myers W A, Hopke P K, Dieterly D and Koops J E 1978 *Phys. Rev. C* **17**(5) 1822–1831 URL `http://link.aps.org/doi/10.1103/PhysRevC.17.1822`
[21] Yalcin P and Kurucu Y 2005 *Applied Radiation and Isotopes* **62** 63 – 67 ISSN 0969-8043 URL `http://www.sciencedirect.com/science/article/pii/S0969804304003653`
[22] Zoller W, Gordon G and Walters W 1969 *Nuclear Physics A* **124** 15 – 33 ISSN 0375-9474 URL `http://www.sciencedirect.com/science/article/pii/0375947469905181`
[23] Raman S and Couch R G 1970 *Phys. Rev. C* **1**(2) 744–746 URL `http://link.aps.org/doi/10.1103/PhysRevC.1.744`
[24] Helmer R and van der Leun C 2000 *Nuclear Instruments and Methods in Physics Research Section A: Accelerators, Spectrometers, Detectors and Associated Equipment* **450** 35 – 70 ISSN 0168-9002 URL `http://www.sciencedirect.com/science/article/pii/S0168900200002527`

Journal of Physics: Conference Series **590** (2015) 012025 doi:10.1088/1742-6596/590/1/012025

# Collective enhancement of nuclear state densities by the shell model Monte Carlo approach

**C Özen[1], Y Alhassid[2] and H Nakada[3]**

[1] Faculty of Engineering and Natural Sciences, Kadir Has University, Istanbul 34083, Turkey

[2] Center for Theoretical Physics, Sloane Physics Laboratory, Yale University, New Haven, CT 06520, USA

[3] Department of Physics, Graduate School of Science,Chiba University, Inage, Chiba 263-8522, Japan

E-mail: cem.ozen@khas.edu.tr

**Abstract.** The shell model Monte Carlo (SMMC) approach allows for the microscopic calculation of statistical and collective properties of heavy nuclei using the framework of the configuration-interaction shell model in very large model spaces. We present recent applications of the SMMC method to the calculation of state densities and their collective enhancement factors in rare-earth nuclei.

## 1. Introduction

Collective states constitute a significant part of the spectra of heavy nuclei up to moderate excitation energies and they are often described by bands that are built on top of intrinsic states. However, the contribution of these collective states to the nuclear state density is difficult to calculate microscopically, and is often described by phenomenological collective enhancement factors [1]. The challenge of computing the collective enhancement factors microscopically can be overcome using the shell model Monte Carlo (SMMC) method [2, 3, 4, 5]. This method is particularly suitable for the calculation of thermal and statistical properties of nuclei.

Here we discuss recent applications of the SMMC method to the isotopic families of even-even samarium and neodymium nuclei [6, 7], which exhibit a crossover from vibrational to rotational collectivity. In particular, we present results for the state densities and their collective enhancement factors.

## 2. The shell model Monte Carlo (SMMC) approach

The SMMC method is based on the Hubbard-Stratonovich (HS) transformation [8] to express the Gibbs operator $e^{-\beta H}$ of a nucleus (described by a Hamiltonian $H$ at inverse temperature $\beta = 1/T$) as a superposition of one-body propagators of non-interacting nucleons moving in external auxiliary fields $\sigma(\tau)$ that depend on imaginary time $\tau$ $(0 \le \tau \le \beta)$

$$e^{-\beta H} = \int D[\sigma] G_\sigma U_\sigma \, . \tag{1}$$

Here $G_\sigma$ is a Gaussian factor and $U_\sigma$ describes a one-body propagator associated with a given set of auxiliary fields $\sigma$. Subsequently, the thermal expectation value of an observable $O$ at inverse temperature $\beta$ can be written in the HS representation as

$$\langle O \rangle = \frac{\mathrm{Tr}\,(Oe^{-\beta H})}{\mathrm{Tr}\,(e^{-\beta H})} = \frac{\int D[\sigma] W_\sigma \Phi_\sigma \langle O \rangle_\sigma}{\int D[\sigma] W_\sigma \Phi_\sigma}\,, \tag{2}$$

where $\langle O \rangle_\sigma = \mathrm{Tr}\,(OU_\sigma)/\,\mathrm{Tr}\,U_\sigma$ is the thermal expectation value of the observable in a given configuration of the auxiliary fields $\sigma$. Since the numbers of neutrons and protons are fixed for a given nucleus, all traces in Eq. (2) are evaluated in the canonical ensemble. Defining a positive-definite function $W_\sigma = G_\sigma |\,\mathrm{Tr}\,U_\sigma|$ and the associated Monte Carlo sign $\Phi_\sigma = \mathrm{Tr}\,U_\sigma/|\,\mathrm{Tr}\,U_\sigma|$, auxiliary-field configurations $\sigma_k$ are sampled according to $W_\sigma$, and the expectation value in (2) is then estimated from $\langle O \rangle \approx \sum_k \langle O \rangle_{\sigma_k} \Phi_{\sigma_k} / \sum_k \Phi_{\sigma_k}$.

## 3. Collective enhancement in the state densities of heavy nuclei

Nuclear collectivity is a ubiquitous phenomenon in heavy nuclei. Various types of collectivity, such as vibrational collectivity and rotational collectivity, are observed and are well described by phenomenological models. However, a microscopic description of nuclear collectivity using the configuration-interaction (CI) shell model is a challenging task. In particular, CI shell model calculations of heavy nuclei require model spaces that are many orders of magnitude larger than model spaces that can be treated by conventional diagonalization methods. The SMMC was successfully applied to describe the rotational character of the rare-earth nucleus $^{162}$Dy [9] in the framework of a truncated spherical shell model space. Recently, using the same approach and model space, the crossover from vibrational to rotational collectivity in families of even-even samarium and neodymium isotopes was also reproduced [6, 7]. Here we discuss the state densities of these nuclei and the associated collective enhancement factors.

The single-particle model space consists of the $50-82$ major shell orbitals plus the $1f_{7/2}$ orbital for protons, and the $80-126$ major shell orbitals plus the $0h_{11/2}$ and $1g_{9/2}$ orbitals for neutrons. The bare single-particle energies of the orbitals were chosen so they reproduce the Woods-Saxon energies in the spherical Hartree-Fock approximation. The effective two-body interaction consists of monopole pairing interaction terms for protons and neutrons, and multipole-multipole interaction with quadrupole, octupole and hexadecupole terms. The calculations were performed using an SMMC code in the proton-neutron formalism [9].

### *3.1. State Densities*

The SMMC method has proven to be a powerful technique for the calculation of state densities [10, 11]. In the SMMC approach, the thermal energy is calculated as the expectation value of the many-body Hamiltonian at an inverse temperature $\beta$. The canonical partition function is computed by integrating the thermal energy with respect to $\beta$, and the state density is obtained by inverting the Laplace transform of the canonical partition function in the saddle-point approximation.

In Fig. 1 we show the SMMC state densities calculated for the even-even $^{148-154}$Sm and $^{144-152}$Nd isotopes. We compare the SMMC results (circles) with the experimental state densities at low energies using level counting data (histograms), and at the neutron resonance energy (when available) assuming the spin cutoff model with rigid-body moment of inertia (triangles). We also show the BBF state densities (solid lines) whose parameters (i.e., the single-particle level density parameter and the backshift parameter) are determined from level counting and neutron resonance data. The SMMC results are also compared with the densities calculated by the finite-temperature Hartree-Fock-Bogoliubov (HFB) approximation (dashed lines). The observed enhancement of the SMMC densities over the HFB densities can be attributed to the presence of collectivity in these nuclei, since collective excitations are included in the SMMC approach but are absent in the HFB densities. The HFB densities also display "kinks" at

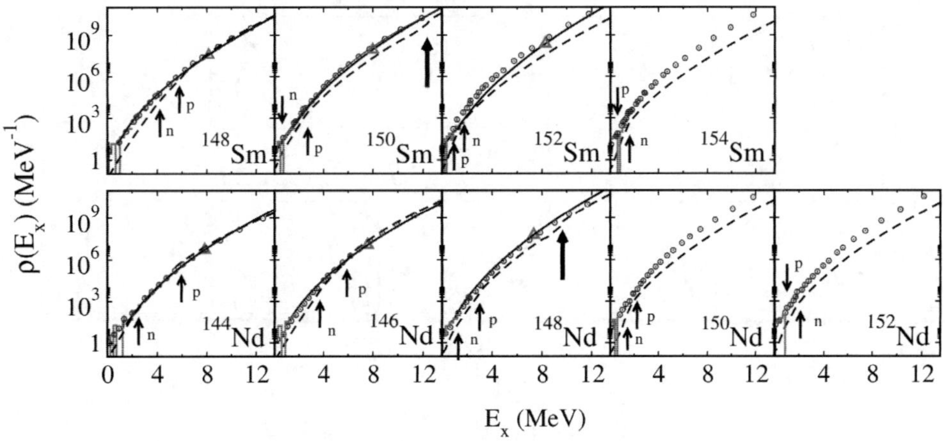


**Figure 1.** State densities in the even-even $^{148-154}$Sm and $^{144-152}$Nd isotopes. The SMMC densities (circles) are compared with level counting data (histograms) and neutron resonance data (triangles) when available. Also shown are the BBF state densities (solid lines), which are parametrized using the experimental data, and the HFB densities (dashed lines). The thin arrows indicate the neutron and proton pairing phase transitions, and the thick arrows indicate the shape phase transitions. Adapted from Refs. [6, 7].

energies that correspond to the phase transitions associated with the breaking of the proton and neutron pairs (thin arrows) and with the transitions from deformed to spherical shapes (thick arrows). These shape transitions occur in all nuclei except $^{148}$Sm, $^{144}$Nd and $^{146}$Nd, which are spherical in their ground states. The shape transitions in $^{152}$Sm, $^{154}$Sm, $^{150}$Nd and $^{152}$Nd occur at excitation energies that are higher than those displayed in Fig. 1. They can be seen, however, in Fig 2.

### *3.2. Collective Enhancement*

The microscopic calculation of collective effects in the state density of heavy nuclei has been a major challenge. Such effects are usually described in terms of phenomenological vibrational and rotational collective enhancement factors [1].

Recently we introduced [6, 7] the ratio of the SMMC and the HFB densities, i.e., $K = \rho_{\rm SMMC}/\rho_{\rm HFB}$, as a microscopic measure of the collective enhancement factor. In Fig. 2, we show $K$ as a function of the excitation energy $E_x$ for the above families of samarium and neodymium isotopes. In the spherical nuclei $^{148}$Sm, $^{144}$Nd, and $^{146}$Nd, collectivity is predominantly vibrational and is observed to be lost completely (i.e., $K \sim 1$) above the pairing transition energies. The other samarium and neodymium isotopes are deformed, and collectivity persists at higher excitation energies than the pairing transition energies where the enhancement is due to rotational collectivity. Above the shape transition energy, the collective enhancement factor decays to $K \sim 1$ since spherical nuclei can no longer support rotational bands.

## 4. Conclusion

We discussed recent SMMC applications to the even-even $^{148-154}$Sm and $^{144-152}$Nd isotopes. The total SMMC state densities were shown to be in very good agreement with experimental data. We also extracted a microscopic measure of the collective enhancement factor defined by the ratio of the SMMC and HFB densities. We observed that the decay of vibrational

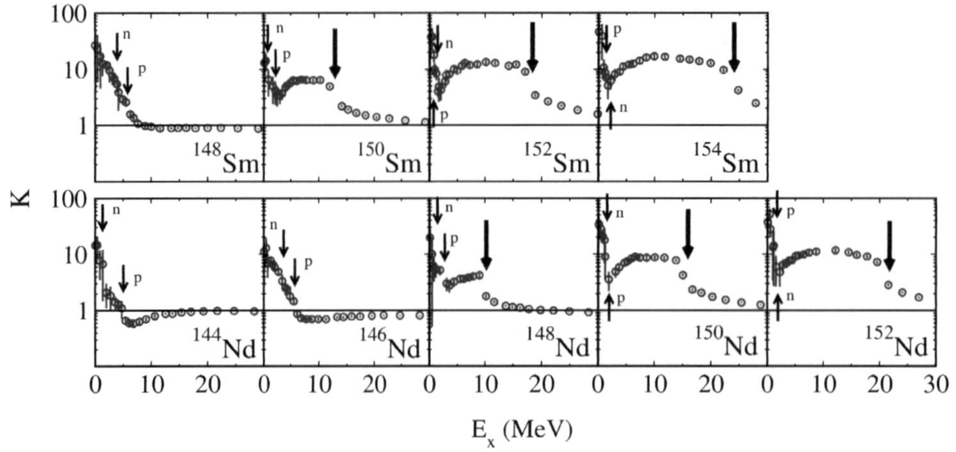


**Figure 2.** Collective enhancement factor K versus excitation energy $E_x$ in the even-even $^{148-154}$Sm and $^{144-152}$Nd isotopes. The pairing and shape transition energies are shown by the thin and thick arrows, respectively. Adapted from Refs. [6, 7].

and rotational collectivity with excitation energy correlates with the pairing and shape phase transitions, respectively.

## Acknowledgments

This work was supported in part by the U.S. Department of Energy Grant No. DE-FG02-91ER40608, and by the Grant-in-Aid for Scientific Research (C) No. 25400245 by the JSPS, Japan. The research presented here used resources of the National Energy Research Scientific Computing Center, which is supported by the Office of Science of the U.S. Department of Energy under Contract No. DE-AC02-05CH11231. It also used resources provided by the facilities of the Yale University Faculty of Arts and Sciences High Performance Computing Center.

## References

[1] Capote R *et al.* 2009 *Nuclear Data Sheets* **110** 3107, and references therein
[2] Lang G H, Johnson C W, Koonin S E and Ormand W E 1993 *Phys. Rev.* C **48** 1518
[3] Alhassid Y, Dean D J, Koonin S E, Lang G H and Ormand W E 1994 *Phys. Rev. Lett.* **72** 613
[4] Koonin S E, Dean D J and Langanke K 1997 *Phys. Rep.* **278** 2
[5] Alhassid Y 2001 *Int. J. Mod. Phys.* B **15** 1447
[6] Özen C, Alhassid Y and Nakada H 2013 *Phys. Rev. Lett.* **110** 042502
[7] Alhassid Y, Özen C and Nakada H 2014 *Nuclear Data Sheets* **118** 233
[8] Hubbard J 1959 *Phys. Rev. Lett.* **3** 77; Stratonovich R L 1957 *Dokl. Akad. Nauk. S.S.S.R.* **115** 1097
[9] Alhassid Y, Fang L and Nakada H 2008 *Phys. Rev. Lett.* **101** 082501
[10] Nakada H and Alhassid Y 1997 *Phys. Rev. Lett.* **79** 2939
[11] Alhassid Y, Liu S and Nakada H 1999 *Phys. Rev. Lett.* **83** 4265

Journal of Physics: Conference Series **590** (2015) 012026 doi:10.1088/1742-6596/590/1/012026

# The effect of proton halo on fusion reactions

**J A Lay, Raj Kumar, and A Vitturi**

Dipartimento di Fisica e Astronomia "Galileo Galilei", Università di Padova, and INFN, Sezione di Padova, via Marzolo, 8, I-35131 Padova, Italy

E-mail: `lay@pd.infn.it`

**Abstract.** In a recent brief report [Kumar *et al.*, Phys. Rev. C **89**, 027601 (2014)] we proposed a method to include the effect of a large break up in fusion reactions with neutron and proton halo nuclei. A stronger enhancement on the total fusion cross section was found for the proton case. However, neither complete nor incomplete fusion was studied and only one target was considered. Here, we revisit the problem in order to address the importance of complete and incomplete fusion. The sensitivity of the cross section to the target is also explored.

## 1. Introduction

Fusion reactions at energies below the barrier is one of the most useful tools to explore the structure of nuclei, since their variation with the energy is known to be driven by the internal degrees of freedom of the two counterparts [1, 2]. This fact has been largerly explored for fusion of heavy-ions, where it is possible to identify the coupled channels driving the behaviour of the cross sections, normally corresponding to rotor or vibrator schemes. These channels have well determined potentials and have been largely studied in comparison with light ions or more generally with weakly bound nuclei.

In the case of fusion with light ions, the large probability of break up imposes the treatment of strongly coupled continuum channels with the consequent inclusion of at least three-body kinematics. This complexity makes almost impossible to single out specific issues from a full continuum calculation, being one of the reasons why there is no general agreement on the effect of the break up and the presence of a charged proton halo on the fusion cross section.

In order to face this problem, we proposed recently a very simplified model consisting of two channels with a schematic description of the break up of the projectile [3]. Within this simple model, a large enhancement of the sub-barrier fusion cross section was found for the case of $^{8}$B on $^{58}$Ni. This reactions was chosen after the experiment by E. F. Aguilera and collaborators [4]. However, this cross section included complete fusion and the fusion of $^{7}$Be, i. e. the fusion of the *core* of the projectile after losing a proton. The particular role of each channel was not explored. Also the sensitivity of the enhancement found to different targets was not considered even though there is an experiment with $^{8}$B on $^{28}$Si by A. Pakou and collaborators [5].

In the present contribution, we continue developing this model to address the role of complete and incomplete fusion following the prescription of [6], as well as the effect of the specific target.

## 2. Reaction framework

We will consider two different halo projectiles representative of the neutron halo case, $^{11}$Be, and the proton halo case, $^{8}$B. Together with the complete fusion of these two projectiles with the

Journal of Physics: Conference Series **590** (2015) 012026 doi:10.1088/1742-6596/590/1/012026

corresponding target, we will include the possibility of break up allowing the remaining core, $^{10}$Be and $^{7}$Be respectively, to fuse afterwards. Following the Austern formalism for coupled-channels [7] we can expand the wavefunction describing the collision of projectile and target in terms of the wavefunctions for the internal state of the projectile $\phi_\beta$ and the radial wave functions $\chi_\beta$ that accounts for the relative motion between projectile and target:

$$\Psi^{(+)} = \Sigma_\beta \frac{\chi_\beta(R)}{R}\phi_\beta. \tag{1}$$

This leads to a set of coupled equations for $\chi_\beta$. In our model case, we will only consider two channels, the incoming channel and one channel representative of the break-up and later fusion without the ejected particle:

$$\begin{aligned} \frac{d^2\chi_1}{dR^2} + \frac{2\mu_1}{\hbar^2}[E_1 - V_1]\chi_1 &= \frac{2\mu_1}{\hbar^2}V_{coup}\chi_2, \\ \frac{d^2\chi_2}{dR^2} + \frac{2\mu_2}{\hbar^2}[E_2 - V_2]\chi_2 &= \frac{2\mu_2}{\hbar^2}V_{coup}\chi_1, \end{aligned} \tag{2}$$

where, in our case, $E_1 = E$, the incoming energy, and $E_2 = E_{bu}$, the energy in the break-up channel.

This energy $E_{bu}$ can be estimated by subtracting the energy needed for break up and the average excitation energy, $\langle E^*\rangle$, in the core-nucleon relative motion, and then sharing the energy between them according to a distant break up scenario. In this way, we consider $E_{bu} = (E - S_{1N} - \langle E^*\rangle)\cdot\frac{A-1}{A}$. $S_{1N}$ is the one neutron or one proton separation energy: $S_{1p} = 0.136$ MeV for $^8$B and $S_{1n} = 0.504$ MeV for $^{11}$Be. In order to estimate $\langle E^*\rangle$ we take the peak energy for the dipole electromagnetic transition probabilities, $\langle E^*\rangle = 0.5$ MeV for $^8$B and $\langle E^*\rangle = 0.4$ MeV for $^{11}$Be.

The total potential for each channel $V_{1,2}(R)$ is given by the sum of Coulomb and a nuclear proximity potential given by Broglia and Winther [8] parameterization. The coupling potential $V_{coup}$ is taken as a derivative Woods Saxon form with the same diffuseness of the proximity potential for the incoming channel and the radius corresponding to the position of the maximum of the Coulomb barrier. The strength is set to a 10% of the strength of the same proximity potential.

Finally, in order to obtain fusion cross section from the coupled-channel equations, we solve them under ingoing-wave boundary conditions (IWBC). Within this approximation the solution is matched with the incoming and outcoming Coulomb wavefunctions obtaining the transmission coefficient for each channel, see [3]. The total transmission probability will be:

$$T = \sum_\beta | T_\beta^2 | = |t_1|^2 + \frac{v_2}{v_1}|t_2|^2 \tag{3}$$

where $v_1$ and $v_2$ are the velocities corresponding to channel 1 and 2.

From here we can compute separate cross section for each channel:

$$\sigma = \sigma_1 + \sigma_2 = \frac{\pi\hbar^2}{2\mu_1 E}\sum_{\ell=0}^{\ell_{max}}(2\ell+1)T_1^\ell(E) + \frac{\pi\hbar^2}{2\mu_1 E}\sum_{\ell=0}^{\ell_{max}}(2\ell+1)T_2^\ell(E), \tag{4}$$

where:

$$T_1 = |t_1|^2, \qquad T_2 = \frac{v_2}{v_1}|t_2|^2, \tag{5}$$

being evaluated for each angular momentum $\ell$.

It is necessary to stress that the cross section for the second channel $\sigma_2$ implies the fusion of all the charge only for the neutron case. In the proton case, the fusion results from the different channels will have different charge.

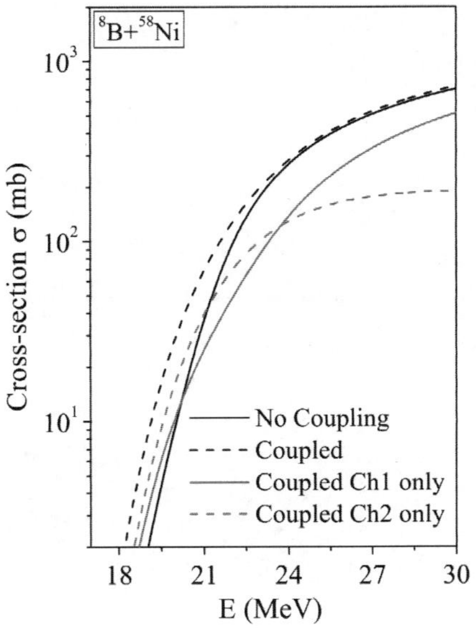


**Figure 1.** Fusion cross section for the system $^{8}B+^{58}Ni$. Black solid line shows the fusion when we do not include the coupling to the break up. Black dashed line is the total cross section with the inclusion of coupling to the break up. This last one is the sum of two components: the cross section in the incoming channel (fusion of $^{8}B$, red solid line) and the cross section in the break up channel (fusion of $^{7}Be$ after ejecting a proton, red dashed line).

## 3. Results

We first attempt the problem of distinguishing the contributions to the total cross section of the two channels. In Fig. 1, we considered the fusion of $^{8}B$ on $^{58}Ni$, the same case as in Ref. [3]. Black solid line shows a case without break up and the black dashed line, the enhancement found for the proton halo where the total cross section is considered. Here we also display the contribution of the two channels. We see in Fig. 1 the cross section for the fusion of the full $^{8}B$ (red solid line) and the cross section of the break up channel, i. e. $^{7}Be$ on $^{58}Ni$ (red dashed line). If we only considered the complete fusion (first channel), we found that the enhancement is only found for really small energies.

On the other hand, we saw that this enhancement can be explained through the definition of an effective $Q$ value arising from the different Coulomb barrier $V_B$ present in each channel:

$$Q_{eff} = (E_{bu} - V_B^2) - (E - V_B^1). \tag{6}$$

For an incoming energy around the Coulomb barrier $E \approx V_B^1$, this $Q$ value can be approximated for the neutron and the proton case:

$$Q_{eff}^{(n)} \simeq -\frac{1}{A_{proj}} V_B^1, \tag{7}$$

$$Q_{eff}^{(p)} \simeq \left(\frac{1}{Z_{proj}} - \frac{1}{A_{proj}}\right) V_B^1. \tag{8}$$

Notice that for the neutron halo $Q_{eff}^{(n)}$ is always negative, whereas $Q_{eff}^{(p)}$ is always positive. Moreover, these values are proportional to the Coulomb barrier. Therefore, the enhancement is expected to be smaller for lighter targets.

Having this in mind, in Fig. 2 we show the reduced cross sections for the fusion of $^{8}B$ (red lines) and $^{11}Be$ (black lines) on a $^{28}Si$ target. Dashed lines represents the case where we include

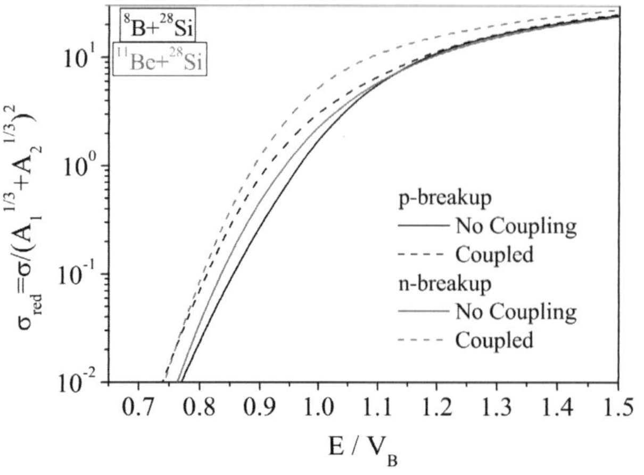


**Figure 2.** Fusion cross section for the systems $^{8}$B+$^{28}$Si and $^{7}$Be+$^{28}$Si. For both reactions, the case without coupling, black and red solid lines respectively, and including the coupling to the break up, black and red dashed lines.

the break up and the solid line, those without break up. Here the proton enhancement reaches the neutron cross section only for really low energies. If we compare with the $^{58}$Ni case, see Fig. 3 in Ref. [3], we found a weaker enhancement for this case as expected with the corresponding Coulomb barriers.

## 4. Conclusions

We have considered here the effect of proton halo in the fusion of the nucleus $^{8}$B on different targets, $^{58}$Ni and $^{28}$Si, using a simplified model of the break up of the halo nucleus. For both targets, a certain enhancement of the fusion cross section is found when we include the coupling to the break up. This enhancement is smaller for the $^{28}$Si case, as expected from a simple analysis of the effective Q value proposed. Here we have compared with a neutron halo case by computing the fusion cross section for $^{11}$Be on $^{28}$Si.

This enhancement is obtained when we considered also the fusion following break up. However, with this simple model we can compute separately the contribution to the fusion of each channel. If we focused on the complete fusion, therefore only considering the fusion from the first channel, it is only enhanced with respect to the case without coupling to the break up for extremely low energies.

## Acknowledgments

This work has been supported by MIUR research fund PRIN 2009TWL3MX. The authors acknowledge L. F. Canto for useful discussions.

## References

[1] Balantekin A B and Takigawa N 1998 *Rev. Mod. Phys.* **70** 77
[2] Hagino K and Takigawa N 2012 *Prog. Theor. Phys.* **128** 1061
[3] Kumar R, Lay J A and Vitturi A 2014 *Phys. Rev. C* **89** 027601
[4] Aguilera E F *et al.* 2011 *Phys. Rev. Lett.* **107** 092701
[5] Pakou A *et al.* 2013 *Phys. Rev. C* **87** 014619
[6] Hagino K, Vitturi A, Dasso C H and Lenzi S M 2000 *Phys. Rev. C* **61** 037602
[7] Austern N, Iseri Y, Kamimura M, Kawai M, Rawitscher G and Yahiro M 1987 *Phys. Rep.* **154** 125
[8] Broglia R A and Winther A 1991 *Heavy Ion Reactions* Frontiers in Physics (Addison-Wesley)

Journal of Physics: Conference Series **590** (2015) 012027 doi:10.1088/1742-6596/590/1/012027

# Monte-Carlo study of the influence of backscattered electrons on the transmission of a mini-orange $\beta$ spectrometer

**Pavel Detistov** [1], **Dimiter L. Balabanski** [2]
[1]INRNE–BAS, 72 Tz. Shousee Blvd, 1784 Sofia, Bulgaria
[2]ELI-NP, IFIN-HH, Reactorului Str 30, 077125 Magurele, Ilfov, Romania
E-mail: detistov@inrne.bas.bg

**Abstract.** This work work is a part of the performance investigation of the recently constructed Mini-Orange beta spectrometer. The spectrometer has eight different configurations using three different magnet shapes and combination of three, four, and six magnet pieces allowing detection of electrons in wide kinetic energy range. The performance of the device is studied using the GEANT4 simulation tool. Evaluation of the device's basic parameters has been made, paying special attention to the backscattering, for which a study of the dependence of this process on the energy and the angle is made.

## 1. Introduction

Mini–orange type $\beta$ spectrometers (MOBS) are a compact instruments for experimental study of conversion electrons in the in-beam measurements. We constructed a MOBS whose magnetic configurations follow the original design suggested at Groeningen [1][2]. The expected improvement is related to the electron detector part. The selected detector was chosen after preliminary investigation reported in [3]. In the present work more detailed study is made with the selected Si particle detector.

The performance of the MOBS depends on the number of detected electrons in the particle detector. Due to the backscattering some part of the electrons are lost. In this work we evaluate them using the GEANT4 simulation tool [4].

## 2. Mini–orange spectrometer description

MOBS is a system consisting of a set of several orange shaped wedges of $SmCo_5$ permanent magnets which are placed around a solid core made of heavy elements, and a particle detector behind. The magnets have been produced in three different shapes, designated as "A", "B" and "C" types. Eight different configurations have been produced in order to cover wide electron energy range. Designation of each configuration was made by the letter for the shape type and the number of magnets. Thus, configuration "A4" has 4 magnets of "A"–type shape. The full set is shown in picture in Figure 1.

Based on a preliminary study [3] a MSQ25-1000 was selected as an electron detector. The active area of the detector is 2500 $mm^2$, it consists of 4 segments with dimensions 24.95 mm

$\times$ 24.95 mm, and active detector thickness of 1.0 mm. According to [5] the thickness of the detector corresponds to a electron range of about 0.55 MeV.

A MOBS is defined by the following parameters: the magnet shape, the number of magnets, the magnetic field strength, the distances between the center line of the magnet set and the detector face (g) or the target face (f), the geometrical shadowing factor (b) defined as the ratio of the angles occupied by the magnets and between the magnet faces, and the transmission (T), which is defined as the ratio of the number of counts in the energy spectra peak to the total number of events.

Using the measured magnetic field map, simulations of the transmission have been calculated and transmission curves have been obtained. Example transmission curves for one of the configurations is shown in Figure 2. From the transmission curves could be seen that the system is not very sensitive to the position of the detector, but is very sensitive to the position of the source.

The evaluated parameters of the MOBS are listed in the Table 1.

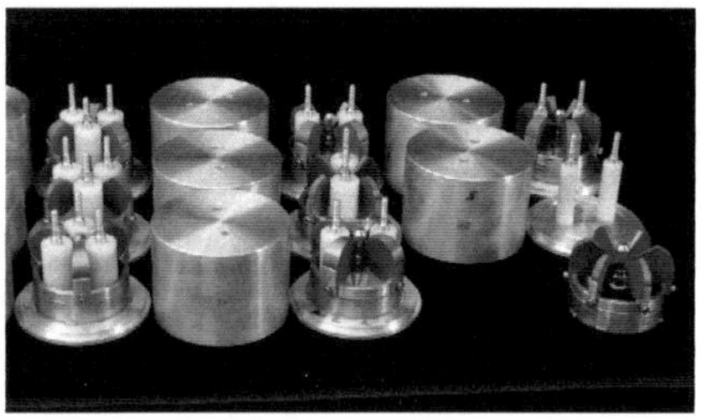

**Figure 1.** The complete set magnet configurations with their protective containers.

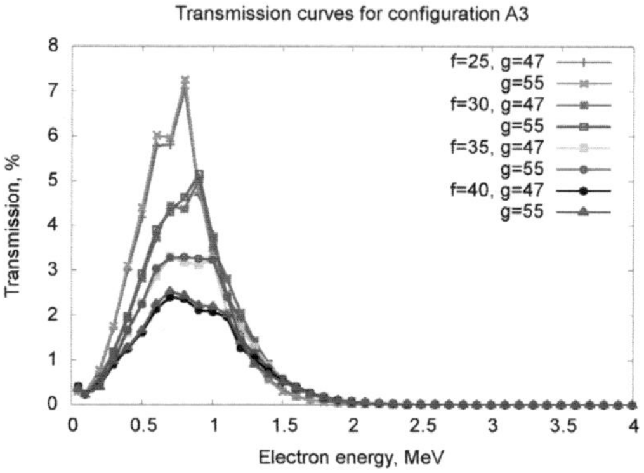


**Figure 2.** Transmission curves for A3, for various distances g and f.

**Table 1.** Parameters of the MOBS. The listed values for the optimal energy ranges are selected for f=25 mm and g=47 mm.

| Parameter | A3 | A4 | A6 | B3 | B4 | B6 | C3 | C4 |
|---|---|---|---|---|---|---|---|---|
| Shadowing factor — b | 0.2 | 0.286 | 0.5 | 0.1 | 0.143 | 0.2 | 0.2 | 0.286 |
| Transmission $T_{max}$, % | 7.3 | 4.2 | 1.15 | 8.4 | 9.8 | 10.8 | 3.4 | 1.36 |
| Optimal energy range, $E_{min}$ MeV | 0.25 | 0.5 | 1.2 | 0.1 | 0.25 | 0.5 | 0.25 | 0.5 |
| Optimal energy range, $E_{max}$ MeV | 1.25 | 1.5 | 2.0 | 1.0 | 1.25 | 1.25 | 1.5 | 1.25 |

For estimation of the detector performance alone a simulation of its efficiency for electrons has been made. The resulting efficiency curves are given in Figure 2. They indicate that there is a gradual drop of the efficiency after around 0.5 MeV. This is a limitation resulting from the detector thickness. In contrast to $\gamma$ -rays, electrons does not travel in the straight lines inside the solid, and their calculated range is a sum of all path they pass inside the material. This is the reason why electrons with higher energies could be detected into a thinner detector.

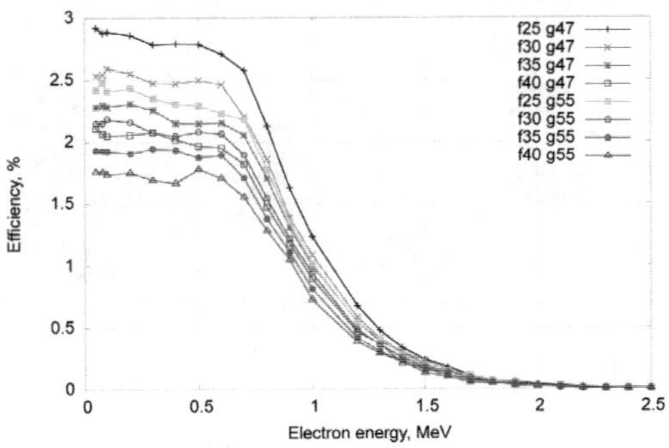


**Figure 3.** Simulated detector efficiency at various f and g distances.

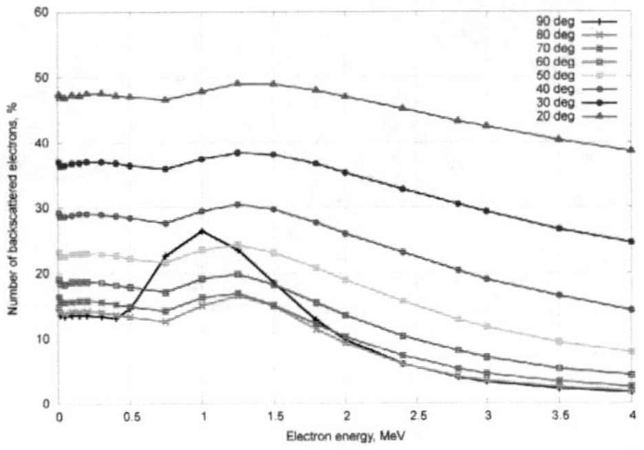


**Figure 4.** Dependence of the number of backscattered electrons on the incident energy and angle.

## 3. Backscattering of electrons from the detector

When electrons bombard a surface part of them scatter back as a result from their interaction with atoms in the solid. These are the backscattered electrons. They scatter either elastically or inelastically. Those who scatter inelastically contribute to the background in the energy spectra of the detector. In both cases, such electrons, though transmitted by the magnetic filter of the MOBS to the detector, are lost and does not contribute to the measurement results, reducing the overall transmission of the system.

Electrons bombard the detector surface at various angles depending on their energy. In Figure 5is shown the area of the detector where electrons hit the detector surface for various electron energies. At the maximum of the transmission value for a given MOBS configuration there is a single spot at the center of the detector. In such a case the angle of the incident electrons is between 40 and 50 degrees. In case when electrons are scattered back towards the system of magnets they are diverted in the direction out of the system, due to the Lorenz force, produced by the magnetic field.

In order to evaluate the amount of backscattered electrons at various incident angles, we made a simulation using monoenergetic electrons which bombard the detector surface at various angles. The results are shown in Figure 4. It could be seen that number of backscattered electrons depends strongly on the angle of incidence, and are not very sensitive to the energy.

Based on this values we could conclude that between 20 and 30 % of electrons backscatter from the surface. This should be valid for all configurations due to the low sensitivity with respect to the energy.

In Figure 6 four energy spectra are shown for different electron energies. From the spectra it could be seen that at lower energies the main contribution is from the backscattered electrons. At higher energies there are other contributions from secondary processes, which are not considered here.

## 4. Conclusions

Using the GEANT4 simulation toolkit an investigation of the MOBS have been made. Transmission curves for all configurations have been obtained for the particular type of particle

Journal of Physics: Conference Series **590** (2015) 012027 doi:10.1088/1742-6596/590/1/012027

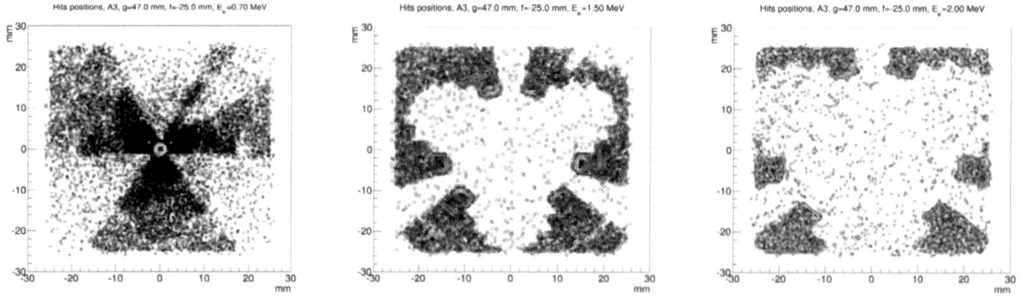


**Figure 5.** Map of the electron hits on the detector surface for magnet configuration A3 at electron energies a) 0.7 MeV, b) 1.5 MeV and c) 2.0 MeV.

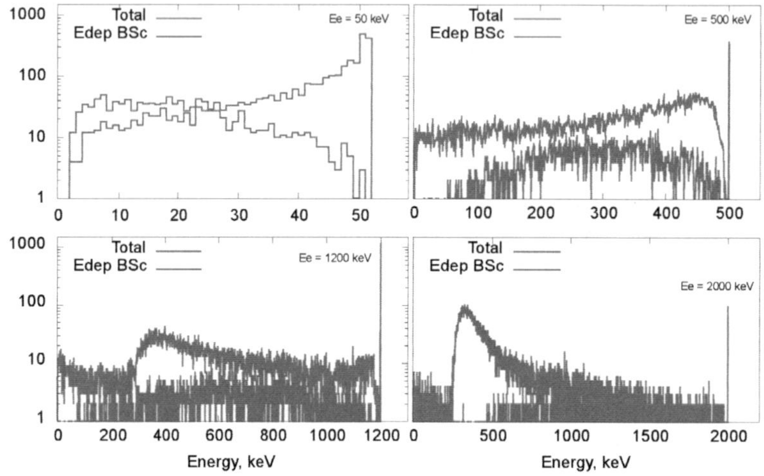


**Figure 6.** Detector spectra for monoenergetic electrons at different incident energies. Red line corresponds to the spectra in the detector and the blue line is the energy deposited by the back scattered electrons.

detector and the recommended energy range for each magnet configuration have been estimated.

Also, backscattering of electrons from the detector surface has been studied. These electrons contribute to the long energy tail in the energy spectra and in this way they reduce the overall transmission of the system.

The work is supported by the DID02/16 contract with the Bulgarian National Science Fund.

## References

[1] Van Klinken J, Feenstra S J, Wisshak K, Faust H 1975 *Nucl. Instr. and Meth.*, **130 (2)** 427
[2] Van Klinken J, Wisshak K, 1972 *Nucl. Instr. and Meth.*, **98 (1)** 1
[3] Detistov P, Balabanski D L 2013 *Proc. of the 32-nd Intern. Workshop on Nucl. Theory, Rila Mount.* **32** 164
[4] Agostinelli S *et al.*, 2003 *Nucl. Instr. and Meth. A* **506(3)** 250
[5] NIST web site: *http://physics.nist.gov/PhysRefData/Star/Text/ESTAR.html*[last visited 7.10.2014]
[6] Guerro L, Blasi N, Saltarelli A 2014 *Nucl. Instr. and Meth. A* **739** 32

Journal of Physics: Conference Series **590** (2015) 012028 doi:10.1088/1742-6596/590/1/012028

# Nuclear level density as a tool for probing the inelastic scattering of $^{6}$He

**Bora Canbula, Halil Babacan**

Department of Physics, Faculty of Arts and Sciences,
Celal Bayar University, 45140 Muradiye, Manisa, Turkey

E-mail: bora.canbula@cbu.edu.tr

**Abstract.** The cross sections are calculated for the both elastic and inelastic scattering of $^{6}$He from $^{12}$C and $^{4}$He. A phenomenological optical potential is used to describe the elastic scattering. $^{4}$He is taken as spherical and inelastic couplings to the first excited states of $^{6}$He and $^{12}$C are described with collective rotational model and coupled-channels method. Deformation lengths for $^{6}$He and $^{12}$C are determined from semi-classical nuclear level density model by using Laplace-like formula for the nuclear level density parameter. The comparison of the predicted and the measured cross sections are presented to test the applicability of nuclear level density model to the light exotic nuclei reactions. Good agreement is achieved between the predicted and measured cross sections.

## 1. Introduction

After the use of radioactive ion beams (RIBs) in the mid-1980s [1], heavy-ion induced reactions, which involve light exotic nuclei, have become a subject of great interest for both theoretical and experimental studies. Nevertheless, understanding the unusual structure of these nuclei, such as halo, skin or Borromean (three-body systems which have no bound two-body subsystems), still remains as a challenge. One of the most studied and also the lightest nucleus among them is $^{6}$He, which is 2n halo ($S_{2n} = 0.975\,\mathrm{MeV}$) [2] and also Borromean ($\alpha + n + n$)[3]. Furthermore, $^{6}$He is of great astrophysical importance because of appearing to be a vital member of the reaction chain that can bridge the instability gaps at $A = 5$ and $A = 8$.

Elastic scattering is the most fundamental and the simplest interaction between the projectile and target nuclei. Thus, at least for stable systems, it is easy to calculate the elastic cross section by using a simple optical potential. With this point of view, it would be a good starting point that employing the optical potential parameters deduced from the elastic scattering data of $^{6}$Li, for describing the elastic scattering of $^{6}$He. However, between the elastic scattering data of $^{6}$Li and $^{6}$He, there are certain differences such as long range absorption, absence of Coulomb rainbow, coupling to inelastic or breakup channels, which require unphysical modification of optical potential parameters to explain. Therefore, when it comes to explaining the elastic cross section of halo nuclei, some other mechanisms should be taken into consideration to deal with such phenomenon. Keeping in mind the low binding energy of $^{6}$He, there is no doubt that the CDCC (continuum-discretized coupled-channels) model is the most appropriate method, but it is still a good choice to combine the phenomenological optical potential with some virtual coupling or dynamic dipole polarization potential, which are surface (derivative) potentials of the usual volume potential, to describe the elastic scattering data of $^{6}$He reasonably well.

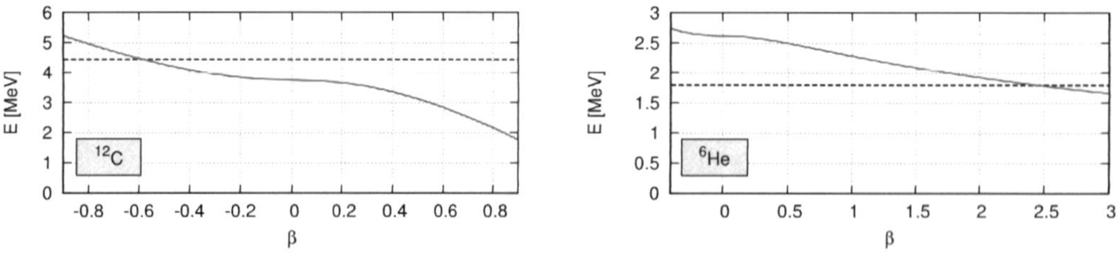


**Figure 1.** (Color online) The energy of the first excited state as a function of deformation parameter. The results are illustrated for $^{12}$C and $^{6}$He in the left and right panels, respectively.

The main idea behind these theoretical efforts is to simulate the absorption from elastic channel due to the surface deformation of $^{6}$He compared to a stable $A = 6$ system. Therefore, one can describe the absorption from elastic channel as inelastic couplings to the low-lying collective states by using coupled-channels method. To use this method the deformation lengths of the projectile and target should be determined from a structure model. Nuclear level density (NLD) could have been one of the candidates but it always has been a problem to describe the low-lying collective states for NLD models [4, 5]. To overcome this problem, recently, we have proposed a Laplace-like formula for the energy dependence of the nuclear level density parameter, which is the main parameter of NLD, including collective effects [6]. Consequently, a significant improvement has been achieved in agreement between the predicted and the observed excited energy levels. In this model, the density of the excited levels are calculated in terms of deformation parameter of nucleus and also a semi-classical approach is employed with a single-particle potential.

## 2. Nuclear level density

The easiest way to determine the nuclear level density of a nucleus is to use a phenomenological model, and one of the most common phenomenological nuclear level density model is Fermi gas model. According to this model, the total nuclear level density is

$$\rho^{\text{tot}}(U) = \frac{1}{12\sqrt{2}\sigma}\frac{\exp[2\sqrt{aU}]}{a^{1/4}U^{5/4}}, \tag{1}$$

**Table 1.** Adjusted optical potential parameters. $r_c = 1.2\,\text{fm}$ is used in all calculations.

| Proj.+Targ. | $E$(MeV) | $V_0$(MeV) | $r_v$ (fm) | $a_v$ (fm) | $W_0$ (MeV) | $r_w$ (fm) | $a_w$ (fm) |
|---|---|---|---|---|---|---|---|
| $^{6}$He + $^{12}$C | 18.0* | 44.789 | 0.738 | 0.867 | 11.284 | 1.227 | 0.696 |
| $^{6}$He + $^{12}$C | 30.0* | 48.775 | 0.931 | 0.745 | 6.991 | 1.310 | 0.890 |
| $^{6}$He + $^{4}$He | 3.8** | 64.800 | 0.910 | 0.660 | 0.500 | 0.900 | 0.650 |
| $^{6}$He + $^{4}$He | 4.2** | 64.000 | 0.910 | 0.660 | 2.200 | 0.900 | 0.650 |
| $^{6}$He + $^{4}$He | 4.7** | 62.800 | 0.910 | 0.660 | 2.400 | 0.900 | 0.650 |
| $^{6}$He + $^{4}$He | 5.1** | 62.000 | 0.910 | 0.660 | 2.600 | 0.900 | 0.650 |
| $^{6}$He + $^{4}$He | 5.4** | 61.200 | 0.910 | 0.660 | 3.000 | 0.900 | 0.650 |
| $^{6}$He + $^{4}$He | 5.8** | 60.800 | 0.910 | 0.660 | 3.200 | 0.900 | 0.650 |

* Energy in lab. system
** Energy in c.m. system

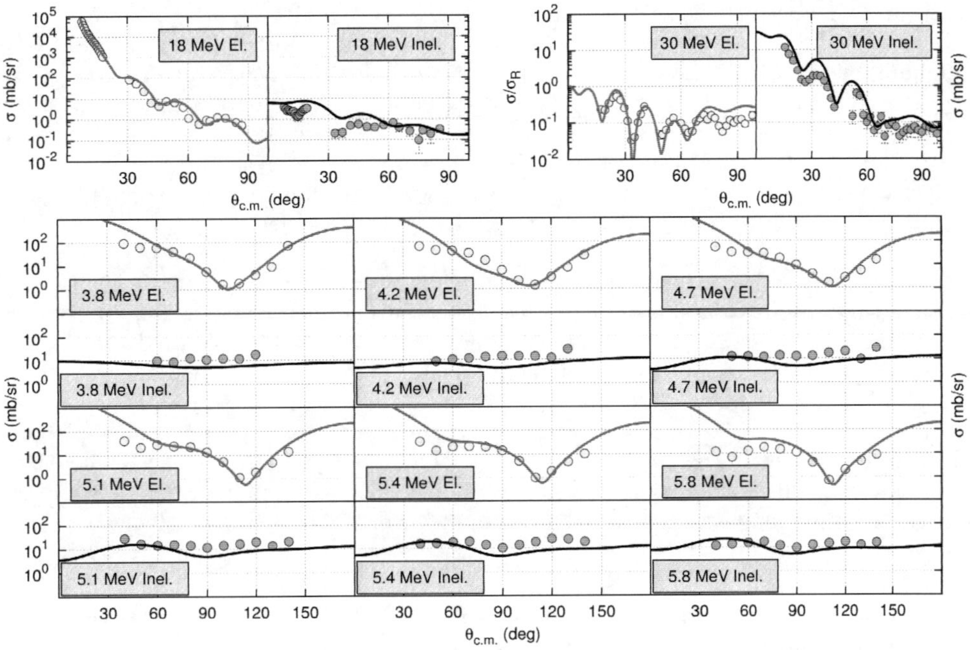


**Figure 2.** (Color online) The results with comparison to experimental data [7–9] for elastic and inelastic scattering of $^6$He on $^{12}$C (in the upper part) target at 18 MeV and 30 MeV laboratory energies and $^4$He (in the lower part) target at 3.8 MeV, 4.2 MeV, 4.7 MeV, 5.1 MeV, 5.4 MeV and 5.8 MeV center-of-mass energies.

where $U$ is effective excitation energy and $\sigma$ is spin cut-off parameter. The main variable of this equation is $a$, which is called as the nuclear level density parameter. In this study, we use the Laplace-like formula [6], which is convenient to describe the low-lying collective levels, for the energy dependence of the nuclear level density parameter. This formula,

$$a(U) = \tilde{a}\left(1 + A_c \frac{S_n}{U} \frac{\exp(-|U - E_0|/\sigma r_c^3)}{\sigma r_c^3}\right), \tag{2}$$

gives the energy dependence of the nuclear level density parameter with a function that depends on the neutron separation energy $S_n$, the energy of the first phonon state $E_0$, and a scale parameter, which is given as $\sigma r_c^3 = \sigma_c^3/\tilde{a}$ in terms of spin cut-off parameter and the asymptotic level density parameter. The so-called collective amplitude $A_c$ is given as the temperature dependent difference between the experimental mass and the theoretical mass calculated with a shape dependent liquid-drop mass formula,

$$A_c = [M_{\text{exp}}(N, Z) - M_{\text{LDM}}(N, Z, \beta)] \frac{\tau_c}{\sinh \tau_c}, \tag{3}$$

where $T_c = \sqrt{S_n/\tilde{a}}$ is defined as some critical temperature and $\tau_c = 2\pi^2 T_c/\hbar\omega$. Further details can be found in Ref. [6]. One can use the level density function to obtain the cumulative number of levels up to certain excitation energy,

$$N_{\text{cum}}(U) = \int_0^U \rho^{\text{tot}}(U) dU. \tag{4}$$

Journal of Physics: Conference Series **590** (2015) 012028 doi:10.1088/1742-6596/590/1/012028

## 3. Results and discussion

In this study, we evaluated the integral (4) with certain deformation parameter values and found the upper limit energy which gives $N_{\text{cum}} = 1$ to find the deformation parameter value of the nucleus for the first excited state. With this procedure we were able to obtain the energy of the first excited state as a function of deformation parameter. The obtained results for $^{12}$C and $^{6}$He are given in Figure 1. In this figure, the red solid line represents the results of our calculation and the black dashed line corresponds the experimentally known value for the energy of the first excited state. As can be seen from the figure, our model predicts the deformation of $^{12}$C as $-0.6$ and $^{6}$He as 2.45.

To do calculations with the coupled-channels method for the inelastic couplings by using the obtained deformation parameters, the optical potential parameters should be adjusted to elastic scattering data first. Adjusted parameters are given in Table 1. We used the Woods-Saxon potential combined with a Coulomb term as optical potential. For $^{6}$He+$^{12}$C system, we found different geometrical parameters for 18 MeV and 30 MeV, but for $^{6}$He+$^{4}$He system we could describe the data at six different energies with the same geometrical parameter values and slowly varying potential depths with the energy. The results of our calculations for these two systems at eight different energies are shown in Figure 2. Good agreement between the predicted and measured cross sections is achieved as seen from figure.

## 4. Conclusions

Theoretical calculations are performed for both elastic and inelastic scattering of $^{6}$He from $^{12}$C and $^{4}$He. These preliminary results, which are given in this paper, show that the nuclear level density can be used as a tool for calculations of inelastic scattering. Obtained deformation parameter of $^{12}$C is in agree with the existing literature [10, 11]. Although the sign of the predicted deformation parameter for $^{6}$He is consistent with some other studies [12], because of its large magnitude it should be reconsidered and a more complicated shape definition should be used to calculate the collective amplitudes. Further details of this study will be published elsewhere.

## Acknowledgments

This work was supported by the Turkish Science and Research Council (TÜBİTAK) under Grant No. 112T566. Bora Canbula acknowledges the support through TÜBİTAK PhD Program fellowship BİDEB-2211 Grant.

## References

[1] Tanihata I, Hamagaki H, Hashimoto O, Shida Y *et al.* 1985 *Phys. Rev. Lett.* **55** 2676
[2] Tanihata I, Hamagaki H, Hashimoto O, Nagamiya S *et al.* 1985 *Phys. Lett. B* **160** 380–384
[3] Bang J, Danilin B, Efros V, Vaagen J *et al.* 1996 *Phys. Rep.* **264** 27–37
[4] Koning A J, Hilaire S and Goriely S 2008 *Nucl. Phys. A* **810** 13–76
[5] Canbula B and Babacan H 2011 *Nucl. Phys. A* **858** 32–47
[6] Canbula B, Bulur R, Canbula D and Babacan H 2014 *Nucl. Phys. A* **929** 54–70
[7] Milin M, Cherubini S, Davinson T, Di Pietro A *et al.* 2004 *Nucl. Phys. A* **730** 285–298
[8] Smalley D, Sarazin F, Nunes F, Brown B *et al.* 2014 *Phys. Rev. C* **89** 024602
[9] Suzuki D, Shore A, Mittig W, Kolata J *et al.* 2013 *Phys. Rev. C* **87** 054301
[10] Nunes F, Thompson I and Johnson R 1996 *Nucl. Phys. A* **596** 171–186
[11] Vermeer W, Esat M, Kuehner J, Spear R *et al.* 1983 *Phys. Lett. B* **122** 23–26
[12] Lee S, Son Y and Park J 2007 *J. Korean Phys. Soc.* **51** 480

Journal of Physics: Conference Series **590** (2015) 012029 doi:10.1088/1742-6596/590/1/012029

# Study of interstrip gap effects and efficiency for full energy detection of Double Sided Silicon Strip Detectors

**D Torresi[1], J Forneris[3], L Grassi[4], L Acosta[2,5], A Di Pietro[2], P Figuera[2], L Grilj[4], M Jakšić[4], M Lattuada[2,6], T Mijatović[4], M Milin[7], L Prepolec[4], N Skukan[4], N Soić[4], D Stanko[7], V Tokić[4], M Uroić[4], M Zadro[4]**

[1]Dipartimento di Fisica e Astronomia, Universit'a di Padova, Italy
[2]INFN Laboratori Nazionali del Sud, Catania, Italy
[3]Dipartimento di Fisica, NIS Centre and CNISM, University of Torino; INFN sez. Torino, Torino, Italy
[4]Ruder Bošković Institute, Zagreb, Croatia
[5]Departamento de Física Aplicada, Universidad de Huelva, Huelva, Spain
[6]Dipartimento di Fisica e Astronomia, Universit di Catania, Catania, Italy
[7]Department of Physics, Faculty of Science, University of Zagreb, Zagreb, Croatia

E-mail: `domenico.torresi@lns.infn.it`

**Abstract.**
In this work is reported a study on the response of double sided silicon strip detectors. In order to investigate the effect of the electrode segmentation on the detector response, two experiments were performed aimed to measure the efficiency for full energy detection. Results show that the efficiency for full energy detection, that is directly related to effective width of the inter-strip region, varies with both detected ion energy and bias voltage. The experimental results are qualitatively reproduced by a simplified model based on the Shockley-Ramo-Gunn framework.

## 1. Introduction

Double Sided Silicon Strip Detectors (DSSSDs) are widely used in high energy and in nuclear physics studies. Their high granularity and their large area make them very suitable to perform, for example, accurate measurements of angular distributions and to study reactions where coincidences between more particles are requested. It is known that the segmentation of the electrodes affects the signal formation for particles whose trajectory crosses an interstrip region [1, 2, 3, 4, 5, 6]. In particular phenomena such as charge sharing and opposite polarity signals can occur. Such phenomena make the efficiency for full energy detection for DSSSDs lower than 100% as one can expect from a single pad detector. Therefore, for the analysis of data obtained by using DSSSDs, it is very important to select the events with the correct full energy and reject interstrip events.

We performed a systematic characterization of the dependence of DSSSDs inter-strip effects on the incident ion type, energy, and polarization voltage. Studies to identify an appropriate

selection procedure of events which allows to maximize the efficiency for the full energy reconstruction have also been performed [2]. Moreover, systematic measurements of the effective width of the inter-strip gap were performed by scanning the front and back inter-strip regions using proton micro-beams at different energies and for different detector bias [1].
In the present paper the results concerning a DSSSD 1000 $\mu$m thick will be discussed. The DSSSD under study is the model W1, manufactured by Micron Semiconductor Ltd. It has an active area of 50×50 mm$^2$, each electrode is divided into 16 parallel strips, and strips of the front and back side are perpendicular each other. Each strip is 3 mm wide and they are separated by a 100 $\mu$m wide $SiO_2$ layer that in the following is called interstrip.

## 2. Measurements of full energy detection efficiency

A first experiment was performed at the Laboratori Nazionali del Sud, Catania, by using $^7$Li and $^{16}$O Tandem beams. The low intensity ( 100 pps) ion beams at energies between 10 and 50 MeV, were directly sent onto the 1000 $\mu$m thick DSSSD placed at 0° with respect to the beam direction. The aim was to study the dependence of the interstrip effects on the beam energy, the ion species and polarization voltages. Moreover the use of a mono-energetic beam allow to easily find a selection procedure of the full energy events.

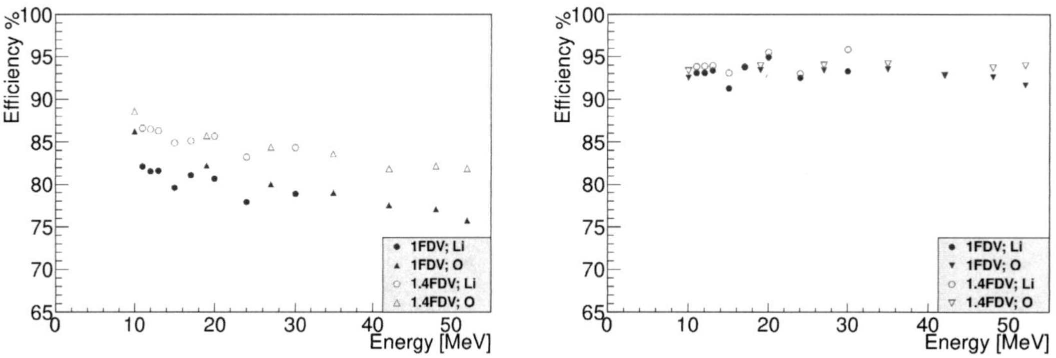


**Figure 1.** Efficiency for full energy detection as a function of the beam energy, for two different bias voltages applied to the detector. In the left panel is shown the efficiency obtained imposing the condition $E_{front} \approx E_{back}$, the continuous line corresponds to the efficiency for the full energy detection calculated using the geometrical interstrip width; in the right panel is shown the efficiency obtained recovering the back interstrip events. See the text for more details.

We observed that not all the particles impinging on the DSSSD are detected with the full energy. In fact particles whose trajectory crosses an interstrip region give rise to signals with a lower amplitude or with an opposite polarity. We divided those events in two classes corresponding to particles crossing the back or the front inter-strip region. For back inter-strip events summing the signals from the two adjacent strips those signals we obtain a full energy signal. Also for front inter-strip events two adjacent front strips give signals, but in this case one of them can have an opposite polarity and the sum of the two signals does not give a full energy signal.

Therefore, the use of such detectors, requires a procedure to select the correct full energy events rejecting the others. A common way to select the full energy events with DSSSDs is to compare the energy measured by the front side $E_{front}$ with the one measured by the back side

$E_{back}$ by imposing $E_{front} \approx E_{back}$, within a given tolerance. By applying this selection to our data we found that the obtained efficiency is slightly dependent on the energy and on the bias applied on the detector as shown in the left panel of Fig. 1 Moreover the efficiency is lower than the one extracted from the geometrical interstrip width. By remembering that for back inter-strip events just charge sharing phenomena with out any charge loss occurs, we proposed another events selection. In this case the energy measured by the front strips is compared to the sum of the energy of two adjacent back strips. In this way all the back inter-strip events are recovered increasing the efficiency and removing the dependence on energy and on bias as shown in right panel of Fig. 1 [2]. The average value of the efficiency for full energy detection is ~94%. The corresponding value of the inter-strip width is larger than the geometrical one.

### 3. Measurements of the effective inter-strip width

Since the efficiency for full energy detection is directly related with the effective inter-strip width, we performed a second experiment with the aim to have a direct measurements of the effective interstrip width for different operating conditions of the detector. The experiment was performed at the Ruder Bošković Institute in Zagreb using the local microbeam facility. The advantage to use a microbeam is that it is possible to know the position of the impinging particle with a precision of few $\mu$m. For each particle impinging on the detector the following information were recorded: the coordinates (x, y) corresponding to the horizontal and the vertical beam position inside the scanning area and the energy signals from all the front and back strips. Proton microbeams of 1.7 and 6 MeV at low intensity (~ 100 pps) were used to irradiate the DSSSD. For each beam energy the measurements were repeated for bias voltages equal to 0.5, 1 and 1.5 times the full depletion voltage (FDV).

In the left panel of fig. 2 is shown the measured energy as a function of the impact position of

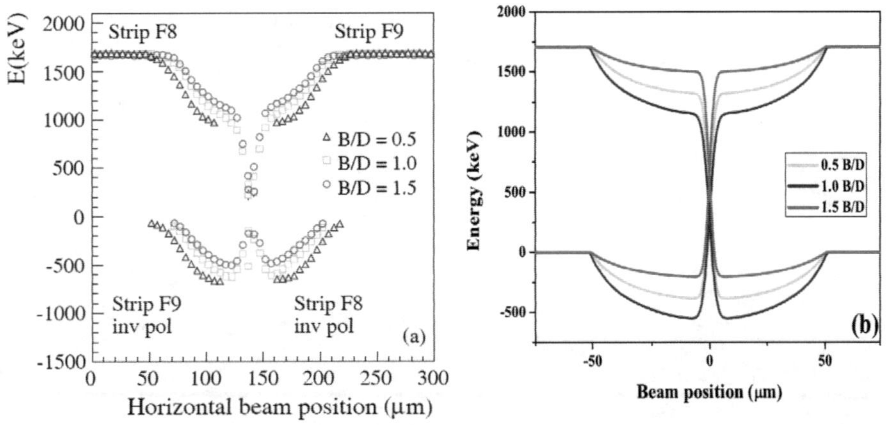


**Figure 2.** Energy as function of position for proton of 1.7 MeV for three different bias voltages. On the left panel are shown experimental data refering to a front interstrip region, on the right panel are shown data coming from simulations (fro more information see the text). B/D indicates the ratio between the applied bias and the full depletion voltage.

the incident proton. From the plot it is possible to distinguish clearly the strip and interstrip regions. When the beam is on one strip the full charge is collected by that strip, as the beam moves over the edge there is a region where both the adjacent strips produce signals that have an amplitude lower than the full energy and that can have opposite polarity. As the beam moves on the next strip the charge is entirely collected by that strip.

In the right side of figure 2 we show the result of a simplified simulation, based on the Shockley-Ramo-Gunn framework, is reported. By assuming a build up of positive charge at the oxide interface in the front inter-strip and of negative charge at the oxide interface in the back inter-strip a satisfactory qualitative reproduction of all the observed inter-strip effects has been obtained [1].

## 4. Summary and conclusions

A systematic study of the dependence of Double Sided Silicon Strip Detectors inter-strip effects on the incident ion type, energy, and polarization voltage was performed. A first measurement was performed by mean of $^{7}$Li and $^{16}$O beams at different energies, showing that the efficiency for full energy detection depends on the energy and type of the detected ion and on the applied bias voltage. A new selection procedure of events that maximize the efficiency for the full energy reconstruction and remove the cited dependences has been proposed. In a second test, direct measurements of the effective width of the interstrip region were performed by mean of a proton micro-beams at different energies and for different detector bias. Results confirmed that both front and back effective inter-strip widths, which are related to the efficacy for full energy detection, can be much larger than the nominal geometric width of the $SiO_2$ zone and depend on the operating conditions. The experimental observations were qualitatively interpreted and described by simplified simulations based on the Shockley-Ramo-Gunn framework. Therefore, for those experiments aiming at an accurate measurements of absolute cross-sections, especially if two or more DSSSD are used in coincidence, efficiency for full energy detection cannot be deduced from the geometric inter-strip width but a characterization of the used DSSSDs is required.

## Acknowledgements

This work has been partly supported by the European Community as an Integrating Activity "Support of Public and Industrial Research Using Ion Beam Technology - SPIRIT" (EC Contract no. 227012) by the European Community FP7 REGPOT project Particle Detectors (EC Contract no. 256783) and by the Italian Ministry of Education and Research (PRIN 2010-2011, project 2010TP-SCSP).

## References

[1] Grassi L *et al.* 2014 *Nucl. Instr. Meth A* **767** 1
[2] Torresi D *et al.* 2013 *Nucl. Instr. Meth A* **713** 1
[3] Poehlsen T *et al.* 2013 *Nucl. Instr. Meth A* **700** 977
[4] Takeda S *et al.* 2007 *Nucl. Instr. Meth A* **579** 044606
[5] Blumenfeld Y *et al.* 1999 *Nucl. Instr. Meth A* **421** 021601
[6] Yorkston J *et al.* 1987 *Nucl. Instr. Meth A* **262** 4097

Journal of Physics: Conference Series **590** (2015) 012030 doi:10.1088/1742-6596/590/1/012030

# Exploring the $^{12}C(^{18}O,^{16}O)^{14}C$ two-neutron transfer reaction at energies far above the Coulomb barrier

**D Carbone[1], F Cappuzzello[1,2], C Agodi[1], A Bonaccorso[3], M Bondì[1,2], M Cavallaro[1], A Cunsolo[1], M De Napoli[4], A Foti[2,4], R Linares[5] and J Lubian[5]**

[1] I.N.F.N. – Laboratori Nazionali del Sud, Catania, Italy
[2] Dipartimento di Fisica e Astronomia, Università di Catania, Catania, Italy
[3] I.N.F.N. – Sezione di Pisa, Pisa, Italy
[4] I.N.F.N. – Sezione di Catania, Catania, Italy
[5] Instituto de Física, Universidade Federal Fluminense, Niteroi, Brazil

E-mail: carboned@lns.infn.it

**Abstract.** The $^{12}C(^{18}O,^{16}O)^{14}C$ two-neutron transfer reaction has been studied at an incident energy of 275 MeV and the ejectiles have been momentum analysed at forward angles by the MAGNEX spectrometer. The identification of the $^{16}O$ ejectiles was obtained by combining a standard $\Delta E$-$E$ technique with the simultaneous measurement of the angle and the position at the focal plane. The $^{14}C$ excitation energy spectra have been obtained and several known bound and resonant states of the residual nuclei have been identified. The results have been compared to a previous measurement of the same reaction at 84 MeV incident energy.

## 1. Introduction

In the last few years, a study of the structure of different nuclei was pursued at the Catania INFN-LNS laboratory by the $(^{18}O,^{16}O)$ two-neutron transfer reaction at 84 MeV on different targets ($^{12}C$,$^{13}C$,$^{9}Be$,$^{11}B$,$^{16}O$), using the MAGNEX spectrometer to detect the ejectiles. Thanks to its high resolution and large acceptance, high quality inclusive spectra were obtained, even in a largely unexplored region above the two-neutron emission threshold in the residual nucleus [1]. New phenomena appeared, such as the dominance of the direct one-step transfer of the two neutrons [2] and the presence of broad resonances at high excitation energy in the $^{14}C$ and $^{15}C$ spectra [3] [4] [5].

Exact finite range coupled reaction channel calculations based on the parameter free Sao-Paulo double-folding optical potential accurately described the measured absolute cross section, for the first time without the need of any arbitrary scaling factor [6]. An interesting feature of the experimental angular distributions is that only the transition to the $^{14}C_{g.s.}$ presented an oscillating pattern characteristic of the expected $L = 0$ angular momentum transfer [6], whereas the other transitions, characterized by $L \neq 0$, did not show such pronounced oscillations. The phenomenon is typical of heavy-ion transfer reactions near the Coulomb barrier [7] and causes a lack of sensitivity for the $L > 0$ at this incident energy. However, there should be a threshold energy, reasonably above the Coulomb barrier, above which the oscillations in the angular distributions for $L \neq 0$ should also appear (good $l$-matching condition).

Recently, the same $^{12,13}C(^{18}O,^{16}O)^{14,15}C$ reactions have been studied at 275 MeV incident energy, by an $^{18}O$ cyclotron beam. The experiment was performed at INFN-LNS using the MAGNEX spectrometer. The aim of the experiment was to study the dynamical evolution of the structures observed in the region at high excitation energy in the spectra and to investigate the energy dependence of the good $l$-matching condition for two-neutron transfer reactions. In the present work, some preliminary results about the $^{12}C(^{18}O,^{16}O)^{14}C$ reaction are presented and compared with the previous study at 84 MeV incident energy.

Journal of Physics: Conference Series **590** (2015) 012030 doi:10.1088/1742-6596/590/1/012030

## 2. The experiment

The $^{12}C(^{18}O,^{16}O)^{14}C$ reaction was measured at the INFN-LNS in Catania using an $^{18}O^{4+}$ beam at 275 MeV accelerated by the K800 Superconducting Cyclotron accelerator. A 63 ± 3 μg/cm$^2$ self-supporting $^{12}C$ was used. The $^{16}O$ ejectiles were momentum analyzed by the MAGNEX large acceptance spectrometer [8] and detected by its focal plane detector [9]. The spectrometer worked in the full acceptance mode (solid angle $\Omega \sim$ 50 msr and momentum range $\Delta p/p \sim 24$ %), thus the total covered angular range of $3° < \theta_{lab} < 18°$ in the laboratory reference frame was obtained by only two runs in which the spectrometer optical axis was located at $\theta_{lab}^{opt} = 7°$, 12°.

The identification of the $^{16}O$ ejectiles was performed by combining two techniques, as described in Ref. [10]. The first step consists in the identification of the atomic number ($Z$) of the ejectiles by the standard $\Delta E$-$E$ technique. An example of the $\Delta E$-$E$ bi-dimensional plot is shown in figure 1 (left panel) with a coarse graphical contour that includes the Oxygen ejectiles. For the mass identification, the correlation between the measurement of the horizontal position at the focal plane ($X_{foc}$) and the residual energy ($E_{resid}$) is exploited, in which the ions are distributed on different curves according to the ratio $m/q^2$. Therefore, it is possible to clearly separate and select the $^{16}O^{8+}$ ejectiles, as shown in figure 1 (right panel). With this identification technique a mass resolution as high as 1/160 is reached [10].

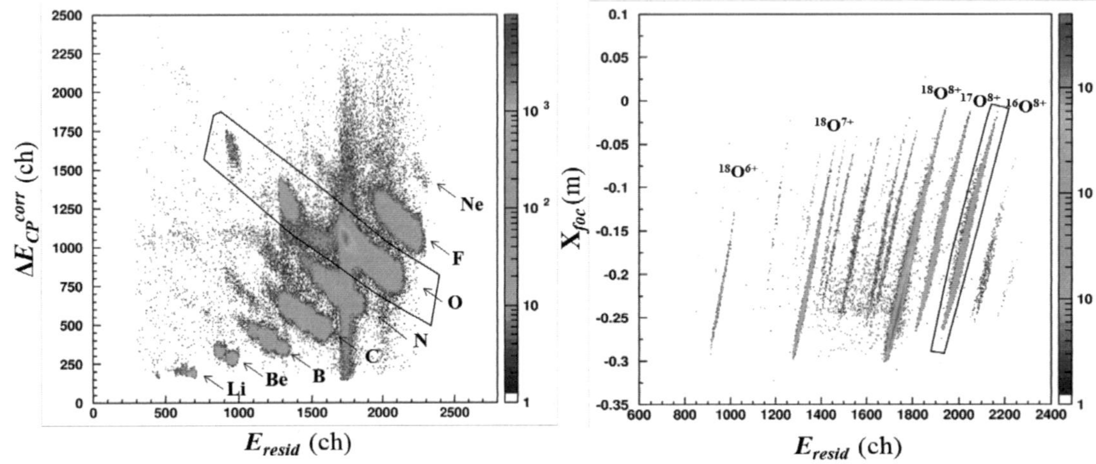


**Figure 1**. Example of the identification plots. (left panel) Typical energy loss $\Delta E_{PC}^{corr}$ vs $E_{resid}$ matrix for the unselected ejectiles detected in the reaction $^{12}C + ^{18}O$ at 275 MeV incident energy at $\theta_{lab}^{opt} = 7°$. The different ion species and a coarse graphical contour on the $^{16}O$ region are indicated. (right panel) Typical $X_{foc}$ vs $E_{resid}$ matrix plotted with the graphical condition indicated in the left panel. The different Oxygen isotopes and a graphical contour selecting the $^{16}O^{8+}$ ejectiles are indicated.

When dealing with large acceptance spectrometers, as MAGNEX, high order aberrations have to be compensated by hardware and/or software techniques. In the present case, a 10$^{th}$ order reconstruction of the scattering angle and momentum modulus was performed, based on the fully algebraic method implemented in MAGNEX [11] [12]. The excitation energies $E_x = Q_0 - Q$ (where is $Q_0$ the ground to ground-state $Q$-value) where then obtained by the application of the relativistic kinematic transformations. An overall energy resolution of about 600 keV (full width at half maximum) was obtained. This was mainly determined by the spectrometer finite energy resolution (1/100), the straggling introduced by the target and the momentum distribution of the beam, as discussed in Ref. [13].

## 3. Discussion

An example of the obtained energy spectrum for the $^{14}C$ residual nucleus is shown in figure 2. In such a spectrum it is possible to recognize some known states of $^{14}C$, namely the ground state ($J^\pi = 0^+$), a peak at ~ 6.9 MeV corresponding to the superposition of the states at $E_x = 6.73$ ($3^-$) and 7.01 ($2^+$) MeV and the states at

$E_x$ = 8.32 ($2^+$), 10.75 ($4^+$), 12.96 ($3^-$) MeV. These states are the same populated in the previous $^{12}C(^{18}O,^{16}O)^{14}C$ reaction explored at 84 MeV incident energy [4] [6], but there are some differences between the two spectra, as a result of the larger incident energy in the present case. First of all, in the region above the two-neutron emission threshold ($S_{2n}$ = 13.122 MeV) a much larger background is observed as a result of the opening of dissipative channels at energies far above the Coulomb barrier. Another effect of the higher incident energy is to favor the population of high spin states. Indeed, in the present $^{14}C$ spectrum the $0^+$ ground state is strongly suppressed and the most populated state is the $4^+$ one at 10.75 MeV.

In the region at $E_x \sim 17$ MeV a large bump is observed, superimposed on the high background. This was observed also in the previous reaction $^{12}C(^{18}O,^{16}O)^{14}C$ studied at 84 MeV [4]. Work is in progress to explain the origin of such a bump, which could reveal the excitation of the Giant Pairing Vibration mode [14] in these experiments.

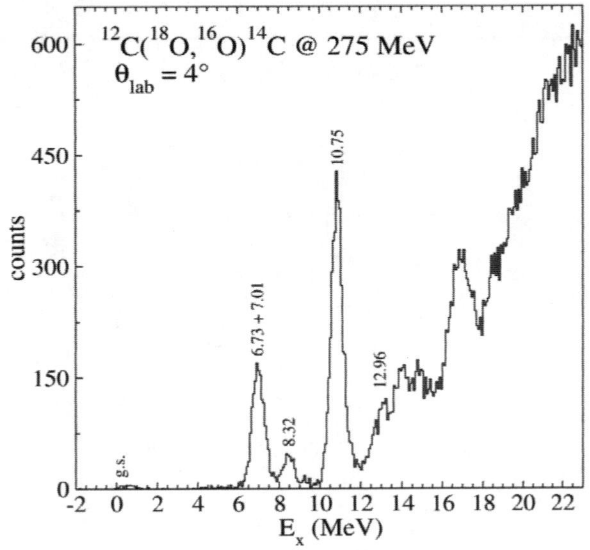


**Figure 2.** Excitation energy spectrum of the $^{12}C(^{18}O,^{16}O)^{14}C$ reaction at 275 MeV incident energy and $\theta_{lab} = 4°$.

**Acknowledgments**
The work of one of us (D.C.) was supported by the Italian Ministry of Education, Universities and Research (MIUR) under the grant "LNS-Astrofisica Nucleare (fondi premiali)".

## 4. References

[1] Cavallaro M *et al.* 2013 *AIP Conf. Proc.* **1529** 269
[2] Carbone D *et al.* 2011 *J. Phys.: Conf. Ser.* **312** 082016
[3] Cavallaro M *et al.* 2011 *AIP Conf. Proc.* **1377** 234
[4] Bondì M, Agodi C, Cappuzzello F, Carbone D, Cavallaro M, Cunsolo A, De Napoli M, Foti A, Nicolosi D and Tropea S 2012 *CERN-Proceedings-2012-002* **1** 139
[5] Cappuzzello F *et al.* 2012 *Phys. Lett. B* **711** 347
[6] Cavallaro M *et al.* 2013 *Phys. Rev. C* **88** 054601
[7] Kahana S and Baltz A 1977 *Adv. Nucl. Phys.* **9** 1
[8] Cappuzzello F, Carbone D, Cavallaro M and Cunsolo A 2011 *MAGNEX: an innovative large acceptance spectrometer for nuclear for nuclear reaction studies, in Magnets: Types, Uses and Safety* (Nova Publisher Inc., New York) p 1-63
[9] Cavallaro M, Cappuzzello F, Carbone D, Cunsolo A, Foti A, Khouaja A, Rodrigues M R D, Winfield J and Bondì M 2012 *Eur. Phys. J. A* **48** 59

[10] Cappuzzello F, Cavallaro M, Cunsolo A, Foti A, Carbone D, Orrigo S E A and Rodrigues M R D 2010 *Nucl. Instr. Methods A* **621** 419
[11] Cappuzzello F, Carbone D and Cavallaro M 2011 *Nucl. Instr. Methods A* **638** 74
[12] Cavallaro M, Cappuzzello F, Carbone D, Cunsolo A, Foti A and Linares R 2011 *Nucl. Instr. Methods A* **637** 77
[13] Cappuzzello F *et al* 2014 *Nucl. Instr. Methods A* **763** 314
[14] Broglia R and Bes D 1977 *Phys. Lett. B* **69** 129

Journal of Physics: Conference Series **590** (2015) 012031 doi:10.1088/1742-6596/590/1/012031

# Reconciliation of mass-asymmetry systematics for incomplete fusion

**Pushpendra P. Singh[1], Abhishek Yadav[2], Vijay R. Sharma[3], R. Kumar[2], Manoj K. Sharma[3], B. P. Singh[4], R. P. Singh[2], S. Muralithar[2], R. K. Bhowmik[2], and R. Prasad[4] for the AMU-IUAC collaboration**

[1]Department of Physics, Indian Institute of Technology Ropar, Rupnagar, PB - 140 001, India
[2]NP-Group, Inter-University Accelerator Center, New Delhi - 110 067, India
[3]Department of Physics, S. V. College, Aligarh, UP - 202 001, India
[4]Acc. Lab., Department of Physics, A. M. University, Aligarh, UP - 202 002, India

E-mail: pps@iitrpr.ac.in

**Abstract.** The onset and strength of incomplete fusion (ICF) has been studied in the framework of Morgenstern's mass-asymmetry systematics. The fraction of ICF has been deduced in $^{12}C+^{169}Tm$ system at energies ranging from 1.02$V_b$ to 1.64$V_b$ ($V_b$ = 54.94 MeV from the analysis of excitation functions (EFs). It has been found that the ICF starts influencing complete fusion at noticeably lower $\beta$-values (i.e., 0.025 or 2.5 % of c) than that proposed by Morgenstern (i.e., $\approx$ 6% of c). The fraction of ICF increases with entrance channel mass-asymmetry for individual projectiles, termed as projectile dependent mass-asymmetry (ProMass-) systematics. The proposed ProMass - systematics has withstood all tests that have been done for fairly large number of systems to verify its validness.

## 1. Introduction

Existence of incomplete fusion (ICF) at low incident energies, where the interactions between two heavy-ions(HIs) are mainly dominated by complete fusion(CF), has gained resurgent interest [1, 2, 3, 4, 5, 6]. Recently, a substantial ICF fraction has been observed at energies as low as $\approx$3-7 MeV/nucleon [7, 8].

Since the very first observation of ICF [9, 10], a variety of studies have been performed to understand the onset and dynamics of ICF [11]. Despite the existing studies (see refs.[2, 7] for detail), the dynamics of ICF at these energies is still not fairly understood, thus continues to be an active area of investigations. The main issues of previous studies were to explore the effects of; (*i*) projectile energy, (*ii*) mass-asymmetry ($\mu_A = A_T/A_{T+P}$), and (*iii*) input angular momenta on the onset and strength of ICF. Some of the important conclusions drawn from the previous studies are; (*i*) at high $\ell$-values ($\ell \geq \ell_{crit}$), projectile breaks up into constituents to provide sustainable input angular momenta, and fuses partly with target nucleus [2, 12, 13], (*ii*) the partially fused composite (PFC)-system forms with less mass/charge and recoil velocity as that of completely fused composite (CFC)-system [6, 14, 15, 16], (*iii*) the fraction of ICF has been found to be large for more mass-asymmetric systems [17]. However, some of the experimental results contradict previous findings. For example, according to Morgenstern's systematics [17], the fraction of ICF is expected to be large for more mass-asymmetric systems at a constant value of $v_{rel}$(i.e., $\sqrt{2(Ecm - Vb)/\mu}$/c). In recent reports [7, 8], the variation of ICF fraction with $\mu_A$ do not found to obey Morgenstern's systematics. While, the data obtained for individual projectiles ($^{12}$C and $^{16}$O) have been fairly explained by the given systematics. This observation may be considered as a supplement to the systematics but rather poorly known, therefore, need to be supported by further measurements. To confirm the findings of ref.[7, 8], and to have better insights into the entrance channel effect on the onset and strength of ICF, the fraction of ICF has been deduced in $^{12}$C+$^{169}$Tm system at energies ranging from 1.02V$_b$ to 1.64V$_b$ (V$_b$ = 54.94 MeV). Findings of present measurement are compared with the previously reported $^{16}$O results [7] to display projectile effect on ICF fraction. A comparison of results from nearby systems, $^{12}$C+$^{128}$Te,$^{165}$Ho,$^{169}$Tm and $^{16}$O+$^{103}$Rh,$^{159}$Tb,$^{169}$Tm [7, 8, 18, ?], is also presented to describe mass-asymmetry (and/or projectile) dependence on the onset and strength of ICF.

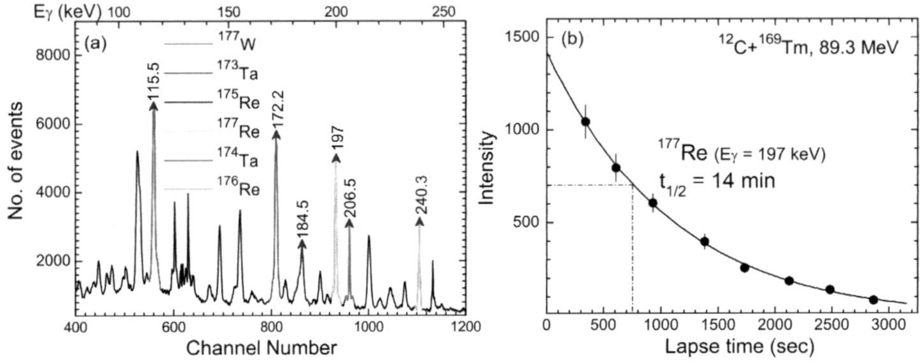


**Figure 1.** (a) Decay $\gamma$-lines of different residues are marked in a spectra obtained at $\approx$89.3 MeV, (b) Decay curve of $^{177}Re$ ($t_{1/2}$ = 14 min) obtained by following 197 keV $\gamma$-line.

## 2. Experimental details and data analysis

Experiments have been performed at IUAC New Delhi using similar experimental methodology and setup as has been used in refs.[7, 8]. However, a brief account of experimental conditions is given here. $^{169}$Tm (abundance = 100%) Targets of thickness $t_m \approx$ 1 mg-cm$^{-2}$ were prepared by rolling technique and the uniformity of each target was verified by $\alpha$-transmission method. The error in thickness was estimated to be $\approx$1%. In ref.[7, 8], energy degradation technique has been employed to cover a wide energy range in single run, and corrections were made for energy spread and beam intensity variations during the data analysis. In this work, the energy spread and beam intensity variations have been avoided by adopting single target irradiation methodology for each projectile energy. Irradiations were carried out using $^{12}$C-beam (E$_{beam}$ = $\approx$ 54-90 MeV, beam current $\approx$20-30 nA). In-situ measurements of $\gamma$-activities were performed off-line using two pre-calibrated HPGe detectors to detect the residues of short half-lives ($t_{1/2}$ $\approx$ 5-10 minutes). The uncertainty in efficiency of detectors is estimated to be $\approx$ 2%. Relevant portion of a $\gamma$-ray spectra obtained at $\approx$89 MeV is shown in Fig.1(a), and a decay curve for $^{177}$Re residue in Fig.1(b). Reaction products have been identified by their characteristic $\gamma$-lines and decay-curve analysis. The most intense $\gamma$-lines have been used for decay-curve analysis and for the production cross-section ($\sigma_{ER}$) measurement [8]. The overall error in $\sigma_{ER}$ is estimated to be $\approx 13\%$.

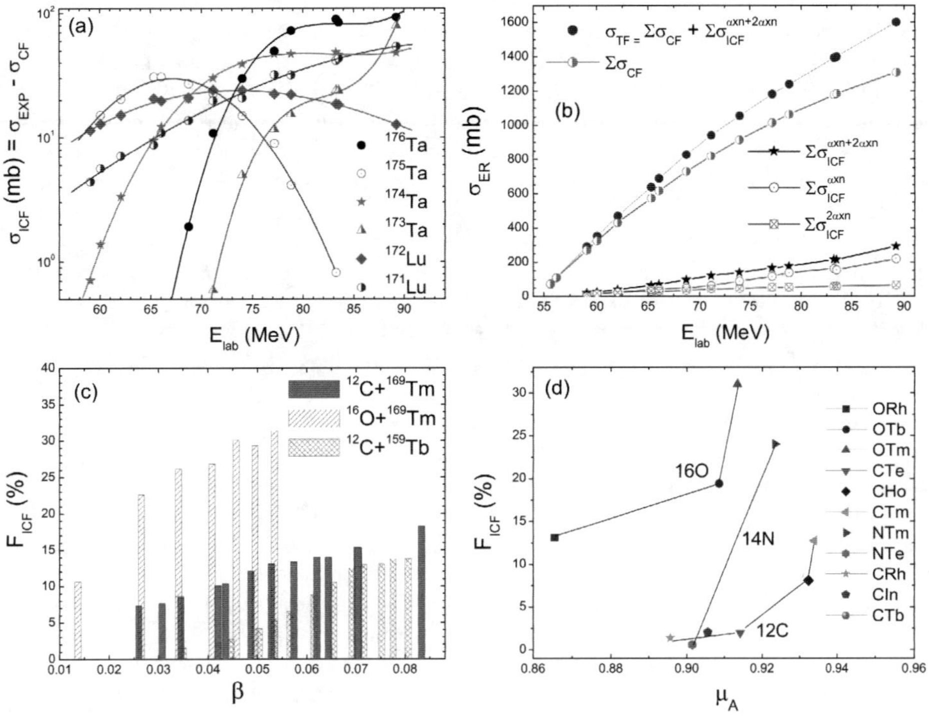


**Figure 2.** ($a$) EFs of individual ICF channels in $^{12}$C+$^{169}$ system, (b) comparision of $\Sigma\sigma_{ICF}$ with $\Sigma\sigma_{CF}$ and $\sigma_{TF}$, (c) the ICF strength function for different systems(see text for description), and ($d$) the values of F$_{ICF}$ as a function of $\mu_A$.

## 3. Results and interpretations

The ICF strength function has been deduced for $^{12}$C,$^{16}$O+$^{169}$Tm systems from the analysis of experimental EFs in the framework of statistical model code PACE4 [6, 7]. Detailed discussion on data reduction procedure can be found elsewhere [6, 7]. Experimentally measured and systematically deduced EFs for individual ICF channels are plotted in Fig.2(a), and the sum of all ICF-channels ($\Sigma\sigma_{ICF}$) is plotted with the sum of all CF-channels ($\Sigma\sigma_{CF}$) and total fusion ($\sigma_{TF}$) in Fig.2(b). The separation between $\Sigma\sigma_{CF}$ and $\sigma_{TF}$ with E$_{lab}$ indicates increasing ICf contribution with energy. In order to better understand the energy dependence, the percentage fraction of ICF (F$_{ICF}$) is plotted with E$_{lab}$ in Fig.2(c), termed as ICF strength function. The ICF strength function defines empirical probability of ICF at different projectile energies. As shown in Fig.2(c), the value of F$_{ICF}$ is found to be $\approx$7% at $\approx$59 MeV, i.e., 1.075Vb (7.5 % above the barrier), and increases smoothly up to ≈18 % at highest measured energy i.e., 1.64 Vb for $^{12}$C+$^{169}$Tm system.

According Morgenstern's systematic [17], ICF contributes significantly above $v_{rel} = \beta = 0.06$ (6 % of c), and the fraction of ICF should increase with entrance channel mass-asymmetry ($\mu_A$). As can be seen in Fig.2(c), the values of $\beta$ are in the range from ≈0.027 (2.7 % of c) to ≈0.084 (8.4 % of c) for $^{12}$C-beam, and from ≈0.014 (1.4 % of c) to ≈0.053 (5.3 % of c) for $^{16}$O-beam. At given value of $\beta$, no significant ICF contribution is expected. But, the results presented here suggest the onset of ICF at relatively lower value of $\beta$ i.e, ≈0.027 ($F_{ICF}$ ≈7 %) in $^{12}$C+$^{169}$Tm system, and at ≈0.014 ($F_{ICF}$ ≈10 %) in $^{16}$O+$^{169}$Tm system. In both cases, the observed value of $F_{ICF}$ is significant even at well below the proposed onset value of $\beta$ (i.e., 6% of c).

Further, the value of $F_{ICF}$ for $^{12}$C is lower than $^{16}$O-projectile. The difference in $F_{ICF}$ for two systems ($^{12}$C,$^{16}$O+$^{169}$Tm) indicates the dependence of $F_{ICF}$ on projectile charge and/or on $\mu_A$. In order to refine this effect, the values of $F_{ICF}$ for nearby systems ($^{12}$C+$^{128}$Te,$^{165}$Ho,$^{169}$Tm and $^{16}$O+$^{103}$Rh,$^{159}$Tb,$^{169}$Tm) are plotted as a function of $\mu_A$ in Fig.2(d) at a constant value of $\beta$=0.053. As shown in this figure, the Morgenstern's systematics does not explain the variation of $F_{ICF}$ with $\mu_A$. However, the value of $F_{ICF}$ increases with $\mu_A$ for individual projectiles ($^{16}$O, $^{12}$C, and $^{14}$N). It is interesting to note that the $^{12}$C+$^{169}$Tm system is a more mass-asymmetric ($\mu_A = 0.9337$) than $^{16}$O+$^{169}$Tm system ($\mu_A = 0.9135$), but the value of $F_{ICF}$ is 18% higher than that observed for $^{12}$C+$^{169}$Tm system. Present observations suggest the inclusion of projectile effect along with the mass-asymmetry of interacting partners to explain low energy ICF and lead to the reconciliation of Morgenstern's systematics.

## 4. Summary and conclusions

This paper briefly summarizes the findings of recent experiments performed to study ICF at energies ≈ 4-7 MeV/nucleon in $^{12}$C,$^{16}$O+$^{169}$Tm systems. It has been found that ICF significantly contributes to total reaction cross-section even at slightly above barrier energies. The percentage fraction of ICF increases with projectile energy and $\mu_A$ for individual projectiles. This suggests the inclusion of projectile type along with the mass asymmetry to explain low energy ICF data and formulate a 'projectile dependent mass-asymmetry (ProMass) systematics'. In order to gain confidence in the proposed systematics a series of similar experiments are in order.

We thank the Director of IUAC New Delhi for extending experimental facilities to perform these measurements. One of the authors (P.P.S.) acknowledges the support and motivation received from Prof. P. K. Raina (Dean A & R) and Prof. M. K. Surappa (Director, IIT Ropar).

## References

[1] Diaz-Torres A 2010 *J. Phys. G: Nucl. Part. Phys.* **37** 075109 and the references therein.
[2] Singh P P *et al.* 2009 *Phys. Lett.* **B671** 20 ; 2009 *Phys. Rev.* C**80**, 064603
[3] Singh D P *et al.* 2010 *Phys. Rev.* C**81** 054607
[4] Gomes P R S *et al.* 2004 *Phys. Lett.* **B601** 20 ; 1984 *Phys. Rev. Lett.* **B53**, 1630
[5] Diaz-Torres A *et al.* 2007 *Phys. Rev. Lett.* **98** 152701
[6] Gupta U *et al.* 2009 *Phys. Rev.* C**80** 024613
[7] Singh P P *et al.* 2008 *Phys. Rev.* C**77** 014607
[8] Gupta U *et al.* 2008 *Nucl. Phys.* **A811** 77
[9] Britt H C and Quinton A R 1961 *Phys. Rev.* **124** 877
[10] Galin J *et al.* 1974 *Phys. Rev.* C**9** 1126 and the references therein.
[11] Gerschel C 1982 *Nucl. Phys.* **A387** 297 and the references therein.
[12] Wilczynski J *et al.* 1980 *Phys. Rev. Lett.* **45** 606 ; 1982 *Nucl. Phys.* **A373**, 109
[13] Barker J H *et al.* 1980 *Phys. Rev. Lett.* **45** 424
[14] Parker D J, Asher J, Conlon T W and Naqib I 1984 *Phys. Rev.* C **30** 143
[15] Udagawa T and Tamura T 1980 *Phys. Rev. Lett.* **45** 1311 and the references therein.
[16] Singh P P 2007 *et al. Eur. Phys. J.* **A34** 29 ; 2008 *Phys. Rev.* C**78** 017602
[17] Morgenstern H *et al.* 1983 *Z. Phys.* **A313** 39 ; 1984 *Phys. Rev. Lett.* **52** 1104
[18] Sharma M K *et al.* 2003 *J. Phys. Soc. Jpn.* **72** 1917 ; Gupta S *et al.* 2000 *Phys. Rev.* C**61** 064613

Journal of Physics: Conference Series **590** (2015) 012032 doi:10.1088/1742-6596/590/1/012032

# Interplay between the pairing and quadrupole interactions in the algebraic realization of the microscopic shell model

**Kalin Drumev, Ana Georgieva**

Institute for Nuclear Research and Nuclear Energy, Bulgarian Academy of Sciences, Sofia 1784, Bulgaria

**Abstract.** We explore the algebraic realization of the Pairing-Plus-Quadrupole Model /PQM/ in the framework of the Elliott's SU(3) Model with the aim to obtain the complementary and competing features of the two interactions through the relation between the pairing and the SU(3) bases. First, we establish a correspondence between the SO(8) pairing basis and the Elliott's SU(3) basis. It is derived from their complementarity to the same LST coupling chain of the shell-model number-conserving algebra. The probability distribution of the SU(3) basis states within the SO(8) pairing states is also obtained and allows the investigation of the interplay between the pairing and quadrupole interactions in the Hamiltonian of the PQM, containing both of them as limiting cases. The description of some realistic N∼Z nuclear systems is investigated in a SU(3)-symmetry-adapted basis within a model space of one and two oscillator shells.

keywords: algebraic models, pairing-plus-quadrupole model, microscopic shell model

## 1. Introduction

Since the early years of the development of the nuclear structure physics, it has been understood that the pairing [1] and the quadrupole-quadrupole interactions [2] are the most important short- and long-range interactions that have to be taken into account in the shell-model description of the nuclear systems [3]. Being with different range of action on the nucleons in the valence shells it is quite clear that these interactions actually influence the behavior of the systems in different parts of the shells. Hence, in some nuclei each of these interactions could reproduce relatively well the observed behavior of the nuclear system, but in most of the cases the study of the relationship between them is of great importance. This is the main motivation for the development of the Pairing-plus-Quadrupole Model /PQM/ [4] for the description of the nuclear excitation spectra. It is most successfully done in the framework of the basic shell model representation of the employed interactions, but the applications to real nuclear systems are rather complicated and cumbersome, due to the enormous dimensionality of the model space in particular for the heavy nuclei. Such a problem is easily avoidable by employing a group-theoretical approach [5], which introduces symmetry principles useful in particular for reducing the basis spaces and in the calculation of matrix elements of transitional operators. In this work we present such an approach by considering the basic interactions of the PQM as invariants of two respective algebras, which reduce the general symmetry of the shell model in a dynamical way. At the same time the two so defined dynamical symmetry chains are both complementary to the Wigner's spin-isospin

Journal of Physics: Conference Series **590** (2015) 012032 doi:10.1088/1742-6596/590/1/012032

$SU_{ST}(4)$ symmetry, which establishes the direct connection between these two limiting cases. The latter allows for the investigation of the competing and complementarity features of the pairing and quadrupole interactions in the description of the realistic nuclear systems in the lower shells up to mass numbers $A \sim 100$.

## 2. Dynamical Symmetries in the Many-Particle Shell-Model Scheme

In this work we explore the reduction of the algebraic realization of the shell model algebra $U(4\Omega) \supset U_{ST}(4) \otimes U(\Omega))$[5] into the spatial $U(\Omega)$ and spin-isospin $U_{ST}(4)$ branches which are complementary. The chain $SU_{ST}(4) \supset SU_S(2) \otimes SU_T(2)$ of the Wigner's supermultiplet model [6] gives the spin $S$ and isospin $T$ of the basis states of the shell model. Further we consider the two possible reductions of the spatial part $U(\Omega)$ to the SO(3) algebra of the angular momentum. The chain $U(\Omega) \supset SU(3) \supset SO_L(3)$ [2] gives the rotational limit of the model with only quadrupole-quadrupole interaction taken into account. The other one through $SO(\Omega)$ which is isomorphic to SO(8), the algebra of the isoscalar and isovector pairing interaction, represents the pairing limit of the shell model algebra. Both these chains are complementary to the spin-isospin $U_{ST}(4)$ algebra. The above chains are unified into the following generalized reduction scheme:

$$\begin{array}{cccccc}
\{1^m\} & & & U(4\Omega) & & \\
 & & & \downarrow & & \\
(\tilde{f}\} & [U(\Omega) & & \otimes & U_{ST}(4)] & \{f\} \\
 & \downarrow & \searrow \alpha & & \downarrow & \\
[\tilde{\mu}] & [SO(\Omega) & [SU(3) & \otimes & SU_{ST}(4)] & \{f'\} \\
(\nu[p]) & \Leftrightarrow SO(8) & (\lambda,\mu) & & \sim SO(6) & [P] \\
\beta & \downarrow & \swarrow K & & \downarrow & \\
L & [SO_L(3) & & \otimes & SU_S(2)] \otimes SU_T(2) & S,T \\
 & & & \downarrow & \downarrow & \\
J & & & SU_J(2) & \otimes SU_T(2) & T
\end{array} \tag{1}$$

Now we obtain the important result that the spatial subalgebra $U(\Omega)$ of the shell-model algebra $U(4\Omega)$ contains two distinct dynamical symmetries defined by the reduction chains: left branch through $SO(\Omega)$ and middle branch through SU(3). Consequently, both chains determine full-basis sets and could be expressed through each other. The basis states labeled by the quantum numbers of the representations of the algebras in the SU(3) chain $|\Psi_R\rangle \equiv |\{f\}\alpha(\lambda,\mu)KL,S;JM\rangle$ are eigenstates of the rotational limit of the model with quadrupole-quadrupole interaction. Correspondingly, the basis states in which the pairing interaction is diagonal [7] are labeled as $|\Psi_P\rangle \equiv |\{f\}\nu[p_1,p_2,p_3]\beta L,S;JM\rangle$. In both types of states $\alpha$, $\beta$ and $K$ give the multiplicity labels of the corresponding reductions. Since the microscopic SU(3) model based on the three-dimensional harmonic oscillator has a well-developed theory, including the Wigner-Racah algebra for the calculation of matrix elements [8] in the SU(3) basis and various successful applications in real nuclei, we choose to expand the states of the pairing basis $|\Psi_P\rangle$ in the set of basis states $|\Psi_R\rangle$, i.e. $|\Psi_P\rangle_i = \sum_j C_{ij}\,|\Psi_R\rangle_j$. Using the above expansion and the diagonalization procedure for the pairing interaction in the SU(3) basis:

$$_i\langle\Psi_P|H_{Pair}|\Psi_P\rangle_i = E_{Pair}(m,i,[P],(ST)) = \sum_{jk} C^*_{ki}C_{ij}.\delta_{kj}\cdot{}_k\langle\Psi_R|H_{Pair}|\Psi_R\rangle_j \tag{2}$$

we obtain numerically the probability $|C_{ij}|^2$ with which the states of the SU(3) basis enter into the expansion of the pairing basis. In this way we actually calculate the transformation brackets between the two chains [9], which is of great use when calculating the matrix elements

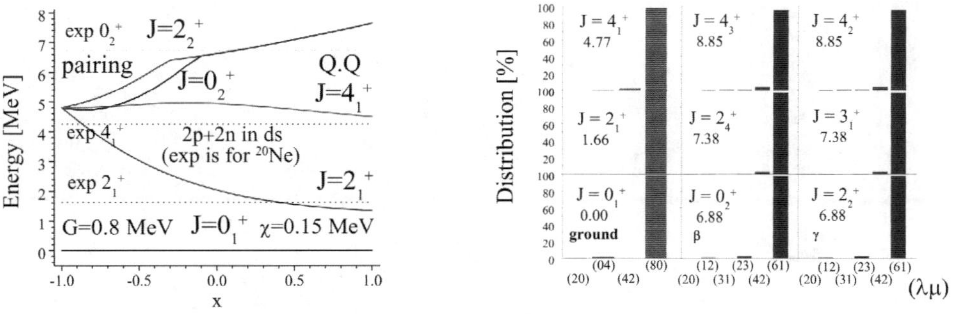


**Figure 1.** Excitation energies (left) and the SU(3) contents of the low-lying states (right) for the system of 2 protons and 2 neutrons (the nucleus $^{20}$Ne) in the *ds* shell.

of different operators in each of the chains. This is important for example for the calculation of transition probabilities. Also, this expansion could help evaluate the importance (weight) of the different SU(3) states, when we need to impose restrictions on the basis because of computational difficulties. The known relations of the SU(3) labels $(\lambda, \mu)$ and the $\beta, \gamma$ shape variables of the geometrical model can be used for the analysis of the deformations of the pairing states, expressed through the respective SU(3) ones.

For the purpose of our investigation we use the Hamiltonian $H = H_0 + V_{res}$ of the PQM [10], where $H_0$ is the harmonic oscillator term and the single-particle interactions and

$$V_{res} = \frac{1}{2}(1-x)\mathbf{G}(S^{\dagger}_{\mu}.S_{\mu} + P^{\dagger}_{\mu}.P_{\mu}) - \frac{1}{2}(1+x)\chi Q.Q, \tag{3}$$

At $x = -1$ we have pure pairing interaction with equal strengths of the isoscalar and isovector terms, and at $x = 1$ the limiting case of pure quadrupole interaction is realized. At $x = 0$ we have both interactions mixed with their respective strengths. This allows us to investigate the influence of these residual interactions on the spectra in real nuclear systems.

## 3. Results and discussion

After presenting in short the algebraic realization of the dynamical symmetries that appear in the microscopic shell model we would now like to exploit their applications in realistic nuclear systems. We start with a real test case for the applications of the theory - the *ds* shell, which is the first one, where both deformation and pairing phenomena play an important role [11], [12]. Our proof-of-case example presents the simple but complete system of 4 particles in the *ds* shell which allows us to study the PQM without any truncation of the model space.

First, we demonstrate the complementarity of the two competing modes by showing the energy spectrum (the left panel of Fig. 1) and the SU(3) contents (the right panel of the same figure) of the first few low-lying energy eigenstates for the system of 2 protons and 2 neutrons in the ds shell ($^{20}$Ne). Different spin-value dominance in the bands is represented by the respective colors - red for $S = 0$ and blue - for $S = 1$. Below the angular momentum labels on the right, the excitation energy in MeV is given. Both $\beta$ and $\gamma$ bands have equal excitation energies and are dominated by the same SU(3) irrep (61).

Next, for the same system, we extend our investigation beyond one oscillator shell and demonstrate to what degree the two-shell eigenfunction can be described by its shell-plus-a-single-orbital and one-shell restrictions. For this purpose, we add the single-particle term to

Journal of Physics: Conference Series **590** (2015) 012032 doi:10.1088/1742-6596/590/1/012032

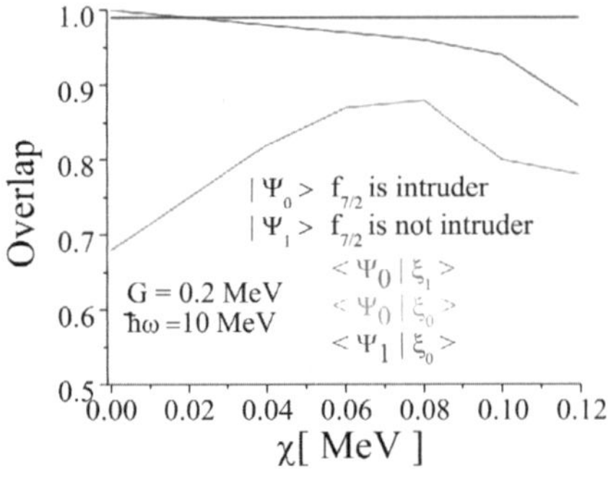


**Figure 2.** The overlap of the eigenstate, obtained in the two-shell model space $(|\Psi_0\rangle, |\Psi_1\rangle)$, with the result in the $ds$ shell plus the $f_{7/2}$ orbital as an intruder $|\xi_1\rangle$ and with the single $ds$-shell result $|\xi_0\rangle$ for the system of 2 protons and 2 neutrons in the $dsfp$ shell at $G = 0.2$ MeV and varying values of the parameter $\chi$. In the $|\Psi_0\rangle$ eigenfunction, the $f_{7/2}$ orbit is considered energetically to belong to the $ds$ shell. In the $|\Psi_1\rangle$ case, $f_{7/2}$ belongs to the $fp$ shell.

the Hamiltonian which accounts for the separation of the orbits in the two adjacent shells. So, on Fig. 2 we demonstrate that the two-shell results $|\Psi_0\rangle, |\Psi_1\rangle$ can be quite well represented at certain interaction strengths by the eigenfunctions in only one shell $|\xi_0\rangle$ or the shell plus a single orbital configuration $|\xi_1\rangle$. The strongest overlap, almost independent of the interaction strengths, is obtained for the case $\langle\Psi_1|\xi_0\rangle$, followed by the $\langle\Psi_0|\xi_1\rangle$, which starts to diminish at higher values of $\chi$.

In conclusion, the algebraic structure of the shell-model algebra $U(4\Omega)$ is investigated to obtain its reductions through the microscopic pairing algebra SO(8), containing both isoscalar $(T = 0, S = 1)$ and isovector $(T = 1, S = 0)$ pairing operators and Elliott's SU(3) algebra. A relation between these chains is established on the basis of the complementarity to the Wigner's spin-isospin $U_{ST}(4)$ symmetry. This elucidates the algebraic structure of an extended Pairing-plus-Quadrupole Model, realized in the framework of the Elliott's SU(3) scheme [5]. This approach is used to study the combined effects of the quadrupole-quadrupole and pairing interactions on the energy spectra of the nuclear systems in a single shell, a shell plus an orbital and two shells.

## Acknowledgments

This work was supported by the Bulgarian National Foundation (DID-02/16/17.12.2009).

## References

[1] Flowers B H 1952 *Proc. Roy. Soc. London*, Ser. A **212** 248
[2] Elliott J P 1958 *Proc. Roy. Soc. London*, Ser. A **245** 128; **245** 562
[3] Bohr A, Mottelson B R and Pines D 1958 *Phys. Rev.* **110** 936
[4] Belyaev S T 1959 *Mat. Fys. Medd. Dan. Vid. Selsk.* **31** No. 11; Kisslinger L S and Sorensen R A 1963 *Rev. Mod. Phys.* **35** 853; Baranger M and Kumar K 1965 Nucl. Phys. **62** 113
[5] Draayer J P 1993 in *Algebraic Approaches to Nuclear Structure: Interacting Boson and Fermion Models*, Contemporary Concepts in Physics VI, edited by R.F. Casten (Harwood Academic, Chur, Switzerland) 423
[6] Wigner E P 1937 *Phys. Rev.* **51** 106
[7] Kota V K B, Castilho Alcaras J A 2006 *Nucl. Phys.* A **764** 181
[8] Akiyama Y and Draayer J P 1973 *Comput. Phys. Commun.* **5** (1973) 405; Bahri C and Draayer J P 1994 *Comput. Phys. Commun.* **83** 59; Draayer J P, Leschber Y, Park S C, and Lopez R 1989 *Comput. Phys. Commun.* **56** 279
[9] Santopinto E, Bijker R and Iachello F 1996 *J. Math. Phys.* **37** 2674
[10] Bahri C, Escher J, Draayer J P 1995 *Nucl. Phys.* A **592** 171
[11] Elliott J P and Harvey M 1963 *Proc. Roy. Soc. London*, Ser. A **272** 557; Elliott J P and Wilsdon C E 1968 *Proc. Roy. Soc. London*, Ser. A **302** 509
[12] Vargas C E, Hirsch J G, Draayer J P 2001 *Nucl. Phys.* A **690** 409

# Analysis of Low Lying Collective States in the Symplectic Interacting Vector Bosons Model

**A. I. Georgieva and V. P. Garistov**

Institute of Nuclear Research and Nuclear Energy, Bulgarian Academy of Sciences bul. Tzarigradsko Chaussee 72, Sofia 1787, Bulgaria

E-mail: anageorg@inrne.bas.bg

**Abstract.** Successful description of both positive and negative parity band configurations and their corresponding band-head's structure is obtained within the framework of the algebraic Interacting Vector Boson Model (IVBM). The analysis of the collective structure of the low lying excited states possessing different parity is carried out together with the results obtained from parabolic distributions of the energies in the space of integer and positive classification parameter (supposedly number of bosons) for large amount of even-even nuclei.

keywords: symplectic boson model, collective bands, energy distribution of states

## 1. Introduction

Contemporary experimental techniques have advanced to the place where information on large sequences of collective states with $L = 0, 2, 4, 6...$ in a given nucleus is now very rich, particularly in the rare-earth and actinide regions [1]. It is a challenge for the theoretical models to achieve a correct interpretation of the data since the observed collective states should be well organized in bands, revealing their structure and origin. Algebraic models play an important role in this respect by relating the structure of the collective states to basis states of a dynamical symmetry group and its subgroup chains. In this regard, the symplectic model provides a natural and very general framework for investigating the nature of collective excitations in many-body systems [2]. The symplectic model incorporates "elementary" excitations into the structure of the collective states. We exploit a phenomenological Interacting Vector Boson Model (IVBM) that yields a rather accurate description of the low-lying spectra of even-even heavy nuclei [3]. The spectrum generating algebra of the model is that of the $Sp(12, R)$ group, which has a very rich subgroup structure [3, 4].

Our aim is to study the behavior of the energies of sequences of collective states with fixed angular momentum $L$ in the spectra of a given even-even nucleus in conjunction with the ordering of these states in collective bands.

This is achieved by employing the algebraic relations between two of the dynamical symmetries of the IVBM, each corresponding to the respective physical interpretation. This gives us from one side possibility to understand the structure of each collective state and from the other side to predict some new unobserved states trough their classification.

## 2. Algebraic framework of the symplectic IVBM

We consider $Sp(12, R)$ – the group of linear canonical transformation in a 12-dimensional phase space – to be the dynamical symmetry group of the model [3, 5]. Its algebra is realized in terms of creation (annihilation) operators $u_m^+(\alpha)(u_m(\alpha) = (u_m^+(\alpha))^\dagger)$, of *two* types of bosons differing by their "pseudospin" projection $\alpha = p = 1/2$ (proton) and $\alpha = n = -1/2$ (neutron) in a 3-dimensional oscillator potential with $m = 0, \pm 1$. All the bilinear products of the creation and annihilation operators of the two vector bosons generate the noncompact symplectic group $Sp(12, R)$ [3]. The following correspondence exists between two of the chains of subalgebras of $sp(12, R)$ [6]:

$$\begin{array}{ccccc} sp(12,R) & \supset & sp(4,R) & \otimes & so(3) \\ \cup & & \cup & & \cap \\ u(6) & \supset & u(2) & \otimes & su(3)\ . \end{array} \tag{1}$$

In the first row of (1) a new possible reduction $sp(12, R) \supset sp(4, R) \otimes so(3)$ of the dynamical symmetry sp(12,R) [5] is introduced. The generators of the $sp(4, R)$ are the set of scalar operators with $L = 0$ from $sp(12, R)$ generators and as such they commute with the components of the angular momentum $L_M$ that generate the $so(3)$ algebra. Hence the eigenvalues of $L$ are used to characterize the representations of $sp(4, R)$ and in this way isolate sets of states with given $L$. The correspondence (1) is a result of the equivalent pseudospin $u(2) \supset sp(4, R)$ algebra in both chains, which is complementary to $su(3) \supset u(6)$. This permits an investigation into the behavior of low-lying collective states with the same angular momentum $L$ with respect to the number of excitations $N$ that build these states [7].

## 3. Energy distribution of low-lying collective bands and states

Because of the correspondence (1) and the relation between the $SU(3)$ and $SU(2)$ second order Casimir operators, the Hamiltonian $H = aN + bN^2 + \alpha_3 T^2 + \beta_3 L^2 + \alpha_1 T_0^2$ and bases $|[N]_6; (\lambda, \mu); K, L, M; T_0\rangle \equiv| (N, T); K, L, M; T_0\rangle$ are equivalent in both chains and the eigenvalues of the Halmitonian $H$

$$E((N,T); KLM; T_0) = aN + bN^2 + \alpha_3 T(T+1) + \alpha_1 T_0^2 + \beta_3 L(L+1). \tag{2}$$

can be used in both cases for evaluation of the energy distribution of states with a fixed $L$ [7] and for calculation of the energies of states from a given band [4]. In the first case, since $L$ is fixed the energies of the collective states depend on the number of phonons (vector bosons) $N$ and is obviously parabolic. All the rest of the quantum numbers defining the states $T$, $T_0$, and $L$ are expressed in terms of $N$ by means of standart reduction procedures [6]. This result confirms

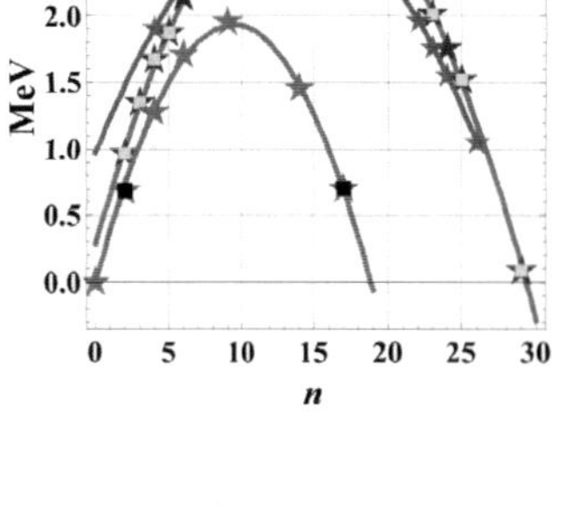


**Figure 1.** Empirical distribution of the experimentally observed energies of the $0^+$ (red curve with red stars) $2^+$ (green curve) and $3^+$ (pink curve with blue stars) states in $^{160}Dy$, with respect to the integer classification parameter $n$ related to the number of bosons, that built the respective states [7].

Journal of Physics: Conference Series **590** (2015) 012033 doi:10.1088/1742-6596/590/1/012033

the empirical investigation of the states with fixed angular momentum [8]. The set of $N_{L_i}$ with minimal value of $\chi^2$ in a multistep fitting procedure determines the distribution of the $L_i^+$ states energies (the parameters of the Hamiltonian) with respect to the number of bosons $N_{L_i}$ that build the states, which is illustrated as an example in fig. 1. Applying this approach two new low-lying excited $0^+$ states in the $^{160}$Dy nucleus were experimentally observed as a result of the theoretical prediction by the parabolic energy distribution of $0^+$ states [9].

Next in the most important application of the $U(6) \subset Sp(12,R)$ limit of the theory we exploit the possibility it affords for describing both even and odd parity bands up to very high angular momentum [4]. This application is related in the first place to the proper identification of the experimentally observed bands with the sequences of basis states from the even and odd representation of $Sp(12,R)$. Here we use the following identification of the experimentally observed ground, excited $(K^\pi = 0^+), \beta-$, $(K^\pi = 2^+), \gamma-$ and octupole $K^\pi = 0^-$ bands. Using

| Experimental states $K^\pi$ | $SU(3)$ Basis states |
|---|---|
| Ground states band $K^\pi = 0^+, \mu_0 = 0$<br>$\beta-$bands $\mu_0 = \lambda_0 \neq 0$-even,$L = 0, 2, 4, ...$ | $(\lambda, \mu) = (\mu_0 + 2i, \mu_0), i = 0, 1, 2, ...,$<br>$L = 2i,\ N = N_0 + L, T = (N_0 + 3L)/2$ |
| $\gamma-$ band $K^\pi = 2^+, \mu_0 = K = 2,$<br>$L = 2, 3, 4, ...$ | $(\lambda, \mu) = (\mu_0 + 2i, \mu_0)$,$i = 0, 1, 2, ...$<br>$L = \mu_0 + i$,$N = N_0 + 2L - 4, T = (L - 1)$ |
| Octupole band $K^\pi = 0^-, \mu_0 = 0,$<br>$L = 1, 3, 5, ...$ | $(\lambda, \mu) = (\mu_0 + 2i, \mu_0), i = 0, 1, 2, ...$,$L = \mu_0 + 2i$<br>$N = N_0 + L - 1, T = \frac{1}{2}(N_0 + L - 1)$ |

the above assignments of the experimentally observed collective bands to sequences of $SU(3)$ basis states and the relations between the quantum numbers labeling the representations of the subalgebras in the reduction scheme (1) we evaluate the parameters of the Hamiltonian $H$ by fitting the energies in (2) to the experimentally observed states in the considered ground, $\beta-$, $\gamma-$ and octupole bands in a given nucleus [4]. In most cases we obtain a very good agreement with the experiment up to very high spins [10]. An example is given on the left panel of fig. 2.

Further from the same assignments we obtain the following dependencies of the energies of the considered collective bands on the numbers of phonon excitations building their band head configurations $N_{0\beta}, N_{0\gamma}$ and $N_{0oct}$, where $N_{0gr} = 0$.

$$E_{gr} = aL + bL^2 + \beta_3 L(L+1), \quad L = N \tag{3}$$

$$E_\beta = a\left(L + N_{0\beta}\right) + b\left(L + N_{0\beta}\right)^2 + \alpha_3 \frac{1}{2}\left(\frac{1}{2}\left(L + N_{0\beta}\right) + 1\right)\left(L + N_{0\beta}\right) + \beta_3 L(L+1) + \alpha_1 \tag{4}$$

$$E_\gamma = a\left(N_{0\gamma} + 2L - 4\right) + b\left(N_{0\gamma} + 2L - 4\right)^2 + \frac{\alpha_3}{2}\left(L - 1\right)\left(2L - 1\right) + \beta_3 L(L+1) + a_1 \tag{5}$$

$$\begin{aligned} E_{oct} = a\left(L + N_{0oct} - 1\right) + b\left(L + N_{0oct} - 1\right)^2 + \tfrac{\alpha_3}{2}\left(\tfrac{1}{2}\left(L + N_{0oct} - 1\right) + 1\right)\left(L + N_{0oct} - 1\right) \\ + \beta_3 L(L+1) + a_1 \end{aligned} \tag{6}$$

Then introducing the obtained parameters in the expressions (3),(4),(5) and (6) we evaluate the dependence of the states with fixed angular momentum $L$ on the numbers $N_{0\beta}, N_{0\gamma}$ and $N_{0oct}$, which define the numbers of phonon excitations that build the band head configurations. See the right panel of fig.3. This dependence defines the energy distribution of the band

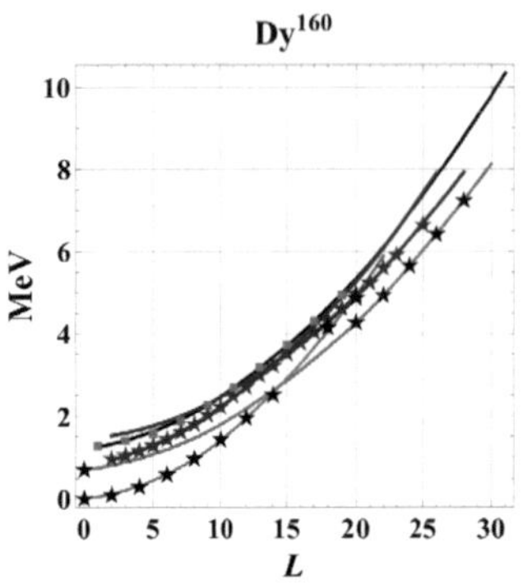


**Figure 2.** Fit of the theoretical (curves) to the experimental (symbols) energies of the ground and $\beta-$ (red) $\gamma-$ (blue) and octupole (black) bands of the ${}^{160}Dy$.

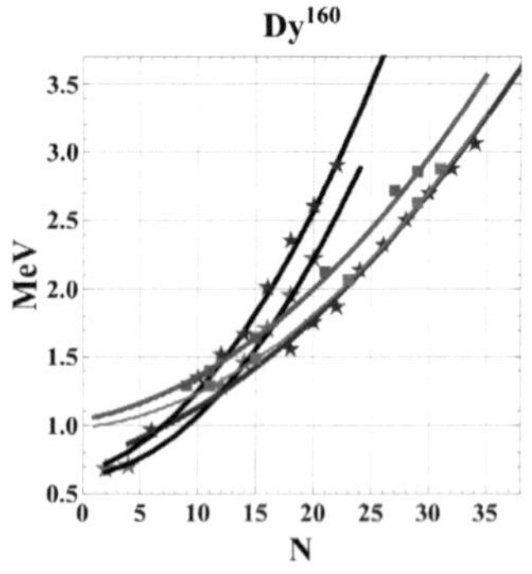


**Figure 3.** Energy distribution of the $0^+$ and $2^+$ states of the $\beta-$bands (black curves), $2^+$ states of the $\gamma-$bands (blue curve), $1^-$ states (red curve) $3^-$ states (brown curve) of the octupole bands in the ${}^{160}Dy$.

head configurations (fig.3) in conjunction with the respective bands' evolution (fig.2) and respectively determines the positions of the bands with respect to each other in the nuclear spectra. The experimental states with the considered $L$ that don't lie on the obtained curves should be considered as having some other type of collective nature that is not included in the IVBM assumptions. The energies corresponding to values of $N_{0\beta}$, $N_{0\gamma}$ and $N_{0oct}$ for which an experimental state is not observed could be considered as predictions for new collective states.

In conclusion we could summarize that in the framework of the IVBM with the present physical assignment of the basis states to the experimentally observed low laying collective bands, their properties for the heavy even-even nuclei from the regions of the rare-earth and the actinides are very well reproduced and interpreted in a rather simple, but physically meaningful way. It is important to proceed systematically in this way, in order to relate the observed phenomena to the microscopic and geometrical structure of the nuclei. The employed group-theoretical approach is a rather convenient tool to achieve these goals.

**Acknowledgments** This work was supported by the Bulgarian National Foundation (DID-02/16/17.12.2009) and the project for collaborative research with JINR, Dubna on the subject "*Nucleus-Nucleus Collisions and Nuclear Properties at Low Energies*".

## References

[1] Level Retrieval Parameters http://iaeand.iaea.or.at/nudat/levform.html;
[2] Rowe D J, 1985 Rep. Prog. Phys. **48** 1419.
[3] Georgieva A I, Raychev P and Roussev R, 1982 J. Phys. G: Nucl. Phys. **8** 1377.
[4] Garistov V P, Georgieva A I and Shneidman T M, 2013 Bulg. J. Phys. **40** 116
[5] Moshinsky M and Quesne C, 1971 J. Math. Phys. **12** 1772.
[6] Vanagas V V, 1988 *Algebraic foundations of microscopic nuclear theory,* Nauka, Moscow, (in russian).
[7] Ganev H G, Garistov V P, Georgieva A I and Draayer J P, 2004 Phys. Rev. **C70** 054317.
[8] Garistov V, 2003 *Proceedings of the XXII International Workshop on Nuclear Theory,* ed. V. Nikolaev, Heron Press Science Series, Sofia, 305.
[9] Adam J et.al. 2014 Bulg. J. Phys. **41**, 10 - 23.
[10] Solnyshkin A A, Garistov V P, Georgieva A, Ganev H and Burov V V 2005 Phys. Rev. **C72**, 064321.

Journal of Physics: Conference Series **590** (2015) 012034 doi:10.1088/1742-6596/590/1/012034

# Fusion hindrance and extra push in fusion reactions with heavy nuclei

**Kouhei Washiyama**

RIKEN Nishina Center, Wako 351-0198, Japan

E-mail: kouhei.washiyama@riken.jp

**Abstract.** We analyze fusion hindrance phenomena involving heavy nuclei using the microscopic time-dependent Hartree-Fock theory combining with a macroscopic equation of motion with a friction term. We find that a barrier structure disappears in the obtained potentials in heavy systems. We show that main contribution to extra-push energy comes from the increase in dynamical potential.

## 1. Introduction

Fusion hindrance has been known by observations that fusion reaction at energies around the Coulomb barrier in heavy systems is strongly hindered compared with that in light- and medium-mass systems [1, 2, 3]. This fusion hindrance partially results in extremely low cross sections of the synthesis of superheavy elements. Before these observations, it was pointed out in Ref. [4, 5] that an extra energy (extra-push energy) is needed to fuse in heavy systems whose so-called effective fissility parameters are greater than a certain value, which can approximately correspond to the charge product of projectile and target nuclei being greater than $1600 \sim 1800$. In such heavy systems, quasi-fission (reseparation without forming a compound nucleus after two nuclei touch) is a dominant process and this can be understood by a geometrical consideration of potential energy landscape. For heavy systems, a conditional saddle point on the potential energy landscape appears inside the touching configuration of two nuclei. Therefore, the system needs to overcome the saddle for fusion. Otherwise, the system will reseparate, leading to quasi-fission. Friction leading to energy dissipation from relative motion to internal excitations becomes strong inside the touching configuration because of large overlap of two nuclei.

Previously, the quasi-fission process has been analyzed by a macroscopic dynamical model based on a Langevin equation [6, 7]. Analysis by microscopic reaction theory such as time-dependent Hartree-Fock (TDHF) theory [8, 9, 10, 11] is only a few [11, 12]. The aim of this work is to understand the origin of the fusion hindrance by using a microscopic reaction model, especially focusing on the properties of nucleus–nucleus potential and energy dissipation. In this work, we employ a method of combining the TDHF theory with a macroscopic equation of motion for directly extracting nucleus–nucleus potential and energy dissipation [13, 14].

## 2. Method

The TDHF theory gives a microscopic self-consistent description through the energy density functional for both static and dynamical properties, which can be applied to nuclei over the

Journal of Physics: Conference Series **590** (2015) 012034 doi:10.1088/1742-6596/590/1/012034

whole nuclear chart. The basic equation for describing time evolution of the dynamics is the so-called TDHF equation for single-particle wave functions $\phi_i(\mathbf{r}, t)$,

$$i\hbar \frac{\partial}{\partial t}\phi_i(\mathbf{r}, t) = \hat{h}[\hat{\rho}]\phi_i(\mathbf{r}, t), \tag{1}$$

where $\hat{h}[\hat{\rho}]$ denotes single-particle Hamiltonian as a functional of one-body density $\hat{\rho}$.

To extract nucleus–nucleus potential and energy dissipation from TDHF evolutions, we use a method of combining microscopic TDHF and macroscopic evolution through a Newton equation, which we proposed in Ref. [13, 14]. This method relies on the assumption that a complicated microscopic mean-field dynamics can be reduced to a one-dimensional macroscopic equation of motion for relative distance $R$ and conjugate momentum $P$ including a friction term,

$$\frac{dR}{dt} = \frac{P}{\mu}, \qquad \frac{dP}{dt} = -\frac{dV}{dR} - \frac{d}{dR}\left(\frac{P^2}{2\mu}\right) - \gamma\frac{dR}{dt}, \tag{2}$$

where $\mu$, $V$ and $\gamma$ denote the reduced mass, nucleus-nucleus potential and friction coefficient, respectively, as a function of relative distance. The friction term describes dissipation from the relative motion to intrinsic degrees of freedom. We would like to note here that our extracted potential is based on neither sudden nor adiabatic approximation. Our extracted potential and friction automatically contain dynamical effects such as dynamical density evolution. We stop the extraction when the overlap of the projectile and target densities becomes significantly large since our extraction method based on the two-body analysis does not properly work with such a large overlap. To compute TDHF evolutions, we use the three-dimensional TDHF code developed by P. Bonche and coworkers [15]. Details of the computations are in Refs. [13, 14].

## 3. Result

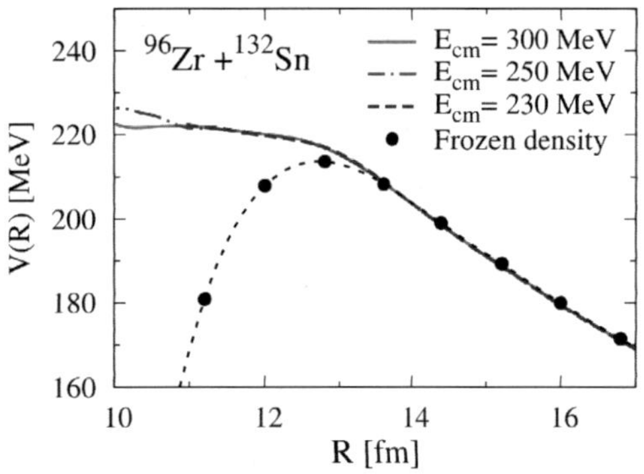


**Figure 1.** Nucleus–nucleus potential as a function of relative distance for the $^{96}\mathrm{Zr} + {}^{132}\mathrm{Sn}$ system. The lines are the results of our method at different $E_{\mathrm{cm}}$ indicated in the figure. The dots are obtained from the frozen density approximation.

First, we show extracted nucleus–nucleus potentials. Figure 1 shows potentials extracted at different center-of-mass energies $E_{\mathrm{cm}}$ for the $^{96}\mathrm{Zr} + {}^{132}\mathrm{Sn}$ system. As a reference, we plot by the filled circles the frozen density potential that is calculated from the same energy density functionals as in TDHF with the projectile and target densities frozen to their ground-state one. Comparing with the frozen density potential, we find an increase in extracted potential at all the energies used here at $R < 13$ fm. In Fig. 2, we investigate in detail the property of extracted potentials for the $^{96}\mathrm{Zr} + {}^{124}\mathrm{Sn}$ system by comparing with those in the $^{40}\mathrm{Ca} + {}^{40}\mathrm{Ca}$ system. We find two significant differences: (1) For the $^{96}\mathrm{Zr} + {}^{124,132}\mathrm{Sn}$ systems, no potential

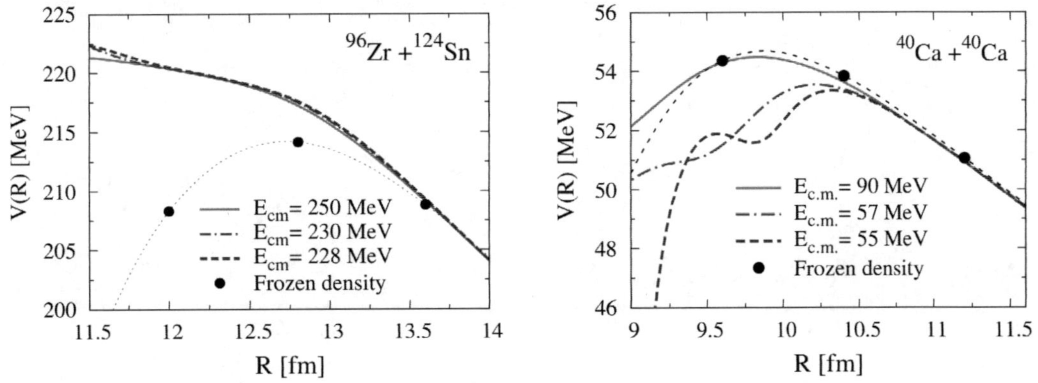


**Figure 2.** Left panel: Zoom around the frozen density barrier for nucleus–nucleus potentials extracted for the $^{96}Zr + ^{124}Sn$ system. Right panel: Same as in the left panel but for the $^{40}Ca + ^{40}Ca$ system.

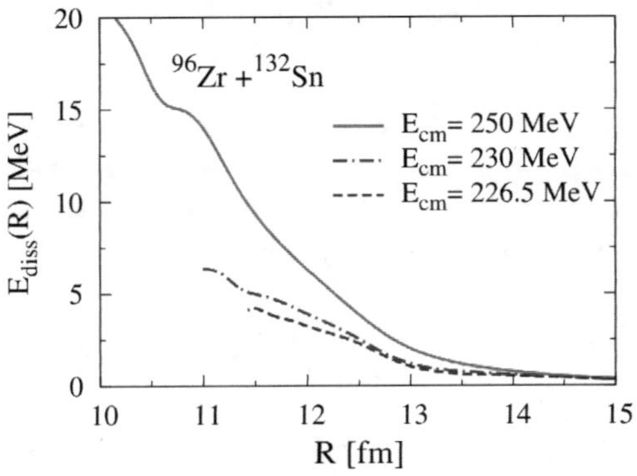


**Figure 3.** Dissipated energy calculated in Eq. (3) for the $^{96}Zr + ^{132}Sn$ system. The lines are the results of our method at different $E_{\rm cm}$ indicated in the panel

barrier is observed in the potentials extracted at any $E_{\rm cm}$ and the potentials monotonically increase as $R$ decreases. For the $^{40}Ca + ^{40}Ca$ system, a barrier is observed as usual. (2) Energy dependence of potential appears in the $^{40}Ca + ^{40}Ca$ system, while it is not significant in the $^{96}Zr + ^{124,132}Sn$ systems around the frozen density barrier at $R \sim 12.8$ fm.

In Fig. 3, we show dissipated energy for the $^{96}Zr + ^{132}Sn$ system. We calculate the dissipated energy through the friction coefficient by

$$E_{\rm diss}[R(t)] = \int_0^t dt' \gamma[R(t')] \left( \frac{dR}{dt} \right)^2 , \tag{3}$$

at time $t$ when we stop the extraction of the friction coefficient. In all the cases, the dissipated energy monotonically increases as nuclei approach to each other.

Finally, we analyze the origin of the fusion hindrance. We first define the extra-push energy $E_{\rm extra}$ of TDHF as the difference between the fusion threshold energy $E_{\rm thres}$ and the frozen density potential barrier $V_{\rm FD}$, $E_{\rm extra} = E_{\rm thres} - V_{\rm FD}$. The fusion threshold energy is determined as the lowest energy for a reaction with the system remaining a compact shape for a sufficiently long time ($\sim 1200\,{\rm fm}/c$) after touching. For the analysis of the extra-push energy, we extract potential and energy dissipation from TDHF trajectories at $E_{\rm thres}$, in which remaining kinetic

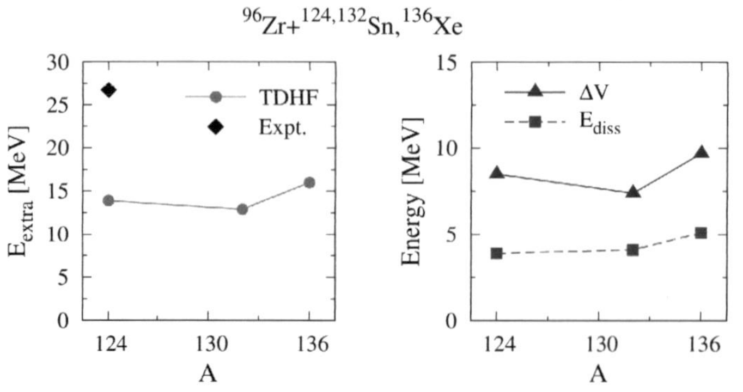


**Figure 4.** Left panel: Extra-push energies extracted from our TDHF method (solid circles) and from experimental data (triangle) [2] for the $^{96}\mathrm{Zr}+{}^{124,132}\mathrm{Sn}$, $^{136}\mathrm{Xe}$ systems. Right panel: increase in potential and dissipated energy extracted from our TDHF method.

energy is smallest at the relative distance where we stop extracting potential. At this relative distance, denoted by $R_{\mathrm{stop}}$, we can identify the extra-push energy as the sum of the dissipated energy $E_{\mathrm{diss}}(R_{\mathrm{stop}})$ in Eq. (3) and the increase in potential $\Delta V$ at $R_{\mathrm{stop}}$ from the frozen density potential barrier, $\Delta V = V(R_{\mathrm{stop}}) - V_{\mathrm{FD}}$. The results are summarized in Fig. 4. It is clear that the contribution from the increase in potential $\Delta V$ is larger than the dissipated energy $E_{\mathrm{diss}}$ for the $^{96}\mathrm{Zr}+{}^{124,132}\mathrm{Sn}$, $^{136}\mathrm{Xe}$ systems. From this finding, we conclude by our microscopic analysis that main contribution to the extra-push energy is from the dynamical increase in potential.

## 4. Summary

We have discussed the origin of the fusion hindrance by using microscopic TDHF combining with a macroscopic equation of motion for extracting dynamical nucleus–nucleus potential and energy dissipation. Dynamical increase in potentials extracted in heavy systems appears, which is different from that in light- and medium-mass systems. We analyze extra-push energies and find that main contribution to the extra-push energy is from the dynamical increase in potential.

**Acknowledgments**

The author is supported by the Special Postdoctoral Researcher Program of RIKEN.

## References

[1] Gaggeler H *et al.* 1984 *Z. Phys. A* **316** 291
[2] Sahm C C *et al.* 1985 *Nucl. Phys. A* **441** 316
[3] Schmidt K H and Morawek W 1991 *Rep. Prog. Phys.* **54** 949
[4] Swiatecki W J 1981 *Phys. Scripta* **24** 113; 1982 *Nucl. Phys. A* **376** 275
[5] Bjornholm S and Swiatecki W J 1982 *Nucl. Phys. A* **391** 471
[6] Zagrebaev V and Greiner W 2005 *J. Phys. G* **31** 825
[7] Aritomo Y, Hagino K, Nishio K and Chiba S 2012 *Phys. Rev. C* **85** 044614
[8] Bonche P, Koonin S E and Negele J W 1976 *Phys. Rev. C* **13** 1226
[9] Flocard H, Koonin S E and Weiss M S 1978 *Phys. Rev. C* **17** 1682
[10] Negele J W 1982 *Rev. Mod. Phys.* **54** 913
[11] Simenel C 2012 *Eur. Phys. J. A* **48** 152
[12] Guo L and Nakatsukasa T 2012 *EPJ Web Conf.* **38** 09003
[13] Washiyama K and Lacroix D 2008 *Phys. Rev. C* **78** 024610
[14] Washiyama K, Lacroix D and Ayik S 2009 *Phys. Rev. C* **79** 024609
[15] Kim K H, Otsuka T and Bonche P 1997 *J. Phys. G* **23** 1267

Journal of Physics: Conference Series **590** (2015) 012035 doi:10.1088/1742-6596/590/1/012035

# Measurements of $^{12}C$ ion fragmentation on thin carbon target from the FIRST collaboration at GSI

**M.Toppi, on behalf of the FIRST collaboration**

Istituto Nazionale di Fisica Nucleare - Laboratori Nazionali di Frascati, Italy; Dipartimento di Fisica, Universitá di Roma Tor Vergata, Italy

E-mail: `marco.toppi@lnf.infn.it`

**Abstract.** The FIRST (Fragmentation of Ions Relevant for Space and Therapy) experiment at GSI laboratory took data in summer 2011, studying the collisions of a $^{12}C$ ion beam with Carbon and Au thin targets. The experiment main purpose is the double differential cross section measurement of the carbon ion fragmentation at energies that are relevant both for tumor therapy and space radiation protection applications (100-1000 MeV/u). The FIRST dataset contains carbon ions collisions on a 3.43 g$\cdot$cm$^{-2}$ carbon target (about 24 M events) and on a 0.96 g$\cdot$cm$^{-2}$ Au target (about 4.5 M events). The SIS (heavy ion synchrotron) was used to accelerate the $^{12}C$ ions at the energy of 400 MeV/u. The preliminary results of differential cross sections measurements as a function of angle and energy for carbon target, in the small angle region ($\theta < 5^{\circ}$), are presented.

## 1. Introduction

The study of the nuclear fragmentation processes occuring in the interaction of highly energetic ions in matter is of great interest in basic research (e.g. to improve the understanding of hadronic showers development in the atmosphere) and in applied physics, in particular in particle therapy and space radiation protection fields. In particle therapy, ion beams slowing down in patient tissues undergo inelastic nuclear reactions and give rise to signficant yields of secondary fragments [1]. A more precise knowledge of the dose map due to the fragments mixed radiation fields with the required accuracy of $\sim 3\%$ is mandatory to achieve a better dose control, expecially when treating tumors nearby organs at risk. Nuclear fragmentation studies are of great interest also in the understanding of the ions component of the galactic cosmic radiation (GCR) effects on human space flight in the energy range of 100-1000 MeV/u (where GCR spectrum has a maximum) [2]. The current particle transport codes, used in particle therapy and space radiation protection applications, are not able to reproduce the measured fragments angular and energy distributions with the accuracy required for space and therapy applications [3]. The main limitation is the lack, or the large related uncertainty, of experimantal data [4]. For these reasons, various recent experiments aimed at double differential fragmentation cross sections measurements as a function of the angle and energy have been performed [5, 6], with a particular focus on the exploration of the energy region of interest for therapeutic application (100-400 MeV/u). The FIRST (Fragmentation of Ions Relevant for Space and Therapy) experiment has been performed at the SIS accelerator of GSI laboratory in Darmstadt with a $^{12}C$ ion beam at 400 MeV/u on carbon and gold targets [7]. Only the results with the carbon target, relative to a data sample of $\sim$ 25 M of detected incoming $^{12}$C ions, are reported here.

## 2. The FIRST detector layout and performances

The FIRST apparatus (figure 1) is composed of several subdetectors optimized for the detection of all the charged fragments produced by the interaction of a $^{12}$C ion with the carbon target up to a maximum aperture angle of 40° with respect to the beam axis. For each subdetector a wide dynamic range is required because the energy released by the fragments ranges from 2 mip (minimum ionizing particle) for protons to 100 mip for carbons. Moreover all the detectors along the carbon beam and the fragments trajectories were designed to minimize the secondary fragmentation inside them. A MC simulation of a $^{12}$C beam at 400 MeV/u on a 3.43 g·cm$^{-2}$ carbon target has been developed with the FLUKA code to optimize the FIRST detector: while most of the fragments go forward with small angles and, with about the same kinetic energy of the carbon beam, a not negligible fraction of the light Z=1,2 fragments are produced with large angles and a wider kinetic energy distribution. In FIRST a $^{12}$C ion, before interacting with the target, impinges on the start counter (SC), a thin scintillator that provides the number of carbon ions, the trigger of the experiment and start for the Time of Flight (ToF) measurement, and goes throught a drift chamber, that measures the direction and the impact point on the target [8]. The fragmentation vertex produced in the interaction with the target is reconstructed with a silicon pixel detector (VD) that has a track angular acceptance of $\theta = \pm 40°$ [9, 10]. The fragments produced with polar angle $\theta < 5°$ enter in the angular acceptance of the ALADIN dipole magnet. The trajectory and the momentum of the fragments are reconstructed using the magnetic bending provided by the ALADIN magnet, by measuring the ToF and matching the fragment position on the ToF-Wall (TW), a large area hodoscope formed of two scintillator layers [11]. The tracking algorithm uses an iterative procedure that matches the VD tracks and the TW hits detected in each event. Special care is needed while performing the extrapolation from the VD to the far away ($\sim$ 6 m) TW. The VD excellent tracking resolution, $<$ 10 µm and $<$ 50 µm in the $(x, y)$ plane and along the beam axis ($z$) respectively, translates in an angular resolution of $\sim$ 1 mrad for each fragment and plays a crucial role in reducing the amount of random combinations made up with wrongly matched tracks and hits (combinatorial

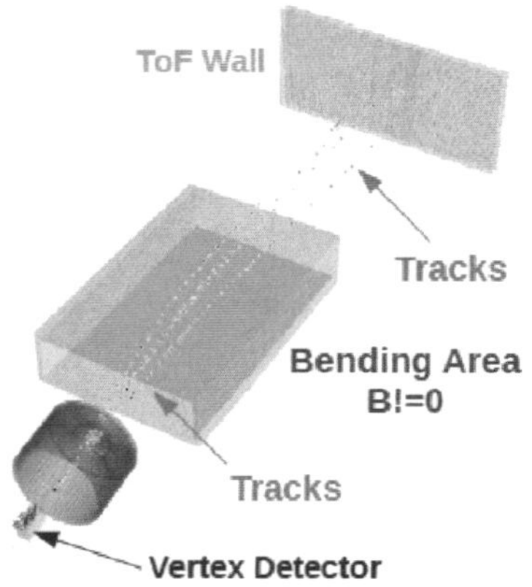


**Figure 1.** Fully reconstructed fragmentation event in the FIRST apparatus.

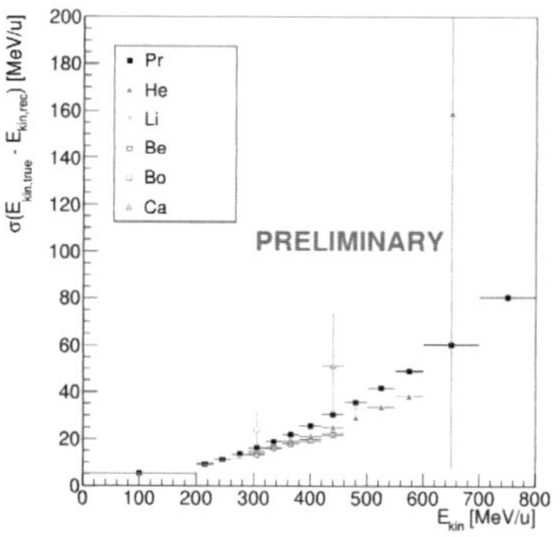


**Figure 2.** Kinetic energy resolution vs kinetic energy for each fragment charge.

background). For each reconstructed global track the charge is measured by the TW and the VD, the ToF is measured by the TW and the SC, the particle path (L) and the rigidity (pc/Z) are determinated by the tracking algorithm, allowing a measurement of the fragment speed $\beta = L/(c \cdot ToF)$. The fragment mass is computed as $Mc^2 = pc/\beta\gamma$ where $\gamma$ is the Lorentz factor and the mass resolution is:

$$\frac{\Delta M}{M} = \sqrt{\left(\frac{\Delta p}{p}\right)^2 + \left(\gamma^2 \frac{\Delta ToF}{ToF}\right)^2} \tag{1}$$

The ToF resolution was measured to be about 800 ps for each fragment charge. The momentum resolution depends essentially on the spatial resolution along $x$ direction of VD (< 10 µm) and TW ($\sim$ 7 mm) and it worsens with higher fragment momentum as shown in figure 2 where the kinetic energy resolution is displayed for the different fragments charges.

## 3. Cross section measurements

The double differential cross section in angle and kinetic energy for the i-th isotope ${}^{Z}_{A}X_i$ with charge Z and mass number A is defined as:

$$\frac{d^2\sigma_i}{d\Omega dE}(\theta, E) = \frac{Y_i(\theta, E)}{N_C \times N_{TG} \times \Delta\Omega \times \Delta E \times \epsilon_{trk}(\theta, E)} \tag{2}$$

where the number of carbon impinging on the target $N_C$ is measured by the SC, the number of

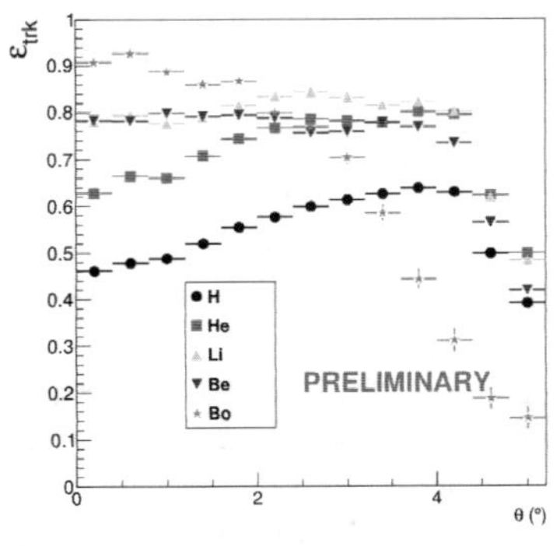


**Figure 3.** Tracking efficiency vs fragment angle $\theta$.

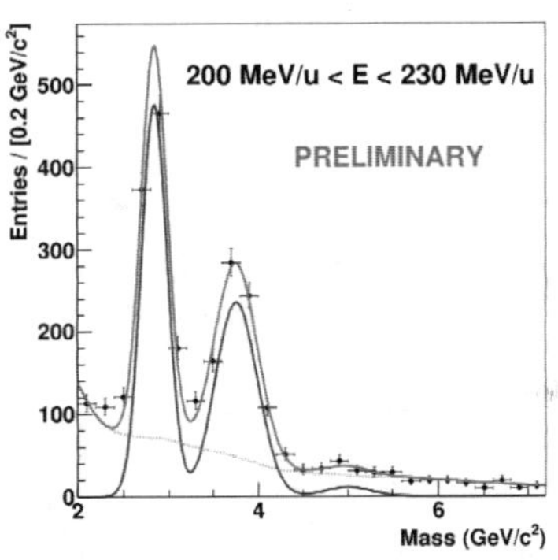


**Figure 4.** Fit result for charge Z=2 mass spectra in an energy bin.

atoms in the target per unit surface $N_{TG}$ is known and $\Delta\Omega$ and $\Delta E$ are the angular and energy phase space factors, accounting for the cuts applied in each $\theta, E$ bin. The study of the tracking efficiencies $\epsilon(\theta, E)$ has been performed on the full MC simulation: for each fragment charge, the efficiency is defined as the ratio between the number of reconstructed tracks and the number of MC generated tracks emerging from the target in the magnet angular acceptance (see figure 3). $Y_i(\theta, E)$ are the fragments yields for a given isotope ${}^{Z}_{A}X_i$ in an angular (energy) bin. The fragment yields $Y_i(\theta, E)$ are measured fitting the reconstructed mass spectra for each charge and angular (energy) bin using an unbinned extended maximum likelihood approach in which the signal and background yields are fitted as well [12]: the signal is modeled with a gaussian PDF and the backgroud PDFs are extracted from the MC simulation. Figure 4 shows the fit result for Z=2 and 200 MeV/u $< E <$ 230 MeV/u where the blue curve is the sum of all the signal gaussian PDFs and the magenta line represents the background PDF. The preliminary single differential cross sections in angle and energy have been obtained for each fragment charge and for each isotope. As an example in figures 5, 6 the angular and energy cross sections for the helium isotopes are shown: the error bars are only accounting for the statistical uncertainty, since the calculation of the systematic uncertainties is still ongoing. A paper containing all the distributions and the tabulated results is in preparation.

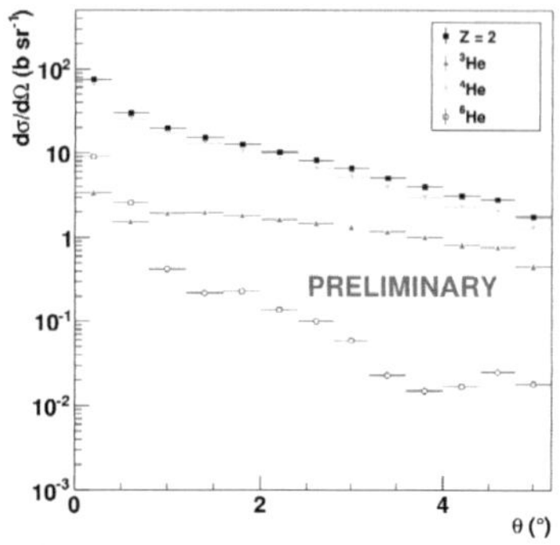


**Figure 5.** Angular fragmentation cross sections for helium isotopes.

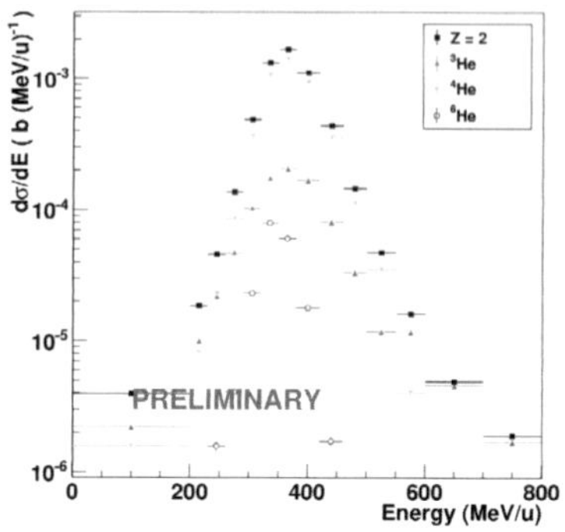


**Figure 6.** Energy fragmentation cross sections for helium isotopes.

## 4. Conclusions

The preliminary differential fragmentation cross sections results in angle and energy for a $^{12}C$ beam at 400 MeV/u on a carbon target in the low angle ($\theta < 5°$) range have been obtained with the FIRST apparatus. Given the very large fragmentation statistics available, the analysis effort is currently focused on the final assessment of the systematic uncertainties related to the PDFs modeling inside the fit, the combinatorial background subtraction and the reconstruction algorithm tuning and efficiency. Future analysis plans foresee the measurements of the full double differential cross sections, together with the benchmarking of MC softwares and analytical algorithms currently used for treatment planning applications.

## References

[1] Schardt D *et al* 1996 *Adv. in Space. Res.* **17** 87-94.
[2] Durante M and Cucinotta F 2011 *Rev. Mod. Phys.* **83** 1245.
[3] Bohlen T *et al* 2010 *Physics in Medicine and Biology* **55** 5833–47.
[4] Patera V 2011 *Nuovo Cimento* C **34** 6.
[5] De Napoli M *et al* 2012 *Physics in Medicine and Biology* **57** 7651-7671.
[6] Dudouet J *et al* 2013 *Physical Review* C **88** 024606.
[7] Pleskac R *et al* 2012 *Nucl. Instrum. Methods Phys. Res.* A **678** 130-138.
[8] Abou-Haidar Z *et al* 2012 *JINST* **7** P02006.
[9] Rescigno R *et al* 2014 *Nucl. Instrum. Methods Phys. Res.* A **767** 34-40.
[10] Spiriti E *et al* 2011 *Nucl. Phys. (Proc. Suppl.)* B **215** 157161.
[11] Introzzi R *et al* 2013 *DOI:10.1109/NSSMIC.2013.682944, Conference: C13-10-26.*
[12] Verkerke W and Kirkby D P 2003 *arXiv:physics* 0306116.

Journal of Physics: Conference Series **590** (2015) 012036 doi:10.1088/1742-6596/590/1/012036

# Lifetime measurements and the high-spin structure of $^{36}$Cl

**S.Aydin[1,2], M. Ionescu-Bujor[3], F. Recchia[2], A. Gadea[4,5], S.M. Lenzi[2], S. Lunardi[2], C.A. Ur[2], D. Bazzacco[2], P.G. Bizzeti[6], A.M. Bizzeti-Sona[6], M. Bouhelal[7], G. de Angelis[4], I. Deloncle[8], E. Farnea[2], A. Gottardo[2,4], F. Haas[9], T. Huyuk[5], H. Laftchiev[10], D. Mengoni[2,11], R. Menegazzo[2], C. Michelagnoli[2], D.R. Napoli[4], E. Sahin[4], P.P. Singh[4], D. Tonev[10] and J.J. Valiente-Dobón[4]**

[1]Aksaray University, Department of Physics, 68100 Aksaray, Turkey
[2]Dipartimento di Fisica e Astronomia dell'Universita' and INFN, Padova, Italy
[3]National Institute for Physics and Nuclear Engineering, Bucharest, Romania
[4]INFN-Laboratori Nazionali di Legnaro, I-46020 Legnaro, Italy
[5]Instituto de Física Corpuscular, CSIC-Universidad de Valencia, Valencia, Spain
[6]Dipartimento di Fisica dell' Universitá and INFN Sezione di Firenze, Firenze, Italy
[7]Laboratoire de Physique Appliquée et Théorique, Université de Tebessa, Algeria
[8]IPNO, IN2P3/CNRS et Université Paris-Sud, Orsay, France
[9]IPHC, IN2P3/CNRS, Université de Strasbourg, Strasbourg, France
[10]Institute for Nuclear Research and Nuclear Energy, BAS, Sofia, Bulgaria
[11]University of the West of Scotland, Paisley, United Kingdom

E-mail: saydin@aksaray.edu.tr

**Abstract.** High-spin states in $^{36}$Cl were populated through the $^{24}$Mg($^{14}$N,2p)$^{36}$Cl reaction at $E(^{14}\mathrm{N})$=31 MeV. Lifetimes have been determined for fifteen states by applying the Doppler shift attenuation method. The results indicated the onset of collectivity in the high spin negative parity states. Large basis shell model calculations have been performed to understand the underlying structure of these states.

## 1. Introduction

Considerable experimental and theoretical efforts have been made in recent years to elucidate the role of particle-hole *sd*-*fp* cross-shell intruder configurations in the structure of the *sd* shell nuclei. At $\beta$ stability, the experimental information, limited previously to low and medium spins [1], was extended at high spins and excitation energies, using heavy ion fusion-evaporation reactions and complex detection arrays [2, 3, 4, 5, 6, 7]. In this region the low-spin structure can be well reproduced by shell model calculations limited to the *sd* main shell with the USD residual interaction introduced by Wildenthal and Brown [8]. At high spins the excitations of particles from the *sd* shell to the *fp* shell have to be taken into account, and the experimental data serve as testing the proposed effective interactions [9, 10, 11]. The coexistence of spherical and deformed states is an interesting feature reported recently in several nuclei of the region [2, 3, 6] and described by large scale shell model calculations involving multi-particle – multi-hole intruder excitations from the *sd* to the *fp* shell. In order to investigate the evolution

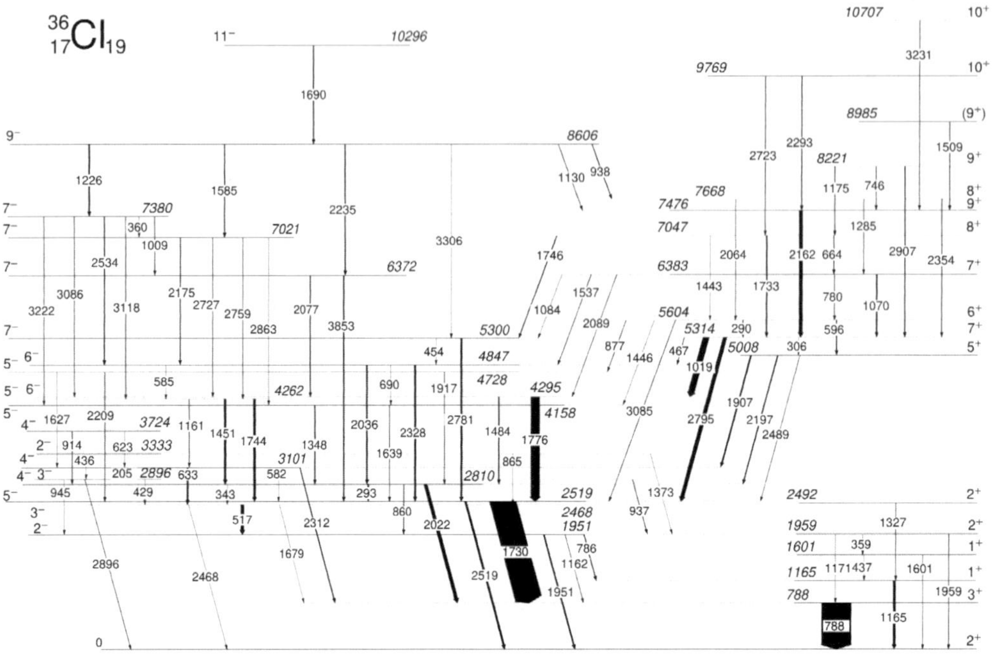


**Figure 1.** High-spin level scheme of $^{36}$Cl.

of collectivity with increasing spins the determination of transition probabilities is of crucial importance. We report here results of lifetime and transition probability investigations for the high spin states in the odd-odd $^{36}$Cl nucleus.

## 2. Experimental results

The high-spin states in $^{36}$Cl were populated via the fusion-evaporation reaction $^{24}$Mg($^{14}$N,2p)$^{36}$Cl with a 31 MeV $^{14}$N beam delivered by the XTU-Tandem accelerator at the INFN-Legnaro National Laboratory, Italy. The 99.7% isotopically enriched $^{24}$Mg target, 1 mg/cm$^2$ thick, was evaporated on a 8 mg/cm$^2$ gold layer. The $\gamma$-rays emitted in the reaction were detected using the 4$\pi$-GASP array [12] composed of 40 Compton-suppressed large volume high-purity Ge detectors arranged in seven rings at different angles with respect to the beam axis. A complex decay scheme, considerably extended at larger spins and excitation energies with respect to previous studies [1], has been established for $^{36}$Cl [5] (see Fig. 1).

The lifetimes of the new high-spin states have been investigated by the Doppler shift attenuation method. For this purpose spectra of interest were created from the asymmetric $\gamma$–$\gamma$ matrices by gating on appropriate lower-lying transitions emitted from stopped nuclei. The analysis has been performed using the LINESHAPE computer code [13]. From lineshape analysis, lifetimes have been derived for 14 new levels in $^{36}$Cl, as well as for the 4295 keV state previously known without measured lifetime. Examples of experimental lineshapes at different angles and the corresponding fits are illustrated in Fig. 2. The investigated states and the derived half-lives are shown in Table 1.

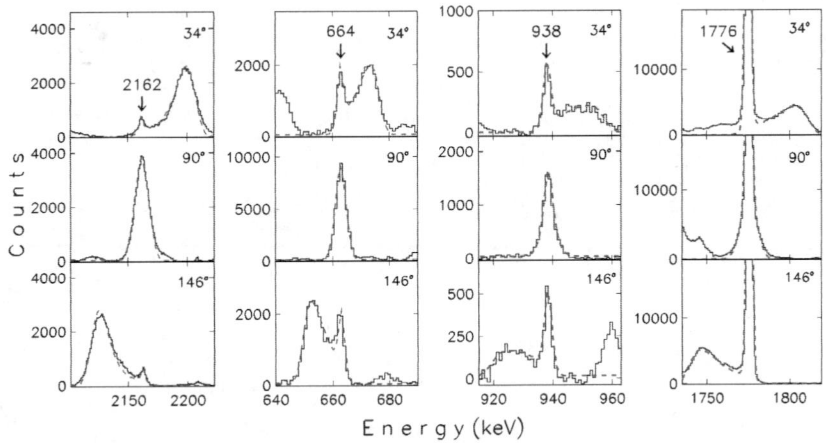


**Figure 2.** Illustrative experimental and fitted (red dashed line) lineshapes in $^{36}$Cl.

## 3. Discussion

Using the measured half-lives and branching ratios (BR), the experimental $B(M1)$ and $B(E2)$ reduced transition probabilities have been derived (see Table 1). Small transition strengths, $B(E2) \approx$ 2-9 W.u., were obtained for all $E2$ transitions, excepting the 1226 and 1690 keV transitions de-exciting the $9^-_1$ and $11^-_1$ states, with values $B(E2)$(1227 keV)=16.9 W.u. and $B(E2)$(1690 keV)=24.2 W.u., respectively. To get insight into the underlying structures, we have performed shell model calculations using two effective interactions developed for this mass region, namely the $sdfp$ interaction [10] and the PSDPF interaction [11]. The PSDPF interaction considers the full $psdpf$ model space with a $^4$He core, and only one nucleon is allowed to jump between major shells. More than one particle-hole excitation to the $fp$ shell is allowed using the $sdpf$ interaction and the model space spanned by the $s_{1/2}, d_{3/2}, f_{7/2}$ and $p_{3/2}$ orbits. The experimental reduced transition probabilities are compared in Table 1 with shell-model calculations. The agreement between experimental data and shell-model predictions is quite satisfactory. The $sdfp$ calculations indicate that the positive-parity states with $J^\pi \geq 5^+$ involve two particle-hole intruder excitations to the $fp$ shell. Negative-parity states are predicted to have one neutron in the $f_{7/2}$ orbit at low spin and the proton in the $sd$ shell, but the excitation of three particles (protons and neutrons) to the upper $fp$ shell becomes relevant already at spin $J^\pi = 7^-$. The large $B(E2)$ value of the $11^-_1 \rightarrow 9^-_1$ is reproduced by the $sdfp$ calculations, in which the involved states are both described by the promotion of three nucleons into the $fp$ shell. On the other hand, the PSDPF calculations, in which only one nucleon is excited in the $fp$ shell, predict a much smaller $B(E2)$ value for this transition. In the $sdfp$ calculations the wave functions of the $7^-$ states involve one neutron excited in the $fp$ shell, with the exception of the $7^-_3$ state where three nucleons are excited. A large $B(E2)$ value for the $9^-_1 \rightarrow 7^-_3$ is correspondingly predicted, while experimentally the $B(E2)$ is largest for the $9^-_1 \rightarrow 7^-_4$ transition. To reproduce the experimental findings one can consider an inversion of states, with the $7^-_4$ observed state described by the calculated $7^-_3$ state, and the $7^-_2$ and $7^-_3$ experimental levels described by the $7^-_4$ and $7^-_2$ calculated states, respectively. The 1226- and 1690-keV $E2$ transitions having enhanced $B(E2)$ values could be the lowest-lying members of a deformed band, similar to that observed recently in the odd-mass $^{35}$Cl nucleus [6]. More experiments are needed to possibly evidence such a deformed structure in the odd-odd $^{36}$Cl nucleus.

**Table 1.** Experimental and calculated reduced transition probabilities $B(\mathrm{M1})$ and $B(\mathrm{E2})$.

| $E_{lev}^{exp}$ (keV) | $T_{1/2}^{exp}$ (ps) | $J_i^\pi$ | $J_f^\pi$ | $E_\gamma^{exp}$ (keV) | $BR$ % | $B(M1)(\mu_N^2)$ exp | PSDPF | *sdfp* | $B(E2)(e^2fm^4)$ exp | PSDPF | *sdfp* |
|---|---|---|---|---|---|---|---|---|---|---|---|
| 5604 | 0.62(14) | $6_1^+$ | $7_1^+$ | 290 | 33(6) | 0.86(25) | | 0.826 | | | |
| | | | $5_1^+$ | 596 | 42(5) | 0.126(32) | | 0.384 | | | |
| 6383 | 0.24(5) | $7_2^+$ | $6_1^+$ | 780 | 21(2) | 0.070(18) | | 0.048 | 60(40) | | 2 |
| | | | $7_1^+$ | 1070 | 51(4) | 0.068(17) | | 0.304 | | | |
| 7047 | 0.34(6) | $8_1^+$ | $7_2^+$ | 664 | 26(3) | 0.102(22) | | 0.098 | | | |
| | | | $6_1^+$ | 1443 | 5(1) | | | | 14(4) | | 12 |
| | | | $7_1^+$ | 1733 | 40(4) | 0.009(3) | | 0.051 | | | |
| 7476 | 0.18(3) | $9_1^+$ | $7_1^+$ | 2162 | 100 | | | | 67(11) | | 68 ($9_2^+$) |
| 7668 | 0.07(2) | $8_2^+$ | $7_2^+$ | 1285 | 26(2) | 0.069(20) | | 0.077 | | | |
| | | | $6_1^+$ | 2064 | 21(1) | | | | 45(14) | | 32 |
| | | | $7_1^+$ | 2354 | 53(3) | 0.023(9) | | 0.054 | 0.05(3) | | 0.51 |
| 8221 | 0.09(2) | $9_2^+$ | $9_1^+$ | 746 | 23(5) | 0.240(72) | | 0.002 | | | |
| | | | $8_1^+$ | 1175 | 23(4) | 0.057(16) | | 0.002 | 57(42) | | 4 |
| | | | $7_1^+$ | 2907 | 54(6) | | | | 16(5) | | 7 ($9_1^+$) |
| 4158 | 0.13(4) | $5_2^-$ | $4_1^-$ | 1348 | 74(6) | 0.091(25) | 0.050 | 0.0002 | 9(4) | 74 | 0.02 |
| | | | $5_1^-$ | 1639 | 26(4) | 0.018(6) | 0.052 | 0.006 | | | |
| 4262 | 0.22(8) | $5_3^-$ | $4_2^-$ | 1161 | 7(1) | 0.008(3) | 0.010 | 0.014 | | | |
| | | | $4_1^-$ | 1451 | 42(1) | 0.025(10) | 0.014 | 0.005 | | | |
| | | | $5_1^-$ | 1744 | 51(3) | 0.017(6) | 0.023 | 0.080 | | | |
| 4295 | 0.23(5) | $6_1^-$ | $4_1^-$ | 1484 | 10(1) | | | | 34(8) | 0.002 | 0.002 |
| | | | $5_1^-$ | 1776 | 90(2) | 0.016(5) | 0.011 | 0.066 | 53(20) | 82 | 17 |
| 4847 | 0.17(4) | $6_2^-$ | $5_3^-$ | 585 | 2.2(1) | 0.024(6) | 0.0001 | 0.103 | 45(25) | 0.02 | 3 |
| | | | $5_2^-$ | 690 | 1.4(1) | 0.010(3) | 0.008 | 0.103 | | | |
| | | | $4_1^-$ | 2036 | 38(1) | | | | 36(9) | 2 | 44 |
| | | | $5_1^-$ | 2328 | 58(2) | 0.009(3) | 0.013 | 0.058 | 5(3) | 2 | 9 |
| 5300 | 0.16(5) | $7_1^-$ | $6_2^-$ | 454 | 8(2) | 0.211(84) | 0.046 | 0.029 | | | |
| | | | $5_1^-$ | 2781 | 92(7) | | | | 20(6) | 15 | 7 |
| 6372 | 0.13(3) | $7_2^-$ | $6_1^-$ | 2077 | 42(2) | 0.014(3) | 0.003 | 0.034 | | | |
| | | | $5_1^-$ | 3853 | 58(4) | | | | 3(1) | 3 | 6 |
| 8606 | 0.55(9) | $9_1^-$ | $7_4^-$ | 1226 | 32(1) | | | | 119(22) | 13 | 95 ($7_3^-$) |
| | | | $7_3^-$ | 1585 | 21(1) | | | | 22(4) | 1 | 20 ($7_2^-$) |
| | | | $7_2^-$ | 2235 | 22(2) | | | | 4(1) | 12 | 9 ($7_4^-$) |
| | | | $7_1^-$ | 3306 | 3(1) | | | | 0.08(3) | 12 | 0.2 |
| 10296 | 0.24(4) | $11_1^-$ | $9_1^-$ | 1690 | 100 | | | | 171(29) | 8 | 116 |

## References

[1] Endt P M 1998 *Nucl. Phys.* A **633** 1, and references therein
[2] Svensson C E *et al.* 2000 *Phys. Rev. Lett.* **85** 2693
[3] Ideguchi E *et al.* 2001 *Phys. Rev. Lett.* **87** 222501
[4] Ionescu-Bujor M *et al.* 2006 *Phys. Rev.* C **73** 024310
[5] Aydin S *et al.* 2012 *Phys. Rev.* C **86** 024320
[6] Bisoi A *et al.* 2013 *Phys.Rev.* C **88** 034303
[7] Aydin S *et al.* 2014 *Phys. Rev.* C **89** 014310
[8] Brown B A and Wildenthal B H 1988 *Ann. Rev. Nucl. Part. Sci.* **38** 29
[9] Warburton E K, Becker J A and Brown B A 1990 *Phys. Rev.* C **41** 1147
[10] Caurier E, Langanke K, Martinez-Pinedo G, Nowacki F and Vogel P 2001 *Phys. Lett.* B **522** 240
[11] Bouhelal M, Haas F, Caurier E, Nowacki F and Bouldjedri A 2011 *Nucl. Phys.* A **864** 113
[12] Rossi Alvarez C 1993 *Nucl. Phys. News* **3** 10
[13] Wells J C and Johnson N R 1991, Report No. ORNL-6689, p. 44

Journal of Physics: Conference Series **590** (2015) 012037 doi:10.1088/1742-6596/590/1/012037

# Exploring the $^{10}$Li structure by the d($^{9}$Li,p)$^{10}$Li transfer reaction

**M. Cavallaro[1*], M. De Napoli[2], F. Cappuzzello[1,3], C. Agodi[1], M. Bondì[1,3], D. Carbone[1], A. Cunsolo[1], B. Davids[4], T. Davinson[5], A. Foti[2,3], N. Galinski[4], R. Kanungo[6], H. Lenske[7], S.E.A. Orrigo[8], C. Ruiz[4] and A. Sanetullaev[5]**
[1] INFN, Laboratori Nazionali del Sud, Italy
[2] INFN, Sezione di Catania, Italy
[3] Dipartimento di Fisica e Astronomia, Università di Catania, Italy
[4] TRIUMF, Vancouver, Canada
[5] Department of Physics and Astronomy, University of Edinburgh, UK
[6] Saint Mary's University, Canada
[7] Institut fur Theoretische Physik, Universitat Gießen,
[8] Instituto de Física Corpuscular, CSIC-Universidad de Valencia, Spain

E-mail: manuela.cavallaro@lns.infn.it

**Abstract.** The $^{9}$Li + $^{2}$H reaction has been investigated at 11 AMeV incident energy at the ISAC II facility (TRIUMF). In the present paper we focus on the one-neutron transfer channel, which potentially holds spectroscopic information on the unbound nucleus $^{10}$Li. The TUDA setup has been used in order to detect and identify the outgoing $^{9}$Li at forward angles and the recoil protons at backward angles. This setup allows to study the $^{10}$Li emitted in the crucial region at forward angles in the center of mass.

## 1. Introduction

Transfer reactions are essential tools to probe selected components of the nuclear wave functions. Thanks to that they have been used in the past and are still crucial for nuclear spectroscopy purposes. With the advent of radioactive beam facilities, the new opportunity to explore nuclear phenomena far from the stability valley has driven a renewed interest to transfer reactions, to be studied in inverse kinematics.

On one hand there is a specific attention to understand how the transfer mechanism is influenced by the reduction of the binding energy of the projectile [1]; on the other hand, the measured spectra could reveal unexpected features due to nucleon correlations which are beyond the mean field description of nuclear structure [2]. Examples of such phenomena have been recently found in light neutron rich nuclei such as the Li isotopes.

In particular, the study of the unbound system $^{10}$Li is of great interest since the knowledge of its structure is a crucial ingredient in the description of the two-neutron halo nucleus $^{11}$Li. Despite the significant amount of experimental information gathered during the last years, the properties of the $^{10}$Li continuum remains unclear, to the extent that even the energy and the spin-parity of the ground state are still controversial [1-4].

* To whom any correspondence should be addressed.

Journal of Physics: Conference Series **590** (2015) 012037 doi:10.1088/1742-6596/590/1/012037

## 2. Past attempts to study the d($^{9}$Li,p)$^{10}$Li reaction

Recently $^{10}$Li has been the subject of different theoretical and challenging experimental studies [3-5], including two attempts to explore the resonant energy spectrum by the d($^{9}$Li,p)$^{10}$Li transfer reaction at 2.35 AMeV incident energy at REX-ISOLDE and at 20 AMeV at NSCL-MSU, respectively.

In the experiment at 2.35 AMeV of Ref. [3], the energy spectrum of $^{10}$Li has been measured up to about 1 MeV excitation energy in an angular range between 100° and 140° in the centre of mass. Due to the low beam intensity of ~5 × 10$^{4}$ pps, a relatively thick target of deuterated polyethylene CD2 (660 μg/cm$^{2}$ thick) was used in order to maximize the yields, which slightly exceeded 100 events in total. The recoiling protons were detected by large area telescopes of position sensitive silicon detectors located between 18° and 80° in the laboratory. At backward laboratory angles, the energy of the protons was too low to be detected, thus excluding in the collected data the crucial region corresponding to forward angles in the centre of mass. Despite the low statistics and the not optimal angular range explored, an excitation energy resolution of about 300 keV (FWHM) was obtained and the authors could draw some conclusions about the role of the $s_{1/2}$ and $p_{1/2}$ neutron orbitals in $^{10}$Li around the neutron emission threshold.

In the NSCL-MSU experiment [4], a similar experimental technique was used but at higher incident energy (20 AMeV). The beam intensity was even lower than in the REX-ISOLDE experiment (~7 × 10$^{3}$ pps) and had considerably worse emittance. The CD2 target was about 2 mg/cm$^{2}$ thick. The recoiling protons were measured at backward angles by a series of large area silicon detectors in coincidence with the $^{9}$Li detected by the S800 spectrometer at forward angles. The obtained energy resolution was about 700 keV and the reconstructed $^{10}$Li spectra were measured up to about 4.5 MeV excitation energy. However, due to the poor statistics (less than 100 counts in total) and energy resolution, not much has been added to our understanding of $^{10}$Li states.

In a recent theoretical study of the $^{10}$Li low energy resonances [2], a state dependent treatment of the pairing interaction, beyond the usual BCS approximation, has proven to be necessary to reproduce the energy spectra measured in the REX-ISOLDE experiment. In addition it has been shown that, despite the complications due to the resonant structure of the final states, the cross sections can be accurately described within the DWBA.

Nevertheless, due to the limitations of the existing data, many of the details predicted by the theory have not been tested. For example only a measurement of the angular distribution at forward angles in the centre of mass would disentangle contributions from *s*, *p* and *d* orbitals in different portions of the energy spectrum. Also the behavior of the observed $s_{1/2}$ and $p_{1/2}$ orbitals should be better addressed, since in both cases the coupling with the $p_{3/2}$ proton generates a doublet of $^{10}$Li states, namely $1^-$, $2^-$ for the $s_{1/2}$ and $1^+$, $2^+$ for the $p_{1/2}$, while experimentally there is no indication of such doublets. In addition, according to the theory, $p_{3/2}$ and $d_{5/2}$ orbitals should generate resonances above 1 MeV, but these have not been observed experimentally, since the MSU data are difficult to be reliably interpreted.

To summarize, many of the key points regarding the $^{10}$Li spectrum are still very uncertain, especially from an experimental point of view. As a consequence the description of the $^{11}$Li halo nucleus is quite deficient, since the binding energy strongly depends on the same n - $^{9}$Li interaction responsible for the $^{10}$Li structure.

## 3. The experiment

We have studied the d($^{9}$Li,p)$^{10}$Li reaction at 11 AMeV incident energy at the TRIUMF laboratory. At this energy the recoiling protons can be detected at backward laboratory angles, thus allowing the exploration of the crucial region at forward angles in the centre of mass.

A $^{9}$Li beam, produced by the ISAC-II facility, impinged on a CD2 target, 126 μg/cm2 thick, evaporated at the INFN-LNS laboratory (Catania). The recoiling protons were detected at backward angles $127° < \theta_{LEDA} < 152°$ by the LEDA array of silicon strip detectors [6], thus allowing the study of

Journal of Physics: Conference Series **590** (2015) 012037 doi:10.1088/1742-6596/590/1/012037

the $^{10}$Li emitted at forward angles. Protons are detected in coincidence with the $^{9}$Li fragments produced from the breakup of the corresponding $^{10}$Li. $^{9}$Li fragments were detected and identified by a ΔE-E telescope of S2 annular Double Sided Silicon Detectors located downstream the target. Fig. 1 shows an example of ΔE-E identification plot where the $^{9}$Li locus is well identified and contoured by a solid line.

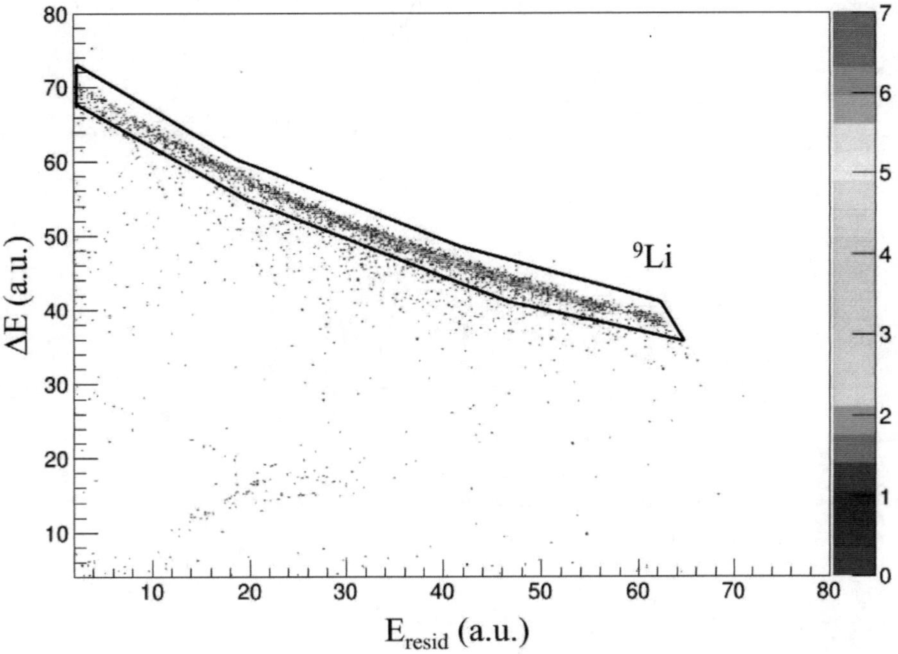


**Figure 1.** Energy loss (ΔE) measured by the first stage of the S2 telescope of annular silicon detectors as a function of the residual energy ($E_{resid}$). The graphical contour indicates the $^{9}$Li events.

An additional benefit to placing the proton detectors at backward scattering angles is that no background was expected from other interactions between the beam and the target.

The $^{10}$Li excitation energy was reconstructed with significant statistics, compared to the previous experiments of refs. [3] and [4], allowing to explore the $^{10}$Li level structure at least in the low excitation energy region. Fig. 2 shows the excitation energy spectrum obtained up to 1.5 MeV in the angular region 8.3°< $\theta_{CM}$ < 16.2°. A centroid $E_0$ = 0.61 MeV and full width at half maximum Γ = 0.62 MeV were extracted for the unbound ground state by fitting the energy spectrum with a lorenzian shape and a background reproducing the three-body phase space. These values are consistent with the presence of a $p_{1/2}$ resonance predicted in the CCBA calculations performed in ref. [3]. Also the attempt to fit the experimental data of ref. [4] with two separate resonances, explored by the authors in the same paper, indicate the possible presence of a structure at that energy. In addition, the calculations of the angle integrated cross section reported in ref. [2], predict a resonance-like structure mainly determined by the $p_{1/2}$ contribution in the same energy region.

The highly segmented detection system available in the present experimental setup, also allowed to measure the angular distributions of the observed resonances at forward angles. The analysis of the angular distribution, which is still in progress, would give some indication about the controversial question of the spin-parity of the $^{10}$Li unbound ground state.

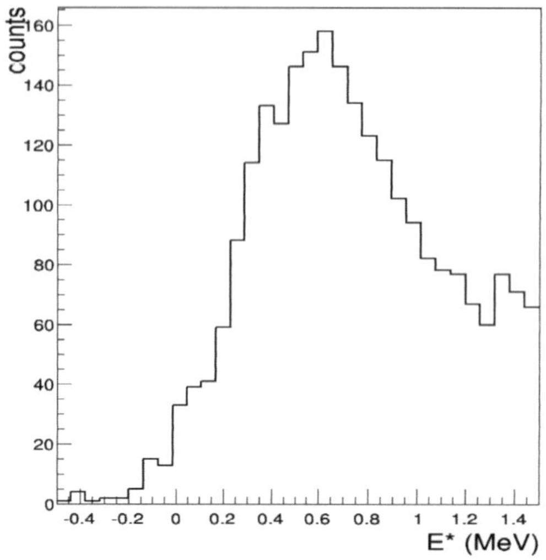


**Figure 2.** $^{10}Li$ reconstructed excitation energy from the $d(^{9}Li,p)^{10}Li$ reaction at 11 AMeV.

[1] Lenske H and Schrieder G 1998 *Eur. Phys. Journ. A* **2** 41
[2] Orrigo S E A and Lenske H 2009 *Phys. Lett. B* **677** 214
[3] Jeppesen H B *et al.* 2006 *Phys. Lett. B.* **677** 449
[4] Santi P *et al.* 2003 *Phys. Rev. C* **67** 024606
[5] Bohlen H G *et al.* 1993 *Z. Phys. A* **344** 381
[6] Davinson T *et al.* 2000 *Nucl. Instr. And Meth. A* **454** 350

Journal of Physics: Conference Series **590** (2015) 012038 doi:10.1088/1742-6596/590/1/012038

# $^{12}C+^{12}C$ and $^{16}O+^{16}O$ fusion reactions at low energies

**M. Assunção**[1] **and P. Descouvemont**[2]

[1] Departamento de Ciências Exatas e da Terra, Universidade Federal de São Paulo, 09972-270, Campus Diadema, São Paulo, Brazil

[2] Physique Nucléaire Théorique et Physique Mathématique, C.P. 229, Université Libre de Bruxelles (ULB), B 1050 Brussels, Belgium

E-mail: massuncao@unifesp.br

**Abstract.**
The $^{12}C+^{12}C$ and $^{16}O+^{16}O$ fusion reactions are investigated in a folding model. We use a multichannel approach involving the $^{12}C(0_1^+, 2^+, 0_2^+, 3^-)$ states. The $^{12}C$ densities (including transition densities) are taken from the RGM calculation of Kamimura. For the nucleon-nucleon interaction, we use the DDM3Y density-dependent interaction. Owing to the explicit presence of inelastic channels, the imaginary part of the optical potential only contains a short-range fusion contribution. The $S$-factor is then virtually insensitive to the precise value, and the model is free of any fitting parameter. From the coupled-channel system, we determine the elastic and fusion cross sections simultaneously. For the $^{16}O+^{16}O$ fusion cross section, we show preliminary results obtained in a single-channel approximation, where the $^{16}O$ densities is obtained from an $\alpha+^{12}C$ cluster calculation.

## 1. Introduction

The $^{12}C+^{12}C$ and $^{16}O+^{16}O$ fusion reactions play an important role in stellar nucleosynthesis [1] and in the evolution of massive stars [2]. The extrapolation of the $S$-factor at low energies is made difficult by the presence of resonances whose interpretation is not clear (see a recent review in [2]). Most fusion calculations to date are performed in a single-channel model, i.e. involving the ground-state only, while the absorption is simulated by a phenomenological imaginary potential [4]. In light systems, however, it is known that inelastic channels may be important and require to be explicitly included in the calculation. In Ref. [3], the authors suggest that mutual excitations play an important role even at low energies, where excited channels are closed. At first sight, this effect may seem surprising since only a single channel is open. It is explained by distortion effects in the wave functions: the cross section is mostly sensitive to the inner part of the wave functions, where closed channels may have a significant amplitude.

The aim of our calculation is to investigate the $^{12}C+^{12}C$ and $^{16}O+^{16}O$ fusion in a folding method [5]. We include the $^{12}C(0_1^+, 2^+, 0_2^+, 3^-)$ and $^{16}O(0_1^+, 0_2^+, 3^-, 2^+, 1^-)$ states and also the corresponding mutual excitations for both fusion reactions. A folding method is performed using the density-dependent M3Y (DDM3Y) potential [7] as nucleon-nucleon (NN) interaction. The $^{12}C$ densities are taken from the RGM values of Kamimura [6], while $^{16}O$ densities are obtained from an $\alpha+^{12}C$ cluster calculation. An important aspect is that our calculation is free of parameter, except for a weak dependence on the absorption potential. It provides a simultaneous description of elastic scattering and of fusion.

Journal of Physics: Conference Series **590** (2015) 012038 doi:10.1088/1742-6596/590/1/012038

## 2. Theoretical framework

In a coupled-channel formalism, the potentials are defined as

$$V_{\alpha_1\alpha_2,\alpha'_1\alpha'_2}(\boldsymbol{r}) = \int\int d\boldsymbol{r}_1 d\boldsymbol{r}_2\, v_{NN}(\boldsymbol{r}-\boldsymbol{r}_1+\boldsymbol{r}_2)\, \rho_1^{\alpha_1\alpha'_1}(\boldsymbol{r}_1)\, \rho_2^{\alpha_2\alpha'_2}(\boldsymbol{r}_2), \tag{1}$$

where $v_{NN}(\boldsymbol{r})$ represents the nucleon-nucleon nuclear interaction, $\boldsymbol{r} = (r, \Omega_r)$ is the relative coordinate, $\rho_k^{\alpha\alpha'}(\boldsymbol{r}_k)$ are the nuclear densities, and labels $\alpha_k$ refer to different states. The same formalism is applied to the Coulomb interaction. In the present work, we include $^{12}$C$(0_1^+, 2^+, 0_2^+, 3^-)$ and $^{16}$O$(0_1^+, 0_2^+, 3^-, 2^+, 1^-)$ states, which means that ten $^{12}$C+$^{12}$C and fifteen $^{16}$O+$^{16}$O channels are introduced in the coupled-channel systems. In practice, the evaluation of the double integral (1) is performed by using Fourier transforms for the nuclear as well as for the Coulomb interactions [5].

As usual, the densities $\rho_k^{\alpha\alpha'}(\boldsymbol{r})$ are expanded in multipoles [6], and the radial wave functions $g_c^{J\pi}(r)$ are obtained from the coupled-channel system

$$-\frac{\hbar^2}{2\mu}\left[\frac{d^2}{dr^2} - \frac{\ell(\ell+1)}{r^2}\right] g_c^{J\pi}(r) + \sum_{c'} V_{cc'}^{J\pi}(r) g_{c'}^{J\pi}(r) = (E - E_1^c - E_2^c) g_c^{J\pi}(r), \tag{2}$$

where $\mu$ is the reduced mass of the system, and $E_i^c$ are the $^{12}$C or $^{16}$O energies. In this equation, label $c$ stands for $c = (\alpha_1, \alpha_2, I, \ell)$, where $I$ is the channel spin and $\ell$ the relative orbital momentum.

In this approach, the absorption to other channels is simulated by adding an imaginary component to the potentials as (see Ref.[12] for details)

$$V_{cc'}^{J\pi}(r) \longrightarrow V_{cc'}^{J\pi}(r) + iW_{cc'}^{J\pi}(r). \tag{3}$$

In the present calculation, the coupled-channel system (2) explicitly includes inelastic channels in a wide energy range. To define the fusion component of the potential, a short-range absorption potential [11] is included as

$$W_{cc'}(r) = -\frac{W_0}{1+\exp((r-R_0)/a)}\delta_{cc'}. \tag{4}$$

The range $R_0$ is chosen smaller than the barrier radius, and this potential acts at short distances only. In this multichannel calculation, the values were taken as $W_0 = 10$ MeV, $R_0 = 3$ fm, and $a = 0.1$ fm, and the same conditions were employed to investigate the elastic-scattering and fusion processes. The calculations are stable within $1-2\%$ when these parameters are modified. An important consequence is that the model is free of parameters, and that all cross sections are obtained without any adjustment.

The coupled-channel system (2) is solved with the $R$-matrix method [9]. In the internal region ($r \leq a$), the radial functions $g_c^{J\pi}(r)$ are expanded over a Lagrange basis [10]. In the external region, they are given by linear combinations of Coulomb functions. The matching provides the collision matrix $\boldsymbol{U}^{J\pi}$. The elastic cross sections are computed from the collision matrices by using standard formulas [4].

## 3. Results and Discussions

The present folding model was first applied to $^{12}$C+$^{12}$C elastic scattering at energies around the Coulomb barrier, where experimental data are available [14]. The idea was to assess the accuracy of the model on elastic scattering, and then apply it to the calculation of fusion cross sections. The procedure was started from a single-channel approximation, and we progressively

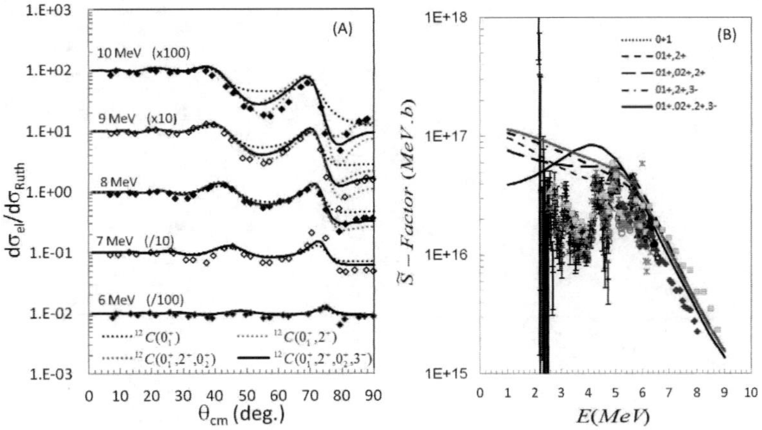


**Figure 1.** (A) Ratios of the elastic and Rutherford cross sections around the Coulomb barrier and (B) modified $S$ factor for increasing numbers of $^{12}$C+$^{12}$C inelastic channels, respectively. Experimental data are taken from Refs. [14, 16]

included additional channels. The comparison between theory and experiment is presented in Fig.1(A).

When the energy increases, and in particular at $E = 10$ MeV, inelastic channels significantly improve the theoretical cross section. At 6 MeV, the physics of the problem is essentially determined by the Coulomb interaction, and the role of the inelastic channels is hardly visible. The most sensitive angular range is beyond $\theta = 70°$, where the single-channel approximation provides a poor fit of the data. Including the $2^+$ state improves the overall agreement, but adding further the $0_2^+$ Hoyle state provides an excellent agreement with the data. Note that good fits can be obtained even in the single-channel approximation [15], but after fitting the imaginary potential to optimize the agreement with experiment.

For the $^{12}$C+$^{12}$C reaction, the fusion cross section is traditionally converted in a modified $S$ factor as $\tilde{S}(E) = \sigma_F(E)E\exp(2\pi\eta + 0.46\,E)$, where $\eta$ is the Sommerfeld parameter and $\sigma_F(E)$ is the fusion cross section [13].

The $^{12}$C+$^{12}$C modified $S$ factor is displayed in Fig.1(B), where the experimental data have been corrected as suggested by Aguilera *et al.* [17]. Above the Coulomb barrier ($\approx 6.5$ MeV) the data are well reproduced by the calculation, and the role of inelastic channels is minor. When the energy decreases, the sensitivity with respect to the number of excited channels is more and more important, in agreement with Ref.[3]. At $E = 1$ MeV, the multichannel calculation provides an enhancement of about a factor of three, in comparison with the single-channel approach.

Preliminary results for the $^{16}$O+$^{16}$O system are shown in Fig.2. We limit the calculations to a single-channel approach, and investigate the sensitivity of the cross sections to a renormalization factor $N_r$. From Fig. 2, we conclude that the elastic cross sections are weakly sensitive to this factor. However the fusion cross section below 10 MeV is underestimated with $N_r = 1$. Using a factor $N_r = 1.3$ provides a reasonable agreement with the data. Multichannel calculations are in progress.

## 4. Conclusion

In the present work, the $^{12}$C+$^{12}$C fusion process was investigated in a multichannel model. The coupling potentials are generated from $^{12}$C densities obtained in a microscopic cluster model.

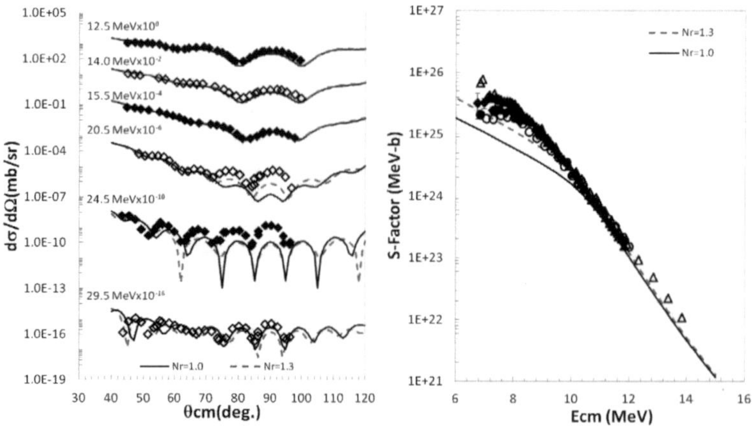


**Figure 2.** (A) $^{16}O+^{16}O$ elastic cross sections at low energies and (B) $S$ factor for $N_r = 1$ and $N_r = 1.3$. Experimental data are from Ref. [18] and Refs. [18, 19].

The calculation does not contain any fitting parameter, and provides simultaneously the fusion and elastic cross sections. Around the Coulomb barrier the elastic data are well reproduced by the model provided that all inelastic channels, and in particular those involving the $0_2^+$ state, are included. The results confirm the conclusion of Ref.[3]. For the $^{16}O+^{16}O$ system, we have shown preliminary results, obtained with a single-channel approximation. Multichannel calculations are in progress.

## Acknowledgments

We acknowledge FAPESP for financial support.

## References

[1] Wiescher M, Käppeler F, Langanke K 2012 *Annu. Rev. Astron. Astrophys.* **50** 165
[2] Pignatari M, *et al.* 2013 *Astrophys. J.* **762** 31
[3] Esbensen H, Tang X and Jiang C L 2011 *Phys. Rev. C* **84** 064613
[4] Canto F and Hussein M S 2013 *Scattering Theory of Molecules, Atoms and Nuclei, World Scientific Publishing*
[5] Satchler G R and Love W G 1979 *Phys. Rep.* **55C** 183
[6] Kamimura M 1981 *Nucl. Phys. A* **351** 456
[7] Kobos A, *et al.* 1982 *Nucl. Phys. A* **384** 65
[8] Ito M, Y.Sakuragi Y, Hirabayashi Y 2001 *Phys. Rev. C* **63** 064303
[9] Descouvemont P and Baye D 2010 *Rep. Prog. Phys.* **73** 036301
[10] Baye D 2006 *Phys. Stat. Sol. (b)* **243** 1095
[11] Rhoades-Brown M, Braun-Munzinger P 1984 *Phys. Lett. B.* **136** 19
[12] Khoa D T and Satchler G 2000 *Nucl. Phys. A* **668** 3
[13] Canto L F,*et al.* 2006 *Phys. Rep.* **424** 1
[14] Treu W, *et al.* 1980 *Phys. Rev. C* **22** 2462
[15] Gasques L R, *et al.* 2005 *Phys. Rev. C* **72** 025806
[16] Assunção M and Descouvemont P 2013 *Phys. Lett. B* **723** 355
[17] Aguilera E F, *et al.* 2006 *Phys. Rev. C* **73** 064601
[18] Kocak G, *et al.* 2010 *Phys. Rev. C.* **81** 024615
[19] Wu S-C, *et al.* 1984 *Nucl. Phys. A* **422** 373

Journal of Physics: Conference Series **590** (2015) 012039 doi:10.1088/1742-6596/590/1/012039

# Half-life measurements of the $^{144}$Pm isotope with $\gamma$-spectroscopy

**C Yalçın[1,2], Gy Gyürky[2], Z Halász[2], T Szücs[2], R T Güray[1], G G Kiss[2], Zs Fülöp[2], N Özkan[1], E Somorjai[2]**

[1] Kocaeli University, Department of Physics, Umuttepe 41380, Kocaeli, Turkey
[2] Atomki, H-4001 Debrecen, POB.51., Hungary

E-mail: caner.yalcin@kocaeli.edu.tr

**Abstract.** The accurate half-life information for reaction products is very crucial in order to measure the cross section of a reaction via the activation method. Precise and accurate cross section measurement of reactions which are resulting in $^{144}$Pm requires the precise half-life value of this isotope. The adopted half-life of $^{144}$Pm has a relatively high uncertainty and conflicting data can be found in the literature. In order to mitigate this controversy and to reduce the uncertainty, the half-life of $^{144}$Pm is measured with gamma spectroscopy. Experimental methods and preliminary results are presented.

## 1. Introduction

In the last decade, a number of cross section measurements were carried out using the activation method for the study of the astrophysical *p*-process (or $\gamma$-process) [1–12]. These cross sections are needed for two purposes: to supply data for the nuclear reaction network [13, 14] related to astrophysical p-process and also to improve our knowledge about theoretical cross section calculations [15, 16] by comparing them with the experimental ones. Experimental aspects of the reaction cross section measurements by the activation methods are discussed e.g. in Ref. [17]. A precise and accurate cross section measurement of a reaction using the activation method requires the precise half-life of the reaction product. One of the recent reaction cross section measurements is on the $^{141}$Pr($\alpha$,n)$^{144}$Pm reaction from Sauerwein et al. from 2011 [7]. This cross section measurement requires precise half-life value of $^{144}$Pm. In the literature the half-life of the $^{144}$Pm isotope is controversial and has a relatively high uncertainty. Half-life values available for this isotope are 377 $\pm$ 16 d [18] and 349 $\pm$ 16 d [19], the uncertainty of these measurements are 4.24% and 4.58% respectively. The recommended half-life of $^{144}$Pm is 363 $\pm$ 14 d [20, 21] which is the weighed average of the two values quoted above. In order to mitigate this controversy and to reduce the uncertainty, the half-life of the $^{144}$Pm isotopes was determined in the present work based on the measurement of the radioactive decay curve of the sources with HPGe detectors. In this paper the experimental methods and preliminary results are presented.

**Table 1.** Main radioactive isotopes in the source. The recommended half-life values are taken from [20, 21] and only the highest relative intensity gamma radiations are tabulated with their intensities in parentheses.

| Isotope | Decay mode | Recommended half-life | Gamma energy (Intensity) [keV (%)] |
|---|---|---|---|
| $^{143}$Pm | $\epsilon$: 100% | 265 ± 7 d | 741.98 (38.5) |
| $^{144}$Pm | $\epsilon$: 100% | 363 ± 14 d | 476.78(43.8), 618.01(98), 696.49(99.490) |
| $^{146}$Pm | $\epsilon$: 66.0% | 5.53 ± 0.05 y | 453.88(65.0), 735.93(22.5) |
| $^{146}$Pm | $\beta^-$: 34.0% | 5.53 ± 0.05 y | 747.24(34.0) |
| $^{148}$Pm | $\beta^-$: 100% | 5.368 ± 0.002 d | 550.27(22.0), 914.85(11.5), 1465.12(22.2) |
| $^{148m}$Pm | $\beta^-$: 95.8% | 41.29 ± 0.11 d | 414.07(18.66), 550.27(94.9), 629.97(89.0), 725.70(32.8), 915.33(17.17), 1013.81(20.3) |
| $^{56}$Co | $\epsilon$: 100% | 77.236 ± 0.026 d | 846.77(99.94), 1037.843(14.05), 1238.288(66.46) |

## 2. Experimental methods

### *2.1. Source Preparation*

The source was produced by the $^{143}$Nd(p,$\gamma$)$^{144}$Pm and $^{144}$Nd(p,n)$^{144}$Pm reactions. For these reactions, a natural neodymium oxides target was prepared onto thin Al foil via physical vapor deposition (PVD) method by using electron beam. The target thickness was determined with weight measurement as 295 ± 15 $\mu$g/cm$^2$. The thickness of the target was also checked by alpha energy loss measurement using the ThiMeT code [22, 23], the results of the two methods were in agreement. The target was irradiated at the cyclotron accelerator of Atomki in Debrecen, Hungary with a 7.5 MeV proton beam of 2.5 $\mu$A intensity for about 5 h. Before the gamma-counting the target was cooled for about 63 days. Because of the natural Nd used for the target preparation, other isotopes of Pm were also produced via (p,$\gamma$) and (p,n) reactions. The composition of the source can be seen in Table 1. The main contaminant beside the Pm isotopes is $^{56}$Co. This isotope is produced via the $^{56}$Fe(p,n)$^{56}$Co reaction on the iron impurities in the Al backing foil.

### *2.2. Gamma counting and data analysis*

In order to decrease the systematic uncertainties of the half-life measurement, the source was counted with two independent measurement systems. The first measurement system is composed of a Canberra HPGe detector with 100% relative efficiency (model GR10024), Canberra High Voltage power supply (model 3105D), Ortec amplifier (model 572A) and Ortec MCA (model ASPEC-927) at Atomki. The second measurement system located at the Nuclear Research Laboratory of Kocaeli University, Turkey is based on an Ortec HPGe detector with 25% relative efficiency (model GEM25P4-70) and an Ortec DSPEC jr 2.0 which incorporates all necessary electronics. Both detectors were covered with 10 cm lead shield to minimize natural background.

The first counting was carried out with the Atomki system. The measurement started 63 days after the irradiation, therefore most of the short-lived contaminants had already decayed out. The spectra were saved every 6 h collecting about 500 spectra in total. A typical gamma spectrum in the relevant energy region is shown in Figure 1. The most prominent peaks in the spectra are from the decay of $^{143,144,146,148}$Pm isotopes and $^{56}$Co. The dead time of the counting system was less than 0.3% from the beginning of the counting. The same source was measured with the Kocaeli system. The measurement there started 261 days after the irradiation, lasted for 92 days and the spectra were stored in every 6 hours.

Spectra were analyzed custom-made peak integration codes as well as with the GammaVision software and a job file prepared for to obtain net peak areas. Half-life values of $^{144}$Pm were determined from the analysis of three gamma peaks (see Table 1) by fitting a linear curve on semi-logarithmic plot of the net areas of the $\gamma$-peaks as a function of time. There were no special method used for the data treatment and none of the data points were excluded for fit. The statistical uncertainty of the half-lives was determined from the fit parameter. The analysis of the systematic uncertainties as well as the comparison of the results at the two counting facilities is still in progress.

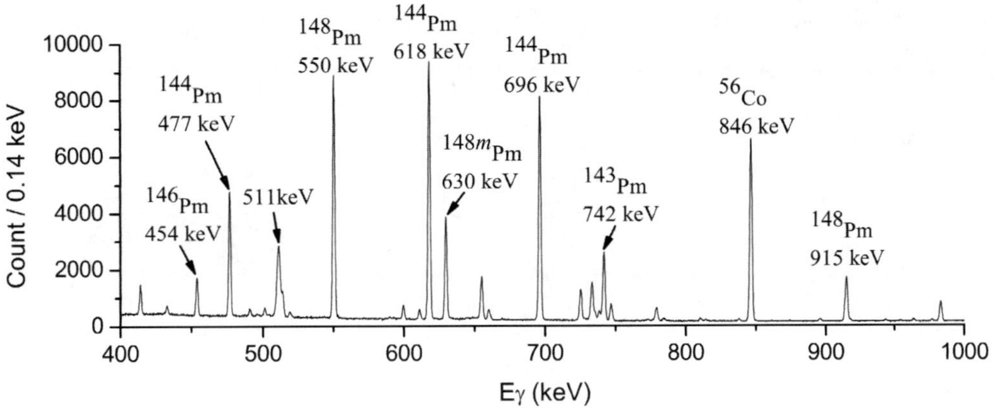


**Figure 1.** Relevant energy region of a typical spectrum.

## 3. Results and Conclusion

According to the preliminary analysis of the currently available data, the obtained half-life for $^{144}$Pm is compatible with previous measurement of Bunney et al. 1964 [19] but significantly lower than the previous measurement of Pagden et al. 1963 [18] and also lower than the recommended value in the literature [20, 21]. The uncertainty of the half-life measurement is about one order of magnitude smaller than that of the previous measurements. For this reason, the $^{141}$Pr($\alpha$,n)$^{144}$Pm reaction cross section results [7] should be revised when the final half-life of $^{144}$Pm becomes available. As a consequence of the shorter half-life value, the cross sections will need to be corrected down by several percent, but taking into account the total uncertainty of the cross sections, the main conclusions of [7] will not change.

The counting of the source is still in progress at the Nuclear Research Laboratory of Kocaeli University, therefore further details about the experiment and the full analysis will be presented in a forthcoming publication.

## Acknowledgments

This work was supported by the Scientific and Technological Research Council of Turkey (TUBITAK), Grants No. 109T585 (under the EUROGENESIS research program) and Grants No. 108T508 and by OTKA grants K101328 and K108459. C Yalçın, acknowledges support through the TUBITAK, under the programme of BIDEB-2219.

## References

[1] Gyürky Gy et al. 2006 *Phys. Rev.* C **74** 025805
[2] Özkan N et al. 2007 *Phys. Rev.* C **75** 025801

[3] Yalçın C et al. 2009 *Phys. Rev.* C **79** 065801
[4] Güray R T et al. 2009 *Phys. Rev.* C **80** 035804
[5] Gyürky Gy et al. 2010 *J. Phys.* G **37** 115201
[6] Kiss G G et al. 2011 *Phys. Lett.* B **695** 419
[7] Sauerwein A et al. 2011 *Phys. Rev.* C **84** 045808
[8] Dillmann I et al. 2011 *Phys. Rev.* C **84** 015802
[9] Palumbo A et al. 2012 *Phys. Rev.* C **85** 028801
[10] Halász Z et al. 2012 *Phys. Rev.* C **85** 025804
[11] Kiss G G et al. 2012 *Phys. Rev.* C **86** 035801
[12] Rauscher T et al. 2012 *Phys. Rev.* C **86** 015804
[13] Woosley S E and Howard W M 1978 *Astrophys. J. Suppl.Ser.* **36** 285
[14] Arnould M and Goriely S 2003 *Phys. Rep.* **384** 1
[15] Rauscher T and Thielemann F -K 2000 *At. Data Nucl. Data Tables* **75** 1
[16] Rauscher T and Thielemann F -K 2001 *At. Data Nucl. Data Tables* **79** 47
[17] Özkan N et al. 2002 *Nucl. Phys.* **A710** 469
[18] Pagden I M H, Jakeways R and Flack F C 1963 *Nucl. Phys.* **48** 555
[19] Bunney L R and Scadden E M 1964 *J. Inorg. Nucl. Chem.* **26** 1795
[20] National Nuclear Data Center Nudat 2 www.nndc.bnl.gov/nudat2/
[21] International Network of Nuclear Structure and Decay Data Evaluators www-nds.iaea.org/nsdd/
[22] Yalçın C this volume
[23] The Thickness Measurement Tools for SRIM (ThiMeT) www.thimet.org

Journal of Physics: Conference Series **590** (2015) 012040 doi:10.1088/1742-6596/590/1/012040

# Coupling to continuum effects in the $^{6,7}$Li $+^{64}$Zn reactions at energies around the Coulomb barrier.

**J. P. Fernández-García[1], M. Zadro[2], A. Di Pietro[1], P. Figuera[1], M. Fisichella[1], M. Lattuada[1,4], A. M. Moro[3] and D. Torresi[1,4]**

[1]INFN, Laboratori Nazionali del Sud, via S. Sofia 62, 1-95123 Catania, Italy
[2] Ruder Bošković Institute, Bijenička cesta 54, HR-10000 Zagreb, Croatia
[3] Departamento de FAMN, Universidad de Sevilla, Apartado 1065, E-41080 Seville, Spain
[4] Dipartamento di Fisica e Astronomia, via S. Sofia 64, I-95123 Catania, Italy

E-mail: `fernandez@lns.infn.it`

**Abstract.**
The elastic scattering angular distributions for the weakly bound nuclei $^6$Li and $^7$Li on $^{64}$Zn were measured in a wide angular range at energies around the Coulomb barrier. In addition, the excitation functions of quasi-elastic scattering at backward angles were measured and the corresponding barrier distributions were derived. The experimental data were analysed within the continuum-discretized-coupled-channel method. In this contribution, we will present a detailed study concerning the effects of the break-up channels in the $^{6,7}$Li $+^{64}$Zn reactions.

## 1. Introduction

Many efforts have been concentrated on the study of nuclear collisions at energies around the Coulomb barrier induced by the halo and stable weakly bound nuclei, such as $^6$Li and $^7$Li (see, e.g. Ref. [1] and references therein). In fact, the coupling to the break-up channel can affect the reaction dynamics and, therefore, strong effects on elastic scattering (e.g. [2–4]) and fusion (e.g. [5, 6]) have been observed.

The barrier distribution method has been proposed as a powerful tool to study the effects of couplings to different reaction channels at near-barrier energies [7–9]. The barrier distribution of quasi-elastic scattering ($D_{qel}$) is defined as [10],

$$D_{qel}(E) = -\frac{d}{dE}\left[\frac{d\sigma_{qel}}{d\sigma_{Ruth}}\right], \tag{1}$$

where $d\sigma_{qel}/d\sigma_{Ruth}$ corresponds to the ratio of the quasi-elastic (QEL) scattering and the Rutherford differential cross sections at a fixed backward angle. QEL scattering is defined as the sum of elastic and inelastic scattering, and all other direct processes. In Refs. [11–14], excitation functions for QEL and/or elastic scattering at backward angles, and the corresponding barrier distribution, have been measured and analyzed for several reactions involving $^{6,7}$Li. The results show that the effects of coupling to break-up channels are large for $^6$Li and smaller for $^7$Li.

Recently, a systematic study of the $^{6,7}$Li + $^{64}$Zn collisions at energies around the Coulomb barrier has been performed [15–17]. In Ref. [15], the elastic scattering of $^6$Li on $^{64}$Zn at energies around the Coulomb barrier has been investigated within the optical model. The energy dependence of the optical potential shows absence of the usual threshold anomaly for the system $^6$Li+$^{64}$Zn. This new kind of anomaly is known as break-up threshold anomaly and it can be understood as an evidence of the effect of the coupling to the break-up channel.

Moreover, for the systems $^{6,7}$Li + $^{64}$Zn the QEL and elastic excitation functions were measured and the corresponding barrier distributions were derived [16]. Coupled-channel calculations, including inelastic excitation of the projectile and target, do not reproduce the experimental data for the $^6$Li system. This result suggests that the effects of couplings to the continuum are important in the reactions induced by $^6$Li. Therefore, to extend the coupled-channel calculations including the continuum states of the projectile and to investigate in detail the effects of such couplings, Continuum-Discretized Coupled-Channel (CDCC) calculations have been performed and presented in this contribution.

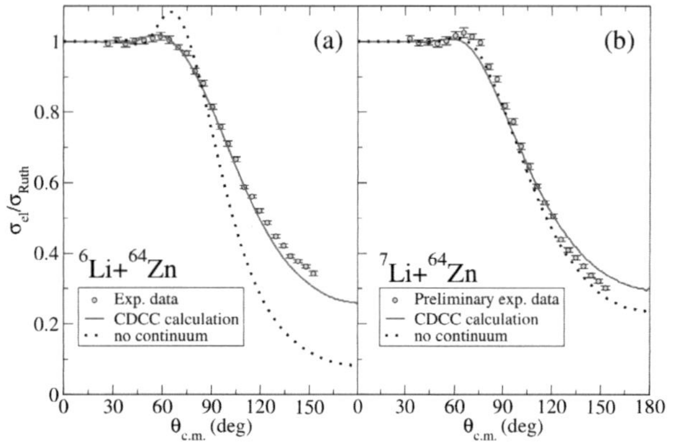


**Figure 1.** (Color online) Elastic scattering angular distribution for the reactions $^6$Li+$^{64}$Zn (a) and $^7$Li+$^{64}$Zn (b) at the same center of mass energy of 13.5 MeV. The CDCC calculations are represented by solid red lines, while the same calculations without the continuum couplings are shown by dotted black lines.

## 2. Experimental set-up and results

The experiments were performed at the Laboratori Nazionali del Sud, INFN, in Catania, Italy. The $^6$Li and $^7$Li beams were produced by the SMP Tandem Van de Graaff accelerator. Elastic-scattering angular distributions were measured at $E_{lab}$=12−22 MeV for the $^6$Li+$^{64}$Zn system and at $E_{lab}$=13−20 MeV for $^7$Li+$^{64}$Zn. The target was a 400 $\mu$g/cm$^2$ thick $^{64}$Zn foil in the $^6$Li+$^{64}$Zn measurement. $^{64}$Zn targets with thicknesses of 90−140 $\mu$g/cm$^2$, made by evaporation onto a carbon backing, were used for the measurements with $^7$Li. Outgoing charged particles were detected and charge identified by an array of five silicon telescopes. Each telescope consisted of a 10 $\mu$m thick $\Delta$E detector and an E detector with a thickness in the range 100−500 $\mu$m. The telescopes were mounted on a rotating plate in the CT2000 scattering chamber (see Ref. [15] for more details).

The measurements of the QEL scattering for the $^{6,7}$Li+$^{64}$Zn systems were performed in the energy range $E_{lab}$=9−20 MeV in steps of 0.5 MeV. In this case, four silicon telescopes were positioned at ±160° and ±170° relative to the beam direction. $^{64}$Zn targets with thicknesses of 140 $\mu$g/cm$^2$ and 70 $\mu$g/cm$^2$, evaporated onto a 15 $\mu$g/cm$^2$ carbon backing, were used for the measurements with $^6$Li and $^7$Li, respectively (more experimental details can be found in Ref. [16]).

Experimental results for the $^6$Li+$^{64}$Zn elastic scattering and for the $^{6,7}$Li+$^{64}$Zn QEL backscattering have already been reported in Refs. [15] and [16], respectively. As an example, the elastic scattering angular distribution at $E_{c.m.}$= 13.5 MeV for $^6$Li [15] is shown in Fig. 1(a), while in Fig. 1(b) the experimental data for $^7$Li at the same energy are presented.

In Ref. [16], in the case of the $^{7}$Li+$^{64}$Zn system, the QEL scattering was defined as the sum of the elastic scattering, inelastic excitation of the $^{7}$Li $1/2^{-}$ and $^{64}$Zn $2^{+}$ states, and 1n transfer to the $^{65}$Zn states. In the case of $^{6}$Li+$^{64}$Zn, the elastic scattering and inelastic excitation of the $^{64}$Zn $2^{+}$ state were included in the QEL scattering. In this paper, we will focus on the elastic backscattering excitation functions and the corresponding contributions to the QEL barrier distributions.

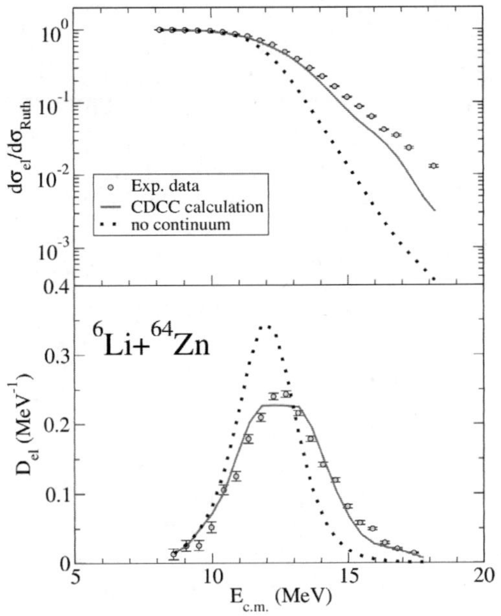


**Figure 2.** (Color online) Elastic excitation function at backward angles (top) and the corresponding barrier distribution (bottom) for the system $^{6}$Li+$^{64}$Zn. The experimental data are represented by the yellow symbols, while the solid red lines correspond to the CDCC calculations. The dotted black lines represent the calculations without the couplings to the continuum states.

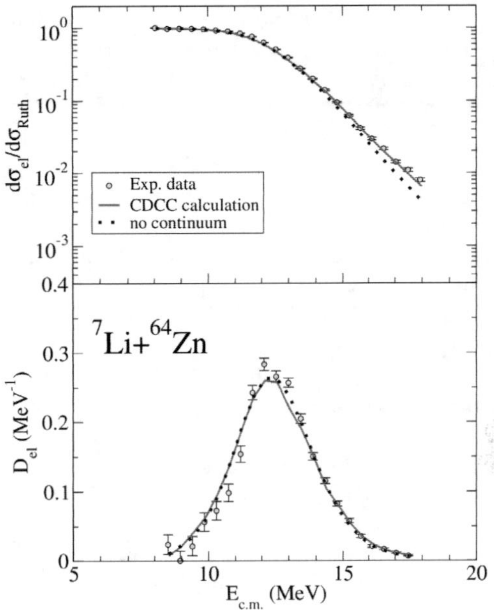


**Figure 3.** (Color online) Elastic excitation function at backward angles (top) and the corresponding barrier distribution (bottom) for the system $^{7}$Li+$^{64}$Zn. The experimental data are represented by the yellow symbols, while the solid red lines correspond to the CDCC calculations. The dotted black lines represent the calculations without the couplings to the continuum states.

## 3. Continuum-Discretized Coupled-Channels calculations

In the last years, the Continuum-Discretized Coupled-Channels (CDCC) method has been applied to describe reactions induced by weakly bound nuclei, using a three-body model of the reaction [18–21]. This formalism uses the coupled channel method to solve the scattering problem. It is known that for weakly bound nuclei it is important to take into account the continuum states of the projectile. These states are included and discretized by means of the so-called *binning* method [22].

In this framework, a three-body model of the reactions $^{6,7}$Li+$^{64}$Zn ($\alpha$+d(t)+$^{64}$Zn) is considered. The $^{6}$Li can be considered as having an $\alpha$+d cluster structure, with a break-up threshold at 1.47 MeV above the ground state. The resonances $3^{+}$, $2^{+}$, $1^{+}$ and the non-resonant states are included in the calculations. On the other hand, the $^7$Li isotope can be described as a $\alpha$+t, with a break-up threshold at 2.47 MeV. The excited state $1/2^-$, the resonances $7/2^-$, $5/2^-$ and the non-resonant states are considered. Both Coulomb and nuclear effects are included. These calculations were performed using the code FRESCO [23].

The calculated elastic scattering angular distributions for both systems are represented in Fig. 1. A good agreement between the prediction of the CDCC calculations and the experimental data is observed. To study the effects of the coupling to the continuum states of the projectile, we have included the calculations without such couplings, represented by the dotted black lines in Fig. 1. These calculations underestimate the experimental data for the $^6$Li case, which suggests a strong coupling to the break-up channel in the $^6$Li system.

In addition, the experimental elastic scattering excitation functions and the corresponding barrier distributions for $^6$Li and $^7$Li are compared with the CDCC calculations in Figs. 2 and 3. The calculations are also compared with those omitting the couplings to the continuum states of the projectile. The results show an important influence of the coupling to the break-up channel in the reactions induced by the $^6$Li nucleus, while in the reaction induced by $^7$Li small changes are observed. As it can be seen in Fig. 2, the effect of the coupling to the break-up channel is to increase the average barrier.

## 4. Conclusions

Experimental results for the $^{6,7}$Li+$^{64}$Zn reactions at energies around the Coulomb barrier have been studied within the CDCC framework. The elastic scattering angular distributions are compared with the calculations, showing that it is necessary to include the couplings to continuum states of the projectile to reproduce the experimental data for the $^6$Li reaction.

Moreover, the elastic excitation functions and the corresponding barrier distributions are well reproduced by the CDCC calculations. The results suggest that the coupling to the breakup channel is more relevant in the $^6$Li induced collisions than in the $^7$Li case. This behavior could be linked to the fact that the break-up threshold of $^6$Li ($S_\alpha$=1.47 MeV) is smaller than that of $^7$Li ($S_\alpha$=2.47 MeV).

## References

[1] Canto L F *et al.* 2006 *Phys. Rep.* **424** 1
[2] Di Pietro A *et al.* 2010 *Phys. Rev. Lett.* **105** 022701
[3] Fernández-García J P *et al.* 2010 *Nucl. Phys. A* **840** 19–38
[4] Cubero M *et al.* 2012 *Phys. Rev. Lett.* **109** 262701
[5] Canto L F *et al.* 2009 *Nucl. Phys. A* **821** 51
[6] Scuderi V *et al.* 2011 *Phys. Rev. C* **84** 064604
[7] Dasgupta M *et al.* 1998 *Annu. Rev. Nucl. Sci.* **48** 401
[8] Lubian J *et al.* 2008 *Phys. Rev. C* **78**(6) 064615
[9] Zagrebaev V I 2008 *Phys. Rev. C* **78**(4) 047602
[10] Timmers H *et al.* 1995 *Nuclear Physics A* **584** 190 – 204
[11] Mukherjee S *et al.* 2009 *Phys. Rev. C* **80**(1) 014607
[12] Otomar D R *et al.* 2009 *Phys. Rev. C* **80**(3) 034614
[13] Zerva K *et al.* 2009 *Phys. Rev. C* **80**(1) 017601
[14] Zerva K *et al.* 2012 *Eur. Phys. J. A* **48** 102
[15] Zadro M *et al.* 2009 *Phys. Rev. C* **80** 064610
[16] Zadro M *et al.* 2013 *Phys. Rev. C* **87** 054606
[17] Di Pietro A *et al.* 2013 *Phys. Rev. C* **87** 064614
[18] Sakuragi Y 1987 *Phys. Rev.* **C35** 2161
[19] Tostevin J A *et al.* 2001 *Phys. Rev. C* **63** 024617
[20] Matsumoto T *et al.* 2003 *Phys. Rev. C* **68** 064607
[21] Beck C *et al.* 2007 *Phys. Rev. C* **75** 054605
[22] Austern N *et al.* 1987 *Physics Reports* **154** 125
[23] Thompson I 1988 *Comp. Phys. Rep.* **7**, *167*

Journal of Physics: Conference Series **590** (2015) 012041 doi:10.1088/1742-6596/590/1/012041

# Nuclear $g$-factor measurement with time-dependent recoil in vacuum in radioactive-beam geometry

**A. Kusoglu[1,2], A.E. Stuchbery[3], G. Georgiev[1], A. Goasduff[1], L. Atanasova[10], D.L. Balabanski[5], M. Bostan[2], M. Danchev[6], P. Detistov[4], K. Gladnishki[6], J. Ljungvall[1], I. Matea[7], D. Radeck[8], C. Sotty[1,9], I. Stefan[7], D. Verney[7] and D.T. Yordanov[7]**

[1] CSNSM, CNRS/IN2P3; Université Paris-Sud, UMR8609, F-91405 ORSAY-Campus, France
[2] Department of Physics Faculty of Science, Istanbul University, Vezneciler/Fatih, 34134, Istanbul, Turkey
[3] Department of Nuclear Physics, RSPE, Australian National University, Canberra ACT 0200, Australia
[4] Institute for Nuclear Research and Nuclear Energy, Bulgarian Academy of Sciences, BG-1784 Sofia, Bulgaria
[5] ELI-NP, IFIN-HH, 077125 Magurele, Romania
[6] Faculty of Physics, St. Kliment Ohridski University of Sofia, 1164 Sofia, Bulgaria
[7] IPN, Orsay, CNRS/IN2P3, Université Paris-Sud, F-91406 Orsay Cedex, France
[8] Institute for Nuclear Physics, University of Cologne, Zülpicher Straße 77, D-50937 Köln, Germany
[9] Instituut voor Kern-en Stralingsfysica, Celestijnenlaan 200d, BE-3001 Leuven (Heverlee), Belgium
[10] Department of Medical Physics and Biophysics, Medical University-Sofia, 1431, Bulgaria

E-mail: kusoglu.asli@gmail.com

**Abstract.**
A modified version of the time-differential recoil-in-vacuum (TDRIV) method, adapted for use with radioactive beams, was applied for the first time to perform a $g$-factor measurement on the first-excited state in $^{24}$Mg ($E_x = 1.369$ MeV, $\tau = 1.97$ ps). A high precision $g$-factor value was obtained using predominantly H-like ions. The results obtained demonstrate the versatility of the new approach and have the precision needed for stringent tests of shell model calculations.

## 1. Introduction

The strong nuclear interaction establishes the distribution and motion of nucleons in the nucleus, and we can probe that distribution with the electromagnetic interaction. Magnetic moments are very sensitive probes to the details of the single-particle configuration of a nuclear state. The $g$ factor reveals which single particle orbits are occupied by the valence nucleons and can be used as a rigorous probe for the proton-neutron character of the nuclear states. The development of radioactive beam facilities requires the development of reliable methods to measure magnetic moments of exotic nuclei produced as radioactive beams.

In order to measure a magnetic moment, a spin oriented ensemble of excited nuclear states (spin $I$, magnetic moment $\mu$) needs to be produced by suitable reaction. One can perturb this ensemble with a magnetic field, either an external one or a hyperfine field ($B$), which

Journal of Physics: Conference Series **590** (2015) 012041 doi:10.1088/1742-6596/590/1/012041

creates a torque that causes a rotation of the spin-oriented ensemble with Larmor frequency ($\omega_L = g\mu_N B/\hbar$) around the magnetic field direction. The rotation angle for a specific time can be determined as $\Delta\theta = \omega_L \cdot t$. This rotation can be observed by measuring the modification of the angular distribution of the associated $\gamma$ rays. If the lifetime of the state of interest is of the order of picoseconds, strong magnetic fields (kTesla) are required to induce a detectable precession. Laboratory magnets cannot provide such strong fields, and therefore one has to rely on hyperfine magnetic fields.

The useful hyperfine magnetic fields are the transient magnetic field (TF) [1] and the free-ion hyperfine fields of ions recoiling in vacuum (RIV) [2, 3]. In the case of the RIV technique, as applied recently to heavy radioactive ion beams [4, 5, 6], the averaging of the hyperfine frequency over the different electronic configurations gives a quasiexponential dependence of the vacuum attenuation factors, $G_k$, as a function of the lifetime of the state of interest. It is not trivial to obtain the exact form of this dependence without knowing the precise charge state distribution and the exact electron configurations involved, therefore, calibration measurements with known $g$ factors are needed. For light nuclei this short-coming of the time-integral RIV can be overcome by using the highly-stripped ions that provide fewer charge-states, as well as by performing time–differential measurements (TDRIV).

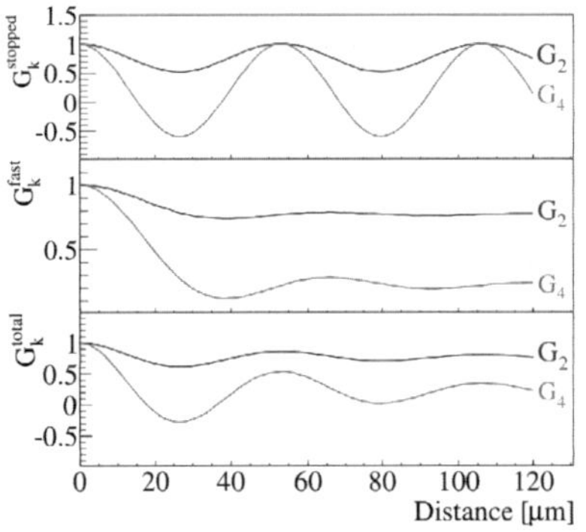


**Figure 1.** Attenuation factors of H-like $^{24}$Mg ions in the conventional technique.

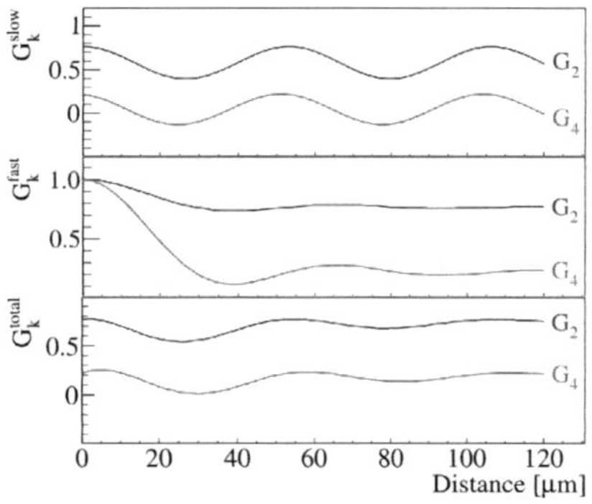


**Figure 2.** Attenuation factors of H-like $^{24}$Mg ions in the modified TDRIV method.

In the conventional TDRIV technique, the interaction time is defined experimentally using a plunger device. After the ions leave the target, and before they reach the stopper, the nuclear and atomic spins $\boldsymbol{I}$ and $\boldsymbol{J}$ precess around the total $\boldsymbol{F}$ with a frequency $\omega_L$. After one period of rotation the spins return to their initial orientation. The precession frequency is proportional to the nuclear $g$ factor and the magnitude of the hyperfine magnetic field at the nucleus. Therefore, although the atomic spins are randomly oriented, the initial angular distribution of $\gamma$ rays is restored after each period of the rotation. At the moment the ions enter the stopper material, the interaction between the atomic and the nuclear spins is 'switched off' and the nuclear spins are 'frozen' to the direction in which they enter the stopper. All these nuclei have therefore experienced the hyperfine interaction for the same time $T$. By changing the position of the stopper foil, we can vary the recoil distance, and thus the interaction time, so the nuclear orientation can be measured time differentially. In TDRIV there are two components of the angular distribution, namely, a 'fast' and a 'stopped' component. As can be seen in Fig. 1, for the $G_k$ stopped component the exact interaction time is well defined and the amplitude of

Journal of Physics: Conference Series **590** (2015) 012041 doi:10.1088/1742-6596/590/1/012041

the observed oscillation pattern is preserved. For the $G_k$ fast component, however, the state of interest has decayed in flight and its interaction time is not well defined, therefore the amplitude of the oscillation pattern decreases with flight time.

Until recently, measurements of $g(2^+)$ were largely confined to stable nuclide. This limitation is being overcome through measurements using radioactive ion beams (RIB). The methods usually applied for moment measurements of short-lived states, the TF and RIV techniques, are also being adapted for radioactive ion beams. TF technique has been developed for use with radioactive ion beams [7, 8, 9]. Here we are focusing on the modification of the TDRIV technique [10] for application with low-$Z$ radioactive beams, and testing it on stable $^{24}$Mg.

A measurement of $g(2^+)$ in $^{24}$Mg with the conventional technique was performed by Horstman *et al.* [2] following the $^{12}$C($^{16}$O,$\alpha\gamma$)$^{24}$Mg reaction. In that measurement only $\sim 15$ % of the Mg ions exiting the target were in H-like charge state. This required the evaluation of complex atomic configurations in order to derive the $g$-factor value. The importance of the use of H-like ions comes from the simplicity of the electron configuration, which allows for a precise determination of the hyperfine field from first principles.

The conventional TDRIV method, in which the beam is stopped at the target, cannot be used for radioactive beams because the build-up of radioactivity becomes prohibitive. To overcome this problem, Stuchbery *et al.* [10] proposed a modification as shown in Fig. 3. The stopper is replaced by a thin foil in order to reset the electron configuration.

The new TDRIV method again has two components of the angular distribution. For the fast component, the state of interest has decayed in flight before the reset foil. After the ion strikes the reset foil several electron exchanges occur, so the charge distribution and electronic configuration is reset randomly. Beyond the reset foil the attenuation factor is just the product of $G_k$(T) and $\bar{G}_k(\infty)$. The nucleus experiences further perturbations identical in effect to those of flight peak for an infinite flight path, which reaches a 'hard core' value. For convenience, we refer to ions that decay between the target and the reset foil as 'fast' and those that decay after the reset foil as 'slow'. As shown in Fig. 2, the TDRIV method does not require that the $\gamma$ rays emitted from fast and slow ions are separated in the observed energy spectrum.

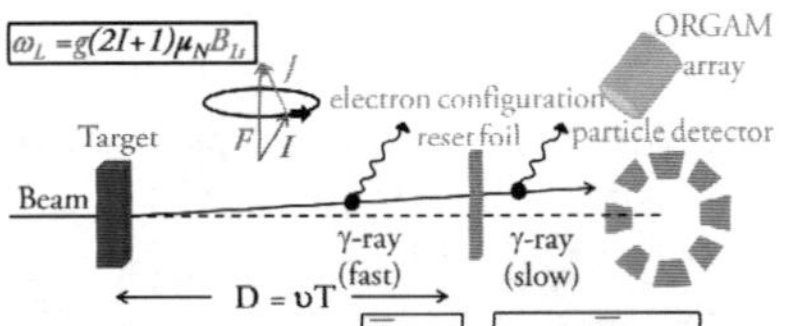


**Figure 3.** Sketch of the experiment.

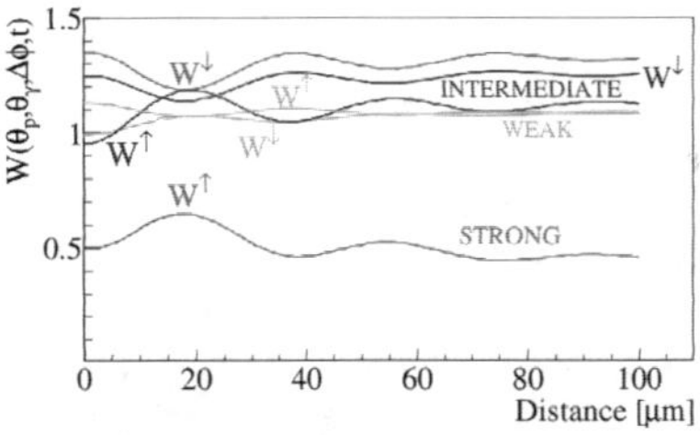


**Figure 4.** Time dependence of particle-$\gamma$ angular correlations for different particle segments and HPGe detectors based on simulations. The red, blue and green lines are respectively strong, intermediate and weak oscillation amplitude.

## 2. Experiment

The experiment was performed at ALTO, Orsay, France. To perform particle-$\gamma$ correlations, the experimental setup comprised of the 13 ORGAM (ORsay GAMma-ray) detectors, surrounding the Orsay plunger (OUPS) [11]. An 8–fold segmented plastic scintillator particle detector was located 61 mm downstream from the target. The breaking of the azimuthal symmetry, due to the use of the 8-fold segmentation, considerably improved the experimental sensitivity. A beam of $^{24}$Mg ions at an energy of 120 MeV was provided by the TANDEM accelerator with about 0.3 pnA intensity and was excited on a stretched foil of $^{93}$Nb with a thickness of 2.4 mg/cm$^2$. The movable, stretched reset foil, positioned downstream of the target, was a 1.7 mg/cm$^2$ thick $^{197}$Au foil. The experiment was performed for a total of four days.

## 3. Analysis

A triggerless data-acquisition system was used for collecting the data. Coincidence events, corresponding to a $\gamma$-ray detected in the ORGAM array and a beam particle detected in the plastic scintillator, were constructed using an universal time stamp and imposing a narrow time gate. The $\gamma$-ray background, originating from radioactive decay and/or reactions, was suppressed by the particle-$\gamma$ coincidence condition.

The data were analyzed by ordering of the amplitude of the oscillations in the time-dependent angular correlations, and whether the $\gamma$-ray intensity should initially increase ($W^{\uparrow}$) or decrease ($W^{\downarrow}$) with the time, as shown in calculated examples in Fig. 4. An $R(t)$ function was constructed for the gamma-ray intensity of the $2^+ \rightarrow 0^+$ transition in $^{24}$Mg formed using a geometric ratio.

$$R(T) = \left( \prod_{i=1}^{n} \frac{W_i^{\uparrow}(T)}{W_i^{\downarrow}(T)} \right)^{1/n} \tag{1}$$

Damped oscillations in $R(t)$ were observed as expected, and the $g$ factor was determined precisely, in agreement with the previous value at the level of one standard deviation.

## 4. Discussion

The new TDRIV method proposed for applications to radioactive beams has been tested using a stable beam giving a precise result. This is the only $g$-factor measurement with the precision needed to test recent shell model predictions [12] that the $g(2^+)$ values should exceed 0.5 by up to about 10% in $N = Z$ nuclei in the *sd* shell. The method is ready for applications to radioactive ion beams.

## Acknowledgments

This work has been supported by the 2214 and 2224-B - TUBITAK (Scientific and Technological Research Council of Turkey) Programmes, the European Community FP7–Capacities–Integrated Infrastructure Initiative–contract ENSAR no 262010, the Australian Research Council grant No. DP0773273, P2I0 (Physique des 2 Infinis et des Origines) laboratory of excellence, and the Scientific Research Projects Coordination Unit of Istanbul University under Project No. 46811.

## References

[1] Benczer-Koller N, Hass M, and Sak J 1980 *Annu. Rev. Nucl. Part. Sci.* **30** 53
[2] Horstman R F *et al* 1975 *Nucl. Phys.* A **248** 291
[3] Goldring G 1982 *in Heavy Ion Collisions* (Vol 3) edited by R Bock (Amsterdam: North-Holland) p 483
[4] Stone N J *et al* 2005 *Phys. Rev. Lett.* **94** 192501
[5] Stuchbery A E and Stone N J 2007 *Phys. Rev.* C **76** 034307
[6] Stuchbery A E *et al* 2013 *Phys. Rev.* C **88** 051304(R)
[7] Kumbartzki G *et al* 2004 *Phys. Lett.* B **591** 213
[8] Davies A D *et al* 2006 *Phys. Rev. Lett.* **96** 112503
[9] Fiori E, Georgiev G, Stuchbery A E *et al* 2012 *Phys. Rev.* C **85** 034334
[10] Stuchbery A E, Mantica P F and Wilson A N 2005 *Phys. Rev.* C **71** 047302
[11] Ljungvall J, Georgiev G *et al* 2012 *Nucl. Instr. Meth.* A **679** 61
[12] Richter W A, Mkhize S, and Brown B A 2008 *Phys. Rev.* C **78** 064302

# Determination of naturally occurring radionuclides in soil samples of Ayrancı, Turkey

**Osman Agar[1,2,*], Canel Eke[1,3], Ismail Boztosun[1,4], M. Emin Korkmaz[2]**

[1]Akdeniz University Nuclear Research and Application Center, Antalya, Turkey
[2]Karamanoglu Mehmetbey University, Physics Department, Karaman, Turkey
[3]Akdeniz University, Faculty of Education, Department of Secondary Science and Mathematics Education, Division of Physics Education, Antalya, Turkey
[4]Akdeniz University, Faculty of Science, Physics Department, Antalya, Turkey

*E-mail: osmanagar@kmu.edu.tr

**Abstract.** The specific activity, radiation hazard index and the annual effective dose of the naturally occurring radioactive elements ($^{238}U$, $^{232}Th$ and $^{40}K$) were determined in soil samples collected from 12 different locations in Ayrancı region by using a NaI(Tl) gamma-ray spectrometer. The measured activity concentrations of the natural radionuclides in studied soil samples were compared with the corresponding results of different countries and the internationally reported values. From the analysis, it is found that these materials may be safely used as construction materials and do not pose significant radiation hazards.

## 1. Introduction

Terrestrial gamma rays and cosmic rays are the two prominent natural sources of external radiation. The first source mainly originates from the primordial radioactive nuclides, originated in the early stages of the formation of the solar system [1]. Natural terrestrial radioactivity arises from the primordial radionuclides, such as, $^{40}K$, and the radionuclides from the $^{238}U$ and $^{232}Th$ and their progenies [2,3]. Terrestrial background radiation represents the main external source of irradiation of the human body. Human beings are also exposed naturally from sources outside their bodies at any time.

The main objective of this study is to determine the activity concentration of natural radionuclides in the soil samples collected from Ayrancı, Turkey and to evaluate the radiological hazard of the natural radioactivity in samples. In the next section, we present the materials and method of the present study. In Section III, we show the result obtained by a NaI(Tl) scintillation detector and Section IV is devoted to our summary and discussion.

## 2. Material and Method

### *2.1 Study Area*

Ayrancı is a district located in south of central Anatolian of Turkey. This study area is between latitudes 37°20' N and 37°22' N and longitudes 33°40' E and 33°42' E. The total urban area of Ayrancı is 2577 km$^2$ and the urban population is approximately 10 600.

### *2.2 Sample collection and preparation*

Soil samples were collected from 12 different points in Ayrancı, Turkey. About half a kilogram of each sample was taken at a depth of 15 – 20 cm and filled into plastic bags and tagged. Each soil sample was air-dried (about 10 d) naturally in the laboratory. Thereafter, these samples were sieved through 2-mm mesh and were transferred to 100 ml beakers. The beakers with the soil samples were stored for more than 30 days to attain secular equilibrium between radium and radon before counting.

*2.3 Experimental techniques*

For measurements of the gamma activity concentrations of naturally occurring radionuclides namely $^{238}U$, $^{232}Th$ and $^{40}K$ in soil samples, a 3" x 3" NaI(Tl) gamma-ray spectrometric system was employed. The detector was surrounded by a special cylindrical lead shield of about 10 cm and a height of 38 cm thickness to reduce the gamma-ray background. The system was calibrated for the gamma energy range 186 keV to 3000 keV. The IAEA gamma-ray spectrometry standard reference sources of known radionuclides with $^{238}U$ (RGU-1), $^{232}Th$ (RGTh-1) and $^{40}K$ (RGK-1) were used to determine the efficiency of the detector for various energies in the prescribed geometry [4]. The soil samples were placed on top of the detector and each sample was counted for 10 000 s. The net area count was calculated by using ScintiVision program.

## 3. Results

Table 1 shows activity concentration of naturally occurring radionuclides in soil samples collected from 12 different points in Ayrancı, Turkey. In soil samples, the average concentrations of $^{238}U$, $^{232}Th$ and $^{40}K$ were found to be 25±14, 50±27 and 228±94 Bq/kg, respectively. The data obtained for the above–cited radionuclides in this study are compared with concentrations from different regions of the world and Turkey in table 2.

Radiation hazard index analysis such absorbed dose rate, radium equivalent activity, annual effective dose and external hazard index is performed to arrive at a better and safer conclusion and calculated according to the equations given in our previous study [5,6].

As can be seen in Fig. 1, the total absorbed dose in the studied area changed from 24.92 to 91.99 nGy $h^{-1}$ with an average value of 52.56 nGy.$h^{-1}$. Whereas, the calculated values for annual effective dose varied from 35.56 to 112.81 $\mu$Sv.$y^{-1}$ with an average value of 51.82 $\mu$Sv.$y^{-1}$. These values are within the permissible dose equivalent limit (460 $\mu$Sv.$y^{-1}$) [2].

Radium equivalent activity in soil samples varied from 53.94 to 204.54 Bq $kg^{-1}$ with mean value of 114.95 Bq $kg^{-1}$ (Fig. 1). All soil samples have Raeq which is commonly used to measure radiation hazards of the cited radionuclides in single quantity lower than the (370 Bq $kg^{-1}$) limit proposed by OECD [7].

**Table 1.** Radioactivity concentrations (in Bq·$kg^{-1}$) in soil samples.

| Sample ID | $^{238}U$ | $^{232}Th$ | $^{40}K$ |
|---|---|---|---|
| S1 | 31.0±16.3 | 107.7±26.6 | 256.5±91.6 |
| S2 | 23.8±14.4 | 60.2±27.5 | 220.4±100.8 |
| S3 | 31.7±12.6 | 21.9±30.3 | 320.7±105.2 |
| S4 | 11.5±13.8 | 52.8±24.2 | 311.1±77.1 |
| S5 | 28.0±15.0 | 49.6±26.2 | 278.3±93.6 |
| S6 | 42.5±14.1 | 40.9±32.6 | 225.1±97.7 |
| S7 | 12.6±11.6 | 51.7±29.8 | 166.3±87.9 |
| S8 | 10.6±13.3 | 32.6±29.4 | 157.8±89.7 |
| S9 | 32.6±16.2 | 75.5±25.9 | 249.6±92.8 |
| S10 | 28.5±16.5 | 54.7±29.7 | 229.5±94.9 |
| S11 | 12.0±14.8 | 21.9±20.2 | 139.1±103.4 |
| S12 | 40.9±16.2 | 34.7±28.1 | 182.0±96.4 |
| Avarage | 25±14 | 50±27 | 228±94 |
| World | 30 | 35 | 400 |

Journal of Physics: Conference Series **590** (2015) 012042 doi:10.1088/1742-6596/590/1/012042

**Table 2**. Activity concentrations of the natural radionuclides in soil samples (in Bq·kg$^{-1}$) from different parts of the world and Turkey

| Region | $^{238}U$ | $^{232}Th$ | $^{40}K$ | References |
|---|---|---|---|---|
| Xiazhuang (China) | 40.2–442 | 32.6–88.1 | 442–913 | [8] |
| Russaifa (Jordan) | 5–1201 | 2–31 | 19–288 | [9] |
| Ibadan (Nigeria) | 10.2–40.7 | 13.3–29.7 | 96.1–336.5 | [10] |
| Buyukeceli | 9.8–258.6 | 11.7–87.6 | 174.8–1949.5 | [11] |
| Çanakkale | 21.39–253.1 | 38.84–160.9 | 583.1–3307 | [12] |
| Karaman | 36.1–106.1 | 12.6–39.5 | 139.9–566.3 | [5] |
| Ayrancı | 11.5–42.5 | 21.9–107.7 | 139.1 – 320.7 | This work |

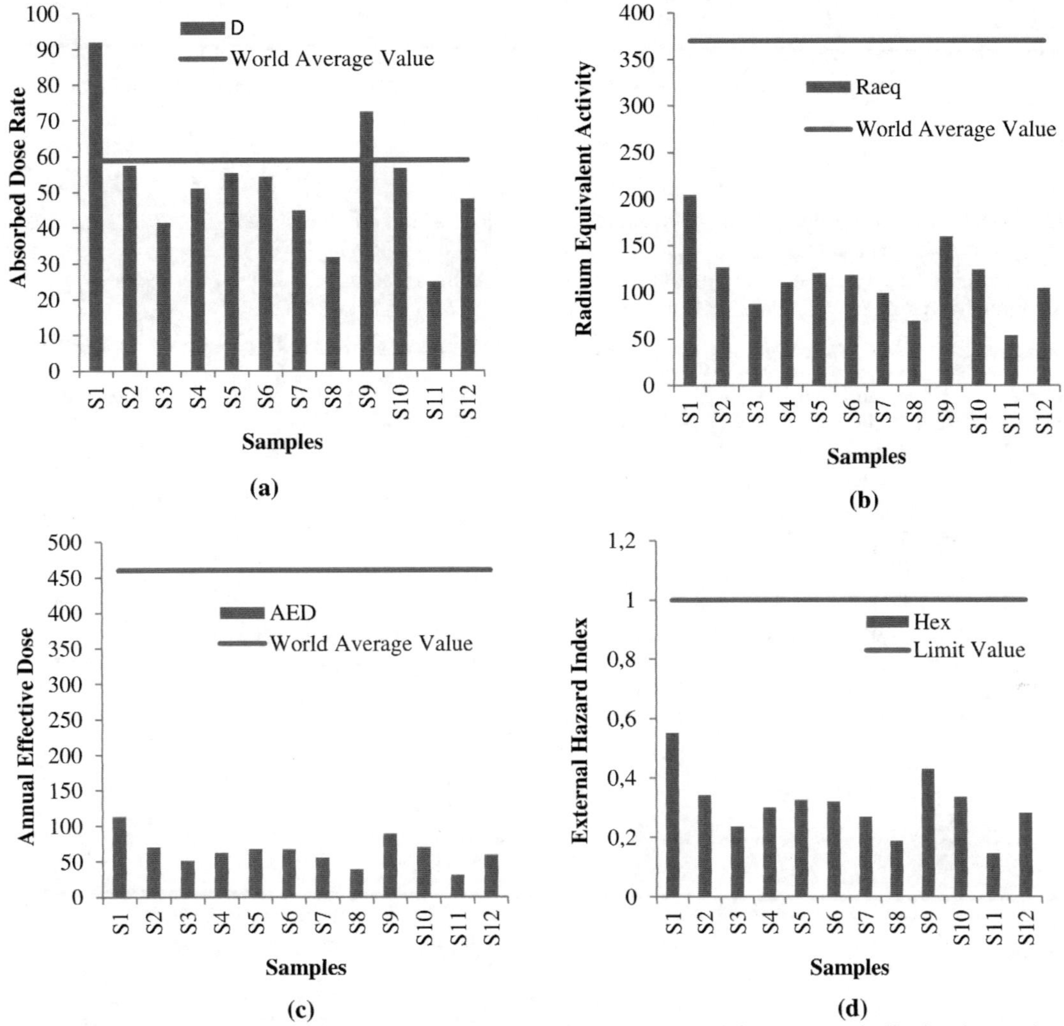


**Fig. 1.** Calculated values of (a) absorbed dose rate, (b) radium equivalent activity, (c) annual effective dose and (d) external hazard index.

The obtained values of external hazard index (Hex) range from 0.39 to 0.64. For the materials to be used in construction, this index should be less than unity. The external radiation hazard in this studied area is less than unity [13].

## 4. Conclusions

In present study, levels of natural radioactivity in soil samples collected from Ayrancı, Turkey is determined using a NaI(Tl) scintillation detector. The activity concentrations of $^{238}U$ and $^{40}K$ are lower than the world wide average, but the activity of $^{232}Th$ is higher than the world wide average. The obtained values of known radiation hazard indices in study area also are calculated and lower than recommended values. The results reveal that the average dose rates do not exceed the average world recommended values and do not pose a significant health hazard. The data presented can be used for comparison in future investigations and may be useful for preparing a radiological map of the area. The results may also be used as a reference data for monitoring possible radioactivity pollutions in future.

## Acknowledgment

This study is supported by Akdeniz University (Project Number:2014.03.0121.006) and Karamanoğlu Mehmetbey University Scientific Research Project (14–M–14).

## References

[1] Rahman S U, Matiullah, Malik F, Rafique M, Anwar J, Ziafat M, Jabbar A 2011 *J. Radioanal. Nucl. Chem.* **287** 647–655
[2] United Nations Scientific Committee on the Effect of Atomic Radiation 2000, United Nations.
[3] Bolca M, Saç M M, Çokuysal B, Karal T, Ekdal E 2007 *Rad. Meas.* **42** 263–270
[4] Ravisankar R, Chandrasekaran A, Vijayagopal P, Venkatraman B, Senthilkumar G, Eswaran P, Rajalakshmi A, 2012 *Radiat. Phy. Chem.* **81** 1789–1795
[5] Agar O, Boztosun I, Korkmaz M E, Özmen S F 2014 *Radiat. Prot. Dosim.* doi:10.1093/rpd/ncu027
[6] Eke C, Boztosun I 2014 *J. Radioanal. Nucl. Chem.* **301** 103–108
[7] OECD 1979 Exposure to radiation from the natural radioactivity in building materials. Report by a Group of Experts of the OECD Nuclear Energy Agency.
[8] Yang Y, Wu X, Jiang Z, Wang W, Lu J, Lin J, Wang L W, Hsia Y 2005 *Appl. Radiat. Isot.* **63** 255–259
[9] Hamideen M S, Sharaf J 2012 *Radiat. Phy. Chem.* **81** 1559–1562
[10] Jibiri N N, Bankole OS 2006 *Radiat. Prot. Dosim.* **118(4)** 453–458
[11] Özmen S F, Boztosun I, Yavuz M, Tunc M R 2014 Radiat. Prot. Dosim. **158 (4)** 461–465
[12] Kam E, Bozkurt A, Ilgar R 2010 Environ. Monit. Assess. **168** 685–690
[13] Al-Sulaiti H, Alkhomashi N, Al-Dahan N, Al-Dosari M, Bradley D A, Bukhari S, Matthews M, Regan P H, Santawamaitre T, 2011 Nucl. Instr. Meth. Phys. Res. A 652 915–919

# Comparison of Different INC Physical Models of MCNPX to Compute Spallation Neutronics of LBE Target

**Seyed Amir Hossein Feghhi [a], Zohreh Gholamzadeh* [b], Claudio Tenreiro[c] and Zahra Alipoor[d]**

[a]Department of Radiation Application, Shahid Beheshti University, G.C, Tehran, Iran
[b]Reactor Research School, Nuclear Science and Technology Research Institute, Tehran, Iran
[c]Department of Physics, Talca University, Talca, Chile
[d]Department of Physics, Zanjan University, Iran.

Author for correspondence (cadmium_109@yahoo.com)

**Abstract**. Spallation particles can utilize in different fields such as neutron scattering studies, external source for burning spent fuel as well as running subcritical reactors. Different computational particle transport codes are widely used to model spallation process into the heavy targets. Among these codes, MCNPX 2.6.0 comprises various intra nuclear cascade models for spallation calculations. Impact of different intra nuclear cascade models on calculation of neutronic parameters of LBE target has been evaluated in this work. Escaped neutron yield, energy deposition and residual nuclei production in the spallation target has been calculated using the physical models. A comparison between the computational and experimental has been carried out to validate the computational data. The simulation data showed there is a good conformity between the obtained data from Bertini/Drenser and Isabel/Drenser. The data achieved by Bertini/Abla and Isabel/Abla models are close to each other for the studied parameters as well. Among the studied models, CEM showed more discrepancies with experimental and other computational data. According to the obtained data, INCL4/Drenser, INCL4/Abla and Isabel/Drenser models can meet more agreements with experimental data.

## 1. Introduction

The spallation target is ideally conceived to be high atomic mass material such as W, Ta, Hg, Pb, LBE (Lead Bismuth Eutectic) and U which are useful to yield high spallation neutron efficiencies and it is well known high density liquid metals like lead and LBE fit the requirement extremely well [1-3]. MCNPX 2.6.0 (Monte Carlo N-Particle transport code) is a general purpose Monte Carlo radiation transport code which involves different Intra-Nuclear Cascade models of Bertini, Isabel, INCL4 (Liège Intra-Nuclear Cascade) and CEM (Cascade-exciton model) for intera nuclear cascade (INC) calculations along with Drenser (associated with RAL (Rutherford Appleton Laboratory) or ORNL (Oak Ridge National Laboratory) fission models) and Abla for de-excitation [4]. Intra-nuclear cascade (INC) models are widely used to describe reactions between nucleon, pion or light ion projectiles with target nuclei. After the cascade an excited remnant nucleus is de-excited using evaporation/fission process. Some extended INC models have been reported in Table 1

**Table 1.**
Intranuclear cascade (INC) models' characteristic [5-8]

| Physics process | Bertini | Isabel | CEM | INCL4 |
|---|---|---|---|---|
| **INC model** | Bertuini INC | Isabel INC | Improved Dubna INC | Cugnon INC |
| **Nuclear density level** | 3 | 16 | 7 | A nucleons - Positions shot in a Saxon-Woods density |
| **Lower energy limit** | 20-150 MeV | 20-150 MeV | ~100 MeV | 200 MeV |
| **Upper energy limit** | 3.5 GeV for Nucl-Nucl 2.5 GeV for Pion-Nucl | 1 GeV | 5 GeV | 2 GeV |
| **The stopping criterium** | Energy cut off | Energy cut off | Optical absorptive potential | Time cut off |

In an accelerator driven system (ADS), high energy particles – usually protons above a few hundreds of MeV – are used to induce spallation reactions in the target. These kinds of reactions have been thoroughly investigated experimentally and theoretically using energetic proton beams by the researchers. Hence, ability study of different INC models introduced in MCNPX 2.6.0 for spallation neutronic calculations has been proposed in this work.

## 2. Material and methods

The spallation process has been modeled using MCNPX 2.6.0 code. A proton beam of 0.4, 0.6, 0.8 and 1 GeV energy has been used respectively to induce spallation reactions in LBE cylindrical target of 15 cm diameter and 60 cm thickness. A proton beam with a Gaussian intensity distribution of 0.2 cm FWHM has been used and the spallation neutron yield has been calculated using different physical models of Bertini/Drenser, Isabel/Drenser, Bertini/Abla, Isabel/Abla, INCL4/Drenser, INCL4/Abla and CEM. The models' effects on the energy deposition in the spallation target due to 1 GeV proton beam have been studied. F6 tally is used to calculate heat deposition into the spallation target. Residual nuclei production calculations have been carried out using the different physical models for the target bombarded by 1 GeV proton beam. Histp card has been used to calculate impurity production in spallation target. Residual masses have been transferred to pico gram (pg) per mA current scale by multiplying the NHTAPE data with A (mass number)×1.036402E-08 (g). To validate the calculations, the neutron yield achieved by the computational code using Bertini/Drenser, Isabel/Drenser, Bertini/Abla, Isabel/Abla, CEM, INCL4/Drenser and INCL4/Abla has been compared to experimental AECL (Atomic Energy of Canada Limited) where a Pb target was irradiated using a 960 MeV proton beam. In addition, the theoretical neutron yields were compared to the experimental data of BNL (Brookhaven National Laboratory) where 0.8, 1.0, 1.2 and 1.4 GeV protons were used. Manipulation of different INC models for residual nuclei calculations has been studied by reviewing of other carried out investigations extracted from the literatures to evaluate the most efficient INC model.

## 3. Result and discussion

### *3.1. Calculation of escaped neutron yield from the spallation target.*

Figure 1 shows neutron yield calculation of the LBE target using different physical models indicated the obtained data using all the investigated physical models fit well each other at 0.4 GeV. Some relative discrepancies are observed at higher energies so that the higher the energy of the proton concludes in the higher the discrepancies between the results of the different models. At an energy of 0.6 GeV, the minimum value belongs to INCL4/Drenser and the maximum value belongs to both

Isabel/Abla and CEM with about 12% relative discrepancy between the maximum and minimum values. Calculations based on the CEM model result in the largest neutron yields for the entire energy span. Isabel/Abla, Bertini/Abla and CEM data have relatively closer agreements than the others at the total investigated energy span whereas the obtained neutron yields were the largest than the others as well. Bertini/Drenser and Isabel/Drenser data resulted in good agreements as well. The lowest values of neutron yield belong to INCL4/Drenser. In case of LBE target some disagreements has been observed between INCL4/Drenser and INCL4/Abla with maximum discrepancy of 5.8% at 1 GeV. (Fig. 1).

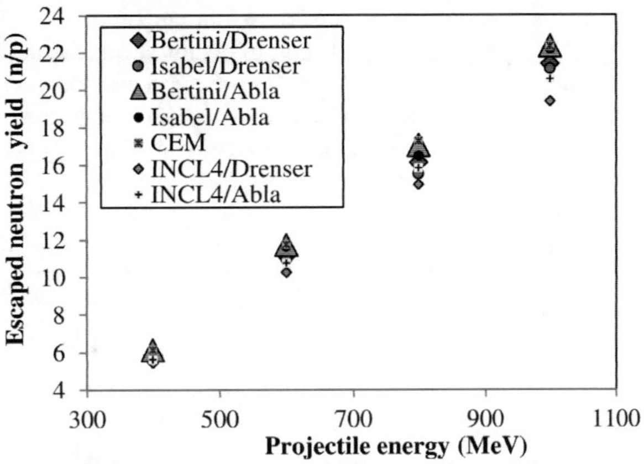


**Figure 1.** Comparison of different INC physical models in escaped neutron yield calculations of LBE target, the target diameter: 15 cm, height: 60 cm.

*3.2. Calculation of deposited energy in the spallation target.*

The energy deposition data achieved by 1 GeV proton irradiation of LBE target showed all the used models are in good conformity with each other in all length of the target (Fig.2).

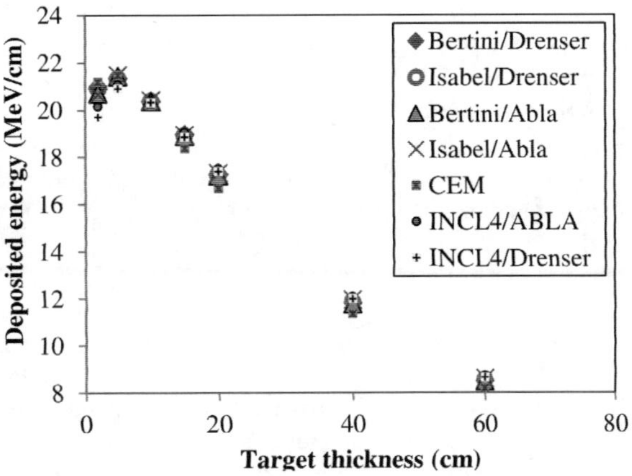


**Figure 2.** Comparison of different INC physical models for calculations of deposited energy in LBE target, the target diameter: 15 cm, height: 60 cm.

*3.3. Calculation of residual nuclei production in the spallation target.*

Calculations of residual nuclei production in a LBE target showed that the data obtained by Bertini/Drenser and Isabel/Drenser are in very good agreement as well as Isabel/Abla and Bertini/Abla result in very well-matched data with each other. The number of residual nuclei calculated using CEM model fits results of Isabel/Abla and Bertini/Abla data very well. These two groups' models (Bertini-Isabel/Drenser, Bertini-Isabel/Abla) have large discrepancies in evaporation as well as fragmentation region and small discrepancies in hot fission region. INCL4/Drenser show noticeably relative discrepancies in both hot fission and evaporation regions compared to the other data; in hot fission the model conclude in underestimated data than all of the other used models and in evaporation region the INCL4/Drenser data are overestimated than the others except Bertini/Drenser and Isabel/Drenser group. In case of INCL4/Abla this behavior is inversed so that in hot fission region the data are overestimated than all the investigated physical models and in evaporation region the obtained data are highly underestimated than all the studied physical models. In fragmentation region, INCL4/Drenser and INCL4/Abla data are close to each other and have fairly good agreements with CEM and Bertini/Abla-Isabel/Abla group. All the obtained data of different physical models have fairly good overlap in spallation region especially above A=175 (Fig.3).

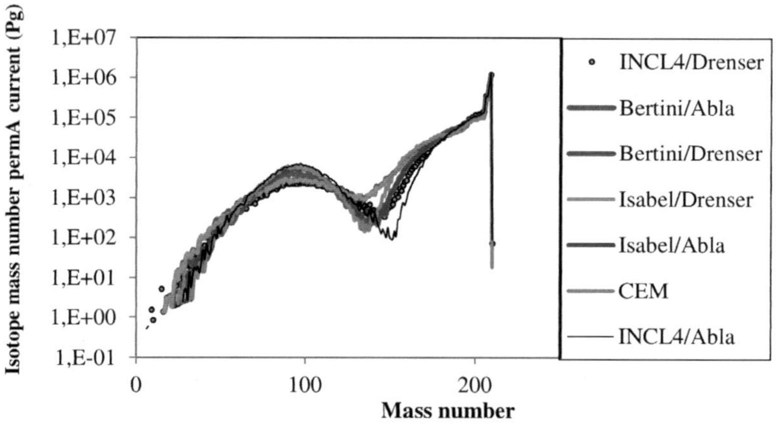


**Figure 3.** Comparison of different INC physical models for calculations of residual nuclei production in LBE target, diameter: 15 cm, height: 60 cm.

*3.4. Comparison of experimental and computational neutron yield.*

To evaluate confidence degree of computational and experimental data, a comparison between experimental neutron yield reported by AECL, BNL and theoretical data obtained by MCNPX 2.6.0 code calculations has been carried out for a Pb target [9-10]. Whereas experimental data on LEB was not available and Pb shows close neutronic behavior to LEB target [1], a comparison between the experimental data and the theoretical neutron yields obtained by the physical models used for a modeled Pb target with the same experimental dimension (10.2×61) has been done. The results showed that the INCL4/Drenser model can obtain more confidence data with experimental data for Pb target. Isabel/Drenser and INCL4/Abla models are also good in describing experimental values properly. The comparisons showed that there is a huge discrepancy between CEM INC model results and experimental data. The average relative discrepancies between experimental and theoretical data for Bertini/Drenser, Isabel/Drenser, Bertini/Abla, Isabel/Abla, CEM, INCL4/Drenser and INCL4/Abla models are 7.67, 5.68, 11.25, 9.13, 11.94, 3.74 and 5.60% respectively, discrepancy=$(\frac{experimental-theoretical)}{experimental})$. As the results show, the code data overestimated experimental data. According to the ACEL report, the experimental results obtained by the foil activation method are believed to be accurate to within ±5%. The computational obtained data in this work had an average uncertainty <0.7%.

### *3.5. Review of some carried out spallation simulations using different INC models.*

Gorse-Pomonti et al. compared residual nuclei production in LBE target (R=10 cm, L=100 cm) irradiated by 1 GeV proton beam of 1 mA current [5]. Their data show there are good agreements between INCL4/Abla and Bertini/Drenser data except within the evaporation region where the Bertini/Drenser data is noticeably larger than the ones from INCL4/Abla. David et al. compared INCL4/Drenser, INCL4/Abla and Isabel/Abla models with experimental data for residual nuclei formation in Pb spallation target bombarded with 1 GeV and 0.8 GeV proton beams respectively. Their results showed that the data obtained using INCL4/Abla and Isabel/Abla models are in reasonable agreement with the experimental data while INCL4/Drenser model did not result in acceptable data especially in the hot fission region with about maximum 80% relative discrepancy. According to David et al. results, it can be seen the evaporation part of the charge distributions are not so well described by INCL4 and Isabel, but Isabel is better [11]. They reported this bad behavior could be due to a lack of excitation energy, since, on the one hand, the number of evaporated particles depends on the excitation energy, and, on the other hand, Isabel left the nucleus with higher excitation energy than INCL4. S. Leray has reported a comparison between Isabel/Drenser, INCL/Drenser and experimental data of residual nuclei formation via 1 GeV proton spallation reactions in Pb target [12]. Results of the investigations by Leray show the two INC models are fitting each other well except within the evaporation region and it is seen there is noticeable discrepancy between them and experimental data especially in the evaporation region and the hot fission region peak. In another report presented by Leray, the comparison between experimental and INCL4/Abla data achieved by irradiation of Pb target using 1 GeV protons shows the used model has acceptable agreements in the hot fission region while some noticeable disagreements has been observed in the spallation region [13].

## 4. Conclusion

In most of the carried out investigations of this work, CEM model presented some data with large relative discrepancy compared to the experimental data and the results of the other considered INC models. Isabel/Drenser and Bertini/Drenser data approximately overlap each other in neutron yield, energy deposition and residual nuclei calculations. Isabel/Abla and Bertini/Abla present roughly well-matched data in the case of the mentioned parameters. The lowest discrepancy with experimental neutron yield can be obtainable using INCL4/Drenser and Isabel/Drenser models. According to the other carried out studies these two models are more successive for residual nuclei and energy deposition calculations as well. However, more complementary benchmarks on wide range of the spallation targets with different material and dimension as well as wide range of projectile energy for spallation process induction in the target can validate the best INC model for such calculations.

## References

[ 1] Feghhi S A H, Gholamzadeh Z, and Tenreiro C 2014 *J of Theoreti.and Appl. Phys.* **8** 1
[ 2] Armbruster P, Benlliure J 2002 Basic Nuclear Data at High and Intermediate Energy for Accelerator-Driven Systems. *NUPECC Report on Nuclear Science in Europe: Impact Applications Interactions, June*
[ 3] Satyamurthy P, Gantayet L M, and Ray A K 2007 Pramana J of phys. 68 343
[ 4] Pelowitz D B, ed., MCNPX User's Manual, Version 2.6.0, Los Alamos National Laboratory report LA-CP-07-1473 (April 2008).
[ 5] Prael R E, Lichtenstein H, User Guide to LCS: The LAHET Code System. LA-UR-89-3014, (1989).
[ 6] Broeders C H M, and Yu. Konobeyev A 2005 *Nucl Inst. and Meth. In Phys Res. B* **234** 387

[ 7] David J C, Boudard A, Fernández-Dominguez B, Leray S and Volant C 2003 Fission within the spallation process. Influence of interanuclear cascade and evaporation modelizations on the fission fragment production. *XVIth International Workshop on Physics of Nuclear Fission, Nov. 7, Obninsk, Russie.*

[ 8] Mashnik S G, and Sierk A J 2001 Recent Developments of the Cascade-Exciton Model of Nuclear Reactions. *International Conference on Nuclear Data for Science and Technology, Oct 7-12, Tsukuba, Japan*

[ 9] Bartholomew G A and Tunnicliffe P R 1996 The AECL study for an intense neutron generator (technical details). *AECL-2600*

[ 10] Kumar V, Kumawat H, Goel U and Barashenkov V S 2006 *Pramana J. of phys.* **60**, 469

[ 11] Gorse-Pomonti D and Russier V 2007 J. of Non-Cryst. Sol. **353** 3600

[ 12] Leray S 2001 Nuclear reaction at high energy. *Workshop on nuclear science and technology, Sep 10-21; France,* LNS0212005.

[ 13] Leray S 2003 HINDAS High-Energy Programme: Main conclusions and implications for spallation neutron sources. *Darmstadt conference, Sep.1-5, Germany.*

# Dependence of activity concentration of natural and artificial radionuclides on depth in soil samples from Antalya in Turkey

**Canel Eke[1,2,*], Osman Agar[1,3] and Ismail Boztosun[1,4]**

[1]Akdeniz Üniversitesi, Nükleer Bilimler Uygulama ve Araştırma Merkezi, 07058, Antalya, Turkey
[2]Akdeniz University, Faculty of Education, Department of Secondary Science and Mathematics Education, Division of Physics Education, 07058, Antalya, Turkey
[3]Karamanoğlu Mehmetbey University, Department of Physics, 70100, Karaman, Turkey
[4]Akdeniz University, Faculty of Science, Department of Physics, 07058, Antalya, Turkey

*E-mail: ceke@akdeniz.edu.tr

**Abstract**. In this study, we attempt to determine the dependence of activity concentration of natural ($^{238}U$, $^{232}Th$, $^{40}K$) and artificial ($^{137}Cs$) radionuclides on depth in soil samples from Antalya in Turkey. Soil samples were collected at different depths (0-10 cm, 10-30 cm, 30-50 cm, 50-70 cm, 70-100 cm and 100-200 cm). Each soil samples were counted by using a high purity Germanium detector (HpGe). For each soil sample, activity concentration, absorbed dose rate, radium equivalent activity, external and internal hazard index were calculated and compared with the published results.

## 1. Introduction

$^{238}U$, $^{232}Th$, $^{40}K$ are natural radioactive sources and $^{137}Cs$, $^{131}I$, $^{90}Sr$ and $^{95}Zr$ are artificial radioactive sources in environment. $^{238}U$, $^{232}Th$, their decay products and $^{40}K$ are the largest contributor to the external dose of the world population [1]. Besides naturally occurring radionuclides, artificial or manmade radioactivity on earth surface is due to anthropogenic activities such as the test of various nuclear weapons, explosions, nuclear power plant leakage and nuclear facility accidents. The isotope $^{137}Cs$ is the most important constituent detected by its gamma radiation in the environment among the class of artificial isotopes released as fission products, such as $^{90}Sr$ and $^{95}Zr$. The importance of this fission product is attributed to its long half-life of 30.17 years, and a high energetic gamma emitter (662 keV) that contribute to people's dose due to the external radiation exposure [2,3].

A wide range of research has been performed to determine the dependence of activity concentration of natural and artificial radionuclides on depth in soil samples [4,5,6,7].

In this paper, our purpose is to determine the dependence of activity concentration of $^{238}U$, $^{232}Th$, $^{40}K$ and $^{137}Cs$ on the depth in soil samples and to calculate absorbed dose rate, annual effective dose equivalent, radium equivalent activity, internal and external hazard index.

## 2. Materials and Methodology

### *2.1. Sample Collection and Preparation*

Six soil samples were collected about 3-4 kg at different depths (0-10 cm, 10-30 cm, 30-50 cm, 50-70 cm and 100-200 cm) from Antalya in Turkey. Each soil samples were sieved 2 mm mesh and dried $130^0$C for 24 h in a oven. Samples were put 1000 ml Marinelli beaker and weighed. They were kept and sealed for 40 days to come to secular radioactive equilibrium between $^{226}$Ra, $^{232}$Th and their decay products.

### *2.2. Gamma-ray Spectrometry Measurement System*

Each soil samples were counted 50000s by using High Purity Germanium Detector (HpGe) (AMATEK-ORTEC (GEM40P4-83). The energy resolution of the detector is 1.85 keV at gamma-ray energy 1332 keV of $^{60}$Co and its relative efficiency is 40% and energy resolution is 768 eV at gamma energy 122 keV of $^{57}$Co. The details can be found in Refs. [8,9,10].

### *2.3. Determination of Activity Concentration and Dose Assessments*

To determine dependence of activity concentration A (Bq/kg) of natural and artificial radionuclides on depth in soil samples from Antalya in Turkey, we used following well known equation [11]:

$$A = \frac{N}{m.t.\varepsilon_0 . P_\gamma} \qquad (1)$$

Where N is number of count, m (kg) is mass of the sample, t (sec) is counting time, $\varepsilon_0$ is detector efficiency and $P_\gamma$ is gamma-ray emission probability.

To calculate absorbed dose rate, annual effective dose equivalent, radium equivalent activity, external and internal hazard index were used by following equations from [12,13,14,15] as shown in table 1.

**Table 1.** Equations for dose assessments

| Dose assessments | Unit | Equation |
|---|---|---|
| Absorbed dose rate (D) | $nGyh^{-1}$ | $D = 0.462A_U + 0.604A_{Th} + 0.0417A_K$ |
| Annual effective dose equivalent (AEDE) | $\mu Svy^{-1}$ | $AEDE = Dx8760x0.7x0.2x10^{-3}$ |
| Radium equivalent activity ($Ra_{eq}$) | $Bqkg^{-1}$ | $Ra_{eq} = A_U + 1.43A_{Th} + 0.077A_K$ |
| External hazard index | | $H_{ex} = \frac{A_U}{370} + \frac{A_{Th}}{259} + \frac{A_K}{4810}$ |
| Internal hazard index | | $H_{in} = \frac{A_U}{185} + \frac{A_{Th}}{259} + \frac{A_K}{4810}$ |

Where $A_U$, $A_{TH}$ and $A_K$ are activity concentration of $^{238}$U, $^{232}$Th and $^{40}$K respectively. 0.462, 0.604 and 0.0417 are dose coefficients from Ref.[12]. Minimum detectable activities (MDA) are 1.43 $Bqkg^{-1}$, 1.07 $Bqkg^{-1}$, 2.19 $Bqkg^{-1}$, 8.82 $Bqkg^{-1}$and 0.41 $Bqkg^{-1}$for $^{214}$Pb (351 keV), $^{214}$Bi (609 keV), $^{228}$Ac (911 keV), $^{40}$K (1460 keV) and $^{137}$Cs (661 keV) respectively.

## 3. Results

Activity concentrations of natural and artificial radionuclides are given in table 2. Results of dose assessments are given in table 3.

Journal of Physics: Conference Series **590** (2015) 012044 doi:10.1088/1742-6596/590/1/012044

**Table 2.** Activity concentration of $^{238}U$, $^{232}Th$, $^{40}K$ and $^{137}Cs$

| Soil depth (cm) | $^{238}U$ (Bqkg$^{-1}$) | | $^{232}$Th (Bqkg$^{-1}$) | | |
|---|---|---|---|---|---|
| | $^{214}$Pb (Bqkg$^{-1}$) (351 keV) | $^{214}$Bi (Bqkg$^{-1}$) (609 keV) | $^{228}$Ac (Bqkg$^{-1}$) (911 keV) | $^{40}$K (Bqkg$^{-1}$) (1460 keV) | $^{137}$Cs (Bqkg$^{-1}$) (661 keV) |
| 0-10 | 8.49 ± 0.08 | 7.79 ± 0.06 | 9.42 ± 0.03 | 54.25 ± 3.28 | <MDA |
| 10-30 | 48.30 ± 0.64 | 43.18 ± 0.46 | 84.10 ± 0.43 | 443.54 ± 50.08 | <MDA |
| 30-50 | 36.34 ± 0.75 | 31.02 ± 0,52 | 58.23 ± 0.13 | 311.71 ± 16.24 | 0.52 |
| 50-70 | 7.82 ± 0.07 | 7.53 ± 0.06 | 10.20 ± 0.75 | 50.24 ± 8.44 | <MDA |
| 70-100 | 4.32 ± 0.04 | 3.96 ± 0.02 | 3.44 ± 0.60 | 17.67 ± 2.19 | <MDA |
| 100-200 | 4.21 ± 0.07 | 3.80 ± 0.05 | 2.90 ± 0.19 | 13.06 ± 0.65 | <MDA |
| Average value | 18.25 | 16.21 | 28.05 | 148.45 | |

MDA: Minimum Detectable Activity

According to Ref.[12] for $^{238}U$ activity concentration, world average value is 33 Bq/kg, for $^{232}Th$, world average value for activity concentration is 45 Bq/kg, for $^{40}K$, world average value for activity concentration is 420 Bq/kg, , the average values of activity concentration of $^{238}U$, $^{232}Th$ and $^{40}K$ in present study is less than the world average value. Activity concentration of $^{214}Pb$, $^{214}Bi$, $^{232}Th$ and $^{40}K$ values are higher than world average value at depth of 10-30 cm and activity concentration $^{214}Pb$ and $^{232}Th$ values are higher than world average value at depth of 50-70 cm. Most of the $^{137}Cs$ activity concentration values are lower than minimum detectable activity.

**Table 3.** Results of dose assessments

| Soil depth (cm) | D (nGyh$^{-1}$) | AEDE(µSvy$^{-1}$) | Ra$_{eq}$ (Bqkg$^{-1}$) | $H_{ex}$ | $H_{in}$ |
|---|---|---|---|---|---|
| 0-10 | 11.71±0.72 | 14.36±0.88 | 25.78±1.58 | 0.069±0.004 | 0.092±0.006 |
| 10-30 | 90.42±10.28 | 110.89±12.60 | 200.15±22.75 | 0.540±0.061 | 0.664±0.075 |
| 30-50 | 63.73±3.54 | 78.16±4.34 | 140.96±7.82 | 0.380±0.021 | 0.471±0.026 |
| 50-70 | 11.80±1.99 | 14.47±2.44 | 26.13±4.40 | 0.070±0.012 | 0.091±0.015 |
| 70-100 | 4.73±0.59 | 5.80±0.72 | 10.42±1.30 | 0.028±0.004 | 0.039±0.005 |
| 100-200 | 4.15±0.22 | 5.08±0.27 | 9.16±0.48 | 0.024±0.001 | 0.035±0.002 |
| Average value | 31.09 | 38.13 | 68.77 | 0.186 | 0.232 |

According to Ref.[12] for absorbed dose rate, world average value is 59 nGy/h, the average value in present study is less than the world average value, for annual effective dose equivalent world average value is 70 µSv/y, the average value in present study is less the than world average value. According to [12] for Radium equivalent activity world average value is 370 Bq/kg, the average value in present study is less than the world average value. $H_{ex}$ and $H_{in}$ values should be less than unity [12,13,16]. In present study, $H_{ex}$ and $H_{in}$ values are lower than unity. Absorbed dose rate and annual effective dose equivalent values are higher than world average value at depths of 10-30 cm and 30-50 cm. Radium equivalent activity values are lower than world average value.

## 4. Conclusion

Dependence of activity concentration of natural and artificial radionuclides on depth was studied in soil samples by using a high purity Germanium detector (HpGe) from Antalya in Turkey. For each soil samples activity concentration, absorbed dose rate, annual effective dose equivalent, radium equivalent activity, internal and external hazard index were calculated. Activity concentration of $^{214}$Pb, $^{214}$Bi, $^{232}$Th and $^{40}$K values are higher than world average value at depth of 10-30 cm and activity concentration $^{214}$Pb and $^{232}$Th values are higher than world average value at depth of 50-70 cm. Most of the $^{137}$Cs activity concentration values are lower than minimum detectable activity. Absorbed dose rate and annual effective dose equivalent values are higher than world average value at depths of 10-30 cm and 30-50 cm. Average values of activity concentration of $^{238}$U, $^{232}$Th and $^{40}$K lower than world average value. Average values of absorbed dose rate, annual effective dose equivalent and radium equivalent lower than world average value. Internal and external hazard index values are lower than unity. Activity concentration of natural and artificial radionuclides can investigate in detail in this area in future and also these results can be useful to compare with future studies.

## References

[1] Nada A, Abd-El Maksoud T M, Abu-Zeid Hosnia M, El-Nagar T, Awad S 2009 *Appl.Radiat.Isot* **67** 643-649
[2] Celik N, Cevik U, Celik A, Koz B 2009 *J. Hazard. Mater.* **162** 146–153
[3] Bara S V, Arora V, Chinnaesakki S, Sartandel S J, Bajwa B S, Tripathi R M, Puranik V D 2012 *J. Radioanal. Nucl. Chem.* **291** 769–776
[4] Shetty P K, Narayana Y, Rajashekara K M 2011 *J. Radioanal. Nucl. Chem.* **290** 159–163
[5] Mabit L, Dornhofer P, Martin P, Toloza A, Zupanc V 2012 Indian J. Pure Appl. Phys. **50** 45-48
[6] Saleh I H 2012 *Turkish J. Eng. Env. Sci.* **36** 236-248
[7] Belivermis M 2012 *Radiat.Prot.Dosim.* **151** **(3)** 511-521
[8] Eke C, Boztosun I 2014 *J. Radioanal. Nucl. Chem.* **301** **(1)** 103-108.
[9] Agar O, Boztosun I, Korkmaz, M E, Ozmen, S F 2014 *Radiat.Prot.Dosim.* DOI:10.1093/rpd/ncu027
[10] Ozmen S F, Boztosun I, Yavuz M, Tunc M R 2014 *Radiat.Prot.Dosim.* **158** **(4)** 461-465.
[11] Agbalagba E O, Avwiri G O, Chad-Umoreh Y E 2012 *J. Environ. Radioact.* **109** 64-70
[12] UNSCEAR 2000, United Nations Scientific Committee on the Effect of Atomic Radiation. Sources and Effects of Ionizing Radiation. Report to General Assembly, with Scientific Annexes, NewYork, United Nations
[13] Uosif M A M, El-Taher A, Abbady A G E 2008 *Radiat.Prot.Dosim.* **131** **(3)** 331-339.
[14] Ravisankar R, Vanasundari K, Chandrasekaran A, Rajalakshmi A, Suganya M, Vijayagopal P,Meenakshisundaram,V 2012 *Appl. Radiat. Isot,* **70** 699-704
[15] NEA-OECD 1979 Exposure to natural radioactivity in building materials. Report by NEA group of Experts of the nuclear energy agency, OECD, Paris, France
[16] Ramasamy V, Sundarrajan M, Paramasivam K, Meenakshisundaram V, Suresh G 2013 *Appl. Radiat. Isot,* **73** 21-31.

Journal of Physics: Conference Series **590** (2015) 012045 doi:10.1088/1742-6596/590/1/012045

# The influence of nuclear deformations on the exotic cluster decay half-lives

**A. Soylu†, O. Bayrak††, S. Evlice†**

†Department of Physics, Nigde University, 51240, Nigde, Turkey

††Department of Physics, Akdeniz University, 07058, Antalya, Turkey

**Abstract.** We systematically study the investigation of the influence of nuclear deformations of the cluster and daughter nuclei on the exotic cluster decay half-lives of heavy nuclei by the WKB method and the Bohr-Sommerfeld quantization condition. Even if the deformations of both cluster and daughter in the half-live values of cluster decays improve the results, considering the deformation of clusters is more efficient than the deformation of daughter for the heavy cluster decay half-live calculations. Moreover, taking into account of angle orientations of daughter and cluster provides a positive contributions to the results as well. The results would be useful for experimental researches in half-lives of exotic decays of some heavy nuclei and radium isotopes.

## 1. Introduction

After Rose and Jones determined the phenomenon of cluster-radioactivity experimentally [1], a new study field in nuclear physics was opened. Rose and Jones have found out that $^{223}$Ra emits a $^{14}$C cluster and their experimental results have been confirmed by Gales and co-workers [2] as well as by Price *et al.* [3]. On the theoretical side, Sandulescu *et al.* made the first theoretical description of the cluster decay [4]. The quantum mechanical tunneling calculations may be done either by fission theory to a larger mass asymmetry or by heavier cluster emitted particle from alpha decay theory. Numerous studies have been performed by using different methods and many papers have been published in order to be able to obtain the half-lives of these exotic decay processes [5–13]. The studies of radium isotopes in the field of exotic cluster decays are also important because their position in the trans-lead region in the nuclear chart and its immediate neighbors as one of the decay fragments comes into play when the cluster emission occurs. Therefore, the systematic study on cluster radioactivity from the $^{210-226}$Ra isotopes was done for the first time and $^{210-226}$Ra isotopes were all alpha instable thereby exhibiting alpha radioactivity [14–16]. In order to investigate the influence of the deformations of half lives of exotic decay clusters, some studies have also been done in the literature [17–19].

In this talk, we present the results of the investigation of the influence of nuclear deformations of the cluster and daughter nuclei on the exotic cluster decay half-lives of the nuclei having the mass numbers between 221 and 242 with Woods-Saxon square potential by using WKB and the Bohr-Sommerfeld quantization condition [18]. Furthermore, we have applied the model in order to get the half-lives of cluster decays of $^{210-226}$Ra which are recent interesting topic [20].

## 2. Model

When we take into account the quadrupole and hexadecapole deformations of both the cluster and the daughter nuclei having the deformations, the effective interaction potential between the daughter and the cluster nuclei can be given the following

$$V_{eff}(r,\theta_c,\theta_d) = V_N(r,\theta_c,\theta_d) + V_C(r,\theta_c,\theta_d) + V_L(r), \tag{1}$$

where $V_N(r,\theta_c,\theta_d)$ is the nuclear potential, $V_C(r,\theta_c,\theta_d)$ is the Coulomb potential and $V_L(r)$ is the centrifugal potential. And, $\theta_c$ and $\theta_d$ represent the orientation angles of cluster and daughter nuclei, respectively. The nuclear radius is deformed as

$$R(\theta_c,\theta_d) = r_0 A_c^{1/3}(1+\beta_2^c Y_{20}(\theta_c)+\beta_4^c Y_{40}(\theta_c)) + r_0 A_d^{1/3}(1+\beta_2^d Y_{20}(\theta_d)+\beta_4^d Y_{40}(\theta_d)), \tag{2}$$

where $A_c$, $\beta_2^c$, $\beta_4^c$ and $Y_{lm}(\theta_c)$ are the atomic mass, quadrupole and hexadecapole deformation parameters and the spherical harmonics of the cluster nuclei and $A_d$, $\beta_2^d$, $\beta_4^d$ and $Y_{lm}(\theta_d)$ are the atomic mass, quadrupole and hexadecapole deformation parameters and the spherical harmonics of the daughter, respectively. The repulsive Coulomb potential can be used in the deformed form [21], and the last term in Eq.(1) is Langer modified centrifugal barrier potential. As we have only focused on the favored cluster decay of the radioactive nuclei, $L = 0$ is used in the calculations. As we know the forms of Coulomb and centrifugal potential very well, the only one unknown term in the effective potential in Eq.(1) is the nuclear interaction potential. In order to find out the influence of deformations on the system effectively, we use the deformed Woods-Saxon square potential as follows

$$V_N(r,\theta_c,\theta_d) = -\lambda(\theta_c,\theta_d)\frac{V_0}{\left[1+exp(\frac{r-R(\theta_c,\theta_d)}{a})\right]^2}, \tag{3}$$

where $\lambda(\theta_c,\theta_d)$, $V_0$, R$(\theta_c,\theta_d)$ and $a$ are normalization parameter determined by Bohr-Sommerfeld quantization condition, depth of the nuclear potentials, nuclear radius and diffusion parameters, respectively. The detailed information about the calculations can be found in Ref. [18]. Turning points are obtained by the numerical solution of the equation $V_{eff}(r,\theta_c,\theta_d) = Q$. The cluster decay width is given by [22],

$$\Gamma = PF\frac{\hbar^2}{4\mu}S, \tag{4}$$

where P is the preformation factor, F is the normalization factor and S is the transition probability of cluster nuclei, respectively. The detailed information about the calculations can be found in Ref. [18]. Then the half life of the cluster decay is calculated by [23, 24]

$$T_{1/2} = \hbar\frac{\ln 2}{\Gamma}. \tag{5}$$

## 3. The Results

We have calculated the heavy-cluster half-lives of the nuclei having the mass numbers 221≤ A ≤242 with WKB method and Bohr-Sommerfeld quantization condition [18]. Comparison between the experimental results and the calculated heavy decay half-lives $\log_{10} T_{1/2}$ for $^{14}$C decay from various parent nuclei for $0^0 - 0^0$ can be shown in Fig. 1. Moreover, the obtained and experimental values of half-lives with deformation of cluster and daughter nuclei for $0^0$-$0^0$, $0^0$-$30^0$, $0^0$-$60^0$, $0^0$-$90^0$ and over all orientation angles for deformation parameters are seen in Fig. 2. It should be noted that the deformation parameters of the cluster and daughter nuclei which have deformations are taken from refs. [23, 25].

Journal of Physics: Conference Series **590** (2015) 012045 doi:10.1088/1742-6596/590/1/012045

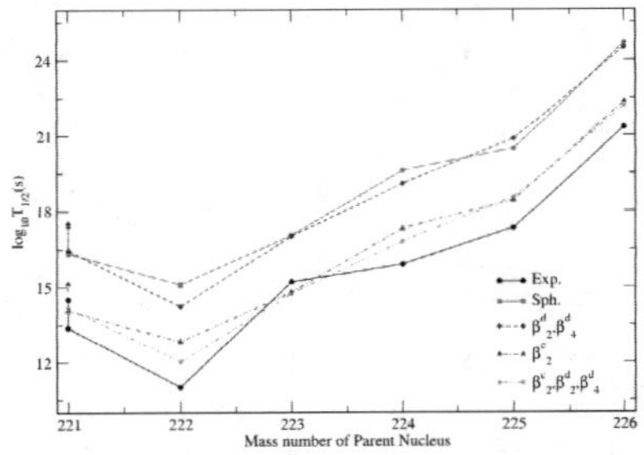


**Figure 1.** Comparison between the experimental results and the calculated heavy decay half-lives $\log_{10} T_{1/2}$ for $^{14}C$ decay from various parent nuclei for $0^0 - 0^0$ [18].

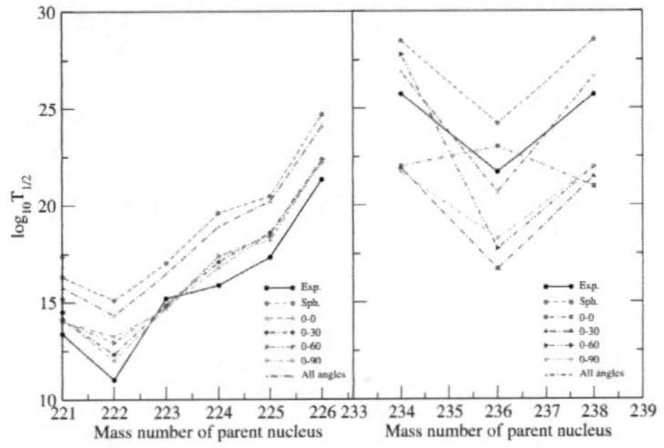


**Figure 2.** Comparison between the experimental results and the calculated heavy decay half-lives $\log_{10} T_{1/2}$ for $^{14}C$ and $^{28}Mg$ decays from various parent nuclei for $0^0 - 0^0$, $0^0 - 30^0$, $0^0 - 60^0$, $0^0 - 90^0$ orientation angles and over all angles [18].

We have also calculated the half-lives of $^4He$, $^8Be$, $^{12}C$, $^{14}C$, $^{16}O$, $^{18}O$, $^{20}O$, $^{22}O$ cluster decays of $^{210-226}Ra$ in order to address whether there is any effect of the deformations on the heavy-cluster half-lives of radium isotopes [20]. The $\log_{10} T_{1/2}$ with Woods-Saxon square and experimental values for **a)** spherical case (sph), **b)** only daughter deformation case **c)** only cluster deformation case **d)** with both daughter and cluster deformations case for $^4He$, $^8Be$, $^{12}C$, $^{14}C$, $^{16}O$, $^{18}O$, $^{20}O$, $^{22}O$ cluster decays of $^{210-226}Ra$ are seen in Fig 3. As seen in these figures, while in the spherical case there are large differences between experimental values and the calculated values for $^{14}C$ decays, taking into account the deformation effects decreases these differences. In other words, considering such kind of deformation effect improves the results effectively and it makes them closer to the experimental values.

## 4. Conclusion

Even if the deformations of both cluster and daughter in the half-live values of exotic cluster decays improve the results, considering the deformation of clusters is more efficient than the deformations of daughter in the heavy cluster decay half-live calculations for heavy nuclei and especially for radium isotopes. Considering the orientations of daughter and cluster in the half-lives calculations improves the results clearly so it would be important to take into account such kind of effects in the theoretical calculations. Searching dominant orientations in the half-lives calculations and finding out the orientation angle is the most compatible one with experimental data would be important to explain the observable in the theoretical models. The model would

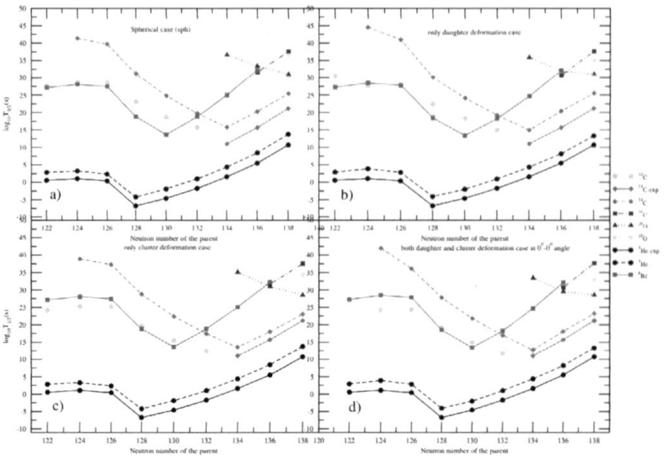


**Figure 3.** The calculated $\log_{10} T_{1/2}$ versus neutron number of the parent nuclei with Woods-Saxon square and experimental values for **a)** Spherical case (sph), **b)** only daughter deformation case, **c)** only cluster deformation case, **d)** both daughter and cluster deformation case for $0^0 - 0^0$ angles for $^4$He, $^8$Be, $^{12}$C, $^{14}$C, $^{16}$O, $^{18}$O, $^{20}$O, $^{22}$O cluster decays of $^{210-226}$Ra [20].

be applied to make some predictions on the unknown experimental half lives of possible cluster decays of Bismuth, Cerium, Barium isotopes as future studies.

## 5. Acknowledgments

This work has been partly supported by the Turkish Science and Research Council (TÜBİTAK) under Project No: 113F225.

[1] Rose H J, Jones G A 1984 *Nature* **307** 245.
[2] Gales S, Hourani E, Hussonnois M, Schapira J P, Stab L, Vergnes M 1984 *Phys. Rev. Lett.* **53** 759.
[3] Price P B, Stevenson J D, Barwick S W, Ravn H L 1985 *Phys. Rev. Lett.* **54** 297.
[4] Sandulescu A, Poenaru D N, Greiner W 1980 *Sov. J. Part. Nucl.* **11** 528.
[5] Sheng Z, Ni D, Ren Z 2011 *J. Phys. G: Nucl. Part. Phys.* **38** 055103.
[6] Buck B, Merchant A C 1989 *J. Phys. G: Nucl. Part. Phys.* **15** 615.
[7] Ibrahim T T, Perez S M, Wyngaardt S M, Buck B, Merchant A C 2012 *Phys. Rev. C* **85** 044313.
[8] Poenaru D N, Gherghescu R A, Greiner W 2011 *Phys. Rev. Lett.* **107** 062503.
[9] Ren Z, Xu C, Wang Z 2004 *Phys. Rev. C* **70** 034304.
[10] Buck B, Merchant A C, Perez S M 1999 *Nucl. Phys. A* **657** 267.
[11] Santhosh K P, Joseph A 2000 *Pramana J. Phys.* **55** 375.
[12] Santhosh K P, Joseph A 2000 *Proc. Int. Nat. Symp. Nucl. Phys. (India)* **43B** 296.
[13] Santhosh K P, Biju R K, Joseph A 2008 *J. Phys. G: Nucl. Part. Phys.* **35**, 085102.
[14] Santhosh K P *et al.* 2009 *Proc. Nat. Symp. Nucl. Phys. (India)* **1** 33.
[15] Santhosh K P *et al.* 2011 *Prc. DAE Symp. Nucl. Phys. (India)* **56** 298.
[16] Santhosh K P, Sahadevan S, Priyanka B, Unnikrishnan M S 2012 *Nuclear Physics A***882** 49.
[17] Sawhney G and Sharma M K, Gupta R K 2011 *Phys. Rev. C* **83** 064610.
[18] Soylu A, Sert Y, Bayrak O, and Boztosun I 2012 *Eur. Phys. J. A* **48** 128.
[19] Santhosh K P, Joseph A 2002 *Pramana, Indian Academy of Sciences* Vol. **59** 679.
[20] Soylu A, and Evlice S 2014 Deformations Effects on Exotic Decays of Radium Isotopes unpublished.
[21] Brink D M and Takigawa N 1977 *Nucl. Phys. A* **279** 159.
[22] Buck B, Merchant A C, and Perez S M 1994 *Phys. Rev. Lett.* **72** 1326.
[23] Denisov V Y and Khudenko A A 2009 *At. Data Nucl. Data Tables* **95** 815.
[24] Xu C and Ren Z 2006 *Phys. Rev. C* **74** 014304.
[25] Belgya T, *et al.* 2006 RIPL-2. IAEA-TECDOC-1506 (IAEA, Vienna, 2006). Available online at http://www-nds.iaea.org/RIPL-2/.

Journal of Physics: Conference Series **590** (2015) 012046 doi:10.1088/1742-6596/590/1/012046

# The effects of Double Folding Cluster Model Potential on some astrophysical reactions

**Azni Abdul Aziz**[1,2,3,*], **Norhasliza Yusof**[1,2], **Muhammad Zamrun Firihu**[4] **and Hasan Abu Kassim**[1,2,5]

[1] Department of Physics, Faculty of Science, University of Malaya, 50603 Kuala Lumpur, Malaysia
[2] Quantum Science Center, Department of Physics, Faculty of Science, University of Malaya, 50603 Kuala Lumpur, Malaysia
[3] Kulliyah of Science, International Islamic University Malaysia, 25200 Kuantan, Pahang, Malaysia
[4] Department of Physics, Halu Oleo University, Kendari, 93232 Sulawesi Tenggara, Indonesia
[5] Institute of Space Science, Universiti Kebangsaan Malaysia, 43600 Bangi, Selangor, Malaysia

E-mail: *azni_abdulaziz@yahoo.com

**Abstract.** The Double Folding Cluster Model Potential is constructed using the $\alpha$ - cluster structure of nuclei. It can be derived by folding an $\alpha - \alpha$ interaction with density distributions of $\alpha$ - clusters inside the projectile and target nuclei. This potential has been successfully tested on elastic scattering data of some selected nuclei. In this work, we are interested to investigate the implications of this potential on astrophysical aspects.

## 1. Introduction

Challenges in determining the accurate nuclear potential motivate the effort in developing various potentials. The most widely used potential is the Woods-Saxon shaped potential given by [1]

$$V_N(r) = \frac{-V_0}{1 + \exp[(r - R_0)/a_0]} + i\frac{-W_0}{1 + \exp[(r - R_W)/a_W]} \tag{1}$$

where the radii $R_0 = r_0(A_P^{1/3} + A_T^{1/3})$ and $R_W = r_W(A_P^{1/3} + A_T^{1/3})$ for the nuclear interaction between a projectile and target nuclei with the mass number $A_P$ and $A_T$ respectively, within their distance of closest approach, $r$. The Akyüz-Winther (A-W) parameterization [1] is often used to determine the parameter for the real part, i.e. the surface diffuseness parameter, $a_0$, the potential depth, $V_0$ and the radius parameter, $r_0$. The double folding potential has become one of the popular microscopic models to calculate the real part of the optical potential. Apart from that, there is also the Double Folding Cluster (DFC) model potential which is based on the $\alpha - \alpha$ interaction folded with the density distributions of $\alpha$ - clusters inside the projectile and target nuclei introduced by Azab et al. [2]. For this potential, the nuclei are assumed to be composed of an integer number $m$ of $\alpha$ particles, i.e., $A = 4m$.

Several studies have shown that the DFC potential has successfully reproduced the differential cross-section of elastic scattering data for a few reactions [2, 3, 4, 5]. Meanwhile, this potential could describe the broad features of the fusion, S-factor and elastic-scattering angular data simultaneously as reported by Kocak et al. [6]. Based on these reports, we are interested to further the study on this type of potential on the $^{12}C+^{12}C$, $^{12}C+^{16}O$ and $^{16}O+^{16}O$ reactions. Since these three reactions are important in stellar evolution and nucleosynthesis, this investigation emphasizes on the effects of this potential on astrophysical aspects.

## 2. Calculation

We present here a short description of the main formula used to calculate the DFC model potential while the detailed description can be found in Azab et al [2]. The DFC can be formulated as an effective $\alpha - \alpha$ interaction, $v_{\alpha\alpha}$ folded with the $\alpha$ - cluster distributions $\rho_{cP}$ and $\rho_{cT}$ for the projectile and target nuclei respectively

$$V_{DFC}(\vec{r}) = \int\int \rho_{cP}(\vec{r}_P)\rho_{cT}(\vec{r}_T)v_{\alpha\alpha}(\vec{s})d\vec{r}_T d\vec{r}_P. \tag{2}$$

The vector $\vec{s} = |\vec{R} + \vec{r}_T - \vec{r}_P|$ while the $\alpha - \alpha$ potential given by $v_{\alpha\alpha} = -122.6225\exp(-0.22r^2)$ is taken from Buck et al. [7].

If $\rho_c$ is the $\alpha$ - cluster distribution function inside the nucleus, then the nuclear matter density distribution function of the nucleus, $\rho_M$ can be related to that of the $\alpha$-particle nucleus, $\rho_\alpha$ as

$$\rho_M(\vec{r}) = \int \rho_c(\vec{r'})\rho_\alpha(|\vec{r} - \vec{r'}|)d\vec{r'}. \tag{3}$$

The matter density distribution of both projectile and target nuclei which can be written in a modified form of the Gaussion shape and the corresponding $\alpha$ density is given by [2]

$$\begin{aligned} \rho_M(\vec{r}) &= \rho_{0M}(1 + \omega r^2)\exp(-\beta r^2) \qquad (4) \\ \rho_\alpha(\vec{r}) &= \rho_{0\alpha}\exp(-\lambda r^2). \qquad (5) \end{aligned}$$

From Eq.(4) and Eq.(5), the $\alpha$-cluster distribution function $\rho_c$ can be obtained by using the Fourier transform [8], on Eq.(3) as

$$\rho_c(\vec{r'}) = \rho_{0c}(1 + \mu r'^2)\exp(-\xi r'^2) \tag{6}$$

where

$$\eta = \lambda - \beta, \qquad \xi = \beta\lambda/\eta, \qquad \mu = \frac{2\omega\lambda^2}{\eta(2\eta - 3\omega)} \tag{7}$$

while $\rho_{0M}$, $\rho_{0\alpha}$ and $\rho_{0c}$ can be obtained from the normalization condition. The parameters used in this work for all nuclei involved are given in Table 1.

Table 1: The nuclear density parameters.

| Nucleus | $\omega$ ($fm^{-2}$) | $\beta(\lambda)$ ($fm^{-2}$) | $\langle r^2\rangle^{1/2}$ (fm) | Ref. |
|---|---|---|---|---|
| $^4He$ | 0 | 0.7024 | 1.461 | [8] |
| $^{12}C$ | 0.4988 | 0.3741 | 2.407 | [9] |
| $^{16}O$ | 0.6457 | 0.3228 | 2.640 | [9] |

The DFC potential is then used to replace the real part in the nuclear potential, $V_N(r)$ to calculate the fusion cross section

$$V_N(r) = N_R V_{DFC(M)}(r) + i\frac{-W_0}{1 + \exp[(r - R_W)/a_W]} \tag{8}$$

Journal of Physics: Conference Series **590** (2015) 012046 doi:10.1088/1742-6596/590/1/012046

with $N_R$ is the normalization constant that can be varied to fit the experimental data. The value $W_0 = 50$ MeV, $r_W = 1.0$ fm, and $a_W = 0.4$ fm are chosen for the imaginary potential parameters to make sure that any differences arise only from the real part of the potential. We only consider the average description of the data by neglecting the resonant oscillation since the main purpose of this work is not to investigate the fusion oscillation especially in the $^{12}$C+ $^{12}$C and $^{12}$C+ $^{16}$O reactions.

## 3. Results

The fusion cross sections for $^{12}$C+$^{12}$C, $^{12}$C+$^{16}$O and $^{16}$O+$^{16}$O reactions have been calculated using the DFC potential. The normalization constant can be adjusted to obtain the best fitting by minimizing the chi-square of the fitting. The results are compared to the A-W potential and tabulated in Table 2. The results show that the DFC potential could produce a fitting as good as the A-W potential or even better.

Table 2: The normalization constant, $N_R$ and the value of chi-square $\chi^2$ of the fitting for A-W and DFC potentials.

| Reaction | Potential | $N_R$ | $\chi^2$ |
|---|---|---|---|
| $^{12}$C+ $^{12}$C | A-W | - | 54.17 |
| | DFC | 1.00 | 42.83 |
| | | 1.17 | 59.41 |
| $^{12}$C+ $^{16}$O | A-W | - | 2.03 |
| | DFC | 1.00 | 2.52 |
| | | 1.09 | 2.00 |
| $^{16}$O+ $^{16}$O | A-W | - | 1.63 |
| | DFC | 1.00 | 6.98 |
| | | 1.35 | 1.42 |

The modified astrophysical S-factor can be defined as $S^*(E) = \sigma(E)E\exp(2\pi\eta - gE)$ where $2\pi\eta = 0.9896Z_1Z_2(\mu/E)^{1/2}$ and $g = 0.46\text{MeV}^{-1}$ [10]. The $S^*(E)$ with $\sigma(E)$ from the DFC potential is compared to the one calculated using A-W potential and depicted in Fig. 1 for $^{12}$C+$^{12}$C reaction. The fusion cross section obtained in this calculation is then used to calculate the thermonuclear reaction rates using the standard formalism [10]

$$\langle\sigma\nu\rangle = \left(\frac{8}{\pi\mu}\right)^{1/2}\frac{1}{(kT)^{3/2}}\int_0^\infty \sigma(E)E\exp\left(-\frac{E}{kT}\right)dE. \quad (9)$$

The ratio of the reaction rates obtained from this work to the rates from Caughlan and Fowler's compilation 1988 (CF88) [11] is then calculated and depicted in Fig. 2. It is shown that the reaction rates obtained from the DFC potential seems lower than the CF88 at very low temperature.

The temperature that is typical of core carbon burning govern by the $^{12}$C+$^{12}$C reaction is within the location of the Gamow peak which is $T \approx 0.85$ GK. For the oxygen burning where the $^{16}$O+$^{16}$O fusion reaction occurs, the typical temperature is in the range of $T = 1.5 - 2.7$ GK, depending on the stellar mass. Although the rates predicted from the DFC potential give large differences at very low temperature, the ratio of the rates is approaching unity at least in the range of the Gamow temperature where the reactions mostly occur.

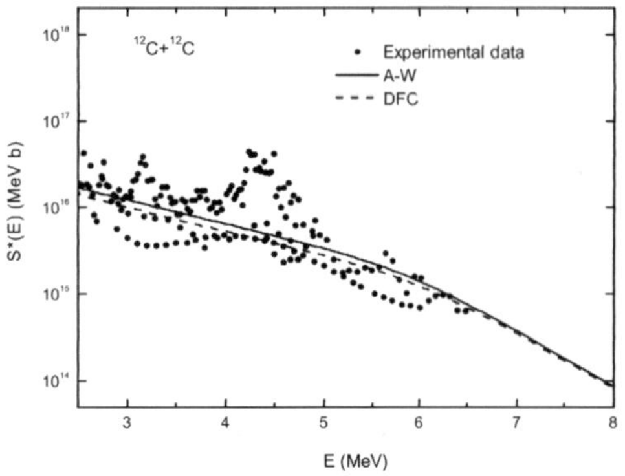


Figure 1: The $S^*(E)$ calculated using $\sigma(E)$ from the DFC potential and the A-W potential for $^{12}$C+$^{12}$C reaction. Experimental data are from [12].

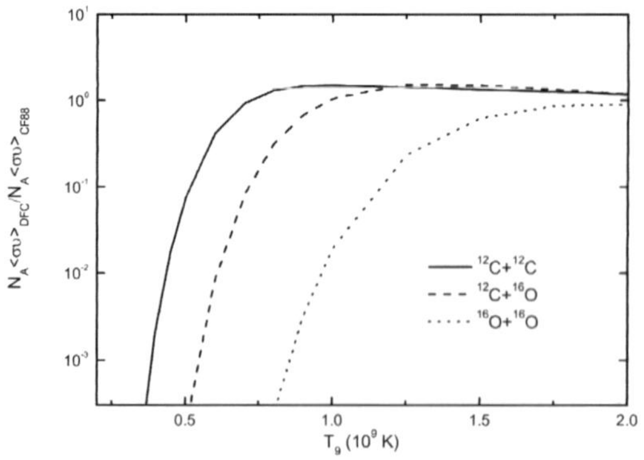


Figure 2: The ratio of the reaction rates calculated using the DFC potential to the CF88 rates for $^{12}$C+$^{12}$C, $^{12}$C+$^{16}$O and $^{16}$O+$^{16}$O reactions.

## 4. Conclusion

The DFC potential can be an alternative potential since it could reproduce the experimental fusion cross section data very well. However, the prediction on the astrophysical aspects gives some significant differences. Hence, further investigation on this potential is required.

## Acknowledgments

We thank University of Malaya for the Postgraduate Research Grant (PPP) (PV089/2011A) and University of Malaya Research Grant (UMRG) Programmes RP006C-13AFR and RP012D-13AFR. A. A. Aziz acknowledges International Islamic University Malaysia and Ministry of Education Malaysia for the Academic Training Scheme (SLAI) scholarship. M. Z. Firihu also acknowledges support from the Directorate General for Higher Education, Ministry of Education and Culture of Republic of Indonesia through Fundamental Research Grant 2012.

## References

[1] Akyuz O and Winter A 1981 Course on Nuclear Structure and Heavy-Ions Reactions *Proc. of the Enrico Fermi School of Physics, 1979* ed Broglia R A, Dasso C H and Ricci R (Amsterdam: North Holland)
[2] Azab Farid M El, Mahmoud Z M M and Hassan G S 2001 *Nucl. Phys.* **A691** 671
[3] Hassanain M A , Ibraheem Awad A, and Azab Farid M El 2008 *Phys. Rev. C* **77** 034601
[4] Hassanain M A, Ibraheem Awad A, Al Sebiey Shikha M M, Mokhtar S R, Zaki M A, Mahmoud Zakaria M M, Behairy K O, and Azab Farid M El 2013 *Phys. Rev. C* **87** 064606
[5] Karakoc M and Boztosun I 2006 *Phys. Rev. C* **73** 047601
[6] Kocak G, Karakoc M, Boztosun I and Balantekin A B 2010 *Phys. Rev. C* **81** 024615
[7] Buck B,Friedrich H and Wheathly C 1977 *Nucl. Phys.* **A275** 246
[8] Satchler G R and Love W G 1979 *Phys. Rep.* **55** 183
[9] Qing-biao S, Da-chun F, and Yi-zhong Z 1991 *Phys. Rev. C* **43** 2773
[10] Claus E Rolfs and William S Rodney 1988 *Cauldrons in the Cosmos* (Chicago: University of Chicago Press)
[11] Caughlan G and Fowler W 1988 *At. Data Nucl. Data Tables* **40** 283
[12] Mazarakis M G and Stephens W E 1973 *Phys. Rev. C* **7** 1280; Aguilera E F et al. 2006 *Phys. Rev. C* **73** 064601; Barrón-Palos L et al. 2006 *Nucl. Phys. A* **779** 318; Spillane T et al. 2007 *Phys. Rev. Lett.* **98** 122501

Journal of Physics: Conference Series **590** (2015) 012047 doi:10.1088/1742-6596/590/1/012047

# Photonuclear Reaction Cross Sections for Gallium Isotopes

**S Akkoyun[1,*], T Bayram[2], SO Kara[3]**
[1]Vocational School of Health Services, Cumhuriyet University, Sivas, Turkey
[2]Department of Physics, Sinop University, Sinop, Turkey
[3]Bor Vocational School, Nigde University, Nigde, Turkey

E-mail: serkan.akkoyun@gmail.com

**Abstract.** The photon induced reactions which are named as photonuclear reactions have a great importance in many field of nuclear, radiation physics and related fields. The cross-sections of neutron (photo-neutron (γ,xn)) and proton (photo-proton (γ,xn)) productions after photon activation have been calculated by using TALYS 1.2 computer code in this study. The target nucleus has been considered gallium which has two stable isotopes, $^{69}$Ga and $^{71}$Ga. Furthermore, the pre-equilibrium and compound process contributions to the total cross-section have been investigated.

## Introduction

Photonuclear reactions have been used in basic and applied sciences in nuclear and radiation physics related fields [1-4]. When a gamma-ray is incident upon a nucleus, the excited nucleus behaves like any compound nucleus with an excitation energy. The most probable decay is neutron emission (γ,n). After this, (γ,2n), (γ,1p) and (γ,2p) reactions take part. Nuclear level and half-life identifications, nucleon binding energy determinations, material analysis, radiation protection applications, dosimetry, absorbed dose assessment, activation analysis, radiation transport analyses, physics of fission and fusion reactors, nuclear waste transmutations and understanding element creations by astrophysical processes can be given as examples to such studies [2,4]. The experimental studies on these reactions have begun in 1934 [5] but there are still lack of existing data. Therefore, systematic studies of these reactions on different nuclei have been needed. The advantages of the reaction by photon activation are determination of the multiple element simultaneously, non-destructive structure of the process, requiring no time-consuming chemical separation procedure, deeper penetrating capability of the photon into the target [6]. The excited nucleus in the target emits particular radiation or photons in order to get rid of its excess energy. Due to the fact that photon do purely electromagnetic interaction with nuclei, the process is non-destructive.

Two main types can be observed when two nuclear system collide and both are very important in order to understand nuclear reaction phenomena. In the compound nucleus reaction, two system collide and a highly excited intermediate system has been created. The compound system has been formed by a sequence of collisions leading to rearrangements of the target nucleus. The excitation energy has been shared by nucleons. After sharing by chance, the overall energy has been localized on one or more or a group of nucleons for it to escape from the compound system. If there is still sufficient energy, further particle emission may occur, otherwise beta or gamma decay will appear. Second type which takes place very quickly according to the compound nucleus reactions has been

[1] To whom any correspondence should be addressed.

named as direct reaction. In this type, there is no intermediate compound system formed. In the compound nucleus formation, the projectile passes into the interior of the nucleus suffering a number of collisions. Therefore, we have expected that the direct reactions are localized in the nuclear surface (projectile interact over the surface of the nucleus) whereas compound nuclear reactions are localized in the interior of the nucleus. For increasing energy or nuclei for which the compound nucleus does not have time to reach thermodynamic equilibrium, pre-equilibrium or direct processes may become significant.

The cross section of the pure compound process is very small with respect to the measured spectra. Additionally, the measured angular distributions are not isotropic as expected isotropic in compound nucleus decays. So, there will be another classification between direct and compound nucleus mechanism. The pre-equilibrium process falls into either category. In these reactions, a nucleon is emitted after a number of collisions have occurred. Pre-equilibrium process [7] contributes to reaction cross-section in the 10 and 200 MeV energy range. In order to describe the pre-equilibrium mechanism, the exciton model is widely used [8]. At any time during the reaction, the nuclear state is characterized by the total energy, the total number of particles above the Fermi surface and total number of holes below the Fermi surface. Particles and holes are referred to as excitons. The complete formalism of the exciton model has been included in the TALYS reaction code [9]. By using the TALYS, it is possible to obtain all the compound nucleus, pre-equilibrium, and direct reactions contributions to the nuclear reactions.

Gallium element has two stable isotopes and both are used in nuclear physics and nuclear medicine. $^{69}$Ga isotope has been used for $^{68}$Ge or $^{68}$Ga generators. The $^{68}$Ga which is a positron emitter has been used as a PET isotope. $^{71}$Ga has been used in the study of behavior of solar neutrinos and in NMR studies. In this work, the photo-neutron and photo-proton cross sections as a function of photon energy with one and two nucleon in out channels have been investigated by using the TALYS 1.2 computer program. The reactions considered were ($\gamma$,1n), ($\gamma$,2n), ($\gamma$,1p) and ($\gamma$,2p). There is very limited photonuclear reaction cross section data in the literature for gallium [10, 11].

In this work, the photo-neutron and photo-proton cross sections of gallium isotopes have been calculated according to the photon energy in the 0.5-30 MeV energy range by using TALYS 1.2 computer program. Default options have been used in the program. The compound nucleus and pre-equilibrium mechanism contributions have been also analyzed for gallium isotopes as done in a systematic study in reference [13]. The results have been compared with the TENDL-2013 [12] databases.

## 2. Results and Discussion

In this work TALYS 1.2 version of the code has been used for the calculations of the ($\gamma$,1n), ($\gamma$,2n), ($\gamma$,1p) and ($\gamma$,2p) reaction cross-sections performed on both $^{69}$Ga and $^{71}$Ga targets separately. Pre-equilibrium and compound reactions cross sections have contributed to the overall cross sections. The energies of the gamma-rays have been between 0.5 and 30 MeV with 0.5 MeV interval. The reaction Q values are -10.3 and -6.6 MeV for $^{69}$Ga($\gamma$,1n)$^{68}$Ga and $^{69}$Ga($\gamma$,1p) $^{68}$Zn reactions, respectively. Besides for $^{71}$Ga isotopes these value are -9.3 and -7.9 MeV for $^{71}$Ga($\gamma$,1n)$^{70}$Ga and $^{71}$Ga($\gamma$,1p)$^{70}$Zn reactions, respectively. Furthermore for two nucleon separations the Q values are -18.6, -16.6, -16.9 and -19.0 MeV for $^{69}$Ga($\gamma$,2n)$^{67}$Ga, $^{69}$Ga($\gamma$,2p) $^{67}$Zn, $^{71}$Ga($\gamma$,2n)$^{69}$Ga and $^{71}$Ga($\gamma$,2p)$^{69}$Zn reactions, respectively. These are the threshold energies for taking place the reactions. As can be seen in the figures that the cross sections are zero below these threshold values.

It is also clear in the Figure 1 that the cross sections for photo-proton reactions has been lower than that of the photo-neutrons because of the suppression by the Coulomb barrier. For instance, for 14 MeV incident photon energy on $^{69}$Ga target, the ($\gamma$,1n) reaction cross section is about 60 mb whilst the ($\gamma$,1p) reaction cross section is only about 3.3 mb for the same energy value. It is clear in Figure 1 that the photo-neutron cross section is almost 2 to 20 times larger than photo-proton cross section in the energy range 10.5 and 30 MeV. Since the threshold energy for the photo-proton reaction is lower than the photo-neutron reaction, there is a probability that occurring photo-proton reaction in the range of

7.5 and 10 MeV where photo-neutron cross is zero. As can be seen in the Figure 1 that after 19 MeV, (γ,2n) reaction become taking place. Until 20.5 MeV the (γ,1n) reaction cross section is larger than that of (γ,2n) reaction. But after this energy value, (γ,2n) become about 1.2 to 5.5 magnitude larger up to 30 MeV. Furthermore (γ,1p) reaction cross section is larger about $6x10^1$ to $9x10^5$ than (γ,2p) cross section. Finally, the (γ,2n) cross section is too larger than that of (γ,2p) cross section about $1x10^3$ to $1x10^7$ magnitude.

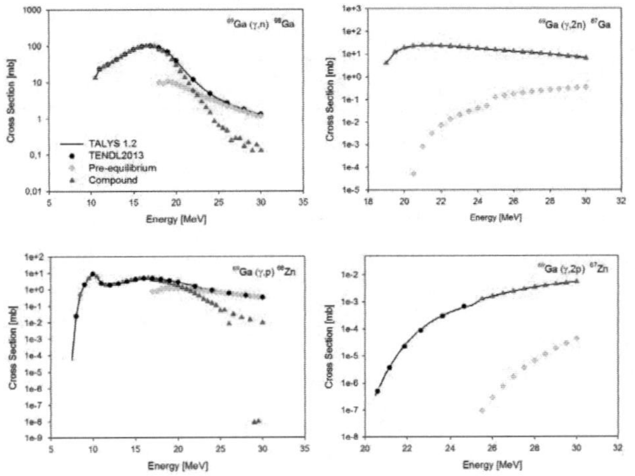


**Figure 1** (γ,1n), (γ,1p), (γ,2n) and (γ,2p) cross sections for $^{69}$Ga nuclei. Comparisons of pre-equilibrium and compound mechanism.

As can be seen in the Figure 1 that the (γ,1p) reaction cross section sharply increases from threshold energy to about 10.5 MeV where (γ,1n) channel opens up. At this value the increase remains slowly and after about 17 MeV, it becomes decrease because of increasing of the new opening channels such as (γ,2p) and (γ,2n). Besides, the (γ,1n) cross section exhibits similar trend. It increases to 17 MeV and decreases from this point where (γ,2p) and other reaction channels open up. The compound mechanism is dominant to about 20 MeV for (γ,n) ans (γ,p) reactions on the $^{69}$Ga target. After this energy value, pre-equilibrium mechanism become dominant to the end. For (γ,2n) and (γ,2p) reactions, the compound mechanism is always dominant.

Similar behavior has been seen for $^{71}$Ga target, but the differences is larger. The (γ,1n) reaction cross section is about 50 mb whilst the (γ,1p) reaction cross section is only about 0.3 mb for the 14 MeV energy value. It is clear in Figure 2 that the photo-neutron cross section is almost 4 to $9x10^2$ times larger than photo-proton cross section in the energy range 9.5 and 30 MeV. Since there is no explicit difference between neutron and proton thresholds, the reaction probabilities take values in the same energy, 9.5 MeV. As can be seen in the Figure 2 that after 17 MeV, (γ,2n) reaction become taking place. Until 19 MeV the (γ,1n) reaction cross section is larger than that of (γ,2n) reaction. But after this energy value, (γ,2n) become about 1.2 to 7.8 magnitude larger up to 30 MeV. Furthermore (γ,1p) reaction cross section is larger about $2x10^3$ to $5x10^5$ than (γ,2p) cross section. Finally, the (γ,2n) cross section is too larger than that of (γ,2p) cross section about $7x10^4$ to $2x10^7$ magnitude. The compound mechanism is dominant to about 20 MeV for (γ,n) ans (γ,p) reactions on the $^{69}$Ga target. After this energy value, pre-equilibrium mechanism become dominant to the end. For (γ,2n) reaction, the compound mechanism is always dominant.

Generally, the cross sections for $^{69}$Ga target is larger than that of the cross sections of $^{71}$Ga for (γ,1n), (γ,1p) and (γ,2p) reactions. This difference is slight for (γ,1n) reaction. But for (γ,2n) reaction, the cross section of the reaction performed on $^{71}$Ga target is larger. The cause of this behaviour can be interpreted through reaction threshold energies. Because the (γ,2n) reaction threshold energy for $^{71}$Ga is larger than that of the $^{69}$Ga, the probability to happen $^{71}$Ga(γ,2n)$^{69}$Ga reaction is higher. The threshold energies are closer to each other for (γ,1n) reactions, because there is a slight difference between the cross sections of the different targets.

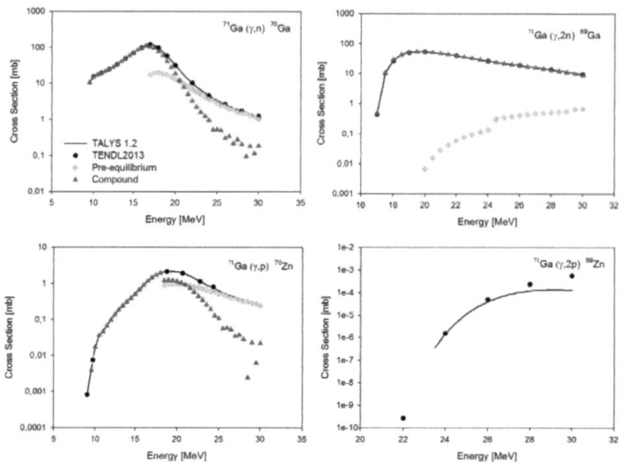


**Figure 2** (γ,1n), (γ,1p), (γ,2n) and (γ,2p) cross sections for $^{71}$Ga nuclei. Comparisons of pre-equilibrium and compound mechanism.

### 3. Conclusions

The stable gallium isotopes have been considered as a target nuclei in photonuclear reaction simulations. According to the results, we have seen that the calculations are in harmony with the TENDL-2013 database. The most probable decay channel is photo-neutron reaction whose cross section is 2 to 20 times larger than that of the photo-proton reaction. By increasing the photon energy, (γ,2n) reaction becomes dominant according to the (γ,n) after about 20 MeV. The pre-equilibrium process is dominant in the high energy region, 20-30 MeV. Besides the compound process contribution to the total cross-section is larger than the others in the 10-20 MeV energy interval.

## References

[1] Schumacher M 1988 *Inst. Phys. Conf. Ser.* No. **88**/*J. Phys. G:Nucl. Phys.* **14** Suppl., S235-S255
[2] Segebade C and Berger A 2008 *Encyclopedia of Analytical Chemistry* R.A. Meyers, John Wiley & Sons Ltd.
[3] Strauch K 1953 *Annu. Rev. Nucl. Sci.* **2**, 105-128
[4] IAEA-TECDOS-1178 2000 *Handbook on photonuclear data for applications Cross-sections and spectra* IAEA in Austria
[5] Chadwick J and Goldhaber M 1934 *Nature* **134** 237-238
[6] Green J Zaijing S Wells D and Maschner H 2011 *AIP Conf. Proc.* **1336** 497-501
[7] Hodgson PE 1981 *Nature* **292** 671-672
[8] Griffin JJ 1966 *Phys. Rev. Lett.* **17** 478-481 (Erratum: *Phys. Rev. Lett.* **19** 57).
[9] http://www.talys.eu/
[10] EXFOR, https://www-nds.iaea.org/exfor/servlet/X4sSearch5
[11] Carlos P Beil H Bergere R Fagot J Lepretre A Veyssiere A Solodukhov GV 1976 *Nucl. Phys. A* **258** 365-387
[12] TENDL-2013, http:// http://www.talys.eu/tendl-2013/
[13] Xu Y Goriely S Koning AJ and Hilaire S 2014 *Phys. Rev. C* **90** 024604

Journal of Physics: Conference Series **590** (2015) 012048 doi:10.1088/1742-6596/590/1/012048

# The Long Range Potential Effect on the $^{20}$O+$^{12}$C System for Near and Sub-barrier Fusion Cross Section

**G. Kocak†, R. Kabak†**

†Department of Physics, Erciyes University, 38039, Kayseri, Turkey

**Abstract.** In this study, we have analyzed the $^{20}$O+$^{12}$C system within the framework of optical model by using miscroscopic folded potential, which is based on M3Y nucleon-nucleon effective interaction and a phenomenological potential. A significant underprediction appeared between experimental data and our results. To solve this problem, one of the important candidate is the use of the long range potential effect. By using the long range potential effect, we have achieved results that are consistent with experimental data for the $^{20}$O+$^{12}$C system fusion cross section.

## 1. Introduction

The studies with neutron-rich radioactive beams provide new insight into production of new isotopes and nuclei. Fusion of very neutron-rich nuclei is also very important in order to determine the observables of neutron stars and for astrophysical interest. There are numerous studies for the near-barrier fusion with neutron-rich nuclei [1–6] and the studies with the exotic nuclei show unexpected behaviors in the barrier region, like several order magnitude enhancement of fusion cross section [7, 8].

One of the most important problem in the sub-barrier fusion studies with very neutron-rich nuclei is the enhancement of fusion cross section at sub-barrier energies. Recently, Rudolph *et al.* [9] measured the $^{20}$O+$^{12}$C system fusion cross section for the near and sub-barrier energies and they have obtained the significant difference between the experimental and theoretical data. In this study, we investigated the fusion cross section for the near and sub-barrier region to explain the unexpected behavior of fusion excitation data. With this purpose, we have used the long range potential effect [10] to explain $^{20}$O+$^{12}$C system fusion cross section for the energies near and below the Coulomb barrier.

## 2. Model

In this study, we have used optical model. In the optical model, the total potential $\mathrm{V_{total}(r)}$ consists of

$$V_{\mathrm{total}}(r) = V_{\mathrm{Nuclear}}(r) + V_{\mathrm{Coulomb}}(r) + V_{\mathrm{Centrifugal}}(r) \tag{1}$$

Coulomb and Centrifugal potentials are well-known. The Coulomb potential [11] due to a charge $Z_P e$ interacting with a charge $Z_T e$ distributed uniformly over a sphere of radius $R_c$ is

$$V_{\mathrm{Coulomb}}(r) = \frac{1}{4\pi\varepsilon_\circ}\frac{Z_P Z_T e^2}{r}, \qquad r \geq R_c, \tag{2}$$

$$= \frac{1}{4\pi\varepsilon_o}\frac{Z_P Z_T e^2}{2R_c}\left(3 - \frac{r^2}{R_c^2}\right), \qquad r < R_c, \tag{3}$$

where $R_c$ is the Coulomb radius, and $Z_P$ and $Z_T$ denote the charges of the projectile $P$ and the target nuclei $T$ respectively. The centrifugal potential is

$$V_{\text{Centrifugal}}(r) = \frac{\hbar^2\, l\,(l+1)}{2\,\mu\, r^2} \tag{4}$$

where $\mu$ is the reduced mass of the colliding pair.

In order to make a comparative study of this reaction, we use two different potentials for the real part of the optical model potential: one is a phenomenological potential and the other one is the microscopic nucleon-nucleon double folding potential.

### *2.1. Phenomenological Potential*

For the phenomenological potential, we used the Woods-Saxon shaped potential. This potential is

$$V_{\text{Nuclear}}(r) = \frac{-V_0}{(1+\exp(r-R_V)/a_V)} + i\frac{-W_0}{(1+\exp(r-R_W)/a_W)} \tag{5}$$

where $V_0 = 198$ MeV, $R_V = r_V(A_P^{1/3} + A_T^{1/3})$ with $r_V = 0,9$ fm, $a_V = 0,46$ fm. The imaginary part of the potential has the same Woods-Saxon volume shape as in Equation 5 and its parameters are $W_0 = 2$ MeV, $r_W = 1,0$ fm and $a_W = 0,8$ fm.

### *2.2. Nucleon-Nucleon Double Folding Potential (NN-DF)*

The NN-DF potential evaluated by using the nuclear matter distributions for projectile $\rho_P$ and target $\rho_T$ nuclei with an effective NN interaction potential ($\nu_{nn}$) is

$$V_{DF}(r) = \int\int \rho_P(r_1)\rho_T(r_1)\nu_{nn}(|\vec{r}+\vec{r}_2-\vec{r}_1|)d^3r_1 d^3r_2 \tag{6}$$

The DF potential which is based upon a realistic effective nucleon-nucleon (NN) interaction (Equation 7), folded with nuclear matter densities of both projectile and target nuclei:

$$\nu_{nn}(r) = 7999\frac{\exp(-4r)}{4r} - 2134\frac{\exp(-2.5r)}{2.5r} + J_{00}(E)\delta(r)MeV \tag{7}$$

The imaginary potential parameters which used in the phenomenological potential have taken for the NN-DF calculations. The code DFPOT [12] has been used for the microscopic DF potential calculation and the code FRESCO [13] has been used to obtain all reaction observables.

## 3. The Results

We have first analyzed the experimental data for the $^{20}$O+$^{12}$C system total fusion cross section with the optical model by using phenomenological Wood-Saxon potential. Our results are shown in fig. 1.

There is large fusion cross section enhancement at the sub-barrier energies that can not explain by the optical model calculation. Therefore, we have performed the more realistic explanation by using microscopic nucleon-nucleon double folding potential to explain this reaction as can be seen in fig. 1.

But, neither the phenomenological Wood-Saxon potential nor the microscopic nucleon-nucleon double folding potential are enough to solve the low energy fusion cross section

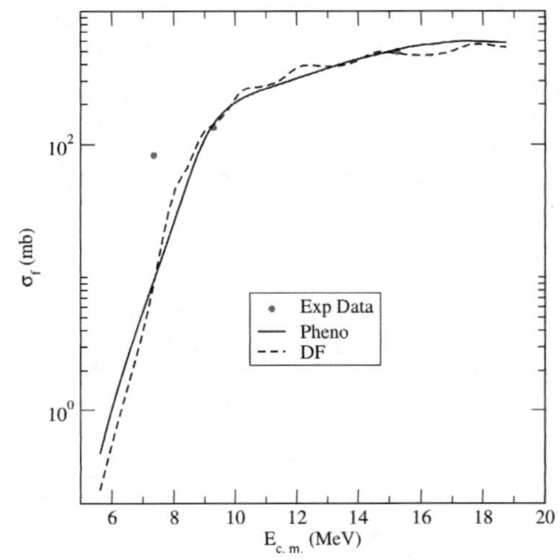


**Figure 1.** The phenomenological and microscopical results for the $^{20}O+^{12}C$ system, the black line gives phenomenological results and the dashed line shows DF results.

| $V_d$ | $r_0$ | $a$ |
|---|---|---|
| MeV | fm | fm |
| 8.0 | 4.3 | 4.0 |

**Table 1.** The parameters of the additional potential.

enhancement problem. The underestimation has been raised between our results and the neutron-rich nuclei fusion cross section in all calculations.

Recently, Dapo *et al.* [10] showed that the importance of long range potential on the low energy observables. They found that the tail of the real potential has a significant effect on the cross section. To understand this relative dominance of the changes induced by long range versus short range of the potential, one can look to the concept of effective theory. In effective theories, a separation of scales is introduced to distinguish effects which occur at low energies (long range) and at high energies (short range). The low energy observables are dominated by the long range of the potential and are not sensitive to the short range [10]. By using this approximation, we have modified our total potential by adding a small additional potential. It has the derivatives of the Wood-Saxon shape and the parameters are shown in Table 1.

As it can be seen from the figure 2, the long range potential has significant effect on the low energy fusion cross section. We have obtained very good agreement with the experimental data by using this effect.

## 4. Conclusion

In this study, we have examined the fusion cross section with very neutron-rich nuclei. The optical model results by using phenomenological and microscopical potential are not enough to explain experimental data. The underestimation has been raised between experimental data and our results. To solve this problem, we have used the long range potential effect and obtained very good results. As a future studies, we will examine the neutron transfer between target and projectile and will compare the long range potential and transfer effect.

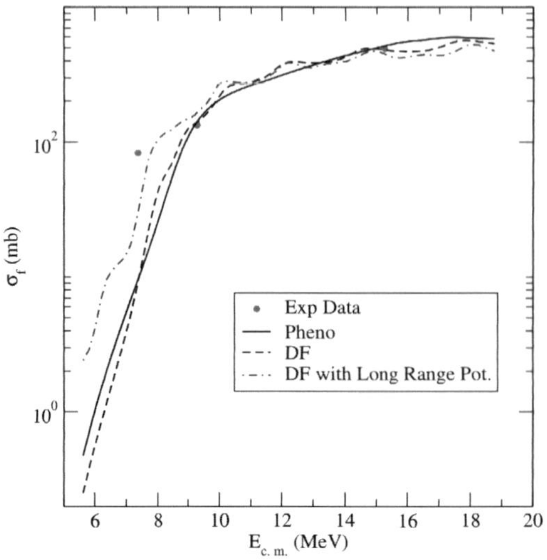


**Figure 2.** The results of the $^{20}O+^{12}C$ system fusion cross section after the inclusion of the long range potential effect

## 5. Acknowledgments
This work was supported by Research Fund of the Erciyes University. Project Number:FBA-2014-5224.

## 6. References
[1] Kolata J J *et al* 1998 *Phys. Rev. Lett.* ***81*** 4580
[2] Fomichev A S *et al* 1995 *Z. Phys. A.* ***351*** 129
[3] Trotta M *et al* 2000 *Phys. Rev. Lett.* ***84*** 2342
[4] Pietro A Di *et al.* 2004 *Phys. Rev. C* ***69*** 044613
[5] Penionzhkevich Yu E *et al* 2006 *Phys. Rev. Lett.* ***96*** 162701
[6] Raabe R *et al* 2004 *Nature* ***431*** 823
[7] Loveland W *et al* 2006 *Phys. Rev. C* ***74*** 064609
[8] Zagrebaev V I, Samarin V V and Greiner W 2007 *Phys. Rev. C* **75**, 035809
[9] Rudolph M J *et al* 2012 *Phys. Rev. C* ***85*** 024605
[10] Dapo H *et al* 2012 *Phys. Rev. C* ***85*** 044602
[11] Satchler G R 1983 *Nucl. Phys.* ***A409*** 3c
[12] Cook J 1982 *Comput. Phys. Commun.* *25* 125
[13] Thompson I J *FRESCO, a coupled-channels code, unpublished*

# Isomeric and ground state energy level measurements of natural tellurium isotopes via (γ,n) reaction

**M Tamkas**[1,2] **O Akcalı**[2] **and A Durusoy**[2]
[1] Istanbul Arel University, Mathematics and Computer Department, TR 34537, Istanbul, Turkey
[2] Yildiz Technical University, Physics Department, Davutpasa Campus, 34220, Istanbul,Turkey

E-mail: makbuletamkas@arel.edu.tr

**Abstract.** We have planned to measure isomeric and ground state energy levels in $^{120}Te(\gamma,n)^{119m,g}Te$, $^{122}Te(\gamma,n)^{121m,g}Te$, $^{128}Te(\gamma,n)^{127m,g}Te$, $^{130}Te(\gamma,n)^{129m,g}Te$ photonuclear reactions of natural tellurium induced by bremsstrahlung photons with end-point energy at 18 MeV. The sample was irradiated in the clinical linear electron accelerator (Philips SLi-25) at Akdeniz University Hospital. The gamma spectrum of the tellurium sample was measured using HP(Ge) semiconductor detector (ORTEC) and multi channel analyzer. We used both MAESTRO (ORTEC) and home made root based gui program (Theia) for data analyzing. The obtained experimental data values are compared with NUDAT energy values.

Keywords: isomeric state, ground state, energy level

## 1. Introduction

There are many nuclear reactions that populate both the ground state and the metastable(isomeric) state of daughter nucleus. Isomeric activities can be induced in many stable nuclides that occur in the region of closed nuclear shells, by gamma activation [1-4]. Gamma activation excites the target nucleus from the ground state to one of its activation energy levels above the metastable isomeric state [5].

The investigation of the properties of excited states in a nuclear reaction such as the energy and spin distributions, the probability of excitation, and different decay modes allow one to acquire important information about interaction mechanism. It is particularly efficient when the excited states last for a adequately long period of time as the isomeric and ground state. It is well known that in the same experimental conditions the isomeric and ground states are formed simultaneously during nuclear reaction process. For this reason, isomeric ratio studies have gained importance from past to the present [1-8]. Today, the isomeric ratios can be determined with high accuracy benefiting from

[1] M Tamkas, Istanbul Arel University, Mathematics and Computer Department, K 34537 Tepekent-Büyükçekmece Istanbul,Turkey

developed experimental measurement equipments and theoretical models. But in the isomeric ratio studies, the determination of ground state and isomeric state energy levels will improve the accuracy of this studies.

The aim of this work, is to determine the isomeric and ground state energy levels of natural tellurium induced by bremsstrahlung photons with end-point energy at 18 MeV. The nuclear reactions considered in this investigation are $^{120}Te(\gamma,n)^{119m,g}Te$, $^{122}Te(\gamma,n)^{121m,g}Te$, $^{128}Te(\gamma,n)^{127m,g}Te$, $^{130}Te(\gamma,n)^{129m,g}Te$ photonuclear reactions.

## 2. Experimental Technique

In our experiment, we used the photo activation technique. Research was performed with the use of bremmstrahlung beam obtained from a clinical linear electron accelerator (cLinac), Philips SLi-25, at the Akdeniz University Hospital.

The cLINAC which is used as the source of bremsthralung photons for phohonucelar reactions is a standard and up-to-date Elekta TM Synergy TM accelerator, SLi-25 [9]. The accelerator's primary electron beam is generated by a gun with an energy of 50 keV. After that the electrons are accelerated in a copper cavity by a 3 GHz (2856MHz to be exact) radio-frequency with peak power of about 5 MW [10]. For an electron energy of about 10 MeV, the typical average electron current is about 300 μA. Figure 1 shows the schematic view of cLinac.

In this experiment, we used 0.7 g of natural tellurium, which is obtained from Sigma-Aldrich Laboratory, with purity of 99.99% in small ore pieces. The sample was irradiated for 50 minutes at 18 MeV end-point energy.

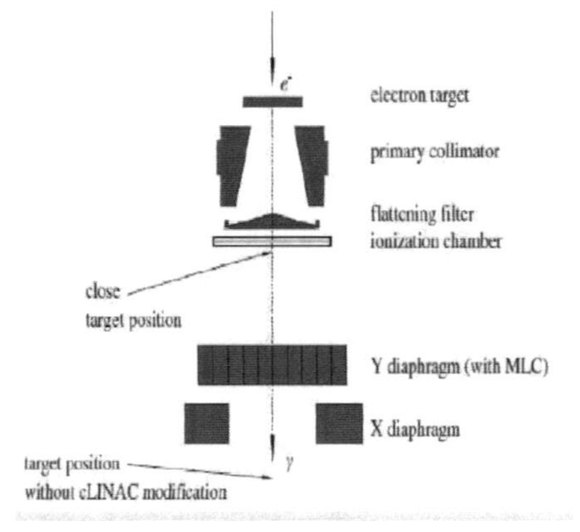


**Figure 1.** Schematic view of the photon beam production at the cLinac

The sample was placed coaxially at 6 cm distance from the High-Purity Germanium (HPGe) detector. Multi spectrum scaling technique is used to collect gamma-ray spectrum with 5 min cooling time and sampling times were adjusted to 3s, 30s, 180s and 900s to sum up to 20 h counting time. Except for the last time interval 200 cumulative spectrums were recorded to be able to correct spectrums due to different half-lives. The detector used was a p-type, coaxial, electrically cooled, HPGe gamma-ray spectrometer AMATEK- ORTEC (GEM40P4-83) 40% relative efficiency. It is connected to a NIM consisting of ORTEC bias supply, 672 spectroscopy amplifier, 927 multi-channel analyzer and a computer. For detector energy calibration, a mixed calibration source supplied by the Çekmece nuclear Research and Training Center emitting gamma rays in the energy range between 47 and 1836 keV was used.

### *2.1. Gamma-Ray Spectra Calculations*

Throughout the data acquisition period Maestro software was used to acquire and analyse the data for consistency. After data acquisition, a home made ROOT (Root 5.26.00) [11] based graphical user interface program named as "Theia" is used to analyse photo peaks and for calculation of the peak parameters (figure 2 and figure 3) Theia designed to enable methodological access to some desired futures of ROOT in a simpler manner while providing Compton background correction for all photo peaks which are assumed to be in Gaussian form.

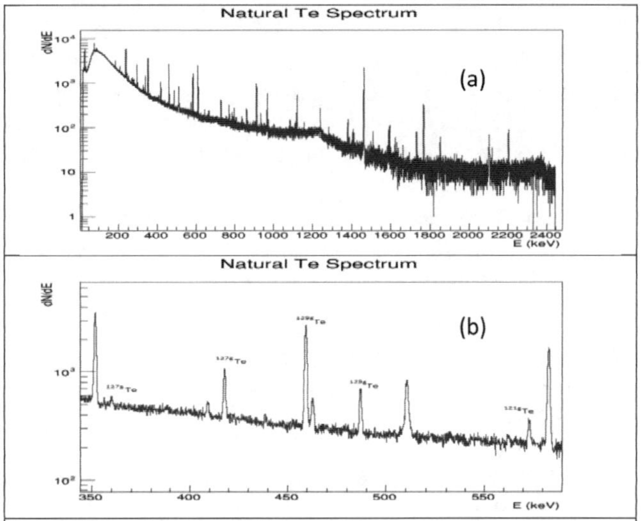


**Figure 2.** Full energy range spectrum of natural Tellurium (γ,n) reaction (a) and the photopeaks of the analyzed region (b)

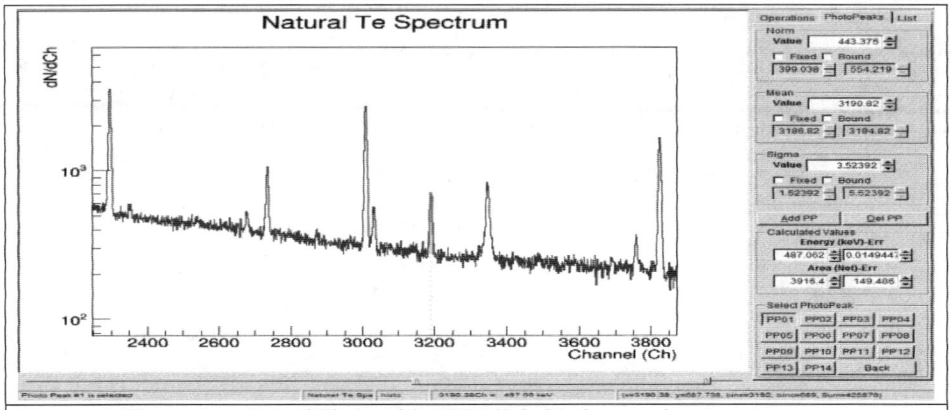


**Figure 3.** The screen shot of Theia with 487.062 keV photopeak parameters

Journal of Physics: Conference Series **590** (2015) 012049 doi:10.1088/1742-6596/590/1/012049

## 3. Conclusion

The obtained experimental data values are compared with NUDAT energy values (table 2). As seen, gamma energy values which was determined by analysis programs Theia and Maestro. Calculated values by both software are very close to NUDAT energy values. All of the planned isotopes could not be determined because only the results of 20 h count could be reached. Our study continues to determine gamma energy levels and half-life of the long half-life isotopes.

**Table 2.** Analysis result of irradiated natural tellurium spectrum

| Reaction Product | Spin | Half Life | Reac.Threshold (MeV) | Gamma Energy (keV) (Err) | Gamma İnt(%)(Err) | Theia Energy (keV)(Err) | Energy Maestro (keV)(Err) |
|---|---|---|---|---|---|---|---|
| $^{121g}$Te | $1^+/2$ | 16.8 d | 17.29 | 573.139(0.011) | 80.4(0.22) | 572.89 (0.04) | 572.95 (0.15) |
| $^{129g}$Te | 3+/2 | 69.5 min | 8.39 | 459.60(0.05) | 7.70(0.06) | 459.272 (0.159) | 459.33 (0.15) |
| | | | | 487.39(0.05) | 1.42(0.10) | 487.062 (0.015) | 487.24 (0.15) |
| $^{127g}$Te | 3+/2 | 9.35 h | 8.75 | 360.30(0.1) | 0.135(0.14) | 360.054 (0.072) | 360.19 (0.15) |
| | | | | 417.90(0.1) | 0.99(0.14) | 417.712 (0.012) | 417.58 (0.15) |

## 4. References


[1] Chertok B T and Booth E C 1965 *Nucl. Phys.* **66** 230.

[2] Booth E C and Brownson J 1967 *Nucl. Phys. A* **98** 529 .

[3] Veres A 1980 *Atom. Energy Rev.* **18** 271 .

[4] Ljubicic A, Pisk K and Logan B A 1981 *Phys. Rev. C* **23** 2238 .

[5] Venkataraman R and Fleming R F 1994 *Nuclear Instruments and Methods in Phys. Research A* **353** (1994) 425-428

[6] Mazur V M,Symochko D M, Derechkey P S and Bigan Z M 2013 *Ukr. J. Phys.* **58** No. 1

[7] Tran D T, An T T, Cuong P V, Khai N t, Vinh N T, Belov A G and Maslov 2011 *J. Rad. Nucl. Chem.* **289** 637-645

[8] Nguyen V D, Pham D K, Kim T T and Tran D T 2007 *J. Korean Phys. Soc.* **50** 417-425

[9] Elekta Digital Accelerator, Technical Training Guide,2003.

[10] Boztosun I *et al.* 2013 *Turk J. Phys* 1-9

[11] Antcheva I *et al.* 2009 *Comp. Phys. Comm.* **180** 2499-2512

# Thickness measurement using alpha spectroscopy and SRIM

**C. Yal¸ın**

Kocaeli University, Department of Physics, Umuttepe 41380, Kocaeli, Turkey

E-mail: caner.yalcin@kocaeli.edu.tr

**Abstract.**

Thickness of a material can be calculated by measuring energy loss through the matter if the stopping power of the material is well known. This method is widely used to measure energy degrader, backing and target thickness for nuclear reaction experiments. However, the ions continuously lose their energy as they move through the matter, and also stopping power changes because of its energy dependent nature. For this reason, when calculating the thickness, one must take into account this effect, especially for a thick material. ThiMeT code takes into account stopping power changes through matter and calculates the accurate thickness by using SRIM. ThiMeT code also can be used for the determination of the required thickness for energy degraders. An experimental method for thickness measurement and calculation with ThiMeT code will be presented.

## . Int ti n

The thin films and self-supported foils are used as energy degraders, backing and targets for the nuclear reactions. The thickness of the energy degraders and targets play important role for the nuclear physics experiments. For example, the energy degraders are needed for light ions to obtain lower energies quickly or to obtain energies which cannot be produced by accelerators [1–4]. The thickness and homogeneity information of the targets are also crucial for the reaction cross section measurement.

Thickness of a material can be calculated by measuring energy loss through matter if the stopping power of the material is well known [5]. The method is very simple and it is based on measurement of the energy loss of ions (charged particles) through the matter. This method is widely used to measure energy degrader, backing and target thicknesses for nuclear reaction experiments [1, 2, 6–8]. The energy loss of a charged particle after it traverses a material of thickness $t$ can be calculated by

$$\Delta E = \int_0^t \frac{dE}{dx} dx \tag{1}$$

Here, it is assumed that the particle range is greater than the material thickness (R>t) and $dE/dx$ is the total stopping power. If t≪R, one may take the stopping power as a constant, and equation (1) is turned into,

$$\Delta E = \left(\frac{dE}{dx}\right)_{E_0} t \tag{2}$$

Journal of Physics: Conference Series **590** (2015) 012050 doi:10.1088/1742-6596/590/1/012050

where $(dE/dx)_{E_0}$ is the total stopping power for the initial energy of the particle. According to equation (2), if the energy loss $\Delta E$ measured and the total stopping power is well known, thickness of material can be easily calculated [5, 9]. But, if the thickness of material is a considerable fraction of the range, stopping power cannot be considered to be a constant because the ions continuously loose their energy as they move through the matter, and also the stopping power changes with the energy (see Figure 1). In this case, in order to calculate an accurate thickness, one must take into account this effect, especially for a thick material. ThiMeT code [10] takes into account energy dependency of the stopping power, and it calculates the accurate thickness by using SRIM stopping power values [11–13].

## . ThiMeT C e

ThiMeT is a code that is a thickness calculation tool for SRIM [10]. SRIM is a group of programs that calculates stopping power and a range of ions into matter using a quantum mechanical treatment of ion-atom collisions [11–13]. SRIM code is based on theoretical formulations and semi-empirical models. There are lots of programs and tables for stopping power values, but for most cases, the SRIM is the best to describe stopping power for all ions at all energies in all targets [18]. Therefore, ThiMeT code uses the SRIM stopping power values. The reliability of the stopping power programs and the tables have been investigated in references [14–18] and references therein.

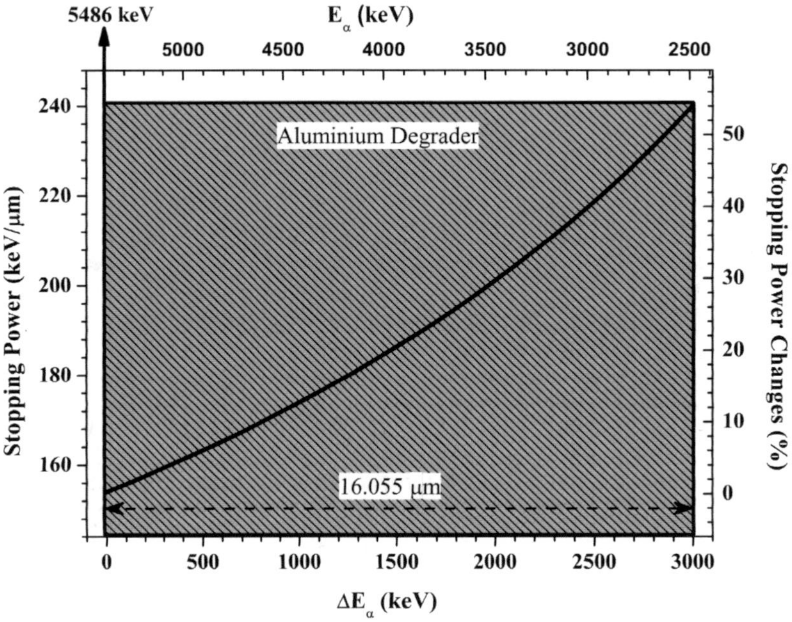


**i e .** Stopping power changes through the Aluminum energy degrader for 5486 keV alphas emitted from $^{241}$Am. Stopping power values were obtained from SRIM-2013 [11]. Upper x-axis shows the alpha energy and lower x-axis shows the energy loss of the alphas. The thickness of the foil is 16.055 $\mu$m and indicated with shaded region in the plot. Right y-axis shows the stopping power difference in percent relative to the value at the incident energy.

Journal of Physics: Conference Series **590** (2015) 012050 doi:10.1088/1742-6596/590/1/012050

ThiMeT code divides the total energy loss into segments, and it calculates the thickness of each segment using the stopping power value at the incident energy of the segment obtained from SRIM. After the first segment, it uses exit energy of the first segment as the incident energy of next segment, and so on. After the last segment, in order to find total thickness, sum of the calculated thicknesses of all segments are taken. Note that ThiMeT code actually uses SR Module in the SRIM, and the nuclear stopping powers of SR Module are slightly different from those in the full tables since these are improved values which will later be incorporated into SRIM [11].

ThiMeT code also can be used for the determination of the required thickness of energy degraders for the nuclear reaction experiments. This is often used for light ions, such as protons, to obtain lower energies quickly [1–4]. Because of the SRIM stopping power for light ions well described above 1 MeV, the accuracy of the calculation is also good. For example, stopping power for proton ions on Al above 1 MeV, the systematic difference between SRIM and experiment is 0.5%, and random error which contains some information about the experimental uncertainty is 0.7% [15].

## . e lt an C n l i n

In order to determine the required thickness of a energy degrader, TRIM simulation can also be used according to the given method in the TRIM help menu and the user manual. This method can be summarized as follows;

(1) estimate the thickness using mean stopping power values of incident and exit energy,
(2) enter this thickness into TRIM and run it,
(3) calculate mean transmission energy from the "Transmit.txt" file of the TRIM,
(4) if the mean transmission energy equals desired exit energy, this energy corresponds to the required thickness,
(5) but if the mean transmission energy dose not equal desired energy, change the thickness entered into TRIM and continue the steps from 2 through 5, and repeat this steps until the mean transmission energy gets a value as close as the desired exit energy.

This methods can be also used for thickness calculation of a material by measuring the energy loss with alpha spectroscopy. Table 1 shows the thickness results of the ThiMeT code and TRIM

**Table .** Calculated ThiMeT (version 1.0) thickness and comparison with TRIM (version SRIM-2013) code for common degrader materials.

| Ions | Material | Incident Energy (keV) | Exit Energy (keV) | TRIM Thickness ($\mu$m) | Calc. Time | ThiMeT Thickness ($\mu$m) | Calc. Time | Diff. (%) |
|---|---|---|---|---|---|---|---|---|
| H | Be | 100000 | 52419 | 35000 | 7 m 34 s | 34986.5 | 7 s | 0.04 |
| H | C | 100000 | 24890 | 35000 | 14 m 10 s | 35034.5 | 12 s | 0.1 |
| H | Al | 100000 | 18644 | 35000 | 15 m 37 s | 35050.4 | 13 s | 0.14 |
| He | Be | 100000 | 17213 | 4000 | 21 m 21 s | 3998.221 | 25 s | 0.04 |
| He | C | 100000 | 18105 | 3000 | 29 m 16 s | 2998.922 | 28 s | 0.04 |
| He | Al | 100000 | 16866 | 3000 | 36 m 10 s | 3000.003 | 25 s | $\approx$0 |
| He | Al | 10000 | 8000 | 17.950 | 54 s | 17.899 | 1 s | 0.28 |
| He | Al | 5486 | 2486 | 16.125 | 1 m 34 s | 16.055 | 1 s | 0.44 |
| He | Al | 3183 | 1184 | 7.86 | 47 s | 7.82 | 1 s | 0.51 |

simulation. The results of ThiMeT code calculation is in agreement with TRIM simulation. But the calculation of the thickness with TRIM simulation takes much longer time, even if, for fastest calculation, "No graphical" and "No auto save" modes are set for TRIM and incident ion numbers are adjusted to 10000. In Table 1, calculation times of the thickness using both TRIM and ThiMeT codes are given, where the TRIM code was run just once with only one input value. For the precise determination of the thickness with TRIM, the TRIM simulation needs to be run more times to obtain desired exit energy. Therefore, determination of the thickness using this iteration method takes much longer time.

The main uncertainties of the thickness calculation depend on the uncertainty of the energy loss measurement and the uncertainty of the stopping power values of SRIM. Because, the energy calibration of the particle detector is crucial for the determination of the energy loss, one should check energy calibration of detector with a pulser and calibrated mix alpha source. Note that the SRIM stopping power and experimental one are not in a good agreement for all ions and targets at all energies. In order to estimate uncertainty of the thickness measurement, one must check the uncertainties of the stopping power values at the energy range of interest from the SRIM web site [11].

ThiMeT code is very useful tool for the calculation of thicknesses and the determination of the required thickness for energy degraders. ThiMeT code calculation is very quick and easy compared to the TRIM calculation. This is the first version of the ThiMeT code. The energy straggling and the uncertainty estimation for thickness calculation will be implemented to the fortcoming version of the ThiMeT code.

**e e en e**

[1] Gyürky Gy et al. 2006 *Phys. Rev.* C **74** 025805
[2] Yalçın C et al. 2009 *Phys. Rev.* C **79** 065801
[3] Anferov V 2003 *Nucl. Instrum. Meth.* A **496** 222
[4] Oberstedt A and Oberstedt S 2007 *Nucl. Instrum. Meth.* A **570** 51
[5] Santry D C and Werner R D 1979 *Nucl. Instrum. Methods* **159** 523
[6] Kiss G G et al. 2011 *Phys. Rev.* C **83** 065807
[7] Kiss G G et al. 2013 *Phys. Rev.* C **88** 045804
[8] Yalçın C this voulume
[9] Tsoulfanidis N 1995 *Measurement and Detection of Radiation* (Washington, DC:Taylor & Francis)
[10] The Thickness Measurement Tools for SRIM (ThiMeT) www.thimet.org
[11] The Stopping and Range of Ions in Matter (SRIM) www.srim.org
[12] Ziegler J F, Biersack J P and Littmark U 1995 *The Stopping and Range of Ions in Solids* (New York:Pergamon Press)
[13] Ziegler J F, Ziegler M D and Biersack J P 2010 *Nucl. Instr. And Meth.* B **268** 1818
[14] Paul H and Schinner A 2003 *Nucl. Instrum. Meth.* B **209** 252
[15] Paul H and Schinner A 2005 *Nucl. Instrum. Meth.* B **227** 461
[16] Paul H and Schinner A 2006 *Nucl. Instrum. Meth.* B **249** 166
[17] Paul H and Schinner A 2012 *Nucl. Instrum. Meth.* B **273** 15
[18] Paul H and Snchez-Parcerisa D 2013 *Nucl. Instrum. Meth.* B **312** 110

Journal of Physics: Conference Series **590** (2015) 012051 doi:10.1088/1742-6596/590/1/012051

# Energy Levels and Half-Lives of Gallium Isotopes Obtained by Photo-Nuclear Reaction

**F Dulger[1, 2], S Akkoyun[3], T Bayram[4], H Đapo[1, 2] and I Boztosun[1, 2]**

[1] Akdeniz University, Department of Physics, 07058, Antalya, Turkey
[2] Akdeniz University, Nuclear Sciences Application and Research Center, 07058, Antalya, Turkey
[3] Cumhuriyet University, Vocational School of Health, Sivas, 58140, Turkey
[4] Sinop University, Department of Physics, 57000, Sinop, Turkey

E-mail: fatihdulger@akdeniz.edu.tr

**Abstract**. We have run an experiment to determine the energy levels and half-lives of Gallium nucleus by using the photonuclear reactions with end-point energy of 18 MeV bremsstrahlung photons, produced by a clinical linear accelerator. As a result of $^{71}$Ga(γ,n)$^{70}$Ga and $^{69}$Ga(γ,n)$^{68}$Ga photonuclear reactions, the energy levels and half-lives of $^{70}$Ga and $^{68}$Ga nuclei have been determined. The results are in good agreement with the literature values.

## 1. Introduction

High-energy electrons can be used to produce bremsstrahlung photons by a linear accelerator. These photons are obtained by the bombardment of a tungsten (W) or tantalum (Ta) bremsstrahlung converter with electrons. The sample is placed downstream behind the converter and thus is exposed to bremsstrahlung photons. Due to activation, the stable nuclei in the sample can be transformed to radioactive ones. A neutron and proton separation is induced by the interaction of bremsstrahlung photons produced by the linear accelerator, *viz.* photo-nuclear reactions. This is called photo-activation technique [1]. Based on this technique, the photonuclear experiments are conducted in two steps: the irradiation and the residual activity measurement. Although there are available different techniques of measurement for the photo-nuclear reaction products, this is the most widely used one.

During a photonuclear reaction, energy of a high energy gamma-ray photon is absorbed fully or partially by the nucleus forcing it into an excited state. From this excited state, the nucleus can emit any particle (p, n, *etc*), provided it has enough energy for such a process to occur. Subsequently, the radiation emitted from the radioactive nuclei is determined by counting it with an appropriate spectrometer. In this way, we can obtain information about the energy levels and half-lives of the created isotopes. This method has a wide range of applications in the medical imaging systems, radioisotope treatment process and radiation protection measures.

So far, many experiments have been performed to determine the energy transitions of stable nuclei. In this area, the most comprehensive study to determine the half-life and the energy levels was measured by Oka *et al.* [2]. They have determined the half-lives of 75 different nuclei and observed (γ,n), (γ,2n), (γ,p), (γ,γ') reactions. In addition, the half-life of $^{68}$Ga was investigated by Oka *et al.* [2]. In this study, as a result of $^{69}$Ga(γ,n)$^{68}$Ga, the half-life of $^{68}$Ga was found to be 68 minute by using scintillator detectors. After the 1970s, this procedure was used for many scientific purposes. In our country, for the first time in 2013, this was done by Akdeniz University Nuclear Sciences Application and Research Center (NUBA) by using the local facilities. In this study, Boztosun *et al.* [3] studied the half-lives and the energy levels of zinc nuclei.

The target nucleus gallium we study in this paper has two stable isotopes $^{69}$Ga and $^{71}$Ga. Their natural abundances are 60.10% and 39.90% for $^{69}$Ga and $^{71}$Ga, respectively. In this paper, our aim is to determine the energy transitions and half-lives of gallium isotopes by using photonuclear reactions. We have used a high-purity germanium detector to obtain more accurate results. Notably, comparing scintillation and HPGe detectors the latter has better resolution while the former has better efficiency. For the measurement of the energy transition, accuracy is the main goal, therefore we have used a HPGe detector for the counting of residuals [4].

In the next section, we describe the experimental methods. Thereafter, we present the results, followed by our conclusions and outlook.

## 2. Experimental

The activation technique based on the detection of residual radioactivity of unstable nuclei produced by the reactions during the irradiation phase. By measuring the activity with a high resolution spectrometer, one obtains characteristic gamma ray spectra of decay in unstable reaction products. Thus, specific reactions can be identified.

In this study, the sample prepared for the experiment was placed 58 cm away from the tungsten converter. Photo-nuclear reactions are obtained by exposing the targets with sufficient energy bremsstrahlung photons. Typically, the irradiation time was 1 or 2 hours; subsequently, counting was performed. Counting time at the HPGe detector after irradiation was optimized depending on the half-life of the radioisotope. After data acquisition, spectra were exported to the gf3 Radware gamma spectroscopy software [5].

Figure 1 shows a schematic of the electronics used in the detection and transmission of the signal from the detector to the computer.

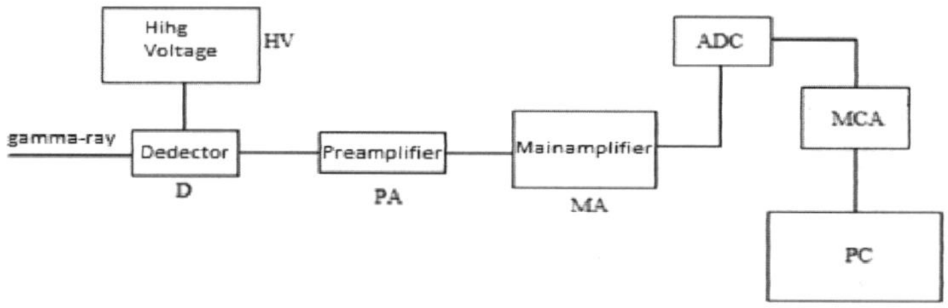


**Figure 1.** Operating sketch of the HPGe detector

## 3. Results

Figure 2 shows the count versus channel spectrum of the irradiated target nuclei measured by HPGe spectrometry [4]. The spectrum analysis is carried out using the gf3 Radware gamma spectroscopy software [5]. After the peak positions are determined, the energy of transition is obtained by energy calibration using several standard sources ($^{60}$Co, $^{54}$Mn, $^{57}$Co, $^{24}$Na, $^{137}$Cs, and $^{133}$Ba). In addition, the time evolution of the counts yields the respective half-lives. Table 1 shows some of resulting half-lives obtained. The left side of Table 1 shows the literature data, in the right side our results for energies and half-lives are presented. We note that all our uncertainty are only statistical.

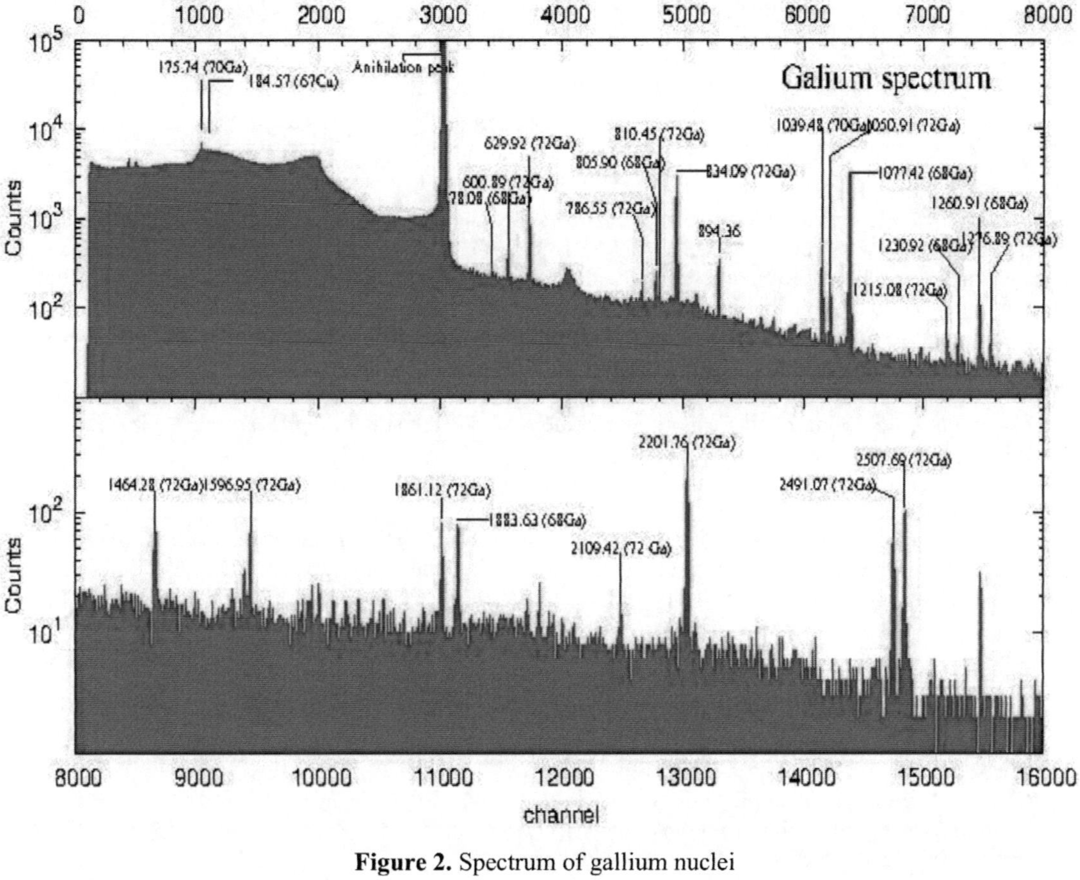


**Figure 2.** Spectrum of gallium nuclei

The results obtained for energy transitions show overall good agreement with the literature data. For a true estimate of the quality of the data, we need to also find a way to handle the systematic deviations. Currently, we are reviewing such procedures. While the uncertainty estimation will certainly change, the peak positions will remain. For the half-lives, the situation is not nearly as good. One can observe about 10 % discrepancy across all results. The probable causes of this discrepancy are the presence of more than one isotope in our sample. For the time being, it would be difficult for us to circumvent this, but we are working on it.

**Table 1.** Results of Our Analysis for $^{71}$Ga and $^{69}$Ga

| | NUDAT [6] | | | | Our Results | | | |
|---|---|---|---|---|---|---|---|---|
| Nucleus | Energy (keV) | err | $T_{1/2}$ (minute) | err | Energy (keV) | Stat. err | $T_{1/2}$ (minute) | Stat. err |
| $^{71}$Ga(γ,n) $^{70}$Ga | 176.17 | 0.02 | 21.14 | 0.03 | 175.74 | 0.02 | 20.10 | 0.05 |
| | 1039.20 | 0.03 | 21.14 | 0.03 | 1039.44 | 0.01 | 20.28 | 0.02 |
| $^{69}$Ga(γ,n) $^{68}$Ga | 805.83 | 0.08 | 67.71 | 0.09 | 805.91 | 0.2 | 65.06 | 0.36 |
| | 1077.34 | 0.05 | 67.71 | 0.09 | 1077.404 | 0.004 | 66.12 | 0.04 |
| | 1230.93 | 0.13 | 67.71 | 0.09 | 1230.92 | 0.08 | 76.2 | 0.47 |
| | 1261.08 | 0.09 | 67.71 | 0.09 | 1260.91 | 0.04 | 77.28 | 0.51 |
| | 1883.16 | 0.06 | 67.71 | 0.09 | 1883.37 | 0.04 | 62.76 | 0.19 |

## 4. Conclusions and Outlook

The aim of this work was to determine the half-lives and the spectra of Ga isotopes through photo-nuclear reaction. In this process, the electrons were absorbed by a tungsten converter inside the c-Linac and a photon beam from a c-Linac with endpoint energy of 18 MeV was obtained. The bremsstrahlung photons with this energy are sufficiently energetic to achieve photonuclear reactions with a Ga target. As a result, several energy transitions of different isotopes showing different half-lives were observed. The energies have shown a good agreement with literature date. However, the half-lives are determined less accurately as expected due to multi-isotope nature of our sample. The particularly interesting point of this experiment is the study of the suitability of a medical linac, used differently from previous studies.

**Acknowledgements**

We acknowledge the financial support of the Scientific and Technological Research Council of Turkey (TÜBİTAK) with a project grant number 114F220 and Cumhuriyet University Scientific Project Unit with a project number SHMYO-007.

## References

[1] Belyshev S S *et al* 2014 NIM **A** 745 133
[2] Oka Y Kato T Nomura K and Saito T 1967 Journal of Nuclear Science and Technology 4 346
[3] Boztosun I *et al* 2014 Turk. J. Phys **38** 1
[4] ORTEC 2012 Maestro Application Software http://www.ortec-online.com/download/MAESTRO.pdf
[5] Radford D C 2000 Notes on the use of the program gf3 http://radware.phy.ornl.gov/gf3/gf3.html
[6] McCuthan E A 2012 Nuclear Data Sheets **113** 1735: Tuli J K 2004 Nuclear Data Sheets **103** 389

Journal of Physics: Conference Series **590** (2015) 012052 doi:10.1088/1742-6596/590/1/012052

# Dynamical Dipole mode in heavy-ion fusion reactions in the $^{192}$Pb mass region

**C. Parascandolo[1], D. Pierroutsakou[2], R. Alba[3], A. Del Zoppo[3], C. Maiolino[3], D. Santonocito[3], C. Agodi[3], V. Baran[4,5,3], A. Boiano[2], M. Colonna[3], R. Coniglione[3], E. De Filippo[6], M. Di Toro[3,7], U. Emanuele[8], F. Farinon[9], A. Guglielmetti[10], M. La Commara[11,2], B. Martin[11,2], C. Mazzocchi[10], M. Mazzocco[1], C. Rizzo[3,7], M. Romoli[2], C. Signorini[1], R. Silvestri[11,2], F. Soramel[1], E. Strano[1], D. Torresi[1], A. Trifirò[8] and M. Trimarchi[8]**

[1] Dip. di Fisica e Astronomia, Università di Padova and INFN, Padova, Italy
[2] INFN - Sezione di Napoli, Napoli, Italy
[3] INFN - Laboratori Nazionali del Sud, Catania, Italy
[4] University of Bucharest, Bucharest, Romania
[5] NIPNE-HH, Magurele, Romania
[6] INFN - Sezione di Catania, Catani, Italy
[7] Dip. di Fisica e Astronomia, Università di Catania and INFN, Catania, Italy
[8] INFN - Gruppo Collegato di Messina and Università di Messina, Messina, Italy
[9] GSI, Darmstadt, Germany
[10] Dip. di Fisica, Università di Milano and INFN - Sezione di Milano, Milano, Italy
[11] Dip. di Scienze Fisiche, Università di Napoli, Napoli, Italy

E-mail: `concetta.parascandolo@na.infn.it`

**Abstract.** The dynamical dipole mode was investigated in the mass region of the $^{192}$Pb compound nucleus, by using the $^{40}$Ca + $^{152}$Sm and $^{48}$Ca + $^{144}$Sm reactions at $E_{lab}$=11 and 10.1 MeV/nucleon, respectively. Both fusion–evaporation and fission events were studied simultaneously for the first time. Our results for evaporation and fission events (preliminary) show that the dynamical dipole mode survives in reactions involving heavier nuclei than those studied previously.

## 1. Introduction

In N/Z asymmetric heavy-ion reactions, it is possible to excite a collective dipole oscillation that can develop along the symmetry axis of the dinuclear system [1, 2, 3]. This oscillation, called "Dynamical Dipole mode" (DD throughout the text), decays emitting prompt dipole $\gamma$-rays, in addition to those coming from the Giant Dipole Resonance (GDR) thermally excited in the hot compound nucleus (CN). The DD radiation presents i) a lower centroid energy than that of a statistical GDR built in a spherical nucleus of similar mass due to the high deformation of the emitting source [2, 3] ii) an anisotropic angular distribution with respect to the beam axis because the oscillation is confined in the reaction plane [4] and iii) a $\gamma$ yield that is predicted to depend on both the beam energy and the reaction dynamics [3].

Experimentally, the existence of the DD mode has been studied in deep inelastic and fusion-evaporation heavy-ion collisions [5, 6, 7, 8, 9]. In these measurements, an excess of $\gamma$-rays was observed in the GDR energy region for a charge asymmetric reaction, with respect to that of a more charge symmetric one forming the same CN at identical conditions [6, 7, 8] or with respect to statistical model calculations [9]. This $\gamma$ excess was attributed to the decay of the predicted DD.

The emission of DD $\gamma$-rays decreases the excitation energy and hence the initial temperature of the nucleus reaching the statistical phase. This cooling mechanism could be suitable to favour the production of super-heavy elements in hot fusion processes. However, TDHF calculations [2] showed that the prompt dipole $\gamma$ yield decreases as the mass of colliding ions increases since the reactions with small nuclei are less damped than those involving more nucleons. In order to understand if this pre-equilibrium effect survives in heavier systems than those studied before and to test its usefulness in super-heavy element production, we decided to study the DD in the mass region of the $^{192}$Pb CN.

## 2. Experimental results for $^{192}$Pb

The experiment was performed by using the $^{40}$Ca($^{48}$Ca) pulsed beam provided by the Superconducting Cyclotron of the Laboratori Nazionali del Sud (LNS, Italy), impinging on a 1 mg/cm$^2$ thick $^{152}$Sm($^{144}$Sm) target at $E_{lab}$ = 440(485) MeV. Both entrance channels populate the same CN, $^{192}$Pb, through a quite different initial dipole moment ranging from 30.6 fm for the $^{40}$Ca + $^{152}$Sm charge asymmetric reaction to 5.3 fm for the $^{48}$Ca + $^{144}$Sm more charge symmetric one. The mass asymmetry of the two entrance channels is very similar, namely 0.22(0.18) for the $^{40}$Ca + $^{152}$Sm ($^{48}$Ca + $^{144}$Sm) system. Furthermore, the formed CN had identical spin distribution: $L_{max} = 74\hbar$ for fusion and $L_{max} = 36\ \hbar$ for fusion-evaporation, according to PACE2 calculations [10] by using a level density parameter of a = A/8, A being the compound nucleus mass, and identical excitation energy, as explained in the following.

The $\gamma$-rays and the light charged particles were detected by using the MEDEA experimental apparatus [12], made of 180 $BaF_2$ scintillators. The discrimination between $\gamma$-rays, light charged particles and neutrons was performed by combining a pulse shape analysis of the $BaF_2$ signal with a time of flight measurement between each scintillator and the radiofrequency signal of the Cyclotron. The fusion-evaporation residues were detected by four position sensitive Parallel Plate Avalanche Counters (PPACs) placed symmetrically around the beam direction at 70 cm from the target at $\theta = 7°$ and subtending 7° in $\theta$. The fission events were selected by detecting the two kinematically coincident fission fragments with position sensitive PPACs, centered at $\theta$ = 52.5° symmetrically around the beam axis at 16 cm from the target covering 22° in both $\theta$ and $\phi$ and allowing the study of $\gamma$-ray - fragment angular correlations. Down-scaled single events together with coincidence events between at least one fired $BaF_2$ scintillator and a PPAC (two PPACs) for evaporation (fission) events were collected during the experiment. The coincidence request eliminated any cosmic ray contamination of the $\gamma$-ray spectra. By using the above trigger there are no normalization factors in the $\gamma$-ray spectra as the double differential $\gamma$ multiplicity is obtained from the ratio of the number of coincidences between $\gamma$-rays and evaporation residues (fission fragments) and the number of single events of evaporation (fission). Preliminary results of the experiment, concerning a partial statistics are shown in [13].

### *2.1. $\gamma$-ray spectra and angular distributions*

The average excitation energy and the average mass of the composite system after pre-equilibrium particle emission were evaluated by studying the energy spectra of the light charged particles (p, $\alpha$) detected in coincidence with evaporation residues. These spectra for the $BaF_2$ rings placed from $\theta = 51.5°$ to $\theta = 159.7°$ were analyzed in the framework of a two moving source scenario: a *slow* source simulating the statistical evaporation from the hot CN and an *intermediate-velocity* (between the CN and the projectile velocity) source related to the pre-equilibrium particles emitted by the composite system before thermalization. The analysis confirmed us that the two considered reactions lead to the formation of a CN with the same average mass at the same average excitation energy. Therefore, as in our previous works [7, 8], all the parameters except the dipole moment were kept identical in the two reactions, so that any difference in their $\gamma$-ray spectra and angular distributions can be attributed to the difference in the entrance channel charge asymmetry. By comparing the center-of-mass double differential $\gamma$-ray spectra of the two reactions for fusion-evaporation and fission events an excess of $\gamma$-rays in the more charge asymmetric reaction was observed, concentrated in the energy range $E_\gamma$ = 8-15 MeV as can be seen in the left-hand side of Figure 1 where the difference between the spectra of the two systems is shown for evaporation (fission) events in the top (bottom). This excess is related to the DD decay and can be reproduced by means of a lorentzian curve folded by the experimental apparatus response function [14] (line in the figure) with a centroid energy $E_{DD}$ = 11 MeV and a width $\Gamma_{DD}$ = 3.5 MeV, for both exit channels. It is interesting to note that $E_{DD}$ is lower than the ground state GDR centroid energy $E_{GDR} \sim 13.5$ MeV for a mass in the region of 192. This result confirms the high deformation of the emitting source, in agreement with expectations [2, 3] and with our previous works [7, 8].

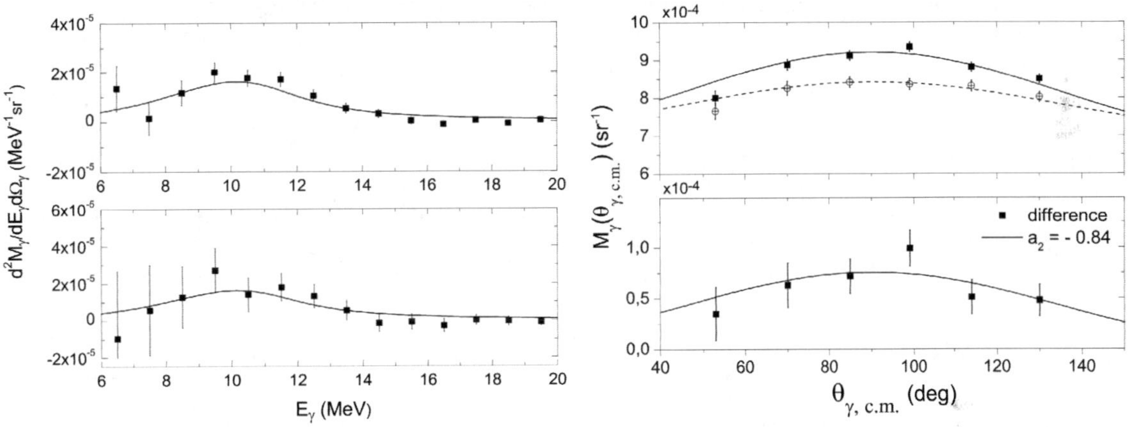


**Figure 1.** (Left-hand side) Difference between the charge asymmetric and charge symmetric reaction center-of-mass $\gamma$-ray spectra for fusion-evaporation (top) and fission (bottom) events. The solid lines in both panels are described in the text. (Right-hand side) Center-of mass angular distribution of the $\gamma$-rays in evaporation events for the two reactions (top) and of their difference (bottom) in the energy interval $10 \leq E_\gamma \leq 14$ MeV corrected by the experimental setup efficiency. The lines are described in the text.

Although such $\gamma$ excess constitutes one of the signatures of the DD radiation, the angular distribution is also an important observable since it gives information about the reaction dynamics and the DD lifetime. This is related to (a) the rotation angular velocity of the dinuclear system during the prompt dipole emission and (b) the instant at which this emission occurs [4]. We display in the right-hand side of Fig. 1 the center-of-mass angular distribution with respect to the beam direction of the $\gamma$-rays detected in coincidence with evaporation residues for the $^{40,48}Ca+^{152,144}Sm$ reactions (top) and for their difference (bottom). The double differential $\gamma$-ray multiplicity was integrated over energy from 10 to 14 MeV, after the subtraction of $(nn)$-bremsstrahlung component, and was corrected by the experimental setup efficiency. The lines in both panels of the above figure describe the angular distribution of the emitted $\gamma$-rays given by the Legendre polynomial expansion $M_\gamma(\theta_\gamma) = M_0[1 + Q_2 a_2 P_2 cos(\theta_\gamma)]$, where $a_2$ is the anisotropy coefficient and $Q_2$ is an attenuation factor for the finite $\gamma$-ray counter, which, for the present geometry, is 0.98 [15]. From a best fit to the data, shown with a solid (dashed) line for the $^{40}$Ca+$^{152}$Sm ($^{48}$Ca+$^{144}$Sm) reaction, we obtained $a_2$ = -0.17 ± 0.05 for the $^{40}$Ca+$^{152}$Sm reaction and $a_2$ = -0.10 ± 0.03 for the $^{48}$Ca+$^{144}$Sm one. The charge asymmetric reaction (squares) displays a more anisotropic angular distribution around 90° than the charge symmetric one (circles). Since we have the same CN, with the same excitation energy and spin distribution, such a difference is related to entrance channel effects.

As a consequence of the above, the experimental angular distribution of the difference (squares in the bottom of the figure) is very anisotropic around 90°. The data can be reproduced with $a_2$ = -0.84 (solid line), obtained within the BNV theoretical calculations as mentioned in the following, that is compatible with an emission from a dipole oscillation along an axis that has performed a small rotation with respect to the beam axis. Although we are not able to evaluate the rotation angle of the DD axis around the beam direction when the DD oscillation is completely damped, due to the large statistical errors, we can confine the $\gamma$-emission time scale at the very beginning of the reaction. This result is in agreement with our previous results [8] for evaporation events corresponding to small impact parameters. By taking into account the DD $\gamma$-ray angular distribution and the response function of the experimental setup, we deduced the DD yield, integrated over energy and over angle: 1.38 ± 0.14*$10^{-3}$ and 1.4 ± 0.4*$10^{-3}$ for evaporation and fission events, respectively, where the quoted errors are statistical. A 3% systematic error in the $BaF_2$ scintillator efficiency introduces a ± 4*$10^{-5}$ error, smaller than the statistical one. We notice that the extracted DD yield is the same, within errors, in evaporation and fission events.

The present experimental findings on the prompt dipole radiation in $^{40}$Ca+$^{152}$Sm reaction were compared with preliminary calculations performed within the BNV transport model framework and based on a collective bremsstrahlung approach of the entrance channel reaction dynamics [3]. These calculations give centroid energy, width and angular distribution of the DD in good agreement with those of the experiment. However, the theoretical $\gamma$ yield overestimates the data, an aspect that should be further investigated.

## References

[1] Chomaz P, Di Toro M and Smerzi A 1993 *Nuc. Phys.* A **563** 509
[2] Simenel C, Chomaz P, and de France G 2001 *Phys. Rev. Let.* **86** 2971
[3] Baran V *et al.* 2001 *Phys. Rev. Let.* 87 182501, 2001 *Nuc. Phys.* A **679** 373
[4] Baran V *et al.* 2009 *Phys. Rev.* C **79** 021603 (R)
[5] Pierroutsakou D *et al.* 2003 *Eur. Phys. Jour.* A **16** 423
[6] Flibotte S *et al.* 1996 *Phys. Rev. Let.* **77** 1448; Amorini F 2004 *et al. Phys. Rev.* C **69** 014608
[7] Pierroutsakou D *et al.* 2003 *Eur. Phys. Jour.* A **17** 71, 2005 *Phys. Rev.* C **71** 054605
[8] Martin B *et al.* 2008 *Phys. Let.* B **664** 47; Pierroutsakou D *et al.* 2009 textitPhys. Rev. C **80** 024612
[9] Corsi A *et al.* 2009 *Phys. Let.* B **679** 197
[10] Gavron A 1980 *Phys. Rev.* C **21** 230
[11] Kelly M P *et al.* 1999 *Phys. Rev. Let.* **82** 3404
[12] Migneco E *et al.* 1992 *Nuc. Inst. Meth.* A **314** 31
[13] Parascandolo C *et al.* 2010 *Nuc. Phys.* A **834** 198c; Pierroutsakou D 2010 *Int. Jour. Mod. Phys.* E **19** 1031
[14] Bellia G *et al.* 1993 *Nuc. Inst. Meth.* A **329** 173
[15] Rose M E 1953 *Phys. Rev.* **91** 610
[16] Baran V, Colonna M, Greco V, Di Toro M 2005 *Phys. Rep.* **410** 33

# Investigation of the Photo-activation Analysis on Concentrations of Various Materials

**A Cesur[1], M Karakoc[1,2], I Boztosun[1,2] and H Đapo[1,2]**
[1] Akdeniz University, Department of Physics, 07058, Antalya, Turkey
[2] Akdeniz University, Nuclear Sciences Application and Research Center – NUBA, 07058 Antalya, Turkey

E-mail: alpcesur@akdeniz.edu.tr

**Abstract.** It is possible to excite atomic nuclei electromagnetically and even separate proton or neutron from a sample which is exposed to gamma rays by a clinical linac. Thus, the determination of the elemental concentration of a sample by photo-activation analysis is a reasonable attempt since it is related to the detailed analysis of the photonuclear reaction of an intended element. In this study, in order to get the proof of principle for photo-activation a titanium dioxide and a 10 kurus Turkish coin samples are irradiated with high energy bremsstrahlung photons at the end-point energy of 18 MeV and their zinc and copper concentrations are determined by a precise measurement with a high purity germanium detector.

## 1. Introduction

The determination of the elemental concentration in a given sample has taken a great deal of interest over the years. There are two ways to achieve this; destructive and non-destructive methods. Non-destructive methods are very important for the analysis of precious elements. These methods are known to be very reliable and accurate. While destructive methods like X-ray fluorescence, laser ablation and radiochemical neutron activation analysis (RNAA) provide accurate results, the analysed samples are often damaged due to the chemical treatment on samples. Neutron activation analysis (NAA) is one of the well-recognised non-destructive method. In general, thermal neutrons from nuclear reactors utilised for activation. However, it has many disadvantages. For instance, the determination of some low-Z elements such as C, N, O, F or several other elements such as Mg, Si, Ca, Ti, Ni, Sr, Y, Zr, Nb, Tl and Pb is not determined accurately or sometimes the determination is impossible at all [1]. In addition, the results may provide only the surface concentrations [2]. Photo-activation analysis (PAA) on the other hand is a complementary method to NAA. In PAA, the target nucleus is activated through photo-induced reactions such as single-neutron or proton knockout, usually at optimal photon energies between 10 and 30 MeV [3]. It is also shown that PAA allows the determination of elements with a better sensitivity compared to NAA [4,5,6].

In this paper, in order to verify the success of PAA method, we aim to investigate the determination of elemental concentration for a given sample by using PAA. In the next section, we present the experimental method and data analysis. Section 3 presents the results of our measurements and section 4 is devoted to our conclusions.

Journal of Physics: Conference Series **590** (2015) 012053 doi:10.1088/1742-6596/590/1/012053

## 2. Experimental

We provide below the details of experimental method in order to determine the elemental concentration in a 10 kurus Turkish coin and titanium dioxide ($TiO_2$) sample by using a c-linac and HPGe detector in our PAA setup.

### 2.1. *Sample and Reference Preparation*

In order to obtain accurate results in PAA, it is important that the sample and reference material have the same geometry. As it is pointed out in the introduction, in contrast to destructive methods, there is no physical or chemical damage to sample under study in PAA method.

### 2.2. *Activation Process*

Samples are activated with a c-linac with an end-point energy of 18 MeV, which is already in use at Akdeniz University Nuclear Sciences Application and Research Center (NUBA). The irradiation surface is adjusted to 55 cm below the tungsten converter. Sample and reference materials are then placed to the central point of irradiation area carefully and exposed to high energy bremsstrahlung photons. Thus, photonuclear reactions are performed.

### 2.3. *Spectrum Measurement and Analysis*

After the irradiation process and the cooling time, the sample and reference material carried to the gamma spectroscopy laboratory safely in a lead container. Gamma-ray spectrometers with a HPGe coaxial detectors with relative efficiencies of 40% and resolution FWHM 1.85 keV for the 1332 keV photons of $^{60}Co$ were used for PAA. Before counting process, a Job Control file is created with detector software Maestro [7]. Thus, spectrum files can be stored in a certain sequence of time. When counting is finished the spectrum is then analysed by gf3 Radware gamma spectroscopy software [8].

### 2.4. *Calculation of Concentration*

Both samples and reference material are exposed to the same amount photon flux during the irradiation process since the beam profile of our c-linac is very well known. It is assumed that all the samples that will be examined are present on earth, so natural abundance of these elements is the same for both sample and reference material [9]. According to these hypothesis, the relative concentrations of the investigated sample calculated as follows [10]:

$$C_S = C_R \frac{M_R}{M_S} \frac{I_S}{I_R} \frac{e^{-\lambda T_{DR}}}{e^{-\lambda T_{DS}}} \frac{1-e^{-\lambda T_{CR}}}{1-e^{-\lambda T_{CS}}} \qquad (1)$$

In this equation, S denotes the sample and R denotes the reference material. $C_R$ and $C_S$ are concentrations, $M_R$ and $M_S$ are total masses, $I_R$ and $I_S$ are the net peak areas, $T_{DR}$ and $T_{DS}$ are cooling times and $T_{CR}$ and $T_{CS}$ are the measurement times in detector of referance and sample respectively, $\lambda$ is the decay constant of the isotope of the element whose concentration is to be determined.

Usually uncertainity related to decay constant, cooling time and measurement times in detector are much smaller than other uncertainties as given in Eq. (1). Hence, these parameters may not be included in error propagation. We will also assume that the mesurement time was recorded without any uncertanities. Error propagation [10] for Eq. (1) is given by:

$$\Delta C_S = C_S \left[ \left(\frac{\Delta C_R}{C_R}\right)^2 + \left(\frac{\Delta M_R}{M_R}\right)^2 + \left(\frac{\Delta M_S}{M_S}\right)^2 + \left(\frac{\Delta I_R}{I_R}\right)^2 + \left(\frac{\Delta I_S}{I_S}\right)^2 \right]^{1/2} \qquad (2)$$

Various elements concentrations in a sample can be obtained by characteristic gamma ray transition energies and also the error propagation *i.e.* reliability of the obtained data could be calculated along with those equations.

Journal of Physics: Conference Series **590** (2015) 012053 doi:10.1088/1742-6596/590/1/012053

In order to find the zinc and copper concentrations in the $TiO_2$ and 10 kurus coin sample, a zinc pill and a 1 kurus coin used as reference materials respectively.

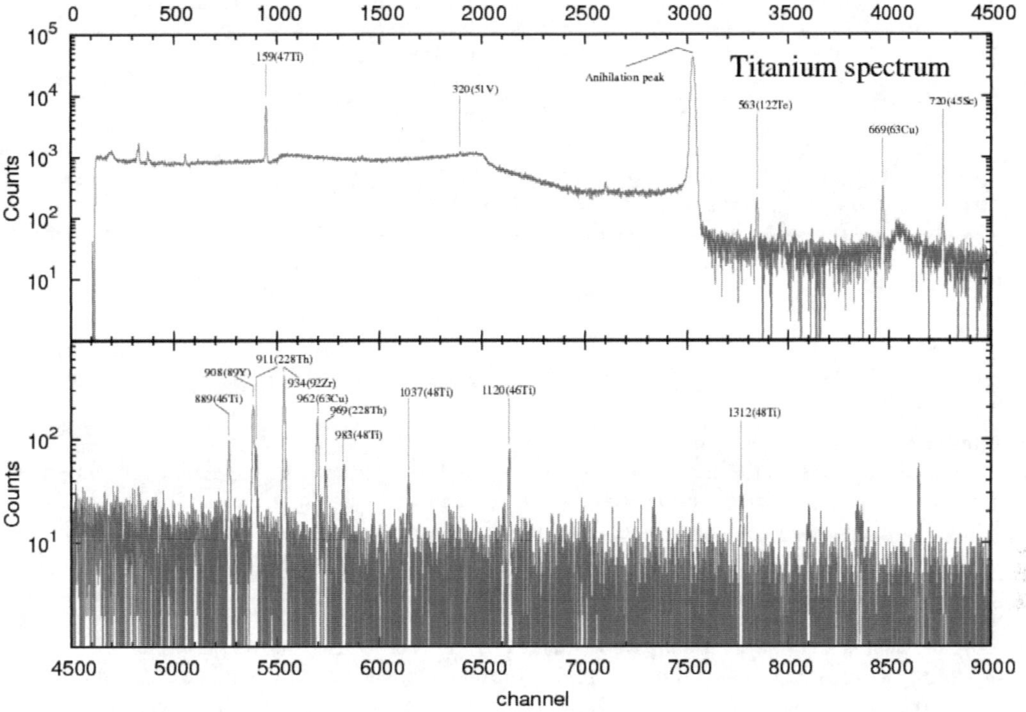


**Figure 1.** Gamma Energy Spectrum of $TiO_2$. $^{64}Zn(\gamma,n)^{63}Zn$ reaction at 669 and 962 keV energies has been observed and analysed.

## 3. Results

As described above, we have irradiated 10 kurus sample together with the reference material by using a c-linac with an end-point energy of 18 MeV and counted the samples with a HPGe detector. The spectrum is shown in Fig. 1. As it can be seen from this figure, we observe not only $(\gamma,n)$ reactions but also $(\gamma,p)$ reactions. However, since we are interested in determining the concentration of zinc, we only use $^{64}Zn(\gamma,n)^{63}Zn$ reaction data of 669 and 962 keV energies. The results of analysis are given in Table 1-2. As it can be seen from the results presented in Tab. 2, the results of concentration analysis for coins are in a good agreement with corresponding literature values. The results for both $TiO_2$ and 10 kurus samples of $^{63}Zn$ at 669 and 962 keV energies are also in a good agreement with each other. Sample concentration values are represented by $C_S$ and their respective uncertainty of the concentration values are represented by $\Delta Cs$.

The reliability of coin results lead us to investigate the zinc concentration in a $TiO_2$ sample as a trace element. Results of this experiment is clearly better owing to the fact that they have smaller uncertainties. It should be noted that the counting time for $TiO_2$ sample was much greater than 10 kurus coin sample. All parameters in Eq. (1) are nearly the same except counting time and so the net peak area. It is then easy to see in order to have minimal uncertainty and more accurate results there should be sufficient amount of counts in an energy peak which should be investigated further.

**Table 1.** Concentrations and their uncertainties in $TiO_2$ sample.

| Name | Isotope | Energy (keV) | $C_R$*(%) | $C_S$(%) | $\Delta C_S$ |
|---|---|---|---|---|---|
| **$TiO_2$** | **$^{63}Zn$** | 669 | 0.333 | 0.269 | 0.013 |
| **$TiO_2$** | **$^{63}Zn$** | 962 | 0.333 | 0.274 | 0.014 |

* *Our Zn reference material*

**Table 2.** Concentrations and their uncertainties in 10 Kr sample.

| Name | Isotope | Energy (keV) | $C_R$*(%) | $C_S$(%) | $\Delta C_S$ |
|---|---|---|---|---|---|
| **10 Kr** | **$^{63}Zn$** | 669 | 29.0 | 22.8 | 3.3 |
| **10 Kr** | **$^{63}Zn$** | 962 | 29.0 | 24.8 | 4.3 |
| **10 Kr** | **$^{63}Cu$** | 1172 | 65.0 | 49.8 | 18.4 |

* *Turkish General Directorate of Mint and Stamp Print House* [11]

## 4. Summary and Conclusions

In this paper, we have shown the application of PAA method to determine the elemental concentration of an intended sample by using a c-linac and a HPGe detector. Therefore, we performed series of experiments on a 10 kurus Turkish coin and $TiO_2$ sample in order to get a proof of principle for photo-activation analysis with our experimental setup. Our findings are in good agreement with the literature values. Through this study, we also show the inherent non-destructive nature of PAA method. This property can used to analyse samples that are too precious to be harmed in any way, like archaeological samples. It is important to note that PAA analysis is used to determine the elemental concentration in a given sample first time in Turkey. Further applications will be made in the future.

In order to improve the results further, counting process has to be handled more carefully since most of the uncertainty in calculations come from the net peak area of the sample and reference material. Effect of dead time, shielding issues and others also have to be taken care of in the future.

**Acknowledgements**
This work is partly supported by the scientific and technological research council of Turkey (TÜBİTAK) with grant number: 114F220.

## References

[1] Randa Z, Kucera J, Mizera J and Frana J 2006 *J. Radioanal. Nucl. Ch.* **271** 589
[2] Green J, Wells D, Benson B, Zaijing S and Maschner H 2011 AIP *Conf. Proc.* **1336** 497
[3] Wells D, Segebade C and Cole P 2010 AIP *Conf. Proc.* **1265** 379
[4] Lutz G J 1971 *Anal. Chem* **43** 93
[5] Randa Z, Spacek B, Kuncir J and Benada J 1981 *Nuclear Information Centre*, Prague
[6] Segebade C, Weise H P and Lutz G J 1988 *Photon Activation Analysis* (Berlin, New York: W. de Gruiter)
[7] ORTEC 2012 Maestro Application Software http://www.ortec-online.com/download/MAESTRO.pdf
[8] Radford D C 2000 Notes on the use of the program gf3 http://radware.phy.ornl.gov/gf3/gf3.html
[9] Boztosun I *et al.* 2014 *Turk. J. Phys.* **38** 1-9
[10] Sun Z J, Wells D, Segebade C and Green J 2011 AIP *Conf. Proc.* **1336** 473
[11] http://www.darphane.gov.tr/tr/content.php?parent_id=182&content_id=735

Journal of Physics: Conference Series **590** (2015) 012054 doi:10.1088/1742-6596/590/1/012054

# Qualitative Elemental Analyses of a Meteorite Sample Found in Turkey by Photo-activation Analysis Method

**C Ertugay[1,2], I Boztosun[1,2], S F Ozmen[1,2] and H Đapo[1,2]**

[1]Akdeniz University, Department of Physics, 07058 Antalya, Turkey
[2]Akdeniz University, Nuclear Sciences Application and Research Center-NUBA, Antalya, Turkey

E-mail : can_ertugay@hotmail.com.tr

**Abstract**. In this paper, a meteorite sample provided from TÜBİTAK National Observatory found in Turkey has been investigated by using a clinical linear accelerator that has endpoint energy of 18 MeV, and a high purity Germanium detector for qualitative elemental analysis within photo-activation analysis method. 21 nuclei ranging from $^{24}$Na to $^{149}$Nd have been identified in the meteorite sample.

## 1. Introduction

The determination or identification of a sample found in nature whether it is a meteorite or not has been studied in detail over the years by using different experimental methods. As it is so far well established, there are two ways to perform elemental analysis of a sample. First type is based on destructive methods. Such methods used for elemental analysis either damage the sample or allow only surface analysis if not damaging the sample. The second one is based on performing the elemental analysis of a sample non-destructively. Non-destructive methods are very important for the analysis of particularly precise samples. As it is known, these non-destructive methods are also very reliable and accurate.

Among these non-destructive methods, X-ray fluorescence (XRF) is based on emission of characteristic secondary X rays from a material excited with high energy X rays or gamma rays. XRF can be applied non-destructively only when it comes to elemental analyses of very thin surface layers, but for a more detailed elemental analysis, XRF cannot be applied non-destructively.

Another non-destructive method is the neutron activation analyses (NAA). NAA is based on neutron capture of a nucleus. Generated radioactive isotopes emit characteristic gamma rays. Then, the elemental analysis can be performed through analyzing the spectrum obtained from these gamma rays. The disadvantage of this method is that it needs a nuclear reactor or a high flux neutron source, and the availability of these facilities is very difficult and expensive.

Another type of non-destructive methods is photo-activation analysis (PAA). In the PAA method, the nuclei are basically activated through the bremsstrahlung photons obtained from an electron accelerator. The generated isotopes emit characteristic gamma rays with certain energies. These gamma photons are counted by a gamma detector. At the end, the nuclei can be identified through the analyzing the gamma spectra.

Journal of Physics: Conference Series **590** (2015) 012054 doi:10.1088/1742-6596/590/1/012054

The bremsstrahlung photons can be obtained through a clinical linear accelerator (c-LINAC) which is mainly used for medicinal purposes, and the spectra can be analyzed through a gamma detector. In this regard, our center has a c-LINAC and a gamma detector. The neutron separation (γ,n) and proton separation (γ,p) occur in the PAA method. But, for heavy elements such as uranium, also the photo-fission reaction (γ,f) may occur [3].

PAA method has been used since 1970s in the world. For the first time in Turkey, it was carried out by Akdeniz University Nuclear Sciences Research and Application Center (NUBA) using local facilities of the country in 2013 [1]. The PAA method we use to perform elemental analysis is quite suitable for elemental analyses of a sample especially when it comes to precise samples such as meteorites mainly because of its non-destructiveness, cheapness, easy accessibility. It also makes an analysis within a reasonable time. It is able to achieve the identification where other methods fail, and it does not include any radioactive waste either. When only small pieces of the meteorite are available, precise methods such as Photo-activation Analysis (PAA) are extremely suitable [2].

Therefore, in this paper, we have performed the qualitative elemental analyses of a meteorite sample found in Turkey by using the PAA method.

In the next section, we explain our experimental method including property of our setup, activation, measurement and analysis process. Section 3 is devoted to our experimental results. In the final section, we present our summary and conclusions.

## 2. Material and Method

In our setup, we have used a cLINAC SLI-25 manufactured by Philips Medical Systems (currently part of Elekta TM Synergy TM) as a bremsstrahlung photon source. The cLINAC primary electron beam is generated by an electron gun with an energy of about 50 keV. The electron gun in SLI-25 is a diode design with a 400 Hz pulse repetition frequency.

The detector used is a high purity Germanium detector (HPGe). It is a p-type, coaxial, electrically cooled HPGe detector, placed in a well shielded cavity. The HPGe detector is connected to a set of Nuclear Instrumentation Modules consisting of ORTEC preamplifier, bias supply, spectroscopy amplifier, analog to digital converter and a computer. Data acquisition was carried out with MAESTRO32 software.

### *2.1. Activation Process*

The sample was placed 60 cm away from irradiation region (the end point of the head of the accelerator). Then, the sample was exposed to bremsstrahlung at endpoint energy with 18 MeV, and the reactions were carried out with 10000 Monitor Unit (MU). Monitor Unit is a measure of the radiation dose, and 100 MU equal approximately 1Gy usually at 100 cm from the source.

### *2.2. Measurement and Analysis of the Spectrum*

When the irradiation ended, after the cooling time, the sample was moved to the gamma spectrometer laboratory in order to count with the HPGe detector. The spectrum analysis recorded by the software program of the detector was performed by the gf3 Radware program, compiling the interactive gamma-ray spectrum data graphically.

## 3. Results

The gamma spectrum of the meteorite sample obtained from analysis with background is shown in Figure 1. As a result of analyses, we present our results in comparison with the literature values taken from National Nuclear Data Center (NUDAT), 21 different nuclei were identified according to energy levels. We also present the type of reactions that have taken place during the irradiation. Table 1 shows all these data. As it is shown in the table, four possible reactions occurred in the experiment.

Journal of Physics: Conference Series **590** (2015) 012054 doi:10.1088/1742-6596/590/1/012054

We have observed (γ,n), (γ,p) and (γ,γ') reactions, and also (n,γ) reaction due to the secondary neutrons produced by the interaction of bremsstrahlung with other materials in the accelerator.

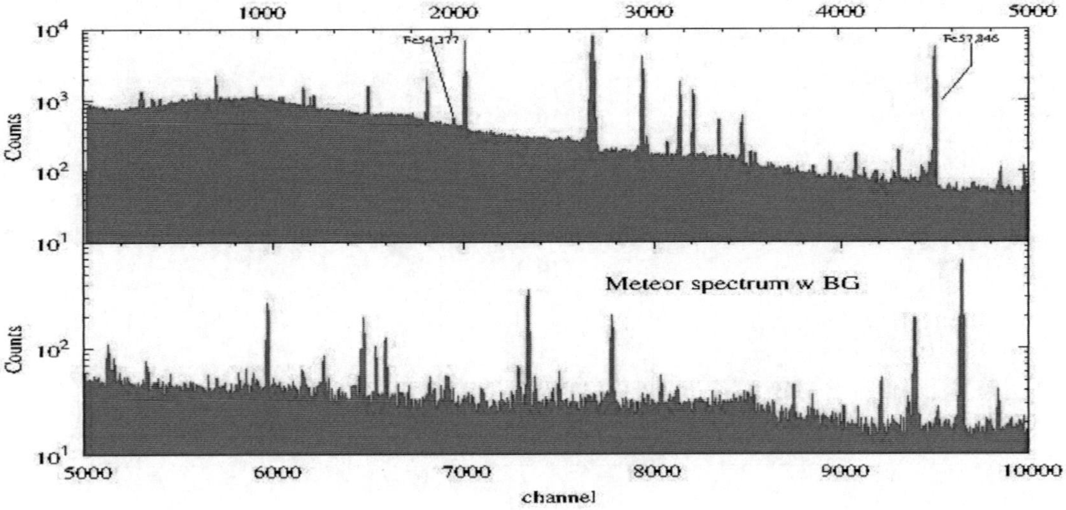


**Figure 1.** Meteorite Spectrum

**Table 1.** The identified nuclei and their energy levels in the meteorite sample obtained from analysis.

| NUCLEUS | Energy (keV) | REACTION |
|---|---|---|
| $^{149}$Nd | 114.45 | |
| $^{149}$Nd | 211.64 | |
| $^{149}$Nd | 270.82 | |
| $^{149}$Nd | 423.81 | |
| $^{74}$As | 635.08 | |
| $^{74}$As | 596.21 | |
| $^{47}$Ca | 1297.41 | |
| $^{57}$Ni | 127.44 | |
| $^{57}$Ni | 1377.9 | |
| $^{57}$Ni | 1758.06 | |
| $^{57}$Ni | 1919.58 | |
| $^{85}$Sr | 151.29 | (γ,n) |
| $^{85}$Sr | 231.98 | |
| $^{53}$Fe | 377.93 | |
| $^{63}$Zn | 670.15 | |
| $^{63}$Zn | 963.34 | |
| $^{58}$Co | 811.07 | |
| $^{54}$Mn | 834.32 | |
| $^{34}$Cl | 1176.95 | |
| $^{34}$Cl | 2127.67 | |
| $^{34}$Cl | 146.61 | |
| $^{69}$Zn | 438.78 | |

| | | |
|---|---|---|
| $^{57}$Mn | 122.07 | |
| $^{76}$As | 559.45 | |
| $^{76}$As | 563.83 | |
| $^{76}$As | 657.44 | |
| $^{76}$As | 665.73 | |
| $^{76}$As | 1213.27 | |
| $^{76}$As | 1216.52 | |
| $^{76}$As | 1229.02 | |
| $^{76}$As | 1787.69 | |
| $^{76}$As | 2096.61 | |
| $^{56}$Mn | 847.13 | (γ,p) |
| $^{56}$Mn | 1810.99 | |
| $^{56}$Mn | 2113.25 | |
| $^{56}$Mn | 2522.96 | |
| $^{56}$Mn | 2657.34 | |
| $^{56}$Mn | 2959.57 | |
| $^{234}$Pa | 1001.45 | |
| $^{60}$Co | 1332.54 | |
| $^{24}$Na | 1368.81 | |
| $^{24}$Na | 2754.01 | |
| $^{42}$K | 1525.24 | |
| $^{87}$Sr | 388.85 | (γ,γ') |
| $^{140}$La | 487.23 | |
| $^{140}$La | 1596.21 | |
| $^{59}$Fe | 1099.58 | (n,γ) |
| $^{38}$Cl | 1642.90 | |
| $^{38}$Cl | 2167.57 | |

## 4. Conclusions and Discussion

In this paper, we have used the PAA method in order to identify the nuclei in a meteorite sample. We have used bremsstrahlung photons obtained from a c-LINAC for irradiation and a HPGe detector has been used for counting. We have reported here preliminary results and a detailed analysis of these measured data including the quantitative results will be further investigated.

## Acknowledgements

This work is partly supported by the scientific and technological research council of Turkey (TÜBİTAK) with grant number: 114F220.

## References

[1] Boztosun I *et al.* 2014 Turk. J. Phys. **38** 1-9
[2] Řanda Z *et al.* 2003 *Journal of Radioanalytical and Nuclear Chemistry* **257** 275
[3] Green J *et al.* 2011 AIP Conf. Proc **1336** 497

Journal of Physics: Conference Series **590** (2015) 012055 doi:10.1088/1742-6596/590/1/012055

# Interaction of $1p$ nuclei: Case of $^{14}N+^{12}C$ Elastic Scattering at 21 0 MeV

**N Burtebayev[1], D Alimov[1,4], I Boztosun[2,3], J Burtebayeva[1], Zh K Kerimkulov[1], M Nassurlla[1,4], N Amangeldy[5], A K Morzabayev[5], S K Sakhiev[5], Sh Hamada[6]**

[1] Institute of Nuclear Physics, 050032, Almaty, Kazakhstan,
[2] Akdeniz University, Faculty of Science, Department of Physics, 07058, Antalya, Turkey,
[3] Akdeniz University, Nuclear Sciences Application and Research Center - NUBA, 07058 Antalya, Turkey,
[4] Kazakh National University, Almaty, Kazakhstan,
[5] Eurasia National University, Astana, Kazakhstan,
[6] Tanta University, Faculty of Science, Tanta, Egypt.

E-mail: boztosun@akdeniz.edu.tr

**Abstract.** Optical model analysis has been conducted for the elastic scattering of $1p$-shell nuclei around the Coulomb barrier energies. We have used both microscopic double-folding and phenomenological potentials for the real part of the complex nuclear potential. The imaginary potential has the shape of phenomenological Wood-Saxon volume. The case $^{14}N+^{12}C$ for $1p$-shell nuclei has been studied in detail and it is noticed that a large normalization of the strength of the double-folding real potential is needed to explain the structure observed in the experimental data. A good agreement between experimental data and theoretical results is obtained for the phenomenological potential case.

## 1. Introduction

The interaction of $1p$-shell nuclei presents a variety of rich structure from clustering effect to nucleon transfer. The reactions around the Coulomb barrier and over the barrier have different diffraction patters. While around the Coulomb barrier, the experimental data presents a Fresnel type, over the Coulomb barrier it changes to Fraunhofer type diffraction pattern. The potential approach in Optical and Coupled-channels models in order to explain the experimental data have certain difficulties to reproduce the certain aspects of the structure observed. The experimental data have a structured peak at forward angles and an oscillatory behavior at intermediate angles while at large angles, the measured cross-section has a backward rise. Theoretical models attempting to explain the experimental data ranging from cluster exchange to parity dependent potential have difficulties to explain the oscillatory structure and backward rise of the data. These theoretical studies have shown that the real part of the complex optical potential should be deep while the imaginary part should be weak, that is a surface transparent potential approach. The backward rise of the experimental data has also related to the $\alpha$-particle exchange between projectile and target nuclei [1, 2, 3, 4, 5, 6, ?, ?, 9, 10, 11, 12].

The experimental measurements for $^{14}N+^{12}C$ nuclear system as an example of $1p$-shell nuclei were performed in the cyclotron DC-60 located in Institute of Nuclear Physics, Almaty-

Journal of Physics: Conference Series **590** (2015) 012055 doi:10.1088/1742-6596/590/1/012055

Kazakhstan [13]. The $^{14}$N ion beam has been accelerated up to energies 1.25, 1.5, and 1.75 MeV/A and then directed to $^{12}$C target nuclei of thickness 31.9 $\mu g/cm^2$. The beam current was nearly equal to 20 $nA$ during the experiment, the angular distribution was measured in a wide range of angles $\approx$ 20-130° in center of mass system. Energy spectra of scattered particles were recorded with a semiconductor silicon surface barrier detector ORTEC company sensitive layer with a thickness of 100 microns. The energy resolution of the registration system was 250-300 keV, which is mainly determined by the energy spread of the primary beam. Only one detector was used in our measurements which detect both $^{14}$N fragment and $^{12}$C residual peaks. Spectrum analysis has been performed using program MAESTRO [14]. The details of experimental measurements can be found in Ref. [15]. As it can be seen from the experimental data, it does not present an oscillatory structure and a fall off at the intermediate angles. A backward rise is not also observed as opposed to similar systems.

Therefore, in order to understand the underlying reaction mechanism, we have conducted a theoretical study of $^{14}$N+$^{12}$C reaction at $E_{Lab}$=21.0 MeV. We have used both optical and coupled channels analysis to explain the elastic scattering data.

In the next section, we introduce our Optical model and potential parameters to explain the observed experimental data. Then, we show the results of optical and coupled-channels model analyzes in section 3. Our conclusion is given in Section 4.

## 2. The Model

In the present calculations, we have used both a phenomenological deep and double-folding real potentials with a volume type imaginary potential [16, 22]. Our total real potential for these cases consists of the nuclear potential, $V_{Nuclear}$, and the Coulomb and centrifugal potentials, $V_{Coul}$, $V_{Cent}$ respectively

$$V_{total}(r) = V_{Nuclear}(r) + V_{Coulomb}(r) + V_{Centrifugal}(r) \tag{1}$$

The phenomenological nuclear potential is assumed to have the square of a Woods-Saxon shape:

$$V_{Nuclear}(r) = \frac{-V_0}{(1 + exp(r - R)/a)^2} \tag{2}$$

where $V_0$=275.0 MeV and R=$r_0({A_a}^{1/3}$+${A_A}^{1/3}$) with $r_0$=0.780 fm and a=1.4 fm.

The Coulomb potential [16] due to a charge $Z_a e$ interacting with a charge $Z_A e$, distributed uniformly over a sphere of radius $R_c$, is also added.

$$V_{Coulomb}(r) = \frac{1}{4\pi\epsilon_o}\frac{Z_a Z_A e^2}{r}, \qquad r \geq R_c \tag{3}$$

$$= \frac{1}{4\pi\epsilon_o}\frac{Z_a Z_A e^2}{2R_c}(3 - \frac{r^2}{R_c^2}), \qquad r < R_c \tag{4}$$

where $R_c$=5.64 fm is the Coulomb radius, and $Z_a$ and $Z_A$ denote the charges of the projectile $a$ and the target nuclei $A$ respectively.

Microscopic nuclear potential we have also used to analyze the experimental data for $^{14}$N+$^{12}$C system is based on the double-dolding model [17]. Double folding potential is calculated by using the nuclear matter distributions of both projectile and target nuclei together with an effective nucleon-nucleon interaction potential ($\nu_{NN}$) are used. Thus, the double folding potential is

$$V_{\rm DF}(\mathbf{r}) = \int d\mathbf{r}_1 \int d\mathbf{r}_2 \rho_P(\mathbf{r}_1)\rho_T(\mathbf{r}_2)\nu_{NN}(r_{12}) \tag{5}$$

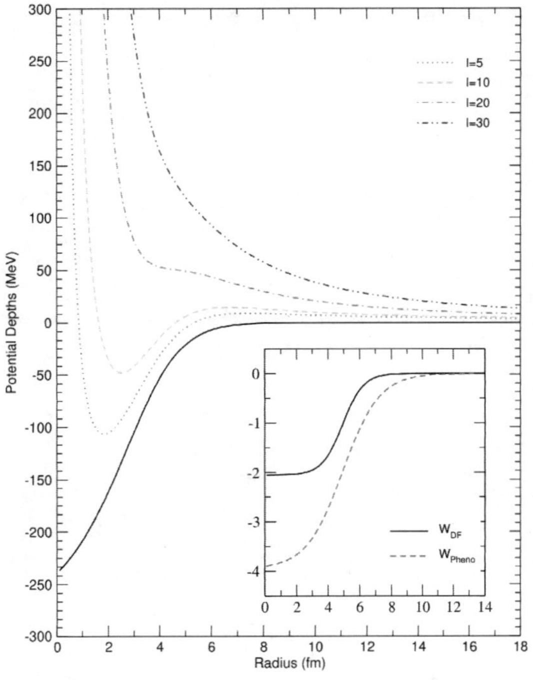


**Figure 1.** Interaction potential between $^{14}N$ and $^{12}C$ for various values of the orbital angular momentum quantum number, $l$. The parameters are given in the text.

where $\rho_P(\mathbf{r}_1)$ and $\rho_T(\mathbf{r}_2)$ are the nuclear matter density of projectile and target nuclei, respectively. Gaussian density distributions (GD) have been used for both nuclei [18, 19, 20] defined as:

$$\rho(r) = \rho_0 \exp\left(-\beta r^2\right) \tag{6}$$

where $\beta$ is adjusted to reproduce the experimental value for the rms radius of the $^{14}N$=2.58 fm and $^{12}C$=2.472 fm [21]. $\rho_0$ values can be obtained from the normalization condition

$$\int \rho(r) r^2 dr = \frac{A}{4\pi} \tag{7}$$

where A is the mass number.

The effective nucleon-nucleon interaction, $\nu_{NN}$, is integrated over both density distributions. Several nucleon-nucleon interaction expressions can be used for the folding model potentials. We have chosen the most common one, the M3Y nucleon-nucleon (Michigan 3 Yukawa) realistic interaction. The M3Y has two form, one corresponds to M3Y-Reid and another is based on the so-called M3Y-Paris interaction [17]. In the present work, we use the former form with the relevant exchange correction term due to the Pauli principle, given by

$$\nu_{NN}(r) = 7999\,\frac{\exp(-4r)}{4r} - 2134\,\frac{\exp(-2.5r)}{2.5r} + J_{00}(E)\delta(r)\text{ MeV}, \tag{8}$$

where $J_{00}(E)$ represents the exchange term, since nucleon exchange is possible between the projectile and the target. $J_{00}(E)$ has a linear energy-dependence and can be expressed as

$$J_{00}(E) = \ 276\ [1 - 0.005\ E_{\text{Lab}}/A_p]\text{ MeV fm}^3 \tag{9}$$

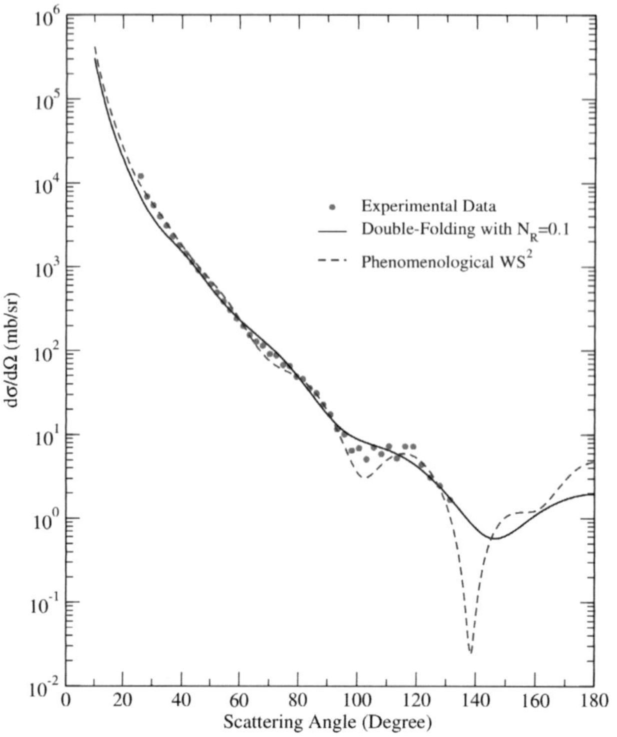


**Figure 2.** The results of Double-folding and phenomenological Woods-Saxon-squared potentials in comparison with the experimental data at 21.0 MeV.

The sum of the nuclear, Coulomb and the centrifugal potentials for both microscopic double-folding and phenomenological Wood-Saxon squared potentials is shown in figure 1 for various values of the orbital angular momentum. The sum of these potential create a pocket in the total effective potential. This pocket is very important for the interference of the barrier and internal waves, which creates the oscillatory structure observed in the cross-section [16].

The imaginary part of the potential [22] was taken as Woods-Saxon volume shaped

$$W(r) = -\frac{W_V}{(1 + exp((r - R)/a))} \tag{10}$$

with $W_V$=3.95 MeV, $r_V$=1.050, $a_V$=1.151 fm. The imaginary potential for the double-folding potential has the same shape but we had to change the parameters in order to obtain a better agreement. Thus, the parameters of the imaginary potential for the double folding calculations are $W_V$=2.05 MeV, $r_V$=1.050, $a_V$=0.651 fm.

The shape of the imaginary potentials for the double-folding and phenomenological potential calculations is shown in the inset of Fig. 1 for $E_{Lab}$=21.0 MeV.

## 3. The Results

The results obtained by using the microscopic double-folding and phenomenological nuclear potentials with the above-described models are shown in Fig. 2 in comparison with the experimental data. As it can be seen from this figure that double folding potential requires a large normalization in order to obtain a reasonable result. Without this normalization, $N_R = 0.1$, we could not get an agreement with the experimental data. The results of double-folding potential calculations are structureless and does not fit the experimental data well at forward angles. The phenomenological Woods-Saxon typed potential has provided a better agreement with the experimental data in comparison with double-folding calculation as seen in the same figure with dashed line. The maximum and minimum observed in the experimental data are in phase with the theoretical results.

## 4. Summary

In this paper, we have conducted an optical model analysis for the elastic scattering of $^{14}$N+$^{12}$C system around the Coulomb barrier energies. We have used both microscopic double-folding and phenomenological potentials for the real part of the complex nuclear potential. It is shown that a large normalization of the microscopic double-folding potential is needed to explain the measured data. The coupling effect may play an important role for a strongly deformed nucleus as $^{12}$C in the coupled-channels calculations. The role of neutron transfer should also be studied in order to clarify the discrepancies observed in the double-folding potential case. Further investigations are being carried out for this and similar systems. For the phenomenological potential case, a good agreement between experimental data and theoretical result is obtained by using a deep real potential.

## Acknowledgments

This work is partly supported by the scientific and technological research council of Turkey (TÜBİTAK) with grant number: 114F220.

[1] Boztosun I and Rae W D M 2001 *Phys. Lett.* **518B** 229
[2] Bremner C *et al.* 2002 *Phys. Rev.* C **66** 034605
[3] Freer M *et al.* 2004 *Phys. Rev.* C **70** 064311
[4] Burtebaev N *et al.* 2005 *Phys. of Atom. Nucl.* **68** 1303
[5] Burtebayev N *et al.* 2010 *Phys. of Atom. Nucl.* **73** 746
[6] Boztosun I and Rae WDM 2001 *Phys. Rev.* C **63** 054607; 2001 *Phys. Rev.* C **64** 054607; 2002 *Phys. Rev.* C **65** 024603
[7] Hamada Sh, Burtebayev N and Amangeldi N 2014 *Int. J. of Mod. Phys.* E **23** 145061
[8] Hamada Sh, Burtebayev N, Gridnev K A and Amangeldi N 2011 *Nucl. Phys.* A. **859** 29
[9] Amar A *et al.* 2011 *Int. J. of Mod. Phys. E-Nucl. Phys.* **20** 980
[10] Freer M *et al.* 2005 *Phys. Rev.* C **71** 047305
[11] Kucuk Y and Boztosun I 2006 *Nucl. Phys.* A **764** 160
[12] Freer M *et al.* 2007 *Phys. Rev.* C **76** 034320
[13] Burtebayev N *et al.* 2014 *WASET* **8** 335
[14] Maestro-32, 2008. Multi-channel analyser software, A65-B32 model. Ortec.
[15] Hamada Sh, Burtebayev N, Gridnev KA and Amangeldi N 2011 *Phys. Scr.* **84** 045201
[16] Boztosun I 2002 *Phys. Rev.* C **66** 024610
[17] Cook J 1982 *Comp. Phys. Comm.* **25** 125
[18] Karakoc M and Boztosun I 2006 *Phys. Rev.* C **73** 047601
[19] Aygun M, Kucuk Y, Boztosun I and Ibraheem A A 2010 *Nucl. Phys.* A **848** 245
[20] Karakoc M and Boztosun I 2006 *Int. J. of Mod. Phys.* E **15** 1317
[21] De Vries H, De Jager C W and De Vries C 1987 *At. Data and Nucl. Data Tables* **36** 495
[22] Kucuk Y, Boztosun I and Topel T 2009 *Phys. Rev.* C **80** 054602

Journal of Physics: Conference Series **590** (2015) 012056 doi:10.1088/1742-6596/590/1/012056

# Optical and Coupled-Channels Description of $^{20}$Ne+$^{16}$O Elastic Scattering

**N Burtebayev[1], M Nassurlla[1,4], D Alimov[1,4], I Boztosun[2,3], J Burtebayeva[1], A K Morzabayev[5], S K Sakhiev[5], S V Artemov[6], J M Mussaev[7]**

[1] Institute of Nuclear Physics, 050032, Almaty, Kazakhstan,
[2] Akdeniz University, Faculty of Science, Department of Physics, 07058, Antalya, Turkey,
[3] Akdeniz University, Nuclear Sciences Application and Research Center - NUBA, 07058 Antalya, Turkey,
[4] Kazakh National University, Almaty, Kazakhstan,
[5] Eurasia National University, Astana, Kazakhstan,
[6] Institute of Nuclear Physics, Tashkent, Uzbekistan,
[7] South Kazakhstan State University, Shymkent, Kazakhstan.

E-mail: boztosun@akdeniz.edu.tr

**Abstract.** The elastic scattering of the $^{20}$Ne+$^{16}$O system has been analyzed with a phenomenological potential approach within the framework of the optical and coupled-channels models at $E_{Lab}$=50.0 MeV. The striking feature of the experimental data is the oscillatory structure at the intermediate angles and a rapid increase at large angles. Optical potentials have difficulty in describing such structures and predict a fall of experimental data around the intermediate angles. In order to explain this structure, we have used a deep real potential with a sum of Woods-Saxon typed surface and volume imaginary potentials. We present that deep real potential with these imaginary potentials explain the oscillatory structure and backward rise observed in the elastic scattering data within both models. It should be pointed out that there is a magnitude problem of the inelastic $2^+$ data for the deformed $^{20}$Ne nucleus that the standard coupled-channels model is unable to predict correctly.

## 1. Introduction

The elastic and inelastic scattering of $4N$-type nuclei have taken a great deal of interest and have been amongst the main sources of information about the structure of complex nuclei over the years [1, 2, 3, 4, 5, 6, 7]. There have been numerous experimental and theoretical investigations to understand the structure and reactions of *sp*-shell and $4N$-type $\alpha$-structured nuclei, and these investigations for the nuclear reactions have displayed a common unexpected feature near $\theta_{CM}$=180° for the elastic and inelastic scattering cross-sections. The scattering of $\alpha$-particles and strongly clusterized heavy ions from light nuclei is an example that one can observe a significant cross section enhancement at backward angles, which in some cases is due to exchange of $\alpha$-particles between the interacting nuclei [8, 9, 10, 11, 12, 13].

The experimental data for the $^{20}$Ne+$^{16}$O system has been measured by Stock *et al.* [14] and it exhibits an oscillatory structure at the intermediate angles and a rapid increase at large angles. Optical potentials proposed to explain the experimental data have failed to describe the oscillatory structure at intermediate angles and also the backward rise within the framework of the optical model. In a recent study, Yang and Li [15] have investigated the $^{20}$Ne+$^{16}$O system in the energy range of $E_{c.m.}$ = 24.5 to 35.5 MeV by using a folding potential based on the four-$\alpha$-particle model for the $^{16}$O nucleus and the $\alpha$+$^{16}$O model for the $^{20}$Ne nucleus. They have obtained a reasonable description of the experimental data for these energies. However, the physical origin of the observed structure is not yet fully understood [16] and interpreted as an exchange of $\alpha$-particle between projectile and target nuclei as Yang and Li [15] underlined.

In this paper, as a part of our further experimental and theoretical research in the future, we first attempt to explain the elastic scattering data of $^{20}$Ne+$^{16}$O system at 50 MeV by using a deep real and imaginary potentials within the framework of optical model. It is known that the inclusion of the excited states of the target nucleus in the coupled-channels calculations may have a significant effect on the scattering. Therefore, we have also conducted coupled-channels calculations to observe this effect.

In the next section, we introduce our Optical model and potential parameters to explain the observed experimental data. Then, we show the results of optical and coupled-channels model analyzes in Section 3. Our conclusion is given in Section 4.

## 2. The Model

The standard Optical model with folding model potentials or with similar phenomenological potentials such as the square of the Woods-Saxon has failed to describe certain aspects of the experimental data, particularly the oscillatory structure observed in the experimental data and the description of the inelastic scattering data. Therefore, in the present calculations, we use a deep real potential with a volume and surface type imaginary potentials [17]. Our total real potential consists of the nuclear potential, $V_{Nuclear}$, and the Coulomb and centrifugal potentials, $V_{Coul}$, $V_{Cent}$ respectively

$$V_{total}(r) = V_{Nuclear}(r) + V_{Coulomb}(r) + V_{Centrifugal}(r) \tag{1}$$

The nuclear potential is assumed to have the square of a Woods-Saxon shape:

$$V_{Nuclear}(r) = \frac{-V_0}{(1 + exp(r - R)/a)^2}. \tag{2}$$

where $V_0$=740.0 MeV and R=$r_0(A_a{}^{1/3}+A_A{}^{1/3})$ with $r_0$=0.680 fm and a=1.36 fm.

The Coulomb potential [17] due to a charge $Z_a e$ interacting with a charge $Z_A e$, distributed uniformly over a sphere of radius $R_c$, is also added.

$$V_{Coulomb}(r) = \frac{1}{4\pi\epsilon_o}\frac{Z_a Z_A e^2}{r}, \qquad r \geq R_c \tag{3}$$

$$= \frac{1}{4\pi\epsilon_o}\frac{Z_a Z_A e^2}{2R_c}(3 - \frac{r^2}{R_c^2}), \qquad r < R_c \tag{4}$$

where $R_c$=5.76 fm is the Coulomb radius, and $Z_a$ and $Z_A$ denote the charges of the projectile $a$ and the target nuclei $A$ respectively.

The sum of the nuclear, Coulomb and the centrifugal potentials is shown in Fig. 1 for various values of the orbital angular momentum. The superposition of the attractive and repulsive potentials results in the formation of a potential pocket. The width and depth of the pocket depend on the orbital angular momentum quantum number for a given nuclear potential. This pocket is very important for the interference of the barrier and internal waves, which creates the oscillatory structure observed in the cross-section [17].

The imaginary part of the potential was taken as the sum of a Woods-Saxon volume and the surface potentials:

$$W(r) = -W_V f(r, R_V, a_V) + 4W_S a_S df(r, R_S, a_S)/dr \tag{5}$$

Journal of Physics: Conference Series **590** (2015) 012056 doi:10.1088/1742-6596/590/1/012056

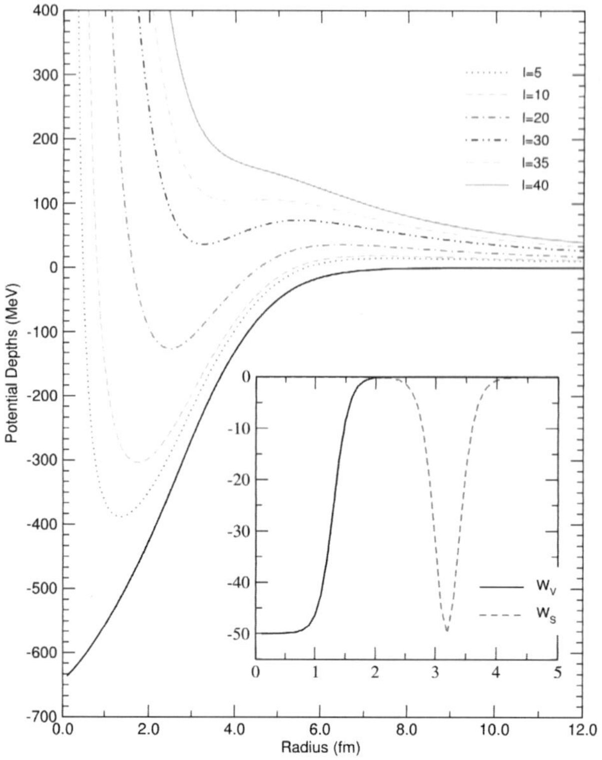


**Figure 1.** Interaction potential between $^{16}O$ and $^{20}Ne$ for various values of the orbital angular momentum quantum number, $l$. The parameters are given in the text.

$$f(r, R, a) = \frac{1}{(1 + exp((r - R)/a))} \tag{6}$$

with $W_V$=50.0 MeV, $r_V$=0.250, $a_V$=0.120 fm and $W_S$=50.0 MeV, $r_S$=0.610, $a_S$=0.140 fm.

The shape of the imaginary potentials are shown in the inset of Fig. 1 for $E_{Lab}$=50.0 MeV. The relative significance of the volume and surface components of the imaginary potential is also studied for this energy. Removing the volume component mainly changes the amplitude of the cross-section at large angles while removing the surface component increases the cross-sections at large angles. The effect of the deep real and imaginary potentials can be understood in terms of the interference between the internal and barrier waves that correspond to a decomposition [17] of the scattering amplitude into two components, the inner and external waves. The inner wave comes from the reflection at the inner face of the total real potential pocket and the external wave comes from the reflection at the outer barrier as the potential pocket is shown in Fig. 1 for the total real potential.

### 3. The Results

Using the above-described Optical model, we have analyzed the elastic scattering data of the $^{20}Ne+^{16}O$ system at 50.0 MeV in the laboratory system. The comparisons between experimental data and the Optical model fits are shown in Fig. 2. In this figure, we have also presented the prediction of the optical potential of Ref. [14]. As it can be seen from this figure, the optical

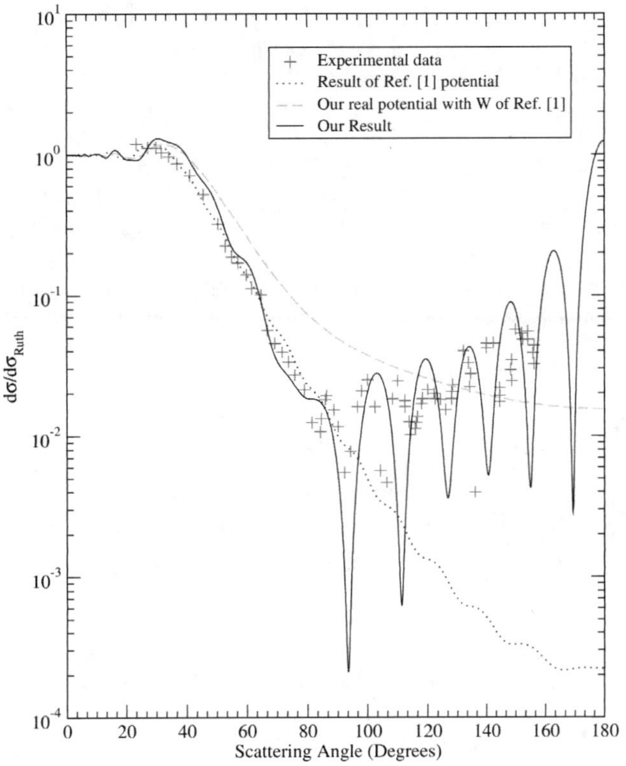


**Figure 2.** The effect of the two imaginary potentials: Dotted line is the prediction of the potential of Ref. [14] while dashed line is the result of our deep real potential with the imaginary potential of Ref. [14]. Solid line is our prediction with deep real potential with the sum of volume and surface imaginary potentials.

potential describes the experimental data at the forward angles up to 90° and then theoretical prediction falls off and unable to reproduce the oscillatory structure and rises at large angles. Therefore, we have used a deep real potential with the imaginary potential of Ref. [14] and it increases the magnitude of the cross-section but it is still unable to produce the oscillatory structure observed in the experimental data. By using the above-described imaginary potentials given in the Model section, we have obtained a reasonable agreement with the experimental data over all angles measured. A good agreement between the experimental data and our prediction is obtained at the forward angles and the oscillatory structure is reproduced at intermediate and backward angles.

In order to see the coupling effect on the scattering observables, we have conducted coupled-channels calculations by deforming the projectile $^{20}$Ne to treat its collective excitation explicitly in the framework of the coupled-channels formalism. It has been assumed that the projectile $^{20}$Ne nucleus has a static quadrupole deformation, and that its rotation can be described in the framework of the collective rotational model. It is therefore taken into account by deforming the real optical potential in the following way

$$R(\theta,\phi) = r_0\ {}^{16}O^{1/3} + r_0\ {}^{20}Ne^{1/3}\ [1 + \beta_2 Y_{20}(\theta,\phi)] \tag{7}$$

Journal of Physics: Conference Series **590** (2015) 012056 doi:10.1088/1742-6596/590/1/012056

where $\beta_2$=-0.23 is the deformation parameter of $^{20}$Ne [18]. This value should be increased in order to fit the magnitude for the $2^+$ data. We have included the first excited state, $2^+$ (1.78 MeV), of the projectile nucleus $^{20}$Ne.

The inclusion of the $2^+$ excited state has an effect on the elastic scattering fits at intermediate and large angles. However, this effect is not substantial and can be altered by changing the parameters of optical potential. Furthermore, the prediction for the $2^+$ state inelastic data has a large magnitude problem. In order to get the magnitude right, we have to increase the deformation parameter enormously. We could not solve this problem but it should be studied in detail by using the $\alpha$-cluster structure of $^{20}$Ne and $^{16}$O as well as by the inclusion of the $\alpha$-transfer in the reaction calculations. This has been a long standing problem that a solution has been proposed by using a second derivative type of coupling potential [1] which should be applied to this reaction.

## 4. Summary

We have shown in this paper that the interaction of $\alpha$-cluster structure type nuclei such as $^{20}$Ne+$^{16}$O system are very sensitive to the shape of the real and imaginary potentials in the surface region. In this paper, we have first shown that the real potential, responsible to describe the scattering observables, should be deep enough to provide a pocket in the total potential for the interference of the external waves, coming from the reflection at the outer barrier and internal waves, coming from the reflection at the inner face of the total real potential pocket. Secondly, the imaginary potential should have two components located inside the real potential as a sum of volume and surface types. Volume one is responsible to account for the flux going to the inelastic channels as excitation while the surface one is responsible to take into account $\alpha$-type cluster exchange occurring at the surface of interaction. Coupled-channels calculations show an effect on the elastic scattering observables but within the standard coupled-channels calculations by using the experimental deformation parameters, it is difficult to describe the excited state data. In particular, the magnitude of the inelastic data is underestimated by a large factor and without increasing the deformation parameter, the experimental data for the inelastic scattering cannot be described.

## Acknowledgments

This work is partly supported by the scientific and technological research council of Turkey (TÜBİTAK) with grant number: 114F220.

[1] Boztosun I and Rae WDM 2001 *Phys. Lett.* **518B** 229
[2] Bremner C *et al.* 2002 *Phys. Rev.* C **66** 034605
[3] Freer M *et al.* 2004 *Phys. Rev.* C **70** 064311
[4] Karakoc M and Boztosun I 2006 *Phys. Rev.* C **73** 047601; 2006 *Int. J. of Mod. Phys.* E **15** 1317
[5] Freer M *et al.* 2005 *Phys. Rev.* C **71** 047305
[6] Kucuk Y and Boztosun I 2006 *Nucl. Phys.* A **764** 160
[7] Freer M *et al.* 2007 *Phys. Rev.* C **76** 034320
[8] Burtebaev N *et al.* 2005 *Phys. of Atom. Nucl.* **68** 1303
[9] Burtebayev N *et al.* 2010 *Phys. of Atom. Nucl.* **73** 746
[10] Burtebayev N *et al.* 2013 *Nucl. Phys.* A **909** 1303
[11] Hamada Sh, Burtebayev N and Amangeldi N 2014 *Int. J. of Mod. Phys.* E **23** 145061
[12] Hamada Sh, Burtebayev N, Gridnev KA and Amangeldi N 2011 *Nucl. Phys.* A **859** 29
[13] Amar A *et al.* 2011 *Int. J. of Mod. Phys. E-Nucl. Phys.* **20** 980
[14] Stock R *et al.* 1976 *Phys. Rev.* C **14** 1824
[15] Yang Yong-Xu and Li Qing-Run 2011 *Phys. Rev.* C **84** 014602
[16] Boztosun I and Rae WDM 2001 *Phys. Rev.* C **63** 054607; 2001 *Phys. Rev.* C **64** 054607; 2002 *Phys. Rev.* C **65** 024603
[17] Boztosun I 2002 *Phys. Rev.* C **66** 024610
[18] Stone N J 2014 *IAEA Nuclear Data Section INDC(NDS)*-0658

# Reaction dynamics studies for the system $^{7}$Be+$^{58}$Ni

**D Torresi[1,2], M. Mazzocco[1,2], L Acosta[3], A Boiano[4], C Boiano[5], A Diaz-Torres[6], N Fierro[1], T Glodariu[7] L Grilj[3], A Guglielmetti[5,8], N Keeley[9], M La Commara[4,10], I Martel[3], C Mazzocchi[11], P Molini[1,2], A Pakou[12], C Parascandolo[1,2], V V Parkar[3], N Patronis[12], D. Pierroutsakou[4], M Romoli[4], K. Rusek[13], A M Sanchez-Benitez[3], M Sandoli [4,10] , C Signorini[1,2], R Silvestri [4,10], F Soramel 1, 2, E Stiliaris[14], E Strano[1,2], L. Stroe[7] , and K. Zerva[12]**

[1]Dipartimento di Fisica e Astronomia, Universit'a di Padova, Italy
[2]INFN - Sezione di Padova, Italy
[3]Departamento de Física Aplicada, Universidad de Huelva, Spain
[4]INFN - Sezione di Napoli, Italy
[5]INFN - Sezione di Milano, Italy
[6]ECT, Villazzano (TN), Italy
[7]NIPNE - Magurele, Romania
[8]Dipartimento di Fisica, Universit'a di Milano, Itay
[9]Department of Nuclear Reactions, INS, Warszawa, Poland
[10]Dipartimento di Scienze Fisiche, Universit'a di Napoli, Italy
[11]Faculty of Physics, University of Warsaw, Warszawa, Poland
[12]Department of Physics and HINP, University of Ioannina, Greece
[13]Heavy Ion Laboratory, University of Warsaw, Warszawa, Poland
[14]epartment of Physics, University of Athens, Greece

E-mail: `domenico.torresi@lns.infn.it`

**Abstract.**
The study of reactions induced by exotic weakly bound nuclei at energies around the Coulomb barrier had attracted a large interest in the last decade, since the features of these nuclei can deeply affect the reaction dynamics. The discrimination between different reaction mechanisms is, in general, a rather difficult task. It can be achieved by using detector arrays covering high solid angle and with high granularity that allow to measure the reaction products and, possibly, coincidences between them, as, for example, recently done for stable weakly bound nuclei [1, 2]. We investigated the collision of the weakly bound nucleus $^{7}$Be on a $^{58}$Ni target at the beam energy of 1.1 times the Coulomb barrier, measuring the elastic scattering angular distribution and the energy and angular distributions of $^{3}$He and $^{4}$He. The $^{7}$Be radioactive ion beam was produced by the facility EXOTIC at INFN-LNL with an energy of 22 MeV and an intensity of $\sim 3\times10^{5}$ pps. Results showed that the $^{4}$He yeld is about 4 times larger than $^{3}$He yield, suggesting that reaction mechanisms other than the break-up mostly produce the He isotopes. Theoretical calculations for transfer channels and compound nucleus reactions suggest that complete fusion accounts for (41±5%) of the total reaction cross section extracted from optical model analysis of the elastic scattering data, and that $^{3}$He and $^{4}$He stripping are the most populated reaction channels among direct processes. Eventually estimation of incomplete fusion contributions to the $^{3,4}$He production cross sections was performed through semi-classical calculations with the code PLATYPUS [3].

## 1. Introduction

The study of reactions induced by stable and unstable weakly bound nuclei at energies around the Coulomb barrier had attracted a large interest in the last decades, since the features of these nuclei can deeply affect the reaction dynamics. Several review papers have been recently published on this topic [1, 2, 4]. The discrimination between different reaction mechanisms is in general a rather difficult task. It can be achieved by using detector arrays covering high solid angle and with high granularity that allow to measure the reaction products and, possibly, coincidences between them. The radioactive and weakly-bound $^7$Be nucleus has been chosen as subject of our investigation since it has a low particle emission threshold ($S_\alpha$=1.586 MeV) and a very well pronounced $^3$He-$^4$He di-nuclear cluster structure. Both of the constituent clusters, that are the main reaction products, are stable charged particles having similar masses. This fact makes easier the study of the interplay of different reaction mechanisms since the detection of the reaction products are easily detectable. In fact, the experimental set-up would be more complex for reactions requiring the detection of neutrons. The $^7$Be nucleus was already studied in two works. In the first study [5] the fusion and transfer-breakup cross sections were measured at five different energies for the system $^7$Be+$^{238}$U. In the second study [6], the total reaction cross section was extracted at five energies measuring the elastic scattering angular distribution for the system $^7$Be+$^{58}$Ni.

In our experiment we measured the elastic scattering angular distributions in order to extract the total reaction cross-section and we were able, for the first time, to detect and identify the $^3$He and $^4$He reaction products for the system $^7$Be+$^{58}$Ni, allowing a deeper understanding of the contribution of direct and indirect reactions at energies below the barrier.

## 2. The experiment

The experiment was performed at the Laboratori Nazionali di Legnaro-INFN in Italy. The $^7$Be beam was produced with the in-flight technique by means of the facility EXOTIC [7]. A $^7$Li primary beam delivered by the XTU tandem accelerator at 34.2 MeV beam energy and with an intensity of about 100 pnA was impinging on a $H_2$ cryogenic gas target. The $^7$Be, produced through the reaction p($^7$Li, $^7$Be)n, was selected by means of a 30°-bending magnet and a 1 $m$ long Wien filter and was delivered at the target position with an energy of 22.3±0.4 MeV, an intensity of 2-3×$10^5$ pps and a purity ∼100%. A second energy 17.7±0.5 MeV, was obtained by inserting a degrader along the beam line. These energy values correspond to approximately +10% and -10% of the nominal Coulomb barrier for the reaction on a 1 mg/cm$^2$ thick $^{58}$Ni target.

The detection system used was DINEX [8]. It consisted of eight Double Sided Silicon Strip Detectors (DSSSDs) arranged in 4 $\Delta$E-E (45+1000 $\mu$m) telescopes placed at angles: +57°, +128°, -61.5°, -132°.

## 3. Result

In Fig. 1 the angular distributions for the scattering of $^7$Be+$^{58}$Ni are shown in linear scale. Due to the beam energy resolution and the energy loss in the $^{58}$Ni target it was not possible to distinguish between the elastic scattering and the inelastic scattering from the first excited states at 0.429 and 1.414 MeV, therefore the scattering cross-section is the quasi-elastic cross-section.

The angular distribution was fitted within the framework of the optical model with the subroutine SFresco of the main code Fresco [3], in order to extract the total reaction cross section. The out-coming theoretical curve is displayed with a solid line in Fig. 1. The total reaction cross section value obtained is 561 mb, that is in fairly good agreement with the data measured at lower beam energies in Ref. [6]. The quasi-elastic scattering data collected at 17.7 MeV beam energy (already measured by Aguilera and collaborators [6]), were substantially used to provide a further cross-check of the data normalization procedure at forward angles, while

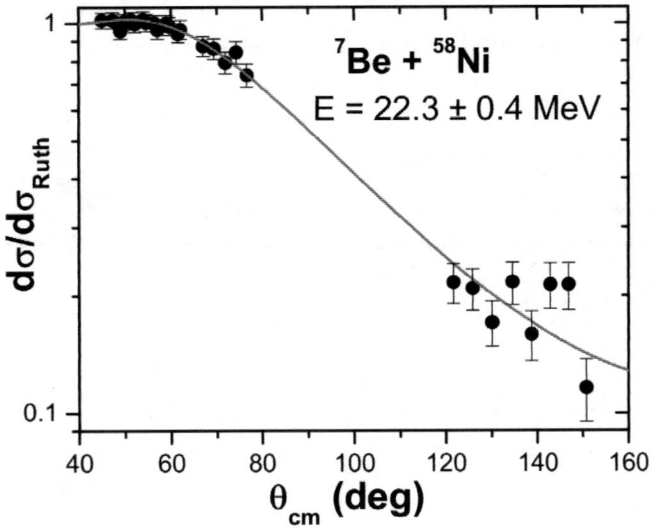


**Figure 1.**

the statistics of the data collected at backward angle was not enough to extract an angular distribution and just an overall ratio-to-Rutherford was evaluated.

A large $^{3,4}$He yields was observed both at forward and backward angles. Several reaction mechanisms can account for the presence of $^3$He and $^4$He ions. Anyway, the only process producing at the same time a $^3$He and $^4$He is the breakup. Since the $^4$He yield is a factor of four (at least) larger than that of $^3$He, we can argue that the exclusive breakup cannot be responsible for the whole $^4$He yields. A possible source of $^4$He is the fusion-evaporation process. We used the code PACE2 to evaluate the expected angular distribution and the multiplicity of $^4$He in the fusion-evaporation of $^7$Be+$^{58}$Ni at 22.3 MeV beam energy. The theoretical curve was normalized to the average value of the measured $^4$He cross section at the most backward angles, corresponding to the range between 40° and 50°. The overall fusion cross section evaluated was 229±29 mb that corresponds to 41±5% of the total reaction cross section extracted from the measurement of the quasi-elastic scattering cross section. However the $^4$He coming from the fusion-evaporation process does not exhausted the total $^4$He yield.

DWBA calculations showed that $^4$He and $^3$He stripping give a significant contribution in the $^3$He and $^4$He yields at forward angles.

## 4. Conclusions

In summary, we have studied the collisions $^7$Be+ $^{58}$Ni at energy close to the Coulomb barrier using a radioactive beam produced at the facility EXOTIC. The measurements of the quasi-elastic cross section angular distribution allowed the extraction of the total reaction cross section, which is in rather good agreement with the data trend individuated in [6]. Fairly large $^3$He and $^4$He production yields were observed at the higher beam energy and for the first time we were able to unambiguously distinguish the two helium isotopes in a $^7$Be-induced experiment.

Detailed theoretical and kinematic calculations together with the exploited similarity with the system $^7$Be+ $^{238}$U suggest that $^3$He and $^4$He are mainly originated by $^4$He and $^3$He stripping, respectively.

## Acknowledgements
This work was partly supported by the Italian MIUR within the project RBFR08P1W2_001 (FIRB 2008).

## References
[1] LF Canto PRS Gomes R D and Hussein M 2006 *Phys. Rep.* **424** 1
[2] Keeley N *et al.* 2006 *Phys. Rep.* **424** 1
[3] Thompson I 1998 *Comput. Phys. Rep.* **2** 167
[4] Mazzocco M 2010 *Int. J. Modern Phys. E* **19** 977
[5] Raabe R *et al.* 2006 *Phys. Rev. C* **74** 044606
[6] Aguilera E *et al.* 2009 *Phys. Rev. C* **79** 021601
[7] Farinon F *et al.* 2008 *Nucl. Instrum. Meth. B* **266** 4097
[8] Sanchez-Benitez A *et al.* 2005 *J. Phys. G* **31** 1935

Journal of Physics: Conference Series **590** (2015) 012058 doi:10.1088/1742-6596/590/1/012058

# Analysis of Elastic Scattering of $^{8}$He+$^{208}$Pb System at around the Coulomb Barrier Energies

**M Direkci[1,2], Y Kucuk[2,3], and I Boztosun[2,3]**

[1] Department of Physics, Bozok University, 66100, Yozgat, Turkey
[2] Department of Physics, Akdeniz University, 07058, Antalya, Turkey
[3] Nuclear Sciences and Application Center, Akdeniz University, 07058, Antalya, Turkey

E-mail: mikaildirekci@yahoo.com

**Abstract.** The elastic scattering angular distribution of $^{8}$He+$^{208}$Pb system is investigated at $E_{lab}$ = 22.0 MeV within the framework of Optical Model by using phenomenological and microscopic potentials. For the phenomenological Optical Model calculations, both real and imaginary parts of the complex nuclear potential have been chosen to have the Wood-Saxon shape. In the microscopic Optical Model calculations, we have used double folding procedure to calculate the real part of optical potential for different kinds of density distributions of $^{8}$He. A comparative study of this system has been conducted for the fist time by using phenomenological and microscopic potentials. It is observed that large imaginary radius value due to the existence of long-range absorption mechanism acting at large distances provides a very good agreement between theoretical results and experimental data with small $\chi^2/N$ values.

## 1. Introduction

By the development of Radioactive-Ion Beam Facilities [1], exotic nuclei especially having large spatial matter radii and neutron skin structure [2] near the drip line has taken a great deal attention in recent years [3, 4, 5, 6]. So far, there have been few studies in determining the radii of the sensitivity of weakly-bound [6, 7] and tightly-bound nuclei near the Coulomb barrier [8, 9]. It is well known that determining the radial sensitivity of the nuclear potential is a considerable task in order to explain the angular distribution of the elastic scattering cross section [8, 9]. In this study, we have investigated the interaction of $^{8}$He nucleus with $^{208}$Pb heavy-stable target at incident energy of laboratory, $E_{lab}$ = 22.0 MeV. There is only one experimental elastic scattering angular distribution data of $^{8}$He+$^{208}$Pb system measured by Marquinez-Duran *et al.* [10] for this system just around the Coulomb barrier energy. For this incident energy limitation, we focus on contribution of extra two neutron ($2n$) to the absorption and also radial sensitivity of optical potential. Hence, instead of determining new optical model parameter set, we used the global optical potential parameter set of $^{6}He$ that was derived by Kucuk *et al.* [11] at low energies to compare the radial sensitivity of imaginary radius $r_w$ and the effect of this imaginary radius to the absorption.

Our aim is to conduct phenomenological and microscopic optical model ($OM$) analysis for the realistic matter density distributions of $^{8}$He. In the next section, we briefly explain the optical models: phenomenological optical model is explained in subsection 2.1 and microscopic one in subsection 2.2. In Section 3, we present our results depending on an extra $2n$ admixture to absorption mechanism of reaction. Our discussion and conclusions are given in section 4.

Journal of Physics: Conference Series **590** (2015) 012058 doi:10.1088/1742-6596/590/1/012058

## 2. Optic Model Analysis

### *2.1. Phenomenological Optical Model*

$^8$He+$^{208}$Pb system has been investigated near the Coulomb barrier energy. We have used the optical model in the theoretical calculations and the total effective potential in our model consists of the Coulomb, centrifugal, and nuclear potentials as follows:

$$V_{total}(r) = V_{nuclear}(r) + V_{Coulomb}(r) + V_{centrifugal}(r) \tag{1}$$

In the total effective potential, the Coulomb and centrifugal potentials are well known. The Coulomb potential [12, 13, 14] due to a charge $Z_P e$ interacting with a charge $Z_T e$ distributed uniformly over a sphere of Radius $R_C$ is given by

$$V_{Coulomb}(r) = \begin{cases} \frac{Z_P Z_T e^2}{r}, & r \geq R_C \\ \frac{Z_P Z_T e^2}{2R_C}(3 - \frac{r^2}{R_C^2}), & r < R_C \end{cases} . \tag{2}$$

where $R_C$ is the Coulomb radius, taken as $R_C = 1.3(A_P^{1/3} + A_T^{1/3})$ fm in the calculations, $Z_P$ and $Z_T$ are defined as the charges of the projectile $P$ and target $T$ respectively. The centrifugal potential is

$$V_{centrifugal}(r) = \frac{\hbar^2}{2\mu}\frac{\ell(\ell+1)}{r^2}, \tag{3}$$

where $\mu$ is the reduced mass. Finally, the complex nuclear potential $V_{nuclear}(r)$ is taken to be the sum of both Woods-Saxon shaped $(W - S)$ real and imaginary potentials given as,

$$V_{nuclear}(r) = -\frac{V_0}{\left[1 + exp(\frac{r-R_v}{a_v}))\right]} - i\frac{W_0}{\left[1 + exp(\frac{r-R_w}{a_w})\right]} \tag{4}$$

Here, $V_0$, $W_0$ are reel and imaginary depth and the nuclear radius are $R_i = r_i(A_P^{1/3} + A_T^{1/3})$ $(i = v$ or $w)$, where $A_P$ and $A_T$ are the masses of projectile and target nuclei and $r_v$ and $r_w$ are the radius parameters of the real and imaginary part of the nuclear potentials respectively. We have used the code FRESCO [15] for the elastic scattering angular distribution calculations. The parameters of real and imaginary potentials are given in Table 1. All potential parameters except the radius of imaginary potential, $r_w$, are taken from Kucuk *et al.* [11]. The experimental data are taken from Ref. [16].

**Table 1.** Optical potential parameters for the $^8$He+$^{20}$Pb system at $E = 22.0$ MeV

| $V_0$ $(MeV)$ | $r_v$ $(fm)$ | $a_v$ $(fm)$ | $W_0$ $(MeV)$ | $r_w$ $(fm)$ | $a_w$ $(fm)$ | $J_V$ $(MeV.fm^3)$ | $J_W$ $(MeV.fm^3)$ | $\sigma_R$ $(mb)$ | $\chi^2/N$ |
|---|---|---|---|---|---|---|---|---|---|
| 153.46 | 0.9 | 0.7 | 9.34 | 1.56 | 0.7 | 153.49 | 45.84 | 1535 | 0.29 |

Journal of Physics: Conference Series **590** (2015) 012058 doi:10.1088/1742-6596/590/1/012058

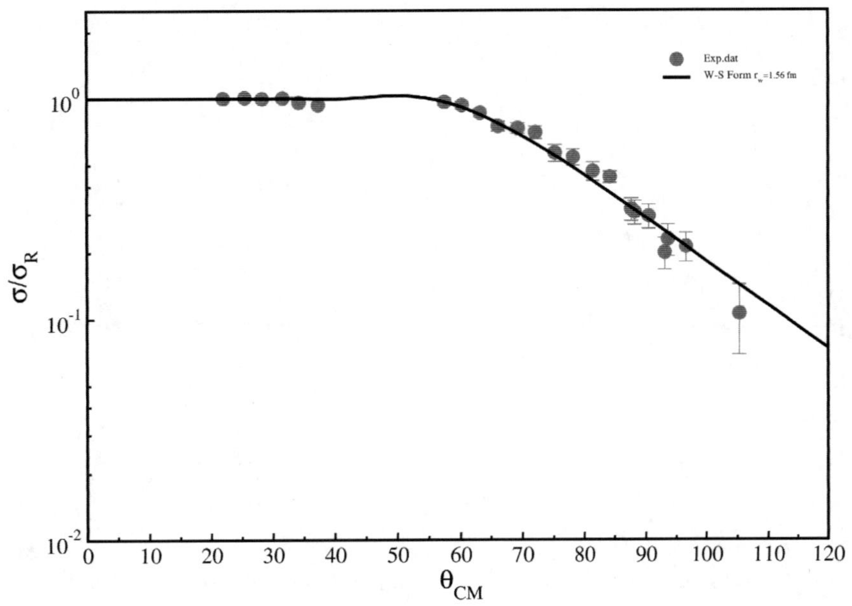


**Figure 1.** The angular distribution of the elastic scattering cross section of the $^8$He+$^{208}$Pb system obtained with $r_w$=1.56 fm with $\chi^2/N$=0.29.

### 2.2. *Microscopic Optical Model*

In this part of theoretical calculations, we determined the real part of the complex $V_{nuclear}(r)$ potential by using the double-folding model. In order to calculate the double-folding potential, the nuclear matter distributions of both projectile $\rho_P$ and target $\rho_T$ nuclei together with an effective nucleon-nucleon interaction potential ($\upsilon_{NN}$) are used. Hence, the double-folding potential is,

$$V_{DF}(\mathbf{r}) = \int d\mathbf{r_1} \int d\mathbf{r_2} \rho_P(\mathbf{r_1}) \rho_T(\mathbf{r_2}) \upsilon_{NN}(\mathbf{r_{12}}) \qquad (5)$$

where $\rho_P$ and $\rho_T$ are the nuclear matter density of projectile and target nuclei respectively [20, 21]. In order to perform a comparative study, we have used three different matter density distributions for the $^8$He ground state. First density distribution of $^8$He structure was obtained by Tanihata *et al.* [17] in which Tanihata calculated $^8$He density by using proton and neutron densities. Second one is obtained by Zhukov *et al.* [2] in which Zhukov applied the cluster-orbital shell-model ($COSMA$) approximation to the $^8$He where $^8$He consist of $\alpha$ core + ($4n$) parts. The last one is Symmetrize Fermi ($SF$) Model which was performed by Lukyanov *et al.* [18] in which $4n$ placed symmetrically around the $\alpha$ core. The detailed explanations of all density distributions of these models used in this study can be found in the relevant papers.

The effective nucleon-nucleon interaction, $\upsilon_{NN}$, is integrated over both density distributions of $\rho_P$ and target $\rho_T$ nuclei. In literature, several nucleon-nucleon interaction expressions is used for the folding model potential calculations. We have chosen the most used one, M3Y nucleon-nucleon (Michigan3Yukawa) realistic interaction [19, 20, 21], that is given by

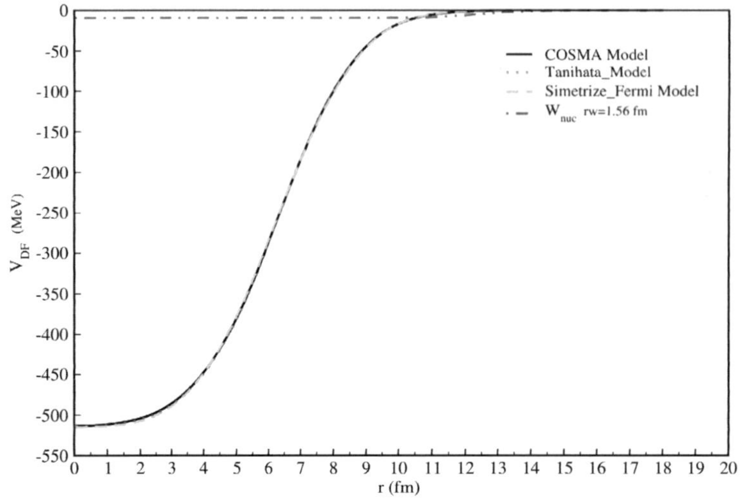


**Figure 2.** The real part of the nuclear potential obtained by three different density distributions by using the double folding model for $^{8}$He. The imaginary part of nuclear potential with W-S form is also shown.

$$v_{NN}(r) = 7999\left[\frac{exp(-4r)}{4r}\right] - 2134\left[\frac{exp(-2.5r)}{2.5r}\right] + J_{00}(E)\delta(r)MeV \tag{6}$$

where $J_{00}(E)$ represents the exchange term since nucleon exchange is possible between the projectile and target. $J_{00}(E)$ has linear energy-dependence and defined as

$$J_{00}(E) = -276\left[1 - 0.005\frac{E_{lab}}{A_P}\right] MeV fm^3 \tag{7}$$

In the microscopical OM calculations, the real part of OM potential has been obtained by using above description of double-folding model, the imaginary potential has been taken as W-S form as mentioned above. Therefore, for the calculation of DF potential case, nuclear potential consists of real and imaginary parts, expressed as given by;

$$V_N(r) = N_R V_{DF}(r) - i\frac{W_0}{\left[1 + exp(\frac{r-R_w}{a_w})\right]} MeV \tag{8}$$

where $N_R$ is the normalization factor and we fixed it as 1.0 for all DF calculations. The parameter set of DF calculations are shown in Table 2 and the double-folding potentials obtained by using above-mentioned three different density distributions together with the imaginary potential with $r_w$=1.56 fm are shown in Fig. 2. As it can be seen from this figure, all density distributions provide almost the same potentials around the Coulomb barrier energies. The angular distribution of elastic scattering of the system obtained by using these density distributions is shown in Fig. 3. For the density distribution of $^{208}$Pb, we has used the density of $RIPL-2$ [22].

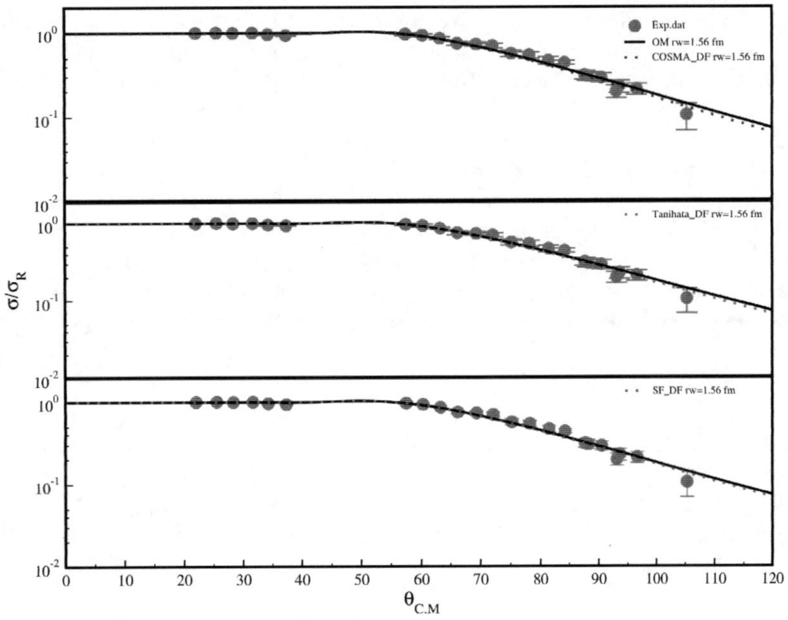


**Figure 3.** The angular distribution of the elastic scattering cross section of the $^8$He+$^{208}$Pb system obtained with $r_w$=1.56 fm with $\chi^2/N = 0.20$.

The real and imaginary potentials have been studied in terms of their volume integrals in our study. We also show the tabulated values of the real and imaginary volume integrals ($J_V$ and $J_W$) per interacting nuclear pair in Table 1 and Table 2 and the volume integrals defined as

$$J_{V,W} = \frac{4\pi}{A_p A_T} \int_0^\infty U_{V,W}(r) r^2 dr \tag{9}$$

where $A_p$ and $A_T$ are the projectile and target masses. Reaction cross-section of these three density distributions are also shown in Table 2. They predict similar values around around 1550 mb.

**Table 2.** The parameters of $^8$He+$^{208}$Pb system and their cross section predictions, $\chi^2/N$ values and volume integrals. $N_R = 1.0$ is fixed for all calculations.

| Density Model | $W_0$ $(MeV)$ | $r_w$ $(fm)$ | $a_w$ $(fm)$ | $J_V$ $(MeV.fm^3)$ | $J_W$ $(MeV.fm^3)$ | $\sigma_R$ $(mb)$ | $\chi^2/N$ |
|---|---|---|---|---|---|---|---|
| COSMA | 9.34 | 1.56 | 0.7 | 417.68 | 45.84 | 1554 | 0.20 |
| Tanihata | 9.34 | 1.56 | 0.7 | 417.76 | 45.84 | 1551 | 0.22 |
| SF | 9.34 | 1.56 | 0.7 | 417.68 | 45.84 | 1546 | 0.23 |

## 3. Results

By using the above-described phenomenological and microscopic potentials, we have investigated the elastic scattering of $^8$He+$^{208}$Pb system at $E_{lab}$ = 22.0 MeV. In order to explain the data, we have to modify the radius of the imaginary potential of Kucuk *et al.* [11] as $r_w = 1.56$ fm. The parameters of the potentials for phenomenological and microscopic potentials are given in Table 1 and Table 2 respectively. The result of phenomenological potential calculations is shown in Fig. 1 and it can be seen from this figure that the agreement is very good over the whole angular range in comparison with the experimental data. The results of microscopic potentials obtained with the three different density distributions are shown in Fig. 3. The results are also in very good agreement with the experimental data. Both phenomenological and microscopic potentials predict similar reaction cross-section values around 1550 mb. Volume integrals of both phenomenological and microscopic potential are also calculated and the values are given in Table 1 and Table 2. The volume integral for the real part of the microscopic potential is much larger than the phenomenological one.

## 4. Discussion and Conclusions

We have studied the elastic scattering of $^{8}$He neutron-skin nucleus from a heavy-stable $^{208}$Pb target. We have used both phenomenological and microscopic potentials to explain the experimental data. Phenomenological potentials have the W-S shape for both real and imaginary parts. Three different density distributions in the double-folding model are used for the real part of microscopic potential calculations. We have observed three different density distributions proposed to define the density of $^{8}$He provide almost the same real potential around Coulomb barrier. We also noticed that the radius of imaginary potential should be large enough to obtain correct absorption. Both phenomenological W-S potential and microscopic calculations with three different density distribution provide a good agreement with the experimental data with phenomenological W-S imaginary potential. The OM calculations for other reaction systems are still in progress.

## Acknowledgments

This work was partly supported by the scientific research projects units of Bozok and Akdeniz Universities.

## References

[1] Ter-Akopian G M *et al.* 2004 *Nucl. Phys.* A **734** 295
[2] Zhukov M V Korsheninnikov A A and Smedberg M H 1994 *Phys. Rev.* C **50** 1
[3] Tanihata I *et al.* 1995 *Prog. Part. Nucl. Phys.* **35** 505
[4] Tanihata I *et al.* 1996 *J. Phys. G: Nucl. Part. Phys.* **22** 157
[5] Thompson I J and Suzuki Y 2001 *Nucl. Phys.* A **693** 424
[6] Jonson B 2004 *Phys. Rep.* **389** 1
[7] Rusek K *et al.* 2004 *Phys. Rev.* C **70** 014603
[8] Mazzocco M *et al.* 2014 *Act. Phys. Pol.* B **45** 2
[9] Biswas M 2008 *Phys. Rev.* C **77** 017602
[10] Marquinez G-Duran *et al.* 2012 *Act. Phys. Pol.* B **43** 2
[11] Kucuk Y, Boztosun I and Topel T 2009 *Phys. Rev.* C **80** 054602
[12] Satchler G R 1983 *Direct Nuclear Reactions* (Oxford,Oxford University Press)
[13] Boztosun I and Rae WDM 2001 *Phys. Lett.* **518B** 229
[14] Boztosun I 2002 *Phys. Rev.* C **66** 024610
[15] Thompson I 1988 *J. Comput. Phys. Rep.* **7** 167
[16] Experimental Nuclear Reaction Data ($EXFOR$) 2011 *Database Version of November 02*
[17] Tanihata I 1992 *Phys. Lett.* B **289** 261
[18] Lukyanov V K *et al.* 2003 arXiv:nucl-t/0311040v1
[19] Satchler G R and Love W G 1979 *Phys. Rep.* **55** 183
[20] Karakoc M and Boztosun I 2006 *Phys. Rev.* C **73** 047601
[21] Karakoc M and Boztosun I 2006 *Int. J. of Mod. Phys.* E **15** 1317
[22] Reference Input Parameter Library ($RIPL-2$), http://www.nds.iaea.org/RIPL-2/

Journal of Physics: Conference Series **590** (2015) 012059 doi:10.1088/1742-6596/590/1/012059

# Photofission of $^{238}$U Nuclei

**S Selek[1], A Çoban[1], I Boztosun[1,2], H Đapo[1,2], M Karakoç[1,2]**

[1]Akdeniz University, Faculty of Science, Physics Department, Antalya, Turkey[2]Akdeniz University Nuclear Application and Research Center, Antalya, Turkey

E-mail: sevdaselek@gmail.com

**Abstract**. The spectrum of the $^{238}$U(γ, f) photofission reaction which is obtained using bremsstrahlung photon beams from a clinical e-linac is analyzed. The following nuclei $^{94}$Y, $^{134}$Te, $^{136}$Pr, $^{142}$Ba, $^{138}$Xe, $^{104}$Tc in the spectrum of our experiment have been identified. The gamma ray energies of the produced nuclei are compared with the literature values. This work is a proof-of-concept experiment which is the first of in the series of our future planned photofission experiments using a clinical e-linac operated with different endpoint energies.

## 1. Introduction

Nuclear fission studies are still attractive and widespread since 1940s. Photofission is one of the way to create nuclear fission products. It is the result of the process which is called as photon-induced nuclear (photonuclear) reaction that is important in a variety of applications such as basic science, activation analyses, preparation of medical isotopes and characterization and transmutation of nuclear wastes [1-7].

Photonuclear reactions are nuclear experiment methods performed with bremsstrahlung photons obtained from the deceleration of fast electrons within the vicinity of target nuclei such as tungsten. Bremsstrahlung photons have a continuous spectrum. The average energy of the fission barrier is about 7-8 MeV. Therefore to obtain photofission products the photons have to have at least this much energy. In this experimental method, the target nuclei can be converted to other nuclei with just one or more proton/neutron ejection or may undergo fission reaction depending on the nature of the target nuclei and the energy of the photons. The experimental yields of the photonuclear reactions can be determined by detection and analysis of the characteristic gamma rays emitted from the product nuclei.

Neutrons are often used as a source to create the fission of $^{235}$U, because it has high fission cross section with neutrons. $^{238}$Uexist in nature with a high abundance has low fission cross section with neutrons. For reaction to take place high flux of fast neutrons are required. Therefore experimental cross section measurement of$^{238}$U is easier for photon induced fission than neutron induced fission [8]. Therefore, this work is a proof-of-concept experiment which is the first of in the series of our future planned photofission experiments with a clinical e-linac operated with different endpoint energies.

## 2. Experimental Procedure

The photofission reaction has been performed using Elekta™ Synergy™ clinical e-linac accelerator as the source of bremsstrahlung photons, its technical documentation can be found in [9].The target nuclei $^{238}$U, inside compound $UO_2(NO_3)_2$, was prepared with appropriate geometry for the irradiation.

The sample target was placed at 58 cm in distance from the photon source and was irradiated for 1 hour with18 MeV end-point energy.

In order to measure the spectra of the $^{238}U$ sample a well-shielded high-purity germanium detector (HPGe) has been used. The sample is placed 15 min after irradiation and counted for two days. The detector used is a p-type, coaxial, electrically cooled, HPGe gamma-ray spectrometer AMATEK-ORTEC (GEM40P4-83) with 40% relative efficiency and 768 eV FWHM at 122 keV for $^{57}Co$ and 1.85 keV FWHM at 1332 keV for $^{60}Co$ [10].

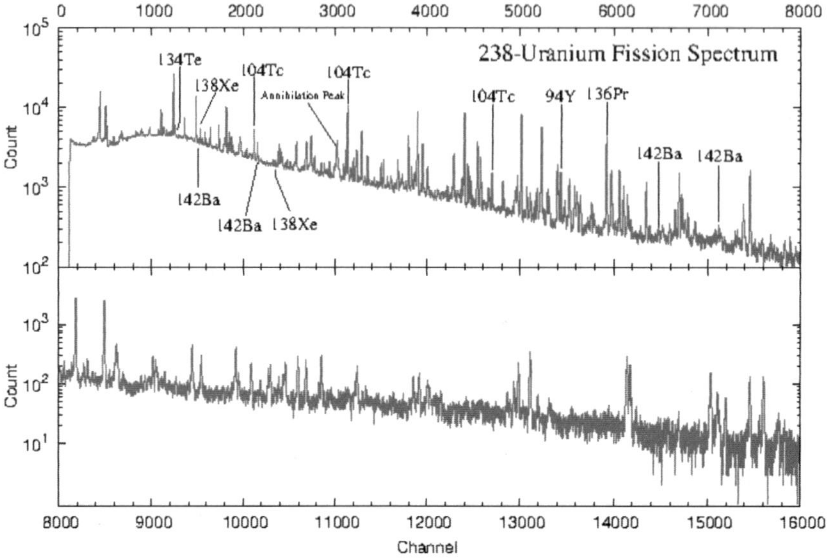


**Figure 1.** Fission spectrum of$^{238}U$ for the irradiated sample after 2 days of counting.

## 3. Results and Discussion

The irradiated $^{238}U$ sample spectrum without background subtractions is shown in Figure 1. This spectrum is analyzed with gf3 RadWare [11] peak-fitting program to identify the photofission fragments. The energy and half lives of the fragment nuclei are determined using the area and the channel of each peak. Then, these obtained values have been used to identify the photofission fragments by comparing them with the values found in NNDC (National Nuclear Data Center) [12].

Analyzing the $^{238}U$ fission spectrum and identifying the fission fragments still continues. In this initial study we were able to show, as proof-of-principle, the concept and the idea behind determining the fission fragments via gamma spectroscopy. However in order to calculate yields of fission fragments more experimental data is necessary. At the very least one more end-point energy of the linac is necessary. At this time we are investigating how to change the end-point energy of medical linac used in the present experiment. After this process is completed detailed amount of the fission yield will be obtained. Then it will be possible to obtain the photofission cross sections using these fission yields.

In conclusion, we have presented the photofission process of $^{238}U$ analysis with gamma spectroscopy and availability of medical linac in nuclear physics experiments. We have positively identified several elements in this analysis. The results of the present study have demonstrated the feasibility of studying photofission in the way that presented in this paper.

Journal of Physics: Conference Series **590** (2015) 012059 doi:10.1088/1742-6596/590/1/012059

**Table 1.** Gamma-ray energies of observed photofission fragments from $^{238}U$ irradiation and values found in the literature (NUDAT). $\sigma_E$ and $\sigma$ are statistical errors.

| | Present study | | *NUDAT* [a] | |
|---|---|---|---|---|
| Element | *E (keV)* | $\sigma_E$ *(keV)* | *E (keV)* | $\sigma$ *(keV)* |
| 134Te | 210.5739 | 0.0162 | 210.465 | 0.016 |
| 94Y | 917.7587 | 0.0094 | 918.74 | 0.05 |
| 136Pr | 999.8549 | 0.0042 | 1000.7 | 0.2 |
| 142Ba | 255.0362 | 0.0428 | 255.300 | 0.012 |
| 142Ba | 364.2100 | 0.0125 | 363.96 | 0.03 |
| 142Ba | 1093.1321 | 0.1388 | 1094.1 | 0.1 |
| 142Ba | 1202.1080 | 0.0895 | 1202.4 | 0.1 |
| 138Xe | 258.2187 | 0.0110 | 258.411 | 0.020 |
| 138Xe | 396.2359 | 0.0806 | 396.513 | 0.010 |
| 104Tc | 357.6323 | 0.0068 | 358.0 | 0.1 |
| 104Tc | 529.4958 | 0.0027 | 530.5 | 0.1 |
| 104Tc | 792.6276 | 0.0118 | 792.5 | 0.1 |

[a]NNDC [12].

## References

[1] Boztosun I, Đapo H, Ozmen S H, Çeçen Y, Karakoç M, Çoban A, Cesur A, Caner T, Bayram E, Keller G B, Güvendi A, Derman M, Kaya D 2014 The results of the first photonuclear reaction performed in Turkey: the zinc example Turkish Journal of Physics **38** 1-9

[2] Wells D P, Segebade C R, Cole P L 2010 Photon Activation Analysis at the Idaho Accelerator Center AIP Conf. Proc. **1265** 379-386

[3] Tickner J, Bencardino R, Roach G 2010 Measurement of activation yields for platinum group elements using Bremsstrahlung radiation with end-point energies in the range 11–14 MeV Nuclear Instruments and Methods in Physics Research B **268** 99-105

[4] Naik H, Suryanarayana S V, Jagadeesan K C, Thakare S V, Joshi P V, Nimje V T, Mittal K C, Goswami A, Venugopal V, Kailas S 2013 An alternative route for the preparation of the medical isotope $^{99}$Mo from the $^{238}$U(γ, f) and $^{100}$Mo(γ, n) reactions J Radioanal Nucl. Chem. **295** 807–816

[5] Gmar M, Jeanneau F, Laine´ F, Makil H, Poumare`de B, Tola F 2005 Assessment of actinide mass embedded in large concrete waste packages by photon interrogation and photofission Applied Radiation and Isotopes **63** 613–619

[6] Dighe P M, Berthoumieux E, Doré D, Laborie J M, Ledoux X, Macary V, Panebianco S, Ridikas D 2009 Delayed gamma studies from photofission of 237Np for nuclear waste characterization Annals of Nuclear Energy **36** 399–403

[7] Carrel F, Agelou M, Gmar M, Lainé F, Loridon J, Ma J-L, Passard C, Poumarède B 2010

Identification and Differentiation of Actinides Inside Nuclear Waste Packages by Measurement of Delayed Gammas IEEE TRANSACTIONS ON NUCLEAR SCIENCE **57** 2862-2871

[8] Prakash H G R, Sanjeev G, Kumar K B V, Kumar H G H, Siddappa K, Nayak B K, Saxena A 2011 Measurement of Photofission Cross-Section of $^{238}U$ Using Microtron Facility International Journal of Modern Physics E **20** 2361–2375

[9] Elekta Digital Accelerator Technical Training Guide 2003 (unpublished)

[10] MAESTRO 2012 URL http://www.ortec-online.com/download/A65-B32-MAESTRO-32-Emulation-Software.pdf

[11] Radford D 1995 ESCL8R and LEVIT8R: Software for interactive graphical analysis of HPGe coincidence data set NIM A **361** 297-305

[12] National Nuclear Data Center (NNDC), http://www.nndc.bnl.gov/

# AUTHOR INDEX


Acosta, L ........ 163, 279
Agar, Osman ........ 215, 225
Agodi, C ........ 167, 195, 257
Akcali, O ........ 245
Akdogan, T ........ 141
Akkoyun, S ........ 237, 253
Alba, R ........ 257
Alhassid, Y ........ 147
Alimov, D ........ 269, 274
Alipoor, Zahra ........ 219
Amangeldy, N ........ 269
Amano, S ........ 13
Andrighetto, A ........ 57
Aprahamian, Ani ........ 87
Arazi, A ........ 69
Artemov, S V ........ 274
Assunção, M ........ 199
Atanasova, L ........ 211
Aydin, S ........ 191
Aziz, Azni Abdul ........ 233
Babacan, Halil ........ 159
Balabanski, D L ........ 211
Balabanski, Dimiter L ........ 25, 155
Balantekin, A B ........ 1
Baran, V ........ 257
Barioni, A ........ 69
Bassato, G ........ 57
Bayrak, O ........ 229
Bayram, E ........ 141
Bayram, T ........ 237, 253
Bazzacco, D ........ 191
Benini, D ........ 57
Benjamim, E A ........ 69
Bermudez, J ........ 57
Bhowmik, R K ........ 171
Bisoffi, G ........ 57
Bizzeti, P G ........ 191
Bizzeti-Sona, A M ........ 191
Boiano, A ........ 257, 279
Boiano, C ........ 279
Bonaccorso, A ........ 167
Bonatsos, Dennis ........ 19
Bondì, M ........ 167, 195
Bostan, M ........ 211
Bouhelal, M ........ 191
Bozkurt, V ........ 141
Bozkurt, Vakkas ........ 105
Boztosun, I ........ 141, 253, 261, 265, 269, 274, 283, 289
Boztosun, Ismail ........ 215, 225
Burtebayev, N ........ 269, 274
Burtebayeva, J ........ 269, 274
Butler, Peter A ........ 93
Calabretta, L ........ 57
Calderolla, M ........ 57
Canbula, Bora ........ 159
Canella, S ........ 57
Caner, T ........ 141
Canto, L F ........ 45, 51, 129
Cappuzzello, F ........ 167, 195
Carbone, D ........ 167, 195
Cavallaro, M ........ 167, 195
Cecen, Y ........ 141
Cesur, A ........ 141, 261
Chiba, S ........ 13
Coban, A ........ 141
Çoban, A ........ 289
Colonna, M ........ 257
Comunian, M ........ 57
Condori, R Pampa ........ 69
Coniglione, R ........ 257
Corradetti, S ........ 57
Cunsolo, A ........ 167, 195
Danchev, M ........ 211
Dapo, H ........ 141, 253, 261, 265, 289
Davids, B ........ 195
Davinson, T ........ 195
De Angelis, G ........ 57, 191
De Faria, P N ........ 69
De Filippo, E ........ 257
De Napoli, M ........ 167, 195
Del Zoppo, A ........ 257
Deloncle, I ........ 191
Descouvemont, P ........ 45, 69, 199
Detistov, P ........ 211
Detistov, Pavel ........ 155
Di Pietro, A ........ 163, 207
Di Toro, M ........ 257
Diaz-Torres, A ........ 279
Dikmen, Erdal ........ 123
Direkci, M ........ 283
Druet, T ........ 45
Drumev, Kalin ........ 175
Dulger, F ........ 253
Durusoy, A ........ 245
Eke, Canel ........ 215, 225
Emanuele, U ........ 257
Ertugay, C ........ 265
Esposito, J ........ 57
Evlice, S ........ 229
Farinon, F ........ 257
Farnea, E ........ 191
Favaron, P ........ 57
Feghhi, Seyed Amir Hossein ........ 219
Fernández-García, J P ........ 207
Fierro, N ........ 279
Figuera, P ........ 163, 207
Filipescu, D M ........ 135
Firihu, Muhammad Zamrun ........ 233
Fisichella, M ........ 207
Forneris, J ........ 163

# AUTHOR INDEX


Foti, A ....167, 195
Freer, Martin....81
Fülöp, Zs....203
Funaki, Yasuro....7
Gadea, A ....191
Galatà, A....57
Galinski, N....195
Garistov, V P ....179
Gasques, L R ....69
Georgiev, G....211
Georgieva, A I ....179
Georgieva, Ana....175
Gheorghe, I....135
Gholamzadeh, Zohreh ....219
Giacchini, M ....57
Gladnishki, K ....211
Glodariu, T ....135, 279
Goasduff, A....211
Gomes, P R S ....129
Goriely, S ....135
Gottardo, A....191
Gramegna, F....57
Grassi, L ....163
Grilj, L....163, 279
Guglielmetti, A ....257, 279
Guimarães, V....69
Güray, R T....203
Gyürky, Gy....203
Haas, F ....191
Hagino, K....117
Halász, Z....203
Hamada, Sh ....269
Hayakawa, T ....13
Horikawa, K ....13
Hussein, M S....45, 51, 129
Huyuk, T....191
Imazaki, K ....13
Ionescu-Bujor, M ....191
Izawa, Y ....13
Jaksic, M....163
Kabak, R....241
Kanungo, R....195
Kara, S O ....237
Karakoc, M....141, 261
Karakoç, M....289
Kassim, Hasan Abu....233
Kaya, D ....141
Keeley, N....279
Kerimkulov, Zh K....269
Kiss, G G....203
Kocak, G ....241
Korkmaz, M. Emin ....215
Kucuk, Y....141, 283
Kumar, R....171
Kumar, Raj....151
Kusoglu, A....211
La Commara, M ....257, 279
Laftchiev, H ....191
Lattuada, M ....163, 207
Lay, J A ....151
Leistenschneider, E ....69
Lenske, H ....195
Lenzi, S M ....191
Lépine-Szily, A....69
Li, D....13
Lichtenthäler, R....69
Linares, R....167
Ljungvall, J ....211
Lollo, M ....57
Lombardi, A....57
Lubian, J ....129, 167
Lui, Y-W ....135
Lunardi, S ....191
Maggiore, M....57
Maiolino, C ....257
Maniero, D ....57
Manzolaro, M ....57
Martel, I ....279
Martin, B....257
Mastinu, P F....57
Matea, I ....211
Mazzocchi, C....257, 279
Mazzocco, M ....257
Mazzocco, M. ....279
Mendes Jr, D R....69
Menegazzo, R....191
Mengoni, D....191
Michelagnoli, C....191
Mijatovic, T....163
Milin, M....163
Minkov, N ....19
Mittig, W ....51
Miyamoto, S....13, 135
Mochizuki, T....13
Molini, P....279
Monetti, A ....57
Morais, M C....69
Morcelle, V....69
Moro, A M ....69, 207
Morzabayev, A K ....269, 274
Moschini, Laura ....39
Muralithar, S ....171
Mussaev, J M....274
Nakada, H ....147
Napoli, D R....191
Nassurlla, M....269, 274
Nyhus, H-T....135
Ogata, K ....13
Orrigo, S E A ....195
Otomar, D R ....129

# AUTHOR INDEX


Özen, C ... 147
Özkan, N ... 203
Ozmen, S F ... 141, 265
Pakou, A ... 279
Parascandolo, C ... 257, 279
Parkar, V V ... 279
Patronis, N ... 279
Pegoraro, R ... 57
Petrellis, D ... 19
Piazza, L ... 57
Pierroutsakou, D ... 257, 279
Pires, K C C ... 69
Pisent, A ... 57
Porcellato, A M ... 57
Prasad, R ... 171
Prepolec, L ... 163
Prete, G ... 57
Quesne, C ... 19
Radeck, D ... 211
Rappold, Christophe ... 105
Recchia, F ... 191
Renstrøm, T ... 135
Rigato, M ... 57
Rizzo, C ... 257
Rodriguez-Gallardo, M ... 69
Romoli, M ... 257, 279
Roncolato, C ... 57
Rossignoli, M ... 57
Rowley, N ... 117
Ruiz, C ... 195
Rusek, K ... 279
Russo, A ... 57
Sahin, D ... 191
Saito, T R ... 141
Saito, Take R ... 105
Sakhiev, S K ... 269, 274
Sanchez-Benitez, A M ... 279
Sandoli, M ... 279
Sanetullaev, A ... 195
Santonocito, D ... 257
Sarchiapone, L ... 57
Savran, Deniz ... 63
Scarduelli, V ... 69
Scarpa, D ... 57
Selek, S ... 289
Sharma, Manoj K ... 171
Sharma, Vijay R ... 171
Shima, T ... 135
Shorto, J M B ... 69
Signorini, C ... 257, 279
Silvestri, R ... 257, 279
Singh, B P ... 171
Singh, P P ... 191
Singh, Pushpendra P ... 171
Singh, R P ... 171
Skukan, N ... 163
Soic, N ... 163
Somorjai, E ... 203
Soramel, F ... 257, 279
Sotty, C ... 211
Soylu, A ... 229
Spitaleri, Claudio ... 75
Stanko, D ... 163
Stefan, I ... 211
Stiliaris, E ... 279
Strano, E ... 257, 279
Stroe, L ... 279
Stuchbery, A E ... 211
Suzuki, Toshio ... 111
Szücs, T ... 203
Takahisa, K ... 135
Tamkas, M ... 245
Tenreiro, Claudio ... 219
Tesileanu, O ... 135
Tokic, V ... 163
Tonev, D ... 191
Toppi, M ... 187
Torresi, D ... 163, 207, 257, 279
Trifiró, A ... 257
Trimarchi, M ... 257
Ur, C A ... 191
Uroic, M ... 163
Utsunomiya, H ... 135
Valiente, J ... 57
Valiente-Dobón, J J ... 191
Vasquez, J ... 57
Verney, D ... 211
Vitturi, A ... 151
Vitturi, Andrea ... 39
Von Neumann-Cosel, Peter ... 99
Washiyama, Kouhei ... 183
Wyss, J ... 57
Yadav, Abhishek ... 171
Yalçin, C ... 203, 249
Yordanov, D T ... 211
Yusof, Norhasliza ... 233
Zadro, M ... 163, 207
Zafiropoulos, D ... 57
Zamora, J C ... 69
Zanella, S ... 57
Zerva, K ... 279
Zilges, A ... 33